T0200565

THEORY OF
ELASTIC STABILITY

Second Edition

Stephen P. Timoshenko
James M. Gere

DOVER PUBLICATIONS
Garden City, New York

Bibliographical Note

This Dover edition, first published in 2009, is an unabridged republication of the second edition of the work, originally published by the McGraw-Hill Book Company, Inc., New York and London, in 1961.

Library of Congress Cataloging-in-Publication Data

Timoshenko, Stephen, 1878–1972.
　　Theory of elastic stability / Stephen P. Timoshenko, James M. Gere. — Dover ed.
　　　　p. cm.
　　Originally published: New York : McGraw-Hill Book, 1961.
　　Includes bibliographical references and index.
　　ISBN-13: 978-0-486-47207-2
　　ISBN-10: 0-486-47207-8
　　1. Elasticity. 2. Strain and stresses. I. Gere, James M. II. Title.

QA931.T54 2009
531'.382—dc22

2009003597

Manufactured in the United States of America
47207812　　2023
www.doverpublications.com

PREFACE TO THE SECOND EDITION

Since the first edition of this book, the subject of stability of structures has steadily increased in importance, especially in the design of metal structures. As a result, many engineering schools now offer courses in this subject, usually as part of a curriculum in applied mechanics. This book is intended primarily to serve the needs of the beginning student of the subject, and the emphasis is on fundamental theory rather than specific applications.

In this second edition the authors have attempted to bring up to date the subject matter of the first edition and at the same time to maintain the presentation which was characteristic of the earlier work. The book begins with the analysis of beam-columns (Chap. 1) and then proceeds to elastic buckling of bars (Chap. 2). The latter chapter has been enlarged to include buckling under the action of nonconservative forces, periodically varying forces, and impact. In addition, the discussion of the determination of critical loads of columns by successive approximations has been expanded. The material on inelastic buckling of bars has been augmented by the introduction of the tangent modulus and placed in a new chapter (Chap. 3). Chapter 4 describes experiments on buckling of bars and is about the same as in the earlier edition, since it was felt that the original material still retains its inherent value.

A new chapter on torsional buckling (Chap. 5) has been added to the book, and the chapter on lateral buckling of beams (Chap. 6) has been extensively revised. Chapter 7 deals with the buckling of rings, curved bars, and arches and contains several additions. The chapters dealing with bending of plates and shells (Chaps. 8 and 10) are substantially unchanged and are included in the book as prerequisites to Chaps. 9 and 11 on buckling of plates and shells. In Chap. 9 (buckling of plates) several new cases of buckling are considered, and some tables for calculating critical stresses have been added. All the important material from the first edition has been retained in this chapter as well as in Chap. 11 (buckling of shells). The material added in Chap. 11 consists of discussion of postbuckling behavior of compressed cylindrical shells and some new material on buckling of curved sheet panels, stiffened cylindrical shells, and spherical shells.

Numerous footnote references are given throughout the text as an aid to the student who wishes to pursue some aspect of the subject further.

The authors take this opportunity to thank Mrs. Thor H. Sjöstrand and Mrs. Richard E. Platt for assistance in preparing the manuscript and reading the proofs for this second edition.

Stephen P. Timoshenko
James M. Gere

PREFACE TO THE FIRST EDITION

The modern use of steel and high-strength alloys in engineering structures, especially in bridges, ships, and aircraft, has made elastic instability a problem of great importance. Urgent practical requirements
have given rise in recent years to extensive investigations, both theoretical and experimental, of the conditions governing the stability of such
structural elements as bars, plates, and shells. It seems that the time
has come when this work, recorded in various places and languages,
often difficult of access to engineers who need it for guidance in design,
should be brought together and put in the form of a book.

The first problems of elastic instability, concerning lateral buckling
of compressed members, were solved about 200 years ago by L. Euler.[1]
At that time the principal structural materials were wood and stone.
The relatively low strength of these materials necessitated stout structural members for which the question of elastic stability is not of primary
importance. Thus Euler's theoretical solution, developed for slender
bars, remained for a long time without practical application. Only with
the beginning of extensive construction of steel railway bridges during
the latter half of the past century did the question of buckling of compression members become of practical importance. The use of steel
led naturally to types of structures embodying slender compression members, thin plates, and thin shells. Experience showed that such structures may fail in some cases not on account of high stresses, surpassing
the strength of material, but owing to insufficient elastic stability of
slender or thin-walled members.

Under pressure of practical requirements, the problem of lateral
buckling of columns, originated by Euler, has been extensively investigated theoretically and experimentally and the limits within which the
theoretical formulas can be applied have been established. However,
lateral buckling of compressed members is only a particular case of
elastic instability. In the modern design of bridges, ships, and aircraft
we are confronted by a variety of stability problems. We encounter
there not only solid struts, but built-up or "lattice-work" columns, and

[1] Leonard Euler's "Elastic Curves," translated and annotated by W. A. Oldfather,
C. A. Ellis, and D. M. Brown, 1933.

vii

tubular members, where there is the possibility of local buckling, as well as buckling as a whole. In the use of thin sheet material, as in plate girders and airplane structures, we have to keep in mind that thin plates may prove unstable under the action of forces in their own planes, and fail by buckling sideways. Thin cylindrical shells, such as vacuum vessels, which have to withstand uniform external pressure, may exhibit instability and collapse at a relatively low stress if the thickness of the shell is too small in comparison with the diameter. The thin cylindrical shell may buckle also under axial compression, bending, torsion, or combinations of these. All such problems are of the utmost importance in the design of airplanes of the modern monocoque type.

In the discussion of these problems and their solutions, it has not been deemed necessary to include an account of the general theory of elastic stability, which finds its appropriate place in books on the theory of elasticity. This book proceeds directly to particular problems showing in each case under what conditions the question of stability calls for consideration. The various methods of solution are presented in connection with the types of problem to which they are best suited. The solutions have in most cases been supplemented by tables and diagrams which furnish values of critical loads and stresses for each particular case. While all available information relevant to a prescribed problem has been given, no attempt has been made to go beyond this into actual design, since it is a field in which other considerations besides rational theory and testing play their parts.

The preliminary knowledge of mathematics and strength of materials taken for granted is that usually covered by our schools of engineering. Where additional mathematical equipment has been found necessary, it is given in the book with the appropriate explanations. To simplify the reading of the book, problems which, although of practical importance, are such that they can be omitted during a first reading are put in small type. The reader may return to the study of such problems after finishing the more essential portions of the book.

Numerous references to papers and books treating stability problems are given in the book. These references may be of interest to engineers who wish to study some special problems in more detail. They give also a picture of the modern development of the theory of elastic stability and may be of some use to graduate students who are planning to take their work in this field.

In the preparation of this book the contents of a previous book[1] dealing with stability problems and representing a course of lectures on the theory of thin plates and shells, as given in several Russian engineering schools, have been freely drawn upon.

[1] "Theory of Elasticity," vol. II, St. Petersburg, Russia, 1916.

To the University of Michigan the author is grateful for financial support obtained from a research fund and used in the preparation of numerical tables and diagrams for this book. He also takes this opportunity to extend thanks to Dr. D. H. Young who read over the complete manuscript and made many valuable suggestions and corrections, to Professors G. H. MacCullough and H. R. Lloyd who read some portions of the manuscript, to Dr. I. A. Wojtaszak and Mr. S. H. Fillion for the checking of equations and numerical tables, to Dr. Wojtaszak for reading proofs, to Miss Reta Morden for the typing of the manuscript, and to Mr. L. S. Veenstra for the preparation of the drawings.

S. Timoshenko

CONTENTS

CONTENTS xiii

7.9 Buckling of a Bimetallic Strip 310
7.10 Lateral Buckling of a Curved Bar with Circular Axis 313

Chapter 8. Bending of Thin Plates 319

8.1 Pure Bending of Plates 319
8.2 Bending of Plates by Distributed Lateral Load 325
8.3 Combined Bending and Tension or Compression of Plates 332
8.4 Strain Energy in Bending of Plates 335
8.5 Deflections of Rectangular Plates with Simply Supported Edges . . . 340
8.6 Bending of Plates with a Small Initial Curvature 344
8.7 Large Deflections of Plates 346

Chapter 9. Buckling of Thin Plates 348

9.1 Methods of Calculation of Critical Loads 348
9.2 Buckling of Simply Supported Rectangular Plates Uniformly Compressed in One Direction 351
9.3 Buckling of Simply Supported Rectangular Plates Compressed in Two Perpendicular Directions 356
9.4 Buckling of Uniformly Compressed Rectangular Plates Simply Supported along Two Opposite Sides Perpendicular to the Direction of Compression and Having Various Edge Conditions along the Other Two Sides . . 360
9.5 Buckling of a Rectangular Plate Simply Supported along Two Opposite Sides and Uniformly Compressed in the Direction Parallel to Those Sides . 370
9.6 Buckling of a Simply Supported Rectangular Plate under Combined Bending and Compression 373
9.7 Buckling of Rectangular Plates under the Action of Shearing Stresses . 379
9.8 Other Cases of Buckling of Rectangular Plates 385
9.9 Buckling of Circular Plates 389
9.10 Buckling of Plates of Other Shapes 392
9.11 Stability of Plates Reinforced by Ribs 394
9.12 Buckling of Plates beyond Proportional Limit 408
9.13 Large Deflections of Buckled Plates 411
9.14 Ultimate Strength of Buckled Plates 418
9.15 Experiments on Buckling of Plates 423
9.16 Practical Applications of the Theory of Buckling of Plates 429

Chapter 10. Bending of Thin Shells 440

10.1 Deformation of an Element of a Shell 440
10.2 Symmetrical Deformation of a Circular Cylindrical Shell 443
10.3 Inextensional Deformation of a Circular Cylindrical Shell 445
10.4 General Case of Deformation of a Cylindrical Shell 448
10.5 Symmetrical Deformation of a Spherical Shell 453

Chapter 11. Buckling of Shells 457

11.1 Symmetrical Buckling of a Cylindrical Shell under the Action of Uniform Axial Compression 457
11.2 Inextensional Forms of Bending of Cylindrical Shells Due to Instability . 461
11.3 Buckling of a Cylindrical Shell under the Action of Uniform Axial Pressure 462
11.4 Experiments with Cylindrical Shells in Axial Compression 468
11.5 Buckling of a Cylindrical Shell under the Action of Uniform External Lateral Pressure 474

NOTATIONS

a, b, c, d Numerical coefficients, distances
A Cross-sectional area
c Distance from neutral axis to extreme fiber of beam
C Torsional rigidity $(C = GJ)$
C_1 Warping rigidity $(C_1 = EC_w)$
C_w Warping constant
D Flexural rigidity of plate or shell $[D = Eh^3/12\ (1 - \nu^2)]$
e Eccentricity, distance from centroid to shear center
E, E_r, E_t Modulus of elasticity, reduced modulus, tangent modulus
F Stress function
g Acceleration of gravity
G Modulus of elasticity in shear
h Thickness of plate or shell, height, distance
I_c, I_o Polar moments of inertia of a plane area with respect to centroid and shear center
I_x, I_y, I_z Moments of inertia of a plane area with respect to x, y, and z axes
I_{xy} Product of inertia of a plane area with respect to x and y axes
J Torsion constant
k Axial load factor for beam-columns $(k^2 = P/EI)$, modulus of elastic foundation, numerical factor
l Length, span
L Reduced length
m, n Integers, numerical factors
m_z Intensity of torque per unit distance along z axis
M Bending moment, couple
M_t Twisting couple or torque
M_x, M_y, M_{xy} Bending and twisting moments per unit distance in plate or shell
n Factor of safety
N Shearing force in beam, normal force
N_x, N_y, N_{xy} Normal and shearing forces per unit distance in middle surface of plate or shell
p, q Intensity of distributed load, pressure
P Concentrated force, axial force in beam-column
P_{cr} Critical buckling load
Q Shearing force in beam, concentrated force
Q_x, Q_y Shearing forces per unit distance in plate or shell
r Radius of gyration, radius of curvature of shell, radius
R Radius, reactive force
s Core radius $(s = Z/A)$, distance
S Axial force

t Thickness, time, temperature

T Work, tensile force

u Axial load factor for beam-columns ($u = kl/2$)

u, v, w Displacements in x, y, and z directions

U Strain energy

v, w Displacements in tangential and radial directions

V Shearing force in beam

w Warping displacement in beam

x, y, z Rectangular coordinates

Z Section modulus ($Z = I/c$)

α, β Angles, numerical factors, ratios, spring constants, coefficients of end restraint

γ Shearing unit strain, weight per unit volume, spring constant, numerical factor

δ Deflection

ϵ Unit normal strain, coefficient of thermal expansion

$\epsilon_x, \epsilon_y, \epsilon_z$ Unit normal strains in x, y, and z directions

$\eta, \lambda, \phi, \chi, \psi$ Amplification factors for beam-columns

θ Angle, angular coordinate, angle of twist per unit length

λ Distance, numerical factor

ν Poisson's ratio

ξ, η, ζ Rectangular coordinates

ρ Radius of curvature

σ Unit normal stress

$\sigma_x, \sigma_y, \sigma_z$ Unit normal stresses in x, y, and z directions

σ_c Average compressive unit stress for columns

σ_{cr} Compressive unit stress at critical load

σ_{ult} Unit stress at ultimate load

σ_W Working unit stress

σ_{YP} Yield-point stress

τ Unit shear stress

$\tau_{xy}, \tau_{yz}, \tau_{sz}$ Unit shear stresses on planes perpendicular to the x, y, and z axes and parallel to the y, z, and x axes

ϕ Angle, angular coordinate, angle of twist of bar

χ Change of curvature in shell

ω Radian frequency of vibration

ω_s Warping function

CHAPTER 1

BEAM-COLUMNS

1.1. Introduction. In the elementary theory of bending, it is found that stresses and deflections in beams are directly proportional to the applied loads. This condition requires that the change in shape of the beam due to bending must not affect the action of the applied loads. For example, if the beam in Fig. 1-1a is subjected to only lateral loads, such as Q_1 and Q_2, the presence of the small deflections δ_1 and δ_2 and slight changes in the vertical lines of action of the loads will have only an insignificant effect on the moments and shear forces. Thus it is possible to make calculations for deflections, stresses, moments, etc., on the basis of the initial configuration of the beam. Under these conditions, and also if Hooke's law holds for the material, the deflections are proportional to the acting forces and the principle of superposition is valid; i.e., the final deformation is obtained by summation of the deformations produced by the individual forces.

Conditions are entirely different when both axial and lateral loads act simultaneously on the beam (Fig. 1-1b). The bending moments, shear forces, stresses, and deflections in the beam will not be proportional to the magnitude of the axial load. Furthermore, their values will be dependent upon the magnitude of the deflections produced and will be sensitive to even slight eccentricities in the application of the axial load. Beams subjected to axial compression and simultaneously supporting lateral loads are known as *beam-columns*. In this first chapter, beam-columns of symmetrical cross section and with various conditions of support and loading will be analyzed.[1]

1.2. Differential Equations for Beam-columns. The basic equations for the analysis of beam-columns can be derived by considering the beam in Fig. 1-2a. The beam is subjected to an axial compressive force P and to a distributed lateral load of intensity q which varies with the distance x along the beam. An element of length dx between two cross sections taken normal to the original (undeflected) axis of the beam is

[1] For an analysis of beams subjected to axial tension see Timoshenko, "Strength of Materials," 3d ed., part II, p. 41, D. Van Nostrand Company, Inc., Princeton, N.J., 1956.

shown in Fig. 1-2b. The lateral load may be considered as having constant intensity q over the distance dx and will be assumed positive when in the direction of the positive y axis, which is downward in this case. The shearing force V and bending moment M acting on the sides of the element are assumed positive in the directions shown.

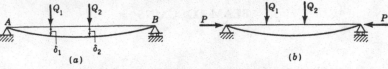

(a) (b)

FIG. 1-1

The relations among load, shearing force V, and bending moment are obtained from the equilibrium of the element in Fig. 1-2b. Summing forces in the y direction gives

$$-V + q\, dx + (V + dV) = 0$$

or
$$q = -\frac{dV}{dx} \tag{1-1}$$

Taking moments about point n and assuming that the angle between the axis of the beam and the horizontal is small, we obtain

$$M + q\, dx\, \frac{dx}{2} + (V + dV)\, dx - (M + dM) + P\frac{dy}{dx}\, dx = 0$$

If terms of second order are neglected, this equation becomes

$$V = \frac{dM}{dx} - P\frac{dy}{dx} \tag{1-2}$$

If the effects of shearing deformations and shortening of the beam axis are neglected, the expression for the curvature of the axis of the beam is

$$EI\frac{d^2y}{dx^2} = -M \tag{1-3}$$

The quantity EI represents the flexural rigidity of the beam in the plane of bending, that is, in the xy plane, which is assumed to be a plane of symmetry. Combining Eq. (1-3) with Eqs. (1-1) and (1-2), we can express the differential equation of the axis of the beam in the following alternate forms:

$$EI\frac{d^3y}{dx^3} + P\frac{dy}{dx} = -V \tag{1-4}$$

and
$$EI\frac{d^4y}{dx^4} + P\frac{d^2y}{dx^2} = q \tag{1-5}$$

Equations (1-1) to (1-5) are the basic differential equations for bending of beam-columns. If the axial force P equals zero, these equations reduce to the usual equations for bending by lateral loads only.

Instead of taking an element dx with sides perpendicular to the x axis (Fig. 1-2b), we can cut an element with sides normal to the deflected axis of the beam (Fig. 1-2c). Since the slope of the beam is small, the normal forces acting on the sides of the element can be taken equal to the axial compressive force P. The shearing force N in this case is related to the shearing force V in Fig. 1-2b by the expression

$$N = V + P \frac{dy}{dx} \tag{a}$$

and instead of Eqs. (1-1) and (1-2) we obtain

$$q = -\frac{dN}{dx} + P\frac{d^2y}{dx^2} \tag{1-1a}$$

and

$$N = \frac{dM}{dx} \tag{1-2a}$$

Equations (1-1a) and (1-2a) can also be derived by considering the equilibrium of the element in Fig. 1-2c. Finally, combining Eq. (1-2a) with Eq. (1-3) yields the equation

$$EI\frac{d^3y}{dx^3} = -N \tag{1-4a}$$

Equation (1-5) remains valid for the element in Fig. 1-2c. Thus we have two sets of differential equations for a beam-column, depending on whether the shearing force is taken on a cross section normal to the deflected or the undeflected axis of the beam.

1.3. Beam-column with a Concentrated Lateral Load. As the first example of the use of the beam-column equations, let us consider a beam of length l on two simple supports (Fig. 1-3) and carrying a single lateral

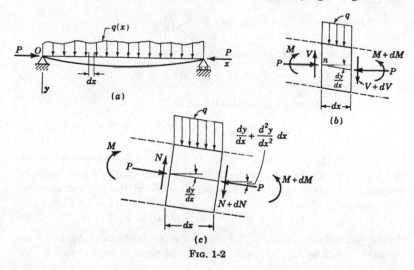

FIG. 1-2

load Q at distance c from the right end.   The bending moments due to the lateral load Q, if acting alone, could be found readily by statics.   How-

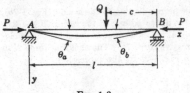

FIG. 1-3

ever, in this case the axial force P causes bending moments which cannot be found until the deflections are determined.   The beam-column is therefore statically indeterminate, and it is necessary to begin by solving the differential equation for the deflection curve of the beam.

The bending moments in the left- and right-hand portions of the beam in Fig. 1-3 are, respectively,

$$M = \frac{Qc}{l} x + Py \qquad M = \frac{Q(l - c)}{l} (l - x) + Py$$

and, therefore, using Eq. (1-3) we obtain

$$EI \frac{d^2y}{dx^2} = - \frac{Qc}{l} x - Py \qquad (a)$$

$$EI \frac{d^2y}{dx^2} = - \frac{Q(l - c)(l - x)}{l} - Py \qquad (b)$$

For simplification the following notation is introduced:

$$k^2 = \frac{P}{EI} \qquad (1\text{-}6)$$

and then Eq. (a) becomes

$$\frac{d^2y}{dx^2} + k^2y = - \frac{Qc}{EIl} x$$

The general solution of this equation is

$$y = A \cos kx + B \sin kx - \frac{Qc}{Pl} x \qquad (c)$$

In the same manner the general solution of Eq. (b) is

$$y = C \cos kx + D \sin kx - \frac{Q(l - c)(l - x)}{Pl} \qquad (d)$$

The constants of integration A, B, C, and D are now determined from the conditions at the ends of the beam and at the point of application of the load Q.   Since the deflections at the ends of the bar are zero, we conclude that

$$A = 0 \qquad C = -D \tan kl \qquad (e)$$

At the point of application of the load Q the two portions of the deflection curve, as given by Eqs. (c) and (d), have the same deflection and a common tangent. These conditions give

$$B \sin k(l - c) - \frac{Qc}{Pl}(l - c)$$

$$= D[\sin k(l - c) - \tan kl \cos k(l - c)] - \frac{Qc}{Pl}(l - c)$$

$$Bk \cos k(l - c) - \frac{Qc}{Pl}$$

$$= Dk[\cos k(l - c) + \tan kl \sin k(l - c)] + \frac{Q(l - c)}{Pl}$$

from which

$$B = \frac{Q \sin kc}{Pk \sin kl} \qquad D = -\frac{Q \sin k(l - c)}{Pk \tan kl} \qquad (f)$$

Substituting into Eqs. (c) and (d) the values of the constants from (e) and (f), we obtain the following equations for the two portions of the deflection curve:

$$y = \frac{Q \sin kc}{Pk \sin kl} \sin kx - \frac{Qc}{Pl} x \qquad 0 \leq x \leq l - c \qquad (1\text{-}7)$$

$$y = \frac{Q \sin k(l - c)}{Pk \sin kl} \sin k(l - x) - \frac{Q(l - c)(l - x)}{Pl} \qquad l - c \leq x \leq l \qquad (1\text{-}8)$$

It is seen that Eq. (1-8) can be obtained from Eq. (1-7) by substituting $l - c$ for c and $l - x$ for x.

By differentiation of Eqs. (1-7) and (1-8) the following formulas, useful in later calculations, are obtained:

$$\frac{dy}{dx} = \frac{Q \sin kc}{P \sin kl} \cos kx - \frac{Qc}{Pl} \qquad 0 \leq x \leq l - c \qquad (1\text{-}9)$$

$$\frac{dy}{dx} = -\frac{Q \sin k(l - c)}{P \sin kl} \cos k(l - x) + \frac{Q(l - c)}{Pl} \qquad l - c \leq x \leq l \qquad (1\text{-}10)$$

$$\frac{d^2y}{dx^2} = -\frac{Qk \sin kc}{P \sin kl} \sin kx \qquad 0 \leq x \leq l - c \qquad (1\text{-}11)$$

$$\frac{d^2y}{dx^2} = -\frac{Qk \sin k(l - c)}{P \sin kl} \sin k(l - x) \qquad l - c \leq x \leq l \qquad (1\text{-}12)$$

In the particular case of a load applied at the center of the beam, the deflection curve is symmetrical and it is necessary to consider only the portion to the left of the load. The maximum deflection in this case is obtained by substituting $x = c = l/2$ in Eq. (1-7), which gives

$$\delta = (y)_{x=l/2} = \frac{Q}{2Pk}\left(\tan \frac{kl}{2} - \frac{kl}{2}\right) \qquad (g)$$

To simplify this equation the following additional notation will be used:

$$u = \frac{kl}{2} = \frac{l}{2}\sqrt{\frac{P}{EI}} \tag{1-13}$$

Then Eq. (g) becomes

$$\delta = \frac{Ql^3}{48EI}\frac{3(\tan u - u)}{u^3} = \frac{Ql^3}{48EI}\chi(u) \tag{1-14}$$

The first factor on the right-hand side of this equation represents the deflection which is obtained if the lateral load Q acts alone. The second factor, $\chi(u)$, gives the influence of the longitudinal force P on the deflection δ. Numerical values of the factor $\chi(u)$ for various values of the quantity u are given in Table A-1 in the Appendix. By using this table, the deflections of the bar can be calculated readily in each particular case from Eq. (1-14).

When P is small, the quantity u is also small [see Eq. (1-13)] and the factor $\chi(u)$ approaches unity. This can be shown by using the series

$$\tan u = u + \frac{u^3}{3} + \frac{2u^5}{15} + \cdots$$

and retaining only the first two terms of this series. It is seen also that $\chi(u)$ becomes infinite when u approaches $\pi/2$. When $u = \pi/2$, we find from Eq. (1-13)

$$P = \frac{\pi^2 EI}{l^2} \tag{1-15}$$

Thus it can be concluded that when the axial compressive force approaches the limiting value given by Eq. (1-15), even the smallest lateral load will produce considerable lateral deflection. This limiting value of the compressive force is called the *critical load* and is denoted by P_{cr}. By using Eq. (1-15) for the critical value of the longitudinal force, the quantity u [see Eq. (1-13)] can be represented in the following form:

$$u = \frac{\pi}{2}\sqrt{\frac{P}{P_{cr}}} \tag{1-16}$$

Thus u depends only on the magnitude of the ratio P/P_{cr}.

To find the slope of the deflection curve at the end of the beam, we substitute $c = l/2$ and $x = 0$ into Eq. (1-9), which gives

$$\begin{aligned}
\left(\frac{dy}{dx}\right)_{x=0} &= \frac{Q}{2P}\left(\frac{1}{\cos kl/2} - 1\right) \\
&= \frac{Ql^2}{16EI}\frac{2(1 - \cos u)}{u^2 \cos u} = \frac{Ql^2}{16EI}\lambda(u)
\end{aligned} \tag{1-17}$$

Again, the first factor is the slope produced by the lateral load Q acting alone at the center of the beam and the second factor represents the effect of the axial load P. Values of the factor $\lambda(u)$ are given in Table A-2 of the Appendix.

By using Eq. (1-11) we obtain the maximum bending moment as follows:

$$M_{max} = -EI \left(\frac{d^2y}{dx^2}\right)_{x=l/2} = \frac{QkEI}{2P} \tan \frac{kl}{2} = \frac{Ql}{4} \frac{\tan u}{u} \qquad (1\text{-}18)$$

The maximum bending moment is obtained in this case by multiplying the bending moment produced by the lateral load by the factor $(\tan u)/u$. The value of this factor, as well as the previous trigonometric factors $\lambda(u)$ and $\chi(u)$, approaches unity as the compressive force becomes smaller and smaller and increases indefinitely when the quantity u approaches $\pi/2$, that is, when the compressive force approaches the critical value given by Eq. (1-15).

1.4. Several Concentrated Loads. The results of the previous article will now be used in the more general case of several lateral loads acting on the compressed beam. Equations (1-7) and (1-8) show that for a given longitudinal force the deflections of the bar are proportional to the lateral load Q. At the same time the relation between deflections and the longitudinal force P is more complicated, since this force enters into the trigonometric functions containing k. The fact that deflections are linear functions of Q indicates that the *principle of superposition*, which is widely used when lateral loads act alone on a beam, can also be applied in the case of the combined action of lateral and axial loads, but in a somewhat modified form. It is seen from Eqs. (1-7) and (1-8) that, if we increase the lateral load Q by an amount Q_1, the resultant deflection is obtained by superposing on the deflections produced by the load Q the deflections produced by the load Q_1, provided the same axial force acts on the bar.

It can be shown that the method of superposition can be used also if several lateral loads are acting on the compressed bar. The resultant deflection is obtained by using Eqs. (1-7) and (1-8) and superposing the separate deflections produced by each lateral load acting in combination with the total axial force. Take the case of two lateral loads Q_1 and Q_2 at the distances c_1 and c_2 from the right support (Fig. 1-4). Proceeding as in the previous article, we find that the differential equation of the deflection curve for the left portion of the beam $(x \leq l - c_2)$ is

$$EI \frac{d^2y}{dx^2} = -\frac{Q_1c_1}{l} x - \frac{Q_2c_2}{l} x - Py \qquad (a)$$

Now consider the loads Q_1 and Q_2 acting separately on the compressed bar, and denote by y_1 the deflections due to Q_1 and by y_2 the deflections caused by Q_2. For these two cases we find the following equations for the deflection curve for the left portion of the beam:

$$EI \frac{d^2 y_1}{dx^2} = - \frac{Q_1 c_1}{l} x - P y_1$$

$$EI \frac{d^2 y_2}{dx^2} = - \frac{Q_2 c_2}{l} x - P y_2$$

By adding these two equations we find

$$EI \frac{d^2(y_1 + y_2)}{dx^2} = - \frac{Q_1 c_1}{l} x - \frac{Q_2 c_2}{l} x - P(y_1 + y_2)$$

It is seen that this equation for the sum of the deflections y_1 and y_2 is the same as Eq. (a) for the deflections obtained when the loads Q_1 and Q_2 were acting simultaneously. The same conclusion also holds for the middle and right-hand portions of the bar. Thus when there are several loads acting on a compressed bar, the resultant deflections can be obtained by superposition of the deflections produced separately by each lateral load acting in combination with the longitudinal force P.

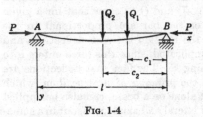

FIG. 1-4

On the basis of this statement we can now write the equation of the deflection curve for any portion of the bar and for any number of lateral loads. Assume that there are n lateral forces $Q_1, Q_2, \ldots, Q_n$ and that their distances from the right support of the beam are $c_1, c_2, \ldots, c_n$, where $c_1 < c_2 < \cdots < c_n$. Then with the use of Eqs. (1-7) and (1-8) for a single lateral load, the deflection curve between the loads Q_m and Q_{m+1} is given by the equation

$$y = \frac{\sin kx}{Pk \sin kl} \sum_{i=1}^{i=m} Q_i \sin kc_i - \frac{x}{Pl} \sum_{i=1}^{i=m} Q_i c_i$$

$$+ \frac{\sin k(l - x)}{Pk \sin kl} \sum_{i=m+1}^{i=n} Q_i \sin k(l - c_i) - \frac{l - x}{Pl} \sum_{i=m+1}^{i=n} Q_i(l - c_i) \quad (1\text{-}19)$$

In the same manner, by using Eqs. (1-9) to (1-12), we can obtain the slope of the deflection curve and the bending moment at any cross section of the beam. Thus, when the method of superposition is used in its modified form, the general problem of calculating deflections for a beam submitted to the action of several lateral loads together with an axial force is solved.

1.5. Continuous Lateral Load. The method of superposition, described in the previous article, can be used in the case of continuous loads also. The formulas of the preceding article can be adapted to this case by replacing summations by integrations. Let us take the case of a *uniform lateral load* of intensity q acting on a compressed bar with hinged ends (Fig. 1-5), and let c denote the variable distance from the right-hand support to an element $q\, dc$ of the continuous load. This element can be considered as an infinitesimally small concentrated force, and the uniform load can be replaced by a system of such infinitesimally small concentrated forces. Then, using Eq. (1-19) and replacing the sum-

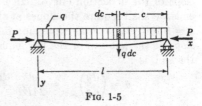

Fig. 1-5

mation from $i = 1$ to $i = m$ by an integration from 0 to $l - x$ and the summation from $i = m + 1$ to $i = n$ by an integration from $l - x$ to l, we obtain

$$y = \frac{\sin kx}{Pk \sin kl} \int_0^{l-x} q \sin kc\, dc - \frac{x}{Pl} \int_0^{l-x} qc\, dc$$
$$+ \frac{\sin k(l - x)}{Pk \sin kl} \int_{l-x}^{l} q \sin k(l - c)\, dc - \frac{l - x}{Pl} \int_{l-x}^{l} q(l - c)\, dc \quad (a)$$

After integration and using notation (1-13), we obtain the following equation of the deflection curve:

$$y = \frac{ql^4}{16EIu^4} \left[\frac{\cos (u - 2ux/l)}{\cos u} - 1 \right] - \frac{ql^2}{8EIu^2} x(l - x) \quad (1\text{-}20)$$

The deflection at the middle of the beam is found by substituting $x = l/2$ into Eq. (1-20). With the aid of some transformations we can express the result in the form

$$\delta = (y)_{x=l/2} = \frac{5ql^4}{384EI} \frac{12(2 \sec u - 2 - u^2)}{5u^4}$$
$$= \frac{5ql^4}{384EI} \eta(u) \quad (1\text{-}21)$$

The first factor on the right-hand side of this equation represents the deflection at the center produced by the lateral load q acting alone, and the second factor $\eta(u)$ shows the effect of the longitudinal compressive force P on the deflection. By expanding sec u in the form of a series, it can be shown that the second factor approaches unity when u approaches zero and increases indefinitely as u approaches $\pi/2$, that is, when P approaches the critical value [Eq. (1-15)]. Thus the effect of the axial load P on the deflection depends on the value of u and hence on the value of the ratio P/P_{cr} [see Eq. (1-16)]. If this ratio is small, the effect of P on the deflection is also small, but as the ratio approaches unity, the effect of P grows indefinitely. The same conclusion is obtained for other kinds of lateral loading. Values of the factor $\eta(u)$ are given in Table A-2 in the Appendix.

By differentiating Eq. (1-20) we can find the general expression for the slope of the deflection curve. In later investigations of bars with fixed ends we shall require the slopes at the ends of the bar. By substituting $x = 0$ into the general expression for the slope, it can be shown that the slope at the left end of the bar, equal to the small angle of rotation θ of the end, is

$$\theta = \left(\frac{dy}{dx}\right)_{x=0} = \frac{ql^3}{24EI}\,\frac{3(\tan u - u)}{u^3} = \frac{ql^3}{24EI}\,\chi(u) \qquad (1\text{-}22)$$

The first factor on the right-hand side is the known formula for the slope at the end of the beam when the uniform load acts alone. The second factor is $\chi(u)$ and represents the effect on the slope of the longitudinal force P. It was shown before that the factor $\chi(u)$ approaches unity when u approaches zero and that it increases indefinitely when u approaches $\pi/2$ and P approaches the critical value.

For calculating the maximum bending moment, Eq. (1-20) for y must be differentiated twice. The maximum bending moment, which in this case is at the center of the span, is

$$M_{max} = -EI\left(\frac{d^2y}{dx^2}\right)_{x=l/2} = \frac{ql^2}{8}\,\frac{2(1 - \cos u)}{u^2 \cos u} = \frac{ql^2}{8}\,\lambda(u) \qquad (1\text{-}23)$$

The same result can be obtained in another way by adding to the moment $ql^2/8$ the moment $P\delta$ due to the longitudinal force. Substituting for δ its value from Eq. (1-21), we obtain

$$M_{max} = \frac{ql^2}{8} + \frac{5ql^4}{384EI}\,\frac{12(2 \sec u - 2 - u^2)}{5u^4}\,P$$

Substituting into this expression $P = k^2EI$ and using notation (1-13), we can bring this result into the same form as Eq. (1-23). The first factor in Eq. (1-23) represents the bending moment produced by the uniform load alone, and the second factor gives the effect of the longitudinal force P on the maximum bending moment. It was mentioned before (see p. 7) that the factor $\lambda(u)$ is near unity for small values of the ratio P/P_{cr} and that it increases indefinitely as P approaches P_{cr}.

The method of superposition in the calculation of deflections can be applied also in the case where the load is distributed along only a portion of the span (Fig. 1-6a). To find, for example, the deflection curve for the portion of the beam to the left of the load, we use Eq. (1-7) for a concentrated load on the span. The deflection produced by one element $q\,dc$ of the total load is obtained by substituting $q\,dc$ for Q in Eq. (1-7). The deflection produced by the total load is then found by integrating between the limits $c = a$ and $c = b$. In this way we obtain the deflection curve for the left portion of the beam in the following form:

$$y = \int_a^b \frac{q\,dc \sin kc}{Pk \sin kl} \sin kx - x \int_a^b \frac{qc\,dc}{Pl} \qquad (b)$$

If it is necessary to find the deflection at any point m under the load (Fig. 1-6a), we use Eq. (1-7) for the load to the right of m and Eq. (1-8) for the load to the left of m. Then the required deflection is

$$y = \int_a^{l-x} \frac{q\,dc \sin kc}{Pk \sin kl} \sin kx - x \int_a^{l-x} \frac{qc\,dc}{Pl}$$
$$+ \int_{l-x}^b \frac{q\,dc \sin k(l-c)}{Pk \sin kl} \sin k(l-x) - \int_{l-x}^b \frac{q\,dc(l-c)(l-x)}{Pl} \qquad (c)$$

When the integrations are carried out as indicated, the equation for the deflection curve under the load is obtained. By substituting $a = 0$ and $b = l$ into that equation we obtain Eq. (1-20) for a uniformly loaded beam.

If q is not constant but is a certain function of c, we can obtain the deflection curve from Eqs. (b) and (c) by substituting for q the given function of c. For instance, in the case shown in Fig. 1-6b, the deflection curve is obtained by substituting $q = q_0 c/l$ into Eq. (c) and also taking $a = 0$, $b = l$.

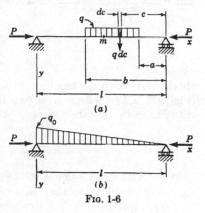

Fig. 1-6

In the preceding examples the deflection curve for a beam-column was found either by using the differential equation (1-3) or by superposing certain known results. An alternate way of determining the deflection curve is to begin with the differential equation (1-5). For example, for the beam shown in Fig. 1-5, carrying a uniform load of constant intensity q, the equation is

$$EI\frac{d^4y}{dx^4} + P\frac{d^2y}{dx^2} = q$$

The general solution of this equation is

$$y = A \sin kx + B \cos kx + Cx + D + \frac{qx^2}{2P} \qquad (d)$$

where A, B, C, and D are constants of integration that must be evaluated from the conditions at the ends of the beam. Since the deflection and bending moment are zero at the ends of the bar, these conditions are

$$y = \frac{d^2y}{dx^2} = 0 \qquad \text{at } x = 0 \text{ and } x = l$$

From the two conditions at $x = 0$ we obtain

$$B = -D = \frac{q}{k^2 P}$$

and the conditions at $x = l$ give

$$A = \frac{q}{k^2 P} \frac{1 - \cos kl}{\sin kl} \qquad C = -\frac{ql}{2P}$$

Substituting the values of the constants into Eq. (d) gives the equation for the deflection curve of the beam-column. When some trigonometric substitutions are made and also Eq. (1-13) is used, this result can be shown to be identical with Eq. (1-20).

1.6. Bending of a Beam-column by Couples. Knowing the solution for a single concentrated force Q (Fig. 1-3) it is not difficult to obtain the equation of the deflection curve for the case when a couple is applied at the end of the beam. For this purpose assume that the distance c in Fig. 1-3 approaches zero and at the same time Q is increasing, so that the product Qc remains finite and equal to M_b. By this means we

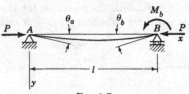

Fig. 1-7

obtain in the limit the bending of the bar by a couple M_b at the right end (Fig. 1-7). The deflection curve is then obtained from Eq. (1-7) by substituting in this equation $\sin kc = kc$ and $Qc = M_b$, giving

$$y = \frac{M_b}{P} \left(\frac{\sin kx}{\sin kl} - \frac{x}{l} \right) \tag{1-24}$$

In our further discussion it will be necessary to have the formulas giving the small angles of rotation θ_a and θ_b of the ends of the bar. These angles are considered as positive when the ends rotate in the direction of positive bending moment as shown in Fig. 1-7. Taking the derivative of Eq. (1-24), we obtain

$$\theta_a = \left(\frac{dy}{dx} \right)_{x=0} = \frac{M_b}{P} \left(\frac{k}{\sin kl} - \frac{1}{l} \right) = \frac{M_b l}{6EI} \frac{3}{u} \left(\frac{1}{\sin 2u} - \frac{1}{2u} \right) \tag{1-25}$$

$$\theta_b = -\left(\frac{dy}{dx} \right)_{x=l} = -\frac{M_b}{P} \left(\frac{k \cos kl}{\sin kl} - \frac{1}{l} \right) = \frac{M_b l}{3EI} \frac{3}{2u} \left(\frac{1}{2u} - \frac{1}{\tan 2u} \right) \tag{1-26}$$

We see that the known expressions $M_b l/6EI$ and $M_b l/3EI$, for the angles produced by the couple M_b acting alone, are multiplied by trigonometric factors representing the influence of the axial force P on the angles of rotation of the ends of the bar. It is not difficult to show that these factors approach unity when u approaches zero and increase indefinitely as u approaches $\pi/2$. In subsequent equations the following notation will be used in order to simplify the expressions:

$$\phi(u) = \frac{3}{u}\left(\frac{1}{\sin 2u} - \frac{1}{2u}\right) \tag{1-27}$$

$$\psi(u) = \frac{3}{2u}\left(\frac{1}{2u} - \frac{1}{\tan 2u}\right) \tag{1-28}$$

Numerical values of these functions are given in Table A-1 in the Appendix.

If two couples M_a and M_b are applied at the ends A and B of the bar (Fig. 1-8a), the deflection curve can be obtained by superposition. From Eq. (1-24) we obtain the deflections produced by the couple M_b.

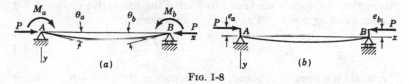

FIG. 1-8

Then by substituting M_a for M_b and $(l - x)$ for x in the same equation, we find the deflections produced by the couple M_a. Adding these results together, we obtain the deflection curve for the case represented in Fig. 1-8a:

$$y = \frac{M_b}{P}\left(\frac{\sin kx}{\sin kl} - \frac{x}{l}\right) + \frac{M_a}{P}\left[\frac{\sin k(l - x)}{\sin kl} - \frac{l - x}{l}\right] \tag{1-29}$$

This type of loading occurs where two eccentrically applied compressive forces P act as shown in Fig. 1-8b. Substituting $M_a = Pe_a$ and $M_b = Pe_b$ in Eq. (1-29), we obtain

$$y = e_b\left(\frac{\sin kx}{\sin kl} - \frac{x}{l}\right) + e_a\left[\frac{\sin k(l - x)}{\sin kl} - \frac{l - x}{l}\right] \tag{1-30}$$

The angles θ_a and θ_b giving the rotation of the ends of the beam in Fig. 1-8a are obtained by using Eqs. (1-25) and (1-26) and notations (1-27) and (1-28). Then, by superposition, we obtain

$$\theta_a = \frac{M_a l}{3EI}\psi(u) + \frac{M_b l}{6EI}\phi(u)$$

$$\theta_b = \frac{M_b l}{3EI}\psi(u) + \frac{M_a l}{6EI}\phi(u) \tag{1-31}$$

Using for $\phi(u)$ and $\psi(u)$ their values from Table A-1, Appendix, the angles θ_a and θ_b can be obtained readily from Eqs. (1-31). These equations will be used frequently in discussing various cases of beams with redundant constraints at the ends.

In the case of two equal couples $M_a = M_b = M_0$, we obtain from Eq. (1-29)

$$y = \frac{M_0}{P \cos{(kl/2)}} \left[\cos\left(\frac{kl}{2} - kx\right) - \cos\frac{kl}{2} \right]$$
$$= \frac{M_0 l^2}{8EI} \frac{2}{u^2 \cos u} \left[\cos\left(u - \frac{2ux}{l}\right) - \cos u \right] \qquad (1\text{-}32)$$

The deflection at the center of the bar is obtained by substituting $x = l/2$, which yields the result

$$\delta = (y)_{x=l/2} = \frac{M_0 l^2}{8EI} \frac{2(1 - \cos u)}{u^2 \cos u} = \frac{M_0 l^2}{8EI} \lambda(u) \qquad (1\text{-}33)$$

The angles at the ends are found by taking the derivative of Eq. (1-32) and substituting $x = 0$. The resulting expression is

$$\theta_a = \theta_b = \left(\frac{dy}{dx}\right)_{x=0} = \frac{M_0 l}{2EI} \frac{\tan u}{u} \qquad (1\text{-}34)$$

The maximum bending moment, which occurs at the middle of the bar, is obtained by using the second derivative of Eq. (1-32), from which

$$M_{\max} = -EI \left(\frac{d^2 y}{dx^2}\right)_{x=l/2} = M_0 \sec u \qquad (1\text{-}35)$$

Equation (1-35) can be used in calculating the maximum bending moment in a bar with eccentrically applied compressive forces (Fig. 1-8b) when both eccentricities are equal. When the longitudinal force P is small in comparison with its critical value (Eq. 1-15), the quantity u is small and sec u can be taken equal to unity; that is, the bending moment can be assumed constant along the length of the bar. As u approaches $\pi/2$ and P approaches P_{cr}, sec u increases indefinitely. At such values of P the slightest eccentricity in the application of the load produces a considerable bending moment at the center of the bar. A discussion of working stresses for such cases will be given in Art. 1.13.

1.7. Approximate Formula for Deflections. In making preliminary design computations, it is frequently useful to have an approximate formula for determining the deflection at the center of a beam-column with simply supported ends. In the preceding articles, equations were derived for the center deflection under three conditions of symmetrical loading (concentrated load at center, uniform load, and two equal end moments). In each case, the deflection is equal to the product of two terms, the first term being the deflection without axial load and the second term being an *amplification factor* which depends upon the value of u and hence upon the ratio P/P_{cr} [see Eq. (1-16)]. The amplification factors for the three cases of loading are $\chi(u)$, $\eta(u)$, and $\lambda(u)$, respectively.

An approximate expression for the amplification factor[1] is

$$\frac{1}{1 - P/P_{cr}} \tag{1-36}$$

This simplified expression can be used with good accuracy, in place of the exact factors $\chi(u)$, $\eta(u)$, and $\lambda(u)$, if the ratio P/P_{cr} is not large. A plot of the amplification factor is given in Fig. 1-9. For values of P/P_{cr} less than 0.6, the error in the approximate expression[2] is less than 2 per cent.

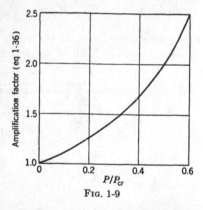

FIG. 1-9

1.8. Beam-columns with Built-in Ends. By using the results of the previous articles and the method of superposition (see Art. 1.4), we can solve various statically indeterminate problems. Take, for instance, the uniformly loaded and compressed beam shown in Fig. 1-10. The uniformly loaded beam is simply supported at end A and built in at end B. The statically indeterminate reactive moment M_b at the support B is obtained from the condition that the tangent to the deflection curve at the built-in end must remain horizontal. Therefore the rotation of the end B produced by the uniform load [Eq. (1-22)] plus the rotation from the action of the moment M_b [found from the second of Eqs. (1-31)] must be zero. In this way we find

$$\frac{ql^3}{24EI} \chi(u) + \frac{M_b l}{3EI} \psi(u) = 0$$

from which
$$M_b = -\frac{ql^2}{8} \frac{\chi(u)}{\psi(u)} \tag{1-37}$$

The negative sign of this result indicates that M_b acts opposite to the direction assumed in Fig. 1-10 and produces bending convex upward. The calculation of the moment M_b can be carried out easily by using Table A-1, Appendix. Having found M_b from Eq. (1-37), we obtain

---

[1] A derivation of this factor is given in Art. 1.11.

[2] Expression (1-36) was presented by Timoshenko in *Bull. Polytech. Inst., Kiev*, 1909.

the deflection curve by superposing on the deflection produced by the uniform load [Eq. (1-20)] the deflection produced by the moment M_b [Eq. (1-24)].

If the uniformly loaded beam has both ends built in (Fig. 1-11), the deflection curve is symmetrical and the moments at the built-in ends are equal ($M_a = M_b = M_0$). The magnitude of these moments is obtained from the condition that the rotation of the ends [Eq. (1-22)] produced by the uniform load is eliminated by the moments acting at the ends [Eq. (1-34)]. Then

$$\frac{ql^3}{24EI}\,\chi(u) + \frac{M_0 l}{2EI}\,\frac{\tan u}{u} = 0$$

from which
$$M_0 = -\frac{ql^2}{12}\,\frac{\chi(u)}{(\tan u)/u} \tag{1-38}$$

Again, the minus sign for the result indicates that the moments are in directions opposite to those assumed in Fig. 1-11. With the end moments

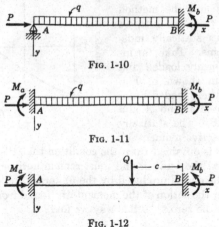

FIG. 1-10

FIG. 1-11

FIG. 1-12

determined, the deflection curve is found by superposing on the deflections produced by the uniform load [Eq. (1-20)] the deflections produced by the two equal moments applied at the ends [Eq. (1-32)]. Similarly, the bending moment at the middle is obtained by superposing the bending moment produced by the uniform load [Eq. (1-23)] and the moment produced by the couples M_0 [see Eqs. (1-35) and (1-38)], which gives

$$(M)_{x=l/2} = \frac{ql^2}{8}\,\lambda(u) - \frac{ql^2}{12}\,\frac{\chi(u)}{(\sin u)/u} = \frac{ql^2}{24}\,\frac{6(u - \sin u)}{u^2 \sin u} \tag{1-39}$$

The trigonometric function appearing on the right-hand side of this equation can be evaluated with the aid of Table A-1, Appendix, by substituting u for $2u$ in the expression for $\phi(u)$.

When the lateral load acting on a beam-column with built-in ends is unsymmetrical, the moments at the ends are found from the conditions that the slopes at the ends are zero. Thus, for the beam shown in Fig. 1-12 the equations for finding the end moments M_a and M_b are

$$
\begin{aligned}
\theta_a &= \theta_{0a} + \frac{M_a l}{3EI} \psi(u) + \frac{M_b l}{6EI} \phi(u) = 0 \\
\theta_b &= \theta_{0b} + \frac{M_a l}{6EI} \phi(u) + \frac{M_b l}{3EI} \psi(u) = 0
\end{aligned}
\tag{1-40}
$$

In these equations, the terms θ_{0a} and θ_{0b} represent the angles of rotation[1] at the ends of the beam with hinged ends when the load Q acts alone and can be found from Eqs. (1-9) and (1-10).

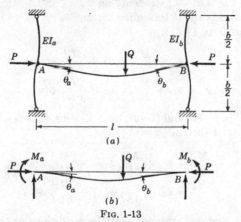

FIG. 1-13

1.9. Beam-columns with Elastic Restraints. As a more general case of a statically indeterminate problem, let us consider a bar with elastically built-in ends. An example of such end conditions is represented in Fig. 1-13a. A laterally loaded beam AB is rigidly connected to vertical bars at A and B and is compressed axially by the forces P. If θ_a and θ_b are the angles of rotation of the ends, there will be couples M_a and M_b at the ends of the beam (see Fig. 1-13b) which we can express in the form

$$
M_a = -\alpha\theta_a \qquad M_b = -\beta\theta_b
\tag{1-41}
$$

The moments and angles of rotation are taken positive in the directions shown in Fig. 1-13. The factors α and β are coefficients defining the

---

[1] The angles of rotation and the moments M_a and M_b are taken positive in the directions shown in Fig. 1-8a.

degree of fixity existing at the ends of the beam and are called *coefficients of end restraint*. The coefficient at one end is equal numerically to the reactive moment at that end when the angle of rotation is equal to unity. The value of the coefficient may vary from zero for a simply supported end to infinity for a built-in end. If, for example, EI_a is the flexural rigidity of the vertical bar at A and the vertical bar is assumed to have hinged ends, the relation between the angle of rotation θ_a and the moment M_a is

$$\theta_a = -\frac{M_a b}{12EI_a}$$

and therefore

$$\alpha = \frac{12EI_a}{b}$$

The angles θ_a and θ_b can now be determined from a consideration of the bending of the bar AB. Again denoting by θ_{0a} and θ_{0b} the angles calculated for hinged ends and determining the angles produced by the couples M_a and M_b from Eqs. (1-31), we obtain

$$\theta_a = \theta_{0a} + \frac{M_a l}{3EI} \psi(u) + \frac{M_b l}{6EI} \phi(u)$$
$$\theta_b = \theta_{0b} + \frac{M_b l}{3EI} \psi(u) + \frac{M_a l}{6EI} \phi(u) \tag{1-42}$$

Finally, from Eqs. (1-41) and (1-42), the following equations for determining the moments at the ends are obtained:

$$-\frac{M_a}{\alpha} = \theta_{0a} + \frac{M_a l}{3EI} \psi(u) + \frac{M_b l}{6EI} \phi(u)$$
$$-\frac{M_b}{\beta} = \theta_{0b} + \frac{M_b l}{3EI} \psi(u) + \frac{M_a l}{6EI} \phi(u) \tag{1-43}$$

When these equations are used, various conditions at the ends of the bar AB can be considered. Taking, for instance, $\alpha = 0$ and $\beta = \infty$, we obtain the case represented in Fig. 1-10, in which the left end of the bar is free to rotate and the right end is rigidly built in. In this case $M_a = 0$ and the moment at the end B, from the second of Eqs. (1-43), is

$$M_b = -\frac{3EI\theta_{0b}}{l\psi(u)} \tag{a}$$

If the bar AB is uniformly loaded, we obtain from Eq. (1-22)

$$\theta_{0b} = \frac{ql^3}{24EI} \chi(u) \tag{b}$$

Substituting expression (b) into Eq. (a), we obtain for the moment at the built-in end the same result as was found previously [see Eq. (1-37)].

By taking $\alpha = \beta = \infty$, we obtain the case of a bar with built-in ends, and Eqs. (1-43) reduce to Eqs. (1-40) of the preceding article. Several applications of Eqs. (1-43) will be given later in discussing stability problems (see Art. 2.3).

1.10. Continuous Beams with Axial Loads. *Continuous Beams on Rigid Supports.* In the case of a continuous beam on rigid supports with both lateral and axial loads, it is advantageous to consider the bending moments at the supports as the statically indeterminate quantities. Let $1, 2, 3, \ldots, m$ denote the consecutive supports; $M_1, M_2, \ldots, M_m$ the corresponding bending moments; $l_1, l_2, \ldots, l_{m-1}$ the span lengths; and $u_1, u_2, \ldots, u_{m-1}$ the corresponding values of the quantity u for each span [from Eq. (1-13)]. The compressive force and the flexural rigidity may vary from one span to the next, but within each span these quantities are assumed constant.

Let us consider any two consecutive spans between the supports $n - 1$, n, and $n + 1$, as shown in Fig. 1-14. The bending moments at the supports are assumed positive in the directions shown in the figure,

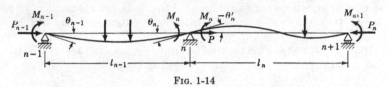

Fig. 1-14

that is, when they cause compression on the top of the beam. The angles of rotation are taken positive when they are in the same directions as the positive bending moments.

The relation among the bending moments M_{n-1}, M_n, and M_{n+1} is obtained from the condition that at the intermediate support n the deflection curves of the two spans have the same tangent. The angle of rotation θ_n of the right end of span $n - 1$ must be equal in magnitude but opposite in sign to the angle of rotation θ'_n of the left end of span n, and therefore

$$\theta_n = -\theta'_n \qquad (a)$$

The angles of rotation are found by considering each span as a simply supported beam subjected to lateral load and to end moments. The expression for θ_n will therefore consist of two parts, the first depending on the lateral load and denoted by θ_{0n} and the second depending on the moments M_{n-1} and M_n and found from Eqs. (1-31). Thus

$$\theta_n = \theta_{0n} + \frac{M_{n-1}l_{n-1}}{6EI_{n-1}} \phi(u_{n-1}) + \frac{M_n l_{n-1}}{3EI_{n-1}} \psi(u_{n-1})$$

A similar expression can be written for the angle θ'_n, and then Eq. (a) becomes

$$\theta_{0n} + \frac{M_{n-1}l_{n-1}}{6EI_{n-1}} \phi(u_{n-1}) + \frac{M_n l_{n-1}}{3EI_{n-1}} \psi(u_{n-1})$$
$$= -\left[\theta'_{0n} + \frac{M_n l_n}{3EI_n} \psi(u_n) + \frac{M_{n+1}l_n}{6EI_n} \phi(u_n) \right]$$

from which

$$M_{n-1}\phi(u_{n-1}) + 2M_n\left[\psi(u_{n-1}) + \frac{l_n}{l_{n-1}} \frac{I_{n-1}}{I_n} \psi(u_n) \right]$$
$$+ M_{n+1} \frac{l_n}{l_{n-1}} \frac{I_{n-1}}{I_n} \phi(u_n) = -\frac{6EI_{n-1}}{l_{n-1}} (\theta_{0n} + \theta'_{0n}) \quad (1\text{-}44)$$

The angles θ_{0n} and θ'_{0n} for any kind of lateral loading can be calculated by the methods explained in Arts. 1.3 to 1.5. Thus Eq. (1-44) contains only three unknown quantities, the moments M_{n-1}, M_n, and M_{n+1}. Writing Eq. (1-44) for each intermediate support of the continuous beam and also using the conditions at the first and last supports, we obtain a sufficient number of equations for calculating all the unknown moments. Equation (1-44) is the *three-moment equation* for a continuous beam with axial loads.

If uniformly distributed loads of intensities q_{n-1} and q_n act on the spans $n - 1$ and n, respectively, we have, from Eq. (1-22),

$$\theta_{0n} = \frac{q_{n-1}l_{n-1}^3}{24EI_{n-1}} \chi(u_{n-1})$$
$$\theta'_{0n} = \frac{q_n l_n^3}{24EI_n} \chi(u_n)$$

and Eq. (1-44) becomes

$$M_{n-1}\phi(u_{n-1}) + 2M_n\left[\psi(u_{n-1}) + \frac{l_n}{l_{n-1}} \frac{I_{n-1}}{I_n} \psi(u_n) \right]$$
$$+ M_{n+1} \frac{l_n}{l_{n-1}} \frac{I_{n-1}}{I_n} \phi(u_n) = -\frac{q_{n-1}l_{n-1}^2}{4} \chi(u_{n-1})$$
$$-\frac{q_n l_n^2}{4} \frac{I_{n-1}}{I_n} \frac{l_n}{l_{n-1}} \chi(u_n) \quad (1\text{-}45)$$

The numerical calculation of bending moments from these equations is greatly simplified by using values of the functions $\phi(u)$, $\psi(u)$, and $\chi(u)$ from Table A-1, Appendix.

Continuous Beams with Supports Not on a Straight Line. If an initially straight compressed bar is held on rigid supports which are not on a straight line, additional bending is introduced. Let $n - 1$, n, and $n + 1$ be three consecutive supports of the continuous beam (Fig. 1-15); h_{n-1}, h_n, and h_{n+1} are the corresponding ordinates to the supports. We assume that the differences between these ordinates are small so that the angle β_n between the two consecutive spans is given with sufficient accuracy by the equation

$$\beta_n = \frac{h_n - h_{n-1}}{l_{n-1}} - \frac{h_{n+1} - h_n}{l_n} \tag{b}$$

The bending moments M_{n-1}, M_n, and M_{n+1}, caused by the differences in elevation of the supports, are now found from the conditions of continuity. Considering the spans $n - 1$ and n as two simple beams, we conclude that the rotation of the ends of these two beams at the common support n must be such as to eliminate the angle β_n. Then by using Eqs. (1-31) we find

$$\frac{M_n l_{n-1}}{3EI_{n-1}} \psi(u_{n-1}) + \frac{M_{n-1} l_{n-1}}{6EI_{n-1}} \phi(u_{n-1}) + \frac{M_n l_n}{3EI_n} \psi(u_n) + \frac{M_{n+1} l_n}{6EI_n} \phi(u_n) = \beta_n$$

from which

$$M_{n-1}\phi(u_{n-1}) + 2M_n \left[\psi(u_{n-1}) + \frac{l_n}{l_{n-1}} \frac{I_{n-1}}{I_n} \psi(u_n) \right]$$
$$+ M_{n+1} \frac{l_n}{l_{n-1}} \frac{I_{n-1}}{I_n} \phi(u_n) = \frac{6EI_{n-1}}{l_{n-1}} \beta_n \tag{1-46}$$

This is the three-moment equation for supports not on the same level. If the positions of all supports are known, β_n can be calculated without difficulty for each intermediate

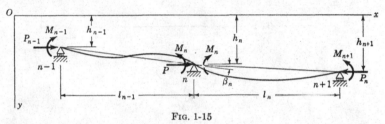

Fig. 1-15

support. The bending moments at the supports are then determined by solution of simultaneous equations (1-46) written for each support. If lateral loads also act on the beam, the three-moment equation is found by combining the right-hand sides of Eqs. (1-44) and (1-46).

Continuous Beams Rigidly Connected to Columns. If the cross sections at the supports of a continuous beam are not free to rotate, being rigidly attached to columns as shown in Fig. 1-16a, the bending moments M_n and M'_n at the two adjacent cross sections to the left and to the right of the support n are not equal. The relation between them is given by the equation of equilibrium of the joint n (Fig. 1-16b):

$$M_n - M'_n + M''_n = 0 \tag{c}$$

Proceeding now as before, and using Eq. (a), we obtain[1]

$$\theta_{0n} + \frac{M'_{n-1} l_{n-1}}{6EI_{n-1}} \phi(u_{n-1}) + \frac{M_n l_{n-1}}{3EI_{n-1}} \psi(u_{n-1})$$
$$= - \left[\theta'_{0n} + \frac{M'_n l_n}{3EI_n} \psi(u_n) + \frac{M_{n+1} l_n}{6EI_n} \phi(u_n) \right] \tag{1-47}$$

[1] The change in axial forces of the beam due to the horizontal reactions of the columns at the supports $n - 1$, n, and $n + 1$ is neglected in this derivation.

or

$$M'_{n-1}\phi(u_{n-1}) + 2M_n\psi(u_{n-1}) + 2M'_n \frac{l_n}{l_{n-1}} \frac{I_{n-1}}{I_n} \psi(u_n)$$

$$+ M_{n+1} \frac{l_n}{l_{n-1}} \frac{I_{n-1}}{I_n} \phi(u_n) = -\frac{6EI_{n-1}}{l_{n-1}} (\theta_{0n} + \theta'_{0n}) \quad (1\text{-}48)$$

Another equation for the same joint is obtained from a consideration of bending of the column. Assuming that the joint n has no lateral displacement, the moment M''_n, representing the action of the joint n on the column (Fig. 1-16a), can be represented by the equation

$$M''_n = \alpha_n \theta_n \qquad (d)$$

where α_n is the coefficient of restraint [see Eq. (1-41)] for the support n. For instance,

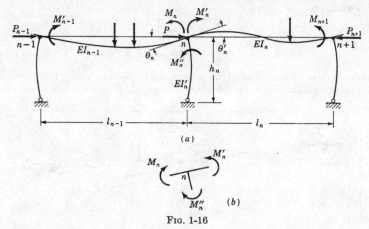

(a)

(b)

Fig. 1-16

in the case of a column hinged at the bottom and having a flexural rigidity EI'_n and a length h_a, we obtain

$$\theta_n = \frac{M''_n h_n}{3EI'_n} \qquad (e)$$

This equation neglects the effect of axial force on bending of the column. From Eq. (e) we now obtain

$$M''_n = \frac{3EI'_n \theta_n}{h_n} \qquad \text{and} \qquad \alpha_n = \frac{3EI'_n}{h_n}$$

Thus α_n can be found for any particular case.

Returning now to Eq. (d) and combining with Eq. (c), we obtain

$$\theta_n = \frac{M''_n}{\alpha_n} = \frac{M'_n - M_n}{\alpha_n}$$

Noting that the left-hand side of Eq. (1-47) is also equal to θ_n, the following additional equation is obtained for each joint:

$$\frac{1}{\alpha_n} (M'_n - M_n) = \theta_{0n} + \frac{M'_{n-1}l_{n-1}}{6EI_{n-1}} \phi(u_{n-1}) + \frac{M_n l_{n-1}}{3EI_{n-1}} \psi(u_{n-1}) \qquad (1\text{-}49)$$

For each intermediate support we can write the two Eqs. (1-48) and (1-49). Thus we have sufficient equations to determine all statically indeterminate moments if the ends of the beam are simply supported. If they are built in, the two additional equations expressing the conditions of fixity of the ends should be added.[1] The solution of these equations is greatly simplified by using Table A-1, Appendix, for the functions $\phi(u)$ and $\psi(u)$.

Another means of analyzing continuous beams and frameworks with combined axial and bending loads is by the method of *moment distribution*. This is done by using values of stiffness factors, carry-over factors, etc., which have been modified to include the axial load effect. Values of these factors are available in graphical and tabular form.[2] With these factors determined, the moment-distribution calculations are carried out in the standard way.

Continuous Beams on Elastic Supports. If the intermediate supports of the beam are elastic, i.e., such that they deflect in proportion to the reactive forces, Eq. (1-46) can still be used. However, if the compressive force P is constant along the length of the beam and the ends are rigidly supported, it is advantageous to take the intermediate reactions as the statically indeterminate quantities. For determining these reactions we use Eq. (1-19), p. 8, and change the previous notations accordingly. The distances of the intermediate supports from the right end of the continuous beam we denote by $c_1, c_2, \ldots, c_n (c_1 < c_2 < c_3 \cdots)$, and the corresponding reactions by

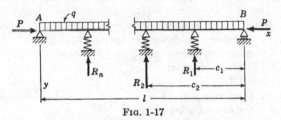

FIG. 1-17

$R_1, R_2, \ldots, R_n$ (Fig. 1-17). The deflection of the continuous beam at any point can be calculated as for a simple beam AB of length l on which are acting a given lateral loading and the unknown reactions $R_1, R_2, \ldots$. Assume that we have found in this way the deflection for any support m. The same deflection can be found in another way, by considering the elasticity of the support. Let α_m be the load necessary to produce a unit deflection of the support m. Then under the action of the pressure R_m, equal to the reaction of the support m, the deflection will be R_m/α_m. Putting this deflection equal to the above calculated deflection, we obtain an equation containing the intermediate reactions $R_1, R_2, \ldots, R_n$. We can write as many equations of this kind as we have intermediate supports, so that there will be enough equations for calculating all intermediate statically indeterminate reactions.

Take as an example the case of a uniform load q distributed along the beam AB (Fig. 1-17). Then the deflection produced by this load is given by Eq. (1-20), and

[1] A very complete discussion of this problem can be found in the book by F. Bleich and E. Melan, "Die gewönlichen und partiellen Differenzengleichungen der Baustatic," Berlin, 1927. See also F. Bleich, "Die Berechnung statisch unbestimmter Tragwerke nach der Methode des Viermomenten satzes," 2d ed., Berlin, 1925.

[2] See B. W. James, Principal Effects of Axial Load on Moment Distribution Analysis of Rigid Structures, *NACA Tech. Note* 534, 1935. The method is also described in Niles and Newell, "Airplane Structures," 3d ed., vol. 2, pp. 120–132, John Wiley & Sons, Inc., New York, 1943.

the deflection produced by the reactions R_1, R_2, . . . is calculated by using Eq. (1-19). Using the notation $x_m = l - c_m$, we obtain then for any support m

$$\frac{ql^4}{16EIu^4}\left[\frac{\cos\left(u - \frac{2ux_m}{l}\right)}{\cos u} - 1\right] - \frac{ql^2}{8EIu^2} x_m(l - x_m)$$

$$- \frac{\sin kx_m}{Pk \sin kl} \sum_{i=1}^{i=m} R_i \sin kc_i + \frac{x_m}{Pl} \sum_{i=1}^{i=m} R_i c_i$$

$$- \frac{\sin k(l - x_m)}{Pk \sin kl} \sum_{i=m+1}^{i=n} R_i \sin k(l - c_i) + \frac{l - x_m}{Pl} \sum_{i=m+1}^{i=n} R_i(l - c_i) = \frac{R_m}{\alpha_m} \quad (1\text{-}50)$$

There will be as many equations of this kind as we have intermediate supports, and all statically indeterminate reactions at these supports can be calculated. Equation (1-50) will be used later in discussing the stability of an elastically supported bar.

Instead of the elastic supports shown in Fig. 1-17, the beam may be supported on a continuous *elastic foundation*. A detailed analysis of this case is given in the book by Hetényi.[1]

1.11. Application of Trigonometric Series.

In studying deflections of a prismatic bar, it is sometimes advantageous to represent the deflection curve in the form of a trigonometric series.[2] In such a case a single mathematical expression holds for the entire length of the beam and it is not necessary to discuss separately each portion of the deflection curve between consecutive loads as was done in Arts. 1.3 and 1.4.

This method of analysis is especially useful in the case of a beam with simply supported ends (Fig. 1-18a). The deflection curve in this case can be represented in the form of a Fourier sine series:

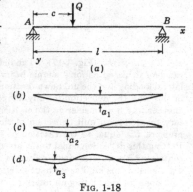

(a)

(b)

(c)

(d)

FIG. 1-18

$$y = a_1 \sin \frac{\pi x}{l} + a_2 \sin \frac{2\pi x}{l}$$
$$+ a_3 \sin \frac{3\pi x}{l} + \cdots \quad (1\text{-}51)$$

Each term of this series satisfies the end conditions of the beam, since each term and its second derivative become zero at the ends ($x = 0$ and $x = l$). Thus the deflections and the bending moments are zero at the ends of the beam. Geometrically, the series (1-51) means that the true deflection

[1] M. Hetényi, "Beams on Elastic Foundation," chap. 6, University of Michigan Press, Ann Arbor, Mich., 1946.

[2] See Timoshenko, Application of Generalized Coordinates to the Solution of Problems on Bending of Bars and Plates, *Bull. Polytech. Inst., Kiev*, 1909 (Russian).

curve of the beam AB can be obtained by superposing sinusoidal curves such as shown in Fig. 1-18b, c, and d. The first term in the series is represented by the curve in Fig. 1-18b; the second term by Fig. 1-18c; etc. The coefficients a_1, a_2, a_3, . . . of the series are the maximum ordinates, or amplitudes, of the consecutive sine curves, and the numbers 1, 2, 3, . . . , with which π is multiplied, indicate the number of half-waves in the sine curves.

It can be rigorously proved that, if the coefficients a_1, a_2, a_3, . . . are properly determined, the series (1-51) can be made to represent any deflection curve with a degree of accuracy which depends upon the number of terms taken.[1] In the following discussion, the coefficients are obtained by considering the strain energy of bending of the beam, which is given by the equation[2]

$$U = \frac{EI}{2} \int_0^l \left(\frac{d^2y}{dx^2}\right)^2 dx \tag{1-52}$$

The second derivative of y with respect to x, from series (1-51), is

$$\frac{d^2y}{dx^2} = -a_1 \frac{\pi^2}{l^2} \sin \frac{\pi x}{l} - 2^2 a_2 \frac{\pi^2}{l^2} \sin \frac{2\pi x}{l} - 3^2 a_3 \frac{\pi^2}{l^2} \sin \frac{3\pi x}{l} - \cdots$$

Substituting in Eq. (1-52), we find that the expression under the integral sign contains terms of two kinds:

$$a_n^2 \frac{n^4 \pi^4}{l^4} \sin^2 \frac{n\pi x}{l} \quad \text{and} \quad 2 a_n a_m \frac{n^2 m^2 \pi^4}{l^4} \sin \frac{n\pi x}{l} \sin \frac{m\pi x}{l}$$

By direct integration it can be shown that

$$\int_0^l \sin^2 \frac{n\pi x}{l} \, dx = \frac{l}{2} \quad \text{and} \quad \int_0^l \sin \frac{n\pi x}{l} \sin \frac{m\pi x}{l} \, dx = 0$$

Hence in expression (1-52) all terms containing the products of coefficients such as $a_m a_n$ vanish and only the terms with squares of those coefficients remain. The expression for strain energy then becomes

$$U = \frac{\pi^4 EI}{4l^3} (a_1^2 + 2^4 a_2^2 + 3^4 a_3^2 + \cdots) = \frac{\pi^4 EI}{4l^3} \sum_{n=1}^{n=\infty} n^4 a_n^2 \tag{1-53}$$

If we give to the beam (Fig. 1-18a) a very small displacement from the position of equilibrium, the change in the strain energy of the beam is equal to the work done by the external load during such a displacement.

[1] The reader is referred to any standard textbook on advanced calculus for a detailed discussion of Fourier series.
[2] See Timoshenko, "Strength of Materials," 3d ed., part I, p. 317, D. Van Nostrand Company, Inc., Princeton, N.J., 1955.

This follows from the principle of virtual displacements, and we shall use it in determining the coefficients of the series (1-51). Small displacements of the beam from the position of equilibrium can be obtained by small variations of the coefficients $a_1, a_2, a_3, \ldots$. If any coefficient a_n is given an increase da_n, we have the term $(a_n + da_n) \sin (n\pi x/l)$ in series (1-51) instead of the term $a_n \sin (n\pi x/l)$. The other terms of the series remain unchanged. Thus the increase da_n in the coefficient a_n represents an additional small deflection of the beam given by the sine curve

$$da_n \sin \frac{n\pi x}{l}$$

superposed upon the original deflection curve. The work done by the external load during this additional deflection can now be calculated. In the case of a single lateral load Q applied at the distance c from the left support (Fig. 1-18a), the point of application of the load undergoes a vertical displacement $da_n \sin (n\pi c/l)$ and the load produces the work

$$Q \, da_n \sin \frac{n\pi c}{l}$$

The change in strain energy [Eq. (1-53)] of the beam due to the increase da_n in a_n is

$$\frac{\partial U}{\partial a_n} \, da_n = \frac{\pi^4 EI}{2l^3} \, n^4 a_n \, da_n$$

Equating the change in strain energy to the work done by the load, we obtain an equation for determining the coefficient a_n:

$$\frac{\pi^4 EI}{2l^3} \, n^4 a_n = Q \sin \frac{n\pi c}{l}$$

from which
$$a_n = \frac{2Ql^3}{\pi^4 EI n^4} \sin \frac{n\pi c}{l}$$

Substituting this expression for the coefficient a_n in the series (1-51), we obtain the equation for the deflection curve in the series form

$$y = \frac{2Ql^3}{\pi^4 EI} \left(\sin \frac{\pi c}{l} \sin \frac{\pi x}{l} + \frac{1}{2^4} \sin \frac{2\pi c}{l} \sin \frac{2\pi x}{l} + \cdots \right)$$

$$= \frac{2Ql^3}{\pi^4 EI} \sum_{n=1}^{n=\infty} \frac{1}{n^4} \sin \frac{n\pi c}{l} \sin \frac{n\pi x}{l} \tag{1-54}$$

Through use of this series, the deflection for any value of x can be calculated.

As an example, let us consider the case in which the load is applied at the center of the span. In order to calculate the deflection under the load, the values $x = c = l/2$ are substituted into Eq. (1-54), which gives

$$\delta = (y)_{x=l/2} = \frac{2Ql^3}{\pi^4 EI}\left(1 + \frac{1}{3^4} + \frac{1}{5^4} + \cdots\right)$$

The series is rapidly converging, and the first few terms give the deflection with a high degree of accuracy. Using only the first term of the series we obtain

$$\delta = \frac{2Ql^3}{\pi^4 EI} = \frac{Ql^3}{48.7 EI}$$

Comparison with the exact solution shows that we obtained 48.7 instead of 48 in the denominator of the expression. Thus the error in using only the first term of the series, instead of the entire series, is about $1\frac{1}{2}$ per cent. This accuracy is sufficient for many practical purposes.

The solution for a single load (Eq. 1-54) having been obtained, other cases of loading can be solved by using the method of superposition. Consider, for instance, a beam carrying a uniformly distributed load of intensity q. Each increment of load, $q\,dc$, at distance c from the left support can be considered as a concentrated load, and the corresponding deflections, which we denote by dy, are obtained from Eq. (1-54) by substituting $q\,dc$ for Q. Then

$$dy = \frac{2q\,dc\,l^3}{\pi^4 EI}\sum_{n=1}^{n=\infty}\frac{1}{n^4}\sin\frac{n\pi c}{l}\sin\frac{n\pi x}{l}$$

Integrating this expression with respect to c, between the limits $c = 0$ and $c = l$, we obtain the deflection curve for the case when the uniform load is distributed along the entire span:

$$y = \frac{4ql^4}{\pi^5 EI}\sum_{n=1,3,5,\ldots}^{n=\infty}\frac{1}{n^5}\sin\frac{n\pi x}{l}$$

Again we obtain a rapidly converging series. Taking, for instance, only the first term and calculating the deflection at the center, we find

$$\delta = \frac{4ql^4}{\pi^5 EI} = \frac{ql^4}{76.5 EI}$$

The exact solution for this case gives

$$\delta = \frac{5ql^4}{384 EI} = \frac{ql^4}{76.8 EI}$$

Thus the error in taking only the first term is less than one-half of 1 per cent in this case.

The representation of the deflection curve in the form of the trigonometric series (1-51) is especially useful in cases where the beam is submitted to the simultaneous action of a lateral load and an axial force. Consider, for example, the beam represented in Fig. 1-19. In determining the coefficients a_1, a_2, . . . , of the series (1-51), we consider as before an infinitely small displacement $da_n \sin(n\pi x/l)$ from the equilibrium deflection curve of the beam. The corresponding change in the strain energy of bending is the same as in the previous case. However, in calculating the work done by the external forces during this displacement, we must consider not only the work $Q\, da_n \sin(n\pi c/l)$ produced by the lateral load, but also the work done by the longitudinal forces P. Any change in the shape of the deflection curve usually results in some displacement of the movable support B, and the force P acting on this support produces

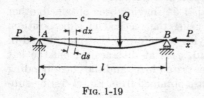

Fig. 1-19

work. Let us consider first the displacement of B which occurs during the deformation of the bar from its initial straight form to the equilibrium curve, shown in Fig. 1-19. This displacement is equal to the difference between the length of the deflection curve and the length of the chord AB if we consider the bar as inextensible. Denoting this displacement by λ and observing that the difference between the length of an element ds of the curve and the corresponding element dx of the chord is equal to

$$ds - dx = dx\sqrt{1 + \left(\frac{dy}{dx}\right)^2} - dx \approx \frac{1}{2}\left(\frac{dy}{dx}\right)^2 dx$$

we obtain[1]

$$\lambda = \frac{1}{2}\int_0^l \left(\frac{dy}{dx}\right)^2 dx \tag{1-55}$$

Substituting in this expression the series (1-51) for y and taking into account that

$$\int_0^l \cos^2\frac{n\pi x}{l}\, dx = \frac{l}{2} \qquad \int_0^l \cos\frac{n\pi x}{l}\cos\frac{m\pi x}{l}\, dx = 0$$

we find

$$\lambda = \frac{\pi^2}{4l}\sum_{n=1}^{n=\infty} n^2 a_n^2 \tag{1-56}$$

---

[1] For a more general expression for λ, in which the effects of initial curvature of the column and eccentricity of the end loading are considered, see T. H. Lin, Shortening of Column with Initial Curvature and Eccentricity and Its Influence on the Stress Distribution in Indeterminate Structures, *Proc. 1st Natl. Congr. Applied Mech.*, ASME, New York, 1952.

If we take now a small displacement from the position of equilibrium by giving to a coefficient a_n an increase da_n, the corresponding small displacement of the hinge B is

$$d\lambda = \frac{\partial \lambda}{\partial a_n} da_n = \frac{\pi^2 n^2}{2l} a_n da_n$$

Equating the change in strain energy of bending to the work done by the external forces during the small displacement $da_n \sin (n\pi x/l)$, we obtain the following equation for determining any coefficient a_n of the series (1-51):

$$\frac{\pi^4 EI}{2l^3} n^4 a_n da_n = Q da_n \sin \frac{n\pi c}{l} + P \frac{\pi^2 n^2}{2l} a_n da_n$$

from which

$$a_n = \frac{2Ql^3}{\pi^4 EI} \sin \frac{n\pi c}{l} \frac{1}{n^2[n^2 - (Pl^2/\pi^2 EI)]}$$

To simplify the equations, we denote by α the ratio of the longitudinal force P to its critical value [Eq. (1-15)]; then

$$a_n = \frac{2Ql^3}{\pi^4 EI} \sin \frac{n\pi c}{l} \frac{1}{n^2(n^2 - \alpha)}$$

Substituting in the series (1-51), we obtain

$$y = \frac{2Ql^3}{\pi^4 EI} \left[\frac{1}{1 - \alpha} \sin \frac{\pi c}{l} \sin \frac{\pi x}{l} + \frac{1}{2^2(2^2 - \alpha)} \sin \frac{2\pi c}{l} \sin \frac{2\pi x}{l} + \cdots \right]$$

$$= \frac{2Ql^3}{\pi^4 EI} \sum_{n=1}^{n=\infty} \frac{1}{n^2(n^2 - \alpha)} \sin \frac{n\pi c}{l} \sin \frac{n\pi x}{l} \quad (1\text{-}57)$$

Comparing this equation with the series (1-54) for the case when only the lateral load Q is acting, we see that each coefficient in the series is increased because of the action of the compressive force P. It is seen also that when P approaches the critical value and α approaches unity, the first term in the series (1-57) increases indefinitely.

It was shown before that the first term of the series (1-51) gives a satisfactory approximation for the deflections of the bar when only lateral load is acting. Then, denoting by δ_0 the maximum deflection of the bar produced by lateral load Q alone, we conclude from the comparison of series (1-54) and (1-57) that, in the case of simultaneous action of lateral load Q and longitudinal compressive force P, the maximum deflection is approximately

$$\delta = \frac{\delta_0}{1 - \alpha} \quad (a)$$

Thus the deflection δ_0 due to lateral load only is increased by the amplification factor $1/(1 - \alpha)$ when an axial load is also present. This amplification factor was discussed previously in Art. 1.7.

Having determined the deflection curve [Eq. (1-57)] for the case of one lateral load Q, we can, without difficulty, obtain the deflections for any kind of lateral loading by using the principle of superposition. In the case of a uniform lateral load on a compressed bar, we substitute $q \, dc$ instead of Q in the series (1-57) and integrate this series by varying c within the limits of the loaded portion of the beam. If the load covers the entire span, the integration limits are 0 and l and we obtain

$$y = \frac{4ql^4}{\pi^5 EI} \sum_{n=1,3,5,\ldots}^{n=\infty} \frac{1}{n^3(n^2 - \alpha)} \sin \frac{n\pi x}{l}$$

Again we have a rapidly converging series, and the first term gives a satisfactory approximation so that a formula analogous to Eq. (a) can also be used in this case and the deflection can be calculated by multiplying the deflection δ_0, produced by lateral load alone, by the factor $1/(1 - \alpha)$. This formula is very accurate for small values of α. With an increase of α, the error of the approximate formula increases also and approaches one-half of 1 per cent when P approaches its critical value.

By moving the load Q to the left support (Fig. 1-19) and making c infinitesimally small, we approach the condition of bending of the bar by a couple Qc applied at the left end. Substituting $\sin (n\pi c/l) = n\pi c/l$ into Eq. (1-57) and using the notation $Qc = M_a$, we obtain the following series giving the deflection curve of a compressed bar bent by a couple at the end:

$$y = \frac{2M_a l^2}{\pi^3 EI} \sum_{n=1}^{n=\infty} \frac{1}{n(n^2 - \alpha)} \sin \frac{n\pi x}{l} \tag{b}$$

If there are two moments M_a and M_b applied at the ends, the deflection curve is obtained by superposing deflections produced by each of the moments. Assuming, for instance, $M_a = M_b = M_0$ and using Eq. (b), we obtain, for the case of two equal moments, the deflection curve

$$\begin{aligned} y &= \frac{2M_0 l^2}{\pi^3 EI} \sum_{n=1}^{n=\infty} \frac{1}{n(n^2 - \alpha)} \sin \frac{n\pi x}{l} \\ &\quad + \frac{2M_0 l^2}{\pi^3 EI} \sum_{n=1}^{n=\infty} \frac{1}{n(n^2 - \alpha)} \sin \frac{n\pi(l - x)}{l} \\ &= \frac{4M_0 l^2}{\pi^3 EI} \sum_{n=1,3,5,\ldots}^{n=\infty} \frac{1}{n(n^2 - \alpha)} \sin \frac{n\pi x}{l} \tag{c} \end{aligned}$$

Since the curve in this case is symmetrical with respect to the center of the beam, the terms with even values of n do not appear in the series (c). The deflection at the middle is

$$\delta = (y)_{x=l/2} = \frac{4M_0 l^2}{\pi^3 EI}\left[\frac{1}{1-\alpha} - \frac{1}{3(9-\alpha)} + \cdots\right] \qquad (d)$$

If the couples at the ends are produced by compressive forces P applied at both ends with the same eccentricity e, we put Pe instead of M_0 in Eq. (d); then

$$\delta = \frac{4e\alpha}{\pi}\left[\frac{1}{1-\alpha} - \frac{1}{3(9-\alpha)} + \cdots\right]$$

Again the series is a rapidly converging one and we can obtain the deflection δ with sufficient accuracy by taking only the first term of the series. Thus

$$\delta = \frac{4e\alpha}{\pi(1-\alpha)} \qquad (e)$$

In the general case where M_a and M_b are not equal, we can always replace them by two equal moments M' of the same sign, equal to $\frac{1}{2}(M_a + M_b)$, and two equal moments M'' of the opposite sign, equal to $\frac{1}{2}(M_a - M_b)$. Only the first two moments produce deflection at the middle. Hence, if compressive forces P are applied at the ends with the eccentricities e_1 and e_2, the deflection

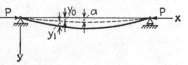

Fig. 1-20

at the middle is obtained from Eq. (e) by substituting[1] $\frac{1}{2}(e_1 + e_2)$ for e.

1.12. The Effect of Initial Curvature on Deflections. When a bar is submitted to the action of lateral load only, a small initial curvature of the bar has no effect on the bending and the final deflection curve is obtained by superposing the ordinates due to initial curvature on the deflections calculated as for a straight bar. However, if there is an axial force acting on the bar, the deflections produced by this force will be substantially influenced by the initial curvature.

Let us consider, as an example, the case in which the initial shape of the axis of the bar is given by the equation (see Fig. 1-20)

$$y_0 = a \sin\frac{\pi x}{l} \qquad (a)$$

Thus the axis of the bar has initially the form of a sine curve with a maximum ordinate at the middle equal to a. If this bar is submitted to the

[1] The eccentricities should be taken positive when they result in positive bending moments at the ends, and negative in the reversed case.

action of a longitudinal compressive force P, additional deflections y_1 will be produced so that the final ordinates of the deflection curve are

$$y = y_0 + y_1 \qquad (b)$$

and the bending moment at any cross section is

$$M = P(y_0 + y_1)$$

Then the deflections y_1 due to deformation are determined in the usual way from the differential equation

$$EI \frac{d^2 y_1}{dx^2} = -P(y_0 + y_1) \qquad (c)$$

or by substituting Eq. (a) for y_0 and using the notation $k^2 = P/EI$, we obtain

$$\frac{d^2 y_1}{dx^2} + k^2 y_1 = -k^2 a \sin \frac{\pi x}{l}$$

The general solution of this equation is

$$y_1 = A \sin kx + B \cos kx + \frac{1}{(\pi^2/k^2 l^2) - 1} a \sin \frac{\pi x}{l} \qquad (d)$$

To satisfy the end conditions ($y_1 = 0$, for $x = 0$ and for $x = l$) for any value of k, we must put $A = B = 0$. Then, by using the previous notation α for the ratio of the longitudinal force to its critical value, we obtain

$$\alpha = \frac{P}{P_{cr}} = \frac{P}{\pi^2 EI/l^2} = \frac{k^2 l^2}{\pi^2} \qquad (1\text{-}58)$$

and

$$y_1 = \frac{\alpha}{1 - \alpha} a \sin \frac{\pi x}{l} \qquad (1\text{-}59)$$

The final ordinates of the deflection curve are

$$y = y_0 + y_1 = a \sin \frac{\pi x}{l} + \frac{\alpha}{1 - \alpha} a \sin \frac{\pi x}{l} = \frac{a}{1 - \alpha} \sin \frac{\pi x}{l} \qquad (1\text{-}60)$$

This equation shows that the initial deflection a at the middle of the bar is magnified in the ratio $1/(1 - \alpha)$ by the action of the longitudinal compressive force. When the compressive force P approaches its critical value and α approaches unity, the deflection ordinates y increase indefinitely.

If the initial shape of the bar is given by a series[1]

$$y_0 = a_1 \sin \frac{\pi x}{l} + a_2 \sin \frac{2\pi x}{l} + \cdots$$

[1] See Timoshenko, *Bull. Soc. Eng. Tech.*, St. Petersburg, 1913 (Russian). See also T. H. Lin, *loc. cit.*

we substitute this expression for y_0 in Eq. (c); then, proceeding as before with each term of the series, we obtain

$$y_1 = \alpha \left(\frac{a_1}{1 - \alpha} \sin \frac{\pi x}{l} + \frac{a_2}{2^2 - \alpha} \sin \frac{2\pi x}{l} + \cdots \right) \tag{1-61}$$

Since α is always less than one and approaches unity when P approaches P_{cr}, the first term in this expression is usually predominant and is seen to coincide with Eq. (1-59).

Solution by Equivalent Lateral Load. The problem of bending of an initially curved bar can be approached in a different way by replacing the effect of the initial curvature on the deflections by the effect of an *equivalent lateral load*. The equivalent lateral load must produce the same bending-moment diagram for a straight bar as the longitudinal forces produce on the initially curved bar when only the initial deflections are taken in calculating the bending moment. Take, for instance, the case in which the initial curvature is given by Eq. (a). The effect of this curvature on deflections of the compressed bar is the same as the effect of a distributed lateral load producing in the bar bending moments $M = Pa \sin \pi x/l$, since the differential equations of the elastic curve (c) are identical for both cases. Then the expression for the intensity q of the equivalent lateral load is obtained by using the known relation between q and M, namely,

$$q = - \frac{d^2 M}{dx^2} = \frac{\pi^2 aP}{l^2} \sin \frac{\pi x}{l}$$

The deflections produced by this lateral load are obtained by using the general method of the previous article. Substituting into the general expression (1-57) the quantity

$$\frac{\pi^2 aP}{l^2} \sin \frac{\pi c}{l} \, dc$$

for Q and integrating from $c = 0$ to $c = l$, we obtain

$$y_1 = \frac{2l^3}{\pi^4 EI} \frac{\pi^2 aP}{l^2} \int_0^l \sin \frac{\pi c}{l} \, dc \sum_{n=1}^{n=\infty} \frac{1}{n^2(n^2 - \alpha)} \sin \frac{n\pi c}{l} \sin \frac{n\pi x}{l}$$

Since

$$\int_0^l \sin \frac{\pi c}{l} \sin \frac{n\pi c}{l} \, dc = 0 \quad \text{when } n \neq 1$$

and

$$\int_0^l \sin^2 \frac{\pi c}{l} \, dc = \frac{l}{2}$$

all the terms of the above integral, except the first, disappear and we finally obtain

$$y_1 = \frac{\alpha}{1 - \alpha} a \sin \frac{\pi x}{l}$$

This expression coincides with Eq. (1-59) previously derived.

As another example, take the case in which the initial deflection curve of the bar (Fig. 1-20) is a parabola:

$$y_0 = \frac{4a}{l^2} x(l - x) \tag{e}$$

The corresponding equivalent lateral load is

$$q = -\frac{d^2(Py_0)}{dx^2} = \frac{8aP}{l^2}$$

After this expression is substituted for q in Eq. (1-20), the expression for the deflection curve becomes

$$y_1 = \frac{2a}{u^2} \left[\frac{\cos (u - 2ux/l)}{\cos u} - 1 \right] - \frac{4a}{l^2} x(l - x) \tag{f}$$

in which u is given by Eq. (1-13). Superposing on these deflections the initial deflections [Eq. (e)], we find the total ordinates of the bent bar:

$$y = y_0 + y_1 = \frac{2a}{u^2} \left[\frac{\cos (u - 2ux/l)}{\cos u} - 1 \right]$$

If the initial shape of the bar consists of two straight-line portions AC and CB, as shown in Fig. 1-21, the equivalent lateral load becomes a concentrated load Q at C, since this gives the same shape of bending-moment diagram as that produced by P in Fig. 1-21. The magnitude of the equivalent load is obtained from the equality of the moments

$$Pa = \frac{Qc(l - c)}{l}$$

from which

$$Q = \frac{Pal}{c(l - c)}$$

Substituting this expression for Q in Eqs. (1-7) and (1-8), we obtain the deflections y_1 for this case.

The Phenomenon of Reversal of Deflections. It is interesting to note that the deflections produced by the compressive force P in an initially curved bar may reverse direction during a continuous increase in the value of P. This occurs because of the nonlinear relation between the deflections and the compressive force. To illustrate this behavior, let us consider that the bar in Fig. 1-22 has the initial curvature

$$y_0 = a_1 \sin \frac{\pi x}{l} + a_2 \sin \frac{2\pi x}{l}$$

If the amplitude a_1 is small in comparison with a_2, the initial curvature will have the shape shown by the solid line in the figure. The deflections produced by the force P are found from Eq. (1-61) as

$$y_1 = \frac{\alpha a_1}{1 - \alpha} \sin \frac{\pi x}{l} + \frac{\alpha a_2}{2^2 - \alpha} \sin \frac{2\pi x}{l} \qquad (g)$$

When the compressive force P is small in comparison with the critical load, the quantity α is also small. Then, since a_1 is small in comparison with a_2, it can be concluded that the second term in Eq. (g) is of greater importance and the deflections are approximately as indicated by the dotted line in Fig. 1-22. At a cross section such as mn the deflection is upward.

Assume now that the force P is gradually increased until it approaches the critical value. Then α approaches unity and the first term in Eq. (g) becomes predominant. The bar now has approximately the deflected shape of a half sine wave, and the direction of the deflection at section mn is downward.

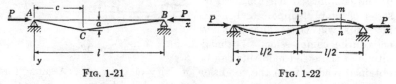

FIG. 1-21　　　　　　　　　FIG. 1-22

Bar with Lateral Loading. If a beam-column with initial curvature is subjected to lateral or end loading, the total deflection is obtained by superposing on the deflections due to curvature, which were discussed above, the deflections due to lateral load calculated for a straight bar. The superposition of these two deflections is justified by the fact that the effect of initial curvature can be replaced by the effect of an equivalent lateral load, and it was shown in Art. 1.4 that the principle of superposition in a modified form holds for all types of lateral loading.

As an example, consider the case of an initially curved bar carrying eccentrically applied end loads P, as shown in Fig. 1-23. The initial deflection y_0 of the bar is assumed to be given by Eq. (a). The eccentrically applied end loads can be replaced by statically equivalent loads consisting of centrally applied loads P and end moments $M_0 = Pe$. Then the total deflection will be the sum of the deflection due to initial curvature [Eq. (1-60)] and the deflection due to the moments M_0. This latter deflection is found from Eq. (1-32) by substituting Pe for M_0. The result of this superposition of deflections is

$$y = \frac{a}{1 - \alpha} \sin \frac{\pi x}{l} + \frac{e}{\cos kl/2} \left[\cos \left(\frac{kl}{2} - kx \right) - \cos \frac{kl}{2} \right] \qquad (h)$$

The maximum deflection occurs at the center of the bar $(x = l/2)$ and is

$$\delta = (y)_{x=l/2} = \frac{a}{1 - \alpha} + e \left(\sec \frac{kl}{2} - 1 \right) \tag{i}$$

The bending moment at any section x of the beam is

$$
\begin{aligned}
M &= P(e + y) \\
&= P \left[\frac{a}{1 - \alpha} \sin \frac{\pi x}{l} + \frac{e}{\cos kl/2} \cos \left(\frac{kl}{2} - kx \right) \right]
\end{aligned} \tag{j}
$$

and the maximum moment at the center of the beam is

$$M_{max} = P \left(\frac{a}{1 - \alpha} + \frac{e}{\cos kl/2} \right) \tag{k}$$

When the principle of superposition is used in this manner, any other case of lateral or end loading on a bar with initial curvature can be analyzed.[1]

Bar with Fixed Ends. If the ends of the compressed bar are fixed instead of simply supported, bending moments will be produced at the ends during compression. The magnitude of these end moments can be

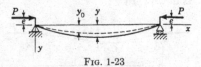

FIG. 1-23

obtained readily from the condition of fixity of the ends. Assume, for example, that the initial curvature of the fixed-end beam is given by Eq. (*a*); that is,

$$y_0 = a \sin \frac{\pi x}{l}$$

If the ends of the bar were free to rotate, the deflection of the bar due to the axial force would be found from Eq. (1-59) and the magnitude of the angle of rotation of each end would be

$$\left(\frac{dy_1}{dx} \right)_{x=0} = \frac{\alpha}{1 - \alpha} \frac{\pi a}{l}$$

To counteract these rotations, moments M_0 must be applied at the ends, and their magnitude can be found [see Eq. (1-34)] from the following equation:[2]

$$\frac{\alpha}{1 - \alpha} \frac{\pi a}{l} + \frac{M_0 l}{2EI} \frac{\tan u}{u} = 0$$

[1] Equations for several specific cases of end loading were worked out by H. K. Stephenson, Stress Analysis and Design of Columns, *Highway Research Board, Proc.* (34th annual meeting), January, 1955.

[2] In this equation M_0 is assumed positive in the direction shown in Fig. 1-8*a*.

from which

$$M_0 = - \frac{\alpha}{1 - \alpha} \frac{2\pi a E I}{l^2} \frac{u}{\tan u} \tag{l}$$

The resultant deflection of the bar is obtained by superposing the deflection due to the moments M_0 [Eq. (1-32)] on the deflection due to curvature [Eq. (1-60)]. In this manner we find that the deflection at the center $(x = l/2)$ is

$$\delta = \frac{a}{1 - \alpha} + \frac{M_0 l^2}{8EI} \frac{2(1 - \cos u)}{u^2 \cos u}$$

or, by using Eq. (l),

$$\delta = \frac{a}{1 - \alpha} \left(1 - \frac{\pi \alpha}{2} \frac{1 - \cos u}{u \sin u} \right) \tag{m}$$

The bending moment at the middle is now obtained from the equation $M = P\delta + M_0$, where M_0 is given by Eq. (l).

In the above discussions we considered in each case a bar compressed by force P applied at the ends. The case of a bar with initial curvature and intermediate axial load has also been considered.[1]

1.13. Determination of Allowable Stresses. In the design of steel beams subjected to the action of lateral loads only, the working stress is ordinarily selected as a certain fraction of the yield-point stress. Thus we have the relation

$$\sigma_W = \frac{\sigma_{YP}}{n} \tag{a}$$

where n is the factor of safety. The cross-sectional dimensions of the beam are so chosen that the maximum stress does not exceed the working stress from Eq. (a).

The same procedure can be used also in cases of the simultaneous action of lateral loads and an axial compressive force provided the longitudinal force remains constant and only the possibility of an increase in the lateral loads must be considered. From the principle of superposition (Art. 1.4) it follows that if the working stress is found from Eq. (a), we obtain a beam of proportions such that the maximum stress becomes equal to the yield-point stress when the lateral loads are increased by the factor n. This assumes that Hooke's law is valid up to the yield point of the material, which is a justifiable assumption for a material such as structural steel.

There are cases, however, in which the longitudinal force increases simultaneously with the lateral loads. The structure shown in Fig. 1-24 is an example of such a case. It is seen that the tensile force T in the

[1] See S. I. Sergev, *Univ. Wash., Eng. Expt. Sta. Bull.* 113, 1945.

wire AC and the corresponding compressive force P in the bar AB increase in the same proportion as the lateral load Q acting on the beam. In such cases the deflections and maximum stresses will increase at a greater rate than the lateral loads, and this fact must be considered in choosing the working stress in order to assure the desired factor of safety. To make the factor of safety equal to n, it is necessary to determine the cross-sectional dimensions of the beam in such a manner that the maximum fiber stress will become equal to the yield-point stress when all loads acting on the beam, including the longitudinal force P, are taken n times greater. This requires the use of a smaller working stress than the value given by Eq. (a).

To illustrate the procedure for selecting the cross-sectional dimensions of such beams, let us consider the case of a column compressed by eccentrically applied loads P (Fig. 1-8b) and assume that both eccentricities are equal to e and that bending occurs in the plane of symmetry of the beam. The maximum bending moment [from Eq. (1-35)] is

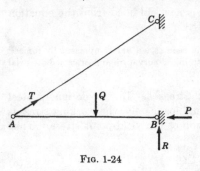

FIG. 1-24

$$M_{max} = Pe \sec \frac{l}{2} \sqrt{\frac{P}{EI}} \qquad (b)$$

Denoting by r the radius of gyration of the cross section and by c the distance from the neutral axis to the extreme fiber, we find the maximum stress at the middle cross section to be

$$\sigma_{max} = \frac{P}{A} + \frac{Mc}{I}$$

$$= \frac{P}{A} \left(1 + \frac{ec}{r^2} \sec \frac{l}{2r} \sqrt{\frac{P}{AE}} \right) \qquad (1\text{-}62)$$

This is the well-known *secant formula* for the maximum stress.

The quantity r^2/c will be denoted by the symbol s and is called the *radius of the core*, since it defines the core of the cross section within which a compressive force can act on a short column without causing tensile stress in any extreme fiber. The core radius is equal to the ratio of the section modulus Z of the cross section to the area, so that[1]

$$s = \frac{Z}{A} = \frac{I}{Ac} = \frac{r^2}{c} \qquad (1\text{-}63)$$

---

[1] The core radius is discussed in Timoshenko, "Strength of Materials," 3d ed., part I, p. 254, D. Van Nostrand Company, Inc., Princeton, N.J., 1955.

In the case of a rectangular cross section of width b and height h, for example, the core radius is

$$s = \frac{bh^2/6}{bh} = \frac{h}{6} \tag{c}$$

Using the notation s for the core radius, we can write the secant formula in the form

$$\sigma_{max} = \frac{P}{A}\left(1 + \frac{e}{s}\sec\frac{l}{2r}\sqrt{\frac{P}{AE}}\right) \tag{1-64}$$

For given dimensions of a column and for a known eccentricity e, the slenderness ratio l/r and the ratio e/s are known, and hence Eq. (1-64) represents the relation between the maximum compressive fiber stress

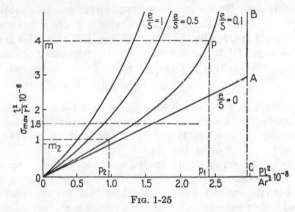

FIG. 1-25

and the average compressive stress $\sigma_c = P/A$. For a given value of the modulus of elasticity E, this relation can be represented graphically by taking Pl^2/Ar^2 as the abscissa and $\sigma_{max}l^2/r^2$ as the ordinate. Three curves of this type are plotted in Fig. 1-25 for the three eccentricity ratios $e/s = 1, 0.5$, and 0.1. The modulus of elasticity is taken as

$$E = 30,000,000 \text{ psi}$$

There is shown also a straight line OA which gives the maximum fiber stress for the case where the load is centrally applied and $e/s = 0$. This line gives the maximum fiber stresses up to the critical value of the average compressive stress. The critical value is indicated in the figure by the vertical line CAB. All curves similar to the three shown in the figure will have this vertical line as their asymptote, since $\sec (l/2r)\sqrt{P/AE}$ in Eq. (1-64) becomes infinitely large when $P = P_{cr}$. As the eccentricity e becomes smaller and smaller, the corresponding curves come closer and closer to the straight lines OA and AB. Having a series of curves like those in Fig. 1-25, we can easily find, for a given column, the maximum fiber stress produced by a given compressive force applied with a known eccentricity.

The same curves can also be used to determine, for a given column and a given eccentricity, the amount of compressive load which can be applied with a given factor of safety. Assume, for instance, that the yield-point stress for the material of the column is 40,000 psi, that the desired factor of safety is 2.5, that $e/s = 0.1$, and that the slenderness ratio $l/r = 100$. Drawing the horizontal line mp for $\sigma_{max} = 40,000$ psi to the intersection point p with the corresponding curve for e/s (Fig. 1-25), we find on the horizontal axis the point p_1 which gives the value of the average stress σ_c which produces a maximum fiber stress equal to the yield-point stress. This value of average stress will be denoted $(\sigma_c)_{YP}$. If the factor of safety is to be 2.5, the allowable average stress should be 0.4 times the value given by point p_1. In Fig. 1-25 this value is indicated by point p_2 where $P/A = 9,700$ psi. The corresponding ordinate to the curve for $e/s = 0.1$, equal to Om_2, gives the maximum fiber stress

$$\sigma_{max.} = 11,400 \text{ psi}$$

which must be taken as the working stress in order to have the desired factor of safety. It is seen that the value of the working stress obtained in this manner is considerably less than the stress of 16,000 psi which would be obtained by using Eq. (a).

In designing an eccentrically loaded column, we begin by assuming probable cross-sectional dimensions. Then, proceeding as explained above, we obtain the safe value of the load that the column can carry. If this load differs substantially from the actual load, the assumed cross-sectional dimensions should be changed and the calculations repeated. Thus, by using the trial-and-error method we can always find satisfactory cross-sectional dimensions for the column.

Instead of using the above curves, we can use the secant formula [Eq. (1-64)] directly in designing eccentrically loaded columns. If P denotes the safe load on the column and n is the factor of safety, then nP is the load at which the maximum fiber stress should become equal to the yield-point stress. Equation (1-64) for this load becomes

$$\sigma_{YP} = \frac{nP}{A}\left(1 + \frac{e}{s}\sec\frac{l}{2r}\sqrt{\frac{nP}{AE}}\right) \qquad (d)$$

We can always solve this equation for P/A by using the trial-and-error method and in this way obtain the safe average stress P/A for a given column. Assuming certain values for σ_{YP}, n, and e/s and using Eq. (d), we can calculate a table of safe values of the average compressive stress

$\sigma_c = P/A$ for various values of the slenderness ratio l/r. This relation between σ_c and l/r can be represented graphically in the form of a family of curves.

To make these curves independent of the factor of safety n, values of $nP/A = (\sigma_c)_{\text{YP}}$ can be plotted against l/r so that the values of the average compressive stress $(\sigma_c)_{\text{YP}}$, at which yielding begins, can be taken directly from the curves and any desired factor of safety can be obtained simply by dividing the ordinates of the curve by the desired value of n. Figure 1-26 represents a set of curves[1] for structural steel and is plotted for $E = 30,000,000$ psi, $\sigma_{\text{YP}} = 36,000$ psi and for values of e/s from 0.1 to 1.0. Having such curves, no difficulty is encountered in determining by trial and error the necessary cross section for an eccentrically compressed bar.[2]

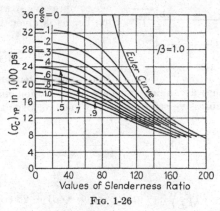

FIG. 1-26

Similar curves can be calculated also for other cases of lateral loading acting on a beam with an axial compressive force.[3] For instance, in the case of a beam with a uniform load it is necessary to use Eq. (1-23) in the same manner as Eq. (1-35) was used above.

In the case of compressive forces applied with two different eccentricities e_a and e_b (Fig. 1-27), we use Eq. (1-30). This case is of practical importance in discussing stresses in compression members of trusses. Because of the rigidity of the joints of a truss, secondary stresses are

---

[1] These curves were calculated by D. H. Young, Rational Design of Steel Columns, *Trans. ASCE*, vol. 101, p. 422, 1936.

[2] A large number of curves of this type, calculated from the secant formula and suitable for design use, were given by H. K. Stephenson and K. Cloninger, Jr., Stress Analysis and Design of Steel Columns, *Texas Eng. Expt. Sta. Bull.* 129, February, 1953.

[3] Several tables of this type were prepared by S. Zavriev, who was the first to develop this idea; see *Mem. Inst. Engrs. Ways of Commun.*, St. Petersburg, 1913.

always present and each compression member is subjected to bending by moments at the ends. If the magnitudes of the bending moments are known from a secondary stress analysis,[1] the maximum stresses in each particular case can be obtained in a manner analogous to that discussed above for the case of two equal end moments.[2] Assuming that e_a is the numerically larger eccentricity, we shall introduce the notation $\beta = e_b/e_a$. The value of β thus varies from $+1$ when the eccentricities are equal and in the same direction to -1 when the eccentricities are equal and in opposite directions.

In the case of comparatively short bars, the maximum fiber stress will occur at the end A, where the eccentricity is larger. The magnitude of this stress is easily calculated from the usual formula for combined compression and bending, and the average compressive stress at which yielding begins is given by the equation

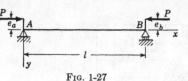

Fig. 1-27

$$(\sigma_c)_{\text{YP}} = \frac{\sigma_{\text{YP}}}{1 + e_a/s} \qquad (e)$$

In the case of slender bars, the maximum stress occurs at an intermediate cross section and the value of the average compressive stress at which yielding begins is given by the equation[3]

$$(\sigma_c)_{\text{YP}} = \frac{\sigma_{\text{YP}}}{1 + (e_a/s)\psi \operatorname{cosec} 2u} \qquad (f)$$

in which

$$2u = kl = l\sqrt{\frac{P_{\text{YP}}}{EI}} \quad \text{and} \quad \psi = \sqrt{\beta^2 - 2\beta \cos 2u + 1}$$

The limiting value of the slenderness ratio l/r, up to which Eq. (e) should be used, is found in each particular case by using the equation

$$\cos^{-1} \beta = \frac{l}{r}\sqrt{\frac{(\sigma_c)_{\text{YP}}}{E}} \qquad (g)$$

which is obtained by equating Eqs. (e) and (f).

The results of calculations made with Eqs. (e), (f), and (g) are represented by the curves in Figs. 1-28 to 1-31. These curves are plotted for structural steel with $E = 30,000,000$ psi and $\sigma_{\text{YP}} = 36,000$ psi. The

[1] See, for example, Timoshenko and Young, "Theory of Structures," pp. 398–403, McGraw-Hill Book Company, Inc., New York, 1945.

[2] This problem was discussed fully by D. H. Young, Stresses in Eccentrically Loaded Steel Columns, *Publ. Intern. Assoc. Bridge Structural Eng.*, vol. 1, p. 507, 1932.

[3] See *ibid.*

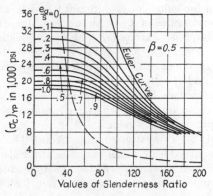

Fig. 1-28

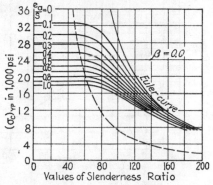

Fig. 1-29

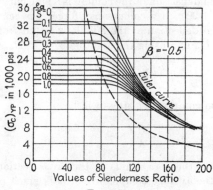

Fig. 1-30

curves are drawn for values of e_a/s from 0.1 to 1.0, and each graph is plotted for a different value of β. With the use of these curves, the value of the average compressive stress $(\sigma_c)_{YP}$ at which yielding begins can be obtained easily. The dotted curves in Figs. 1-28 to 1-30 are obtained from Eq. (g) and represent the dividing line between the ranges of applicability of Eqs. (e) and (f).

When the eccentricity e_a approaches zero, the average stress $(\sigma_c)_{YP}$ for a short column is equal to σ_{YP}. For slender columns with larger l/r the value of $(\sigma_c)_{YP}$ approaches the value P_{cr}/A, where P_{cr} is the critical load given by Eq. (1-15). This latter curve is labeled "Euler curve" in the figures, since the critical load is also known as the Euler load.

In a similar manner we can determine the safe load for the case of an initially curved and compressed column. Take, for instance, the case

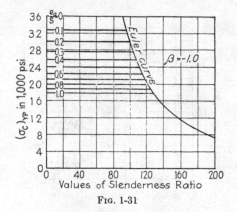

Fig. 1-31

in which the initial deflection of the column is a sine curve $y_0 = a \sin \pi x/l$ as shown in Fig. 1-20. If a compressive force P is centrally applied at the ends, the total deflection at the middle is found from Eq. (1-60) by substituting $x = l/2$ and using the notation $\alpha = P/P_{cr}$, which gives

$$\delta = \frac{a}{1 - P/P_{cr}}$$

and the maximum compressive stress will be

$$\sigma_{max} = \frac{P}{A}\left(1 + \frac{a}{s}\frac{1}{1 - P/P_{cr}}\right) \qquad (h)$$

in which s denotes the core radius of the cross section. Denoting as before by $(\sigma_c)_{YP}$ the average compressive stress, which produces a maximum fiber stress equal to the yield-point stress, the equation for determining $(\sigma_c)_{YP}$, from Eq. (h), becomes

$$\sigma_{YP} = (\sigma_c)_{YP}\left[1 + \frac{a}{s}\cfrac{1}{1 - \cfrac{(\sigma_c)_{YP}}{\pi^2 E}\cfrac{l^2}{r^2}}\right] \tag{i}$$

This is a quadratic equation for $(\sigma_c)_{YP}$ which can be solved for any values of the ratios a/s and l/r.

Having determined the average compressive stress $(\sigma_c)_{YP}$ at which yielding begins, the allowable average compressive stress for an initially

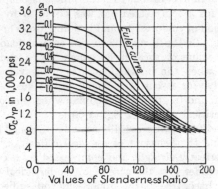

FIG. 1-32

curved column can be obtained simply by dividing this stress by the desired factor of safety. In Fig. 1-32 are given curves[1] for the average compressive stress at which yielding begins, calculated for structural steel with $E = 30,000,000$ psi and $\sigma_{YP} = 36,000$ psi. The curves are drawn for values of a/s from 0.1 to 1.0. Through the use of these curves, the allowable compressive load for a column with a given initial curvature and for any desired factor of safety can be calculated by trial and error. It should be noted that the curves in Fig. 1-32 are very similar to the curves in Fig. 1-26 for corresponding values of a/s and e/s.

[1] These curves are from D. H. Young, Rational Design of Steel Columns, *Trans. ASCE*, vol. 101, p. 422, 1936.

CHAPTER 2

ELASTIC BUCKLING OF BARS AND FRAMES

2.1. Euler's Column Formula. In the previous chapter the value of the critical load for a compressed bar was obtained by considering the simultaneous action of compressive and bending forces or by assuming an initial curvature. In the former case the critical load was found by determining the value of axial force which would cause large lateral deflections even when the lateral load itself was very small [see Eq. (1-15)]. Similarly, for a bar with small initial curvature the lateral deflections were found to grow without limit when the compressive force approached the critical value [see Eq. (1-60)].

The critical load for a compressed bar can be obtained in another manner by considering the behavior of an *ideal column*, which is assumed initially to be perfectly straight and compressed by a centrally applied load. Let us consider first the case of a slender, ideal column built in vertically at the base, free at the upper end and subjected to an axial force P (Fig. 2-1a).[1] The column is assumed to be perfectly elastic, and the stresses do not exceed the proportional limit. If the load P is less than the critical value, the bar remains straight and undergoes only axial compression. This straight form of elastic equilibrium is *stable*, which means that if a lateral force is applied and a small deflection produced, the deflection disappears when the lateral force is removed and the bar returns to its straight form. If P is gradually increased, a condition is reached in which the straight form of equilibrium becomes *unstable* and a small lateral force will produce a deflection which does not disappear when the lateral force is removed. The *critical load* (or Euler load) is then defined as the axial force which is sufficient to keep the bar in such a slightly bent form (Fig. 2-1b).

The critical load can be calculated by using the differential equation of

---

[1] This is the case which was solved originally by Leonhard Euler and published in the appendix, "De curvis elasticis," of his book "Methodus inveniendi lineas curvas maximi minimive proprietate gaudentes," Lausanne and Geneva, 1744. An English translation of the appendix was published in *Isis*, vol. 20, no. 58, p. 1, November, 1933 (reprinted in Bruges, Belgium). For a more complete historical discussion see Timoshenko, "History of Strength of Materials," pp. 30–36, McGraw-Hill Book Company, Inc., New York, 1953.

the deflection curve (see Art. 1.2). When the coordinate axes are taken as indicated in Fig. 2-1b and also the column is assumed to be in a slightly deflected position, the bending moment at any cross section mn is

$$M = -P(\delta - y)$$

and the differential equation (1-3) becomes

$$EI \frac{d^2y}{dx^2} = P(\delta - y) \tag{2-1}$$

Since the upper end of the column is free, it is apparent that buckling of the bar will occur in the plane of minimum flexural rigidity, which we assume is a plane of symmetry. This minimum value of EI is used in

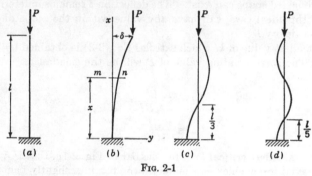

FIG. 2-1

Eq. (2-1). Using the previous notation

$$k^2 = \frac{P}{EI}$$

we can write Eq. (2-1) in the form

$$\frac{d^2y}{dx^2} + k^2y = k^2\delta \tag{a}$$

The general solution of this equation is

$$y = A \cos kx + B \sin kx + \delta \tag{b}$$

in which A and B are constants of integration. These constants are determined from the following conditions at the built-in end of the bar:

$$y = \frac{dy}{dx} = 0 \qquad \text{at } x = 0$$

These two conditions are fulfilled if

$$A = -\delta \qquad B = 0$$

and then
$$y = \delta(1 - \cos kx) \tag{2-2}$$

The condition at the upper end of the bar requires that

$$y = \delta \qquad \text{at } x = l$$

which is satisfied if

$$\delta \cos kl = 0 \tag{c}$$

Equation (c) requires that either $\delta = 0$ or $\cos kl = 0$. If $\delta = 0$, there is no deflection of the bar and hence no buckling (Fig. 2-1a). If $\cos kl = 0$, we must have the relation

$$kl = (2n - 1) \frac{\pi}{2} \tag{2-3}$$

where $n = 1, 2, 3, \ldots$. This equation determines values of k at which a buckled shape can exist. The deflection δ remains indeterminate and, for this ideal case, can have any value within the scope of small-deflection theory.[1]

The smallest value of kl which satisfies Eq. (2-3) is obtained by taking $n = 1$. The corresponding value of P will be the smallest critical load, and we have

$$kl = l \sqrt{\frac{P}{EI}} = \frac{\pi}{2}$$

from which

$$P_{cr} = \frac{\pi^2 EI}{4l^2} \tag{2-4}$$

This is the smallest critical load for the bar in Fig. 2-1a, that is, it is the smallest axial force which can maintain the bar in a slightly bent shape. The quantity kx in Eq. (2-2) varies in this case from 0 to $\pi/2$, and the shape of the deflection curve is therefore as shown in Fig. 2-1b.

Substituting $n = 2, 3, \ldots$ into Eq. (2-3), we obtain for the corresponding values of the compressive force

$$P_{cr} = \frac{9\pi^2 EI}{4l^2} \qquad P_{cr} = \frac{25\pi^2 EI}{4l^2} \cdots$$

The quantity kx in Eq. (2-2) varies in these cases from 0 to $3\pi/2$, from 0 to $5\pi/2, \ldots$, and the corresponding deflection curves are shown in Fig. 2-1c and d. For the shape shown in Fig. 2-1c a force nine times larger than the smallest critical load is necessary, and for the shape in Fig. 2-1d a force twenty-five times larger is required. Such forms of buckling can be produced by using a very slender bar and by applying external constraints at the inflection points to prevent lateral deflection. Otherwise these forms of buckling are unstable and have little practical meaning because the structure develops large deflections when the load reaches the value given by Eq. (2-4).

[1] Note that the differential equation (2-1) is based upon an approximate expression for curvature and is valid only for small deflections.

The critical loads for columns with some other end conditions can be obtained from the solution of the preceding case. For example, in the case of a bar with hinged ends (Fig. 2-2) it is evident from symmetry that each half of the bar is in the same condition as the entire bar of Fig. 2-1. Hence the critical load for this case is obtained by substituting $l/2$ for l in Eq. (2-4), which gives

$$P_{cr} = \frac{\pi^2 EI}{l^2} \tag{2-5}$$

The case of a bar with hinged ends is probably assumed in practice more frequently than any other; it is called the *fundamental case* of buckling of a prismatic bar.

If the bar has both ends built in (Fig. 2-3), there are reactive moments that prevent the ends of the column from rotating during buckling.

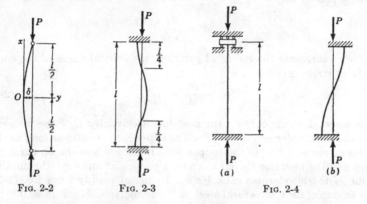

FIG. 2-2   FIG. 2-3   FIG. 2-4

These end moments and the axial compressive forces are equivalent to forces P applied eccentrically as shown in the figure. Inflection points are located where the line of action of P intersects the deflection curve, because at these points the bending moments are zero. The inflection points and the mid-point of the span divide the bar into four equal regions, each of which is in the same condition as the bar in Fig. 2-1b. Hence the critical load for a column with built-in ends is found by substituting $l/4$ for l in Eq. (2-4), which gives

$$P_{cr} = \frac{4\pi^2 EI}{l^2} \tag{2-6}$$

As a final example, consider the column shown in Fig. 2-4a. This bar is free to displace laterally at the upper end but is guided in such a manner that the tangent to the elastic curve remains vertical. At the lower end the column is built in. Since there is a point of inflection at the center of the bar (Fig. 2-4b), the critical load is found by substituting $l/2$ for l in Eq. (2-4), and thus it is seen that Eq. (2-5) holds for this case also.

In each of the preceding cases it was assumed that the column was free to buckle in any direction, and hence EI represents the smallest flexural rigidity. If a column is constrained in such a manner that buckling is possible in one principal plane only, then EI will represent the flexural rigidity in that plane.

It was assumed in the previous discussion that the bar was very slender, so that the maximum compressive stresses which occurred during buckling remained within the proportional limit of the material. Only under these conditions will the preceding equations for the critical loads be valid. To establish the limit of applicability of these formulas, let us consider the fundamental case (Fig. 2-2). Dividing the critical load from Eq. (2-5) by the cross-sectional area A of the bar and letting

$$r = \sqrt{\frac{I}{A}}$$

where r represents the radius of gyration, the critical value of the compressive stress is

$$\sigma_{cr} = \frac{P_{cr}}{A} = \frac{\pi^2 E}{(l/r)^2} \tag{2-7}$$

This stress depends only on the modulus of elasticity E of the material and on the slenderness ratio l/r. The expression is valid as long as the stress σ_{cr} remains within the proportional limit. When the proportional limit and the modulus E are known for a particular material, the limiting value of the slenderness ratio l/r can be found readily from Eq. (2-7). For example, for a structural steel with a proportional limit of 30,000 psi and $E = 30,000,000$ psi, we find the minimum l/r from Eq. (2-7) to be about 100. Consequently, the critical load for a bar of this material, having hinged ends, can be calculated from Eq. (2-5) if l/r is greater than 100. If l/r is less than 100, the compressive stress reaches the proportional limit before buckling can occur and Eq. (2-5) cannot be used. The question of the buckling of bars compressed beyond the proportional limit is discussed in the next chapter.

Equation (2-7) can be represented graphically by the curve ACB in Fig. 2-5, where the critical stress is plotted as a function of l/r. The curve approaches the horizontal axis asymptotically, and the critical stress approaches zero as the slenderness ratio increases. The curve is also asymptotic to the vertical axis but is applicable in this region only as long as the stress σ_{cr} remains below the proportional limit of the material. The curve in Fig. 2-5 is plotted for the structural steel mentioned above, and point C corresponds to a proportional limit of 30,000 psi. Thus only the portion BC of the curve can be used.

Referring now to the cases represented in Figs. 2-1a and 2-3 and proceeding as for a bar with hinged ends, we find the following expressions for the critical stresses:

$$\sigma_{cr} = \frac{\pi^2 E}{(2l/r)^2} \qquad \sigma_{cr} = \frac{\pi^2 E}{(l/2r)^2}$$

It is seen that in these two cases equations analogous to Eq. (2-7) for the fundamental case can be used in calculating the critical stress. These equations are obtained from Eq. (2-7) by substituting in place of the actual length l of the bar a *reduced length* L. Thus we can write in general

$$\sigma_{cr} = \frac{\pi^2 E}{(L/r)^2} \tag{2-8}$$

In the case of a prismatic bar with one end built in and the other end free, the reduced length is twice the actual length ($L = 2l$). In the case

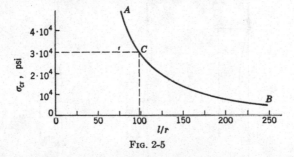

FIG. 2-5

of a bar with both ends built in, the reduced length is half the actual length ($L = l/2$). Thus the results obtained for the fundamental case can be used for other cases of buckling of bars by using the reduced length instead of the actual length of the bar.

2.2. Alternate Form of the Differential Equation for Determining Critical Loads. In the preceding article it was shown that the critical load for an ideal column could be found by beginning with the differential equation (1-3), which expresses the curvature of the bar in terms of the bending moment. An alternate method is to begin with Eq. (1-5). Since in determining critical loads of buckled bars the lateral load vanishes, the differential equation for the column is

$$EI \frac{d^4 y}{dx^4} + P \frac{d^2 y}{dx^2} = 0$$

or, substituting $k^2 = P/EI$,

$$\frac{d^4 y}{dx^4} + k^2 \frac{d^2 y}{dx^2} = 0 \tag{2-9}$$

The general solution of this equation is

$$y = A \sin kx + B \cos kx + Cx + D \tag{2-10}$$

The constants in this equation and the value of the critical load are found from the end conditions of the bar. A number of particular cases will now be considered.

Column with Hinged Ends. In the case of a bar with hinged ends (Fig. 2-6a) the deflection and the bending moment are zero at the ends, and hence we have the conditions

$$y = \frac{d^2y}{dx^2} = 0 \qquad \text{at } x = 0 \text{ and } x = l$$

Applying these conditions to the general solution [Eq. (2-10)] gives

$$B = C = D = 0 \qquad \sin kl = 0$$

and therefore

$$kl = n\pi \tag{a}$$

This equation determines the values of the critical load and for $n = 1$ gives the previous result, Eq. (2-5). The shape of the deflection curve is given by the equation

$$y = A \sin kx = A \sin \frac{n\pi x}{l} \tag{b}$$

where the constant A represents the undetermined amplitude of the deflection. For the lowest critical load ($n = 1$) the buckled shape is shown in Fig. 2-6a. For $n = 2$,

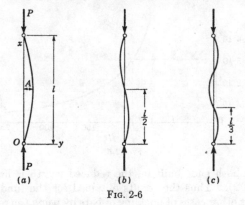

FIG. 2-6

3, . . . higher values of the critical load are obtained from Eq. (a) and the corresponding buckled shapes are shown in Figs. 2-6b and c.

Column with One End Fixed and the Other Free. For the bar shown in Fig. 2-1a, fixed at the base and free at the upper end, the conditions at the lower end are

$$y = \frac{dy}{dx} = 0 \qquad \text{at } x = 0$$

At the free end ($x = l$) the bending moment and shearing force must be zero. Referring to Eqs. (1-3) and (1-4), Art. 1.2, we see that these conditions mean that

$$\frac{d^2y}{dx^2} = 0 \qquad \text{at } x = l$$

$$\frac{d^3y}{dx^3} + k^2 \frac{dy}{dx} = 0 \qquad \text{at } x = l$$

From the conditions at the lower end of the bar we obtain

$$B = -D \qquad C = -Ak$$

and the last two conditions give

$$A \sin kl + B \cos kl = 0 \qquad C = 0$$

Therefore we finally conclude that $C = A = 0$ and

$$\cos kl = 0 \qquad kl = \frac{(2n - 1)\pi}{2}$$

which agrees with Eq. (2-3) of the preceding article.

Column with One End Fixed and the Other Pinned. This case is illustrated in Fig. 2-7, where the lower end of the bar is built in and the upper end is hinged. When lateral buckling occurs, a reactive force R is developed at the pinned end. The direction of this reaction is determined by noting that it must oppose the reactive moment at the built-in end. The end conditions for this column are

$$y = \frac{dy}{dx} = 0 \qquad \text{at } x = 0$$

$$y = \frac{d^2y}{dx^2} = 0 \qquad \text{at } x = l$$

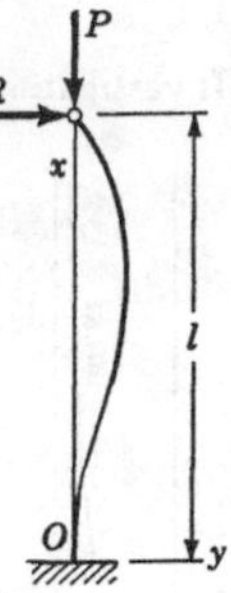

Using these conditions with the general solution (2-10) gives the following equations for the constants:

$$R + D = 0$$
$$Ak + C = 0$$
$$Cl + D = 0$$
$$A \sin kl + B \cos kl = 0$$

Fig. 2-7

All four of these equations will be satisfied by taking $A = B = C = D = 0$, in which case the deflection [see Eq. (2-10)] vanishes and we have the straight form of equilibrium. In order to have the possibility of a buckled shape of equilibrium, we need a solution of the equations other than the trivial one. Solving for A in terms of B from the first three equations and substituting into the last equation gives

$$-B \frac{\sin kl}{kl} + B \cos kl = 0$$

and hence,

$$\tan kl = kl \tag{2-11}$$

Thus, in order to get a curved shape of equilibrium satisfying the end conditions of the bar, the transcendental equation (2-11) must be satisfied.

To solve Eq. (2-11) a graphical method is useful.[1] The curves in Fig. 2-8 represent $\tan kl$ as a function of kl. These curves are asymptotic to the vertical lines $kl = \pi/2$, $3\pi/2$, . . . since for these values of kl, $\tan kl$ becomes infinite. The roots of Eq. (2-11) are represented by the intersection points of the above curves with the straight line $y = kl$. The smallest root, corresponding to point A, is

$$kl = 4.493$$

and the corresponding critical load is

$$P_{cr} = \frac{20.19EI}{l^2} = \frac{\pi^2EI}{(0.699l)^2} \tag{2-12}$$

[1] Also, solutions of Eq. (2-11) are tabulated in Jahnke and Emde, "Tables of Functions," 4th ed., p. 30 of Addenda, Dover Publications, New York, 1945.

Thus the critical load is the same as for a bar with hinged ends having a reduced length equal to $0.699l$ [see Eq. (2-8)].

Fixed-end Column. If both ends of the bar are fixed (Fig. 2-9a), the end conditions are

$$y = \frac{dy}{dx} = 0 \qquad \text{at } x = 0 \text{ and } x = l$$

These conditions give the following equations for determining the constants in Eq. (2-10):

$$\begin{aligned}
B + D &= 0 \\
Ak + C &= 0 \\
A \sin kl + B \cos kl + Cl + D &= 0 \\
Ak \cos kl - Bk \sin kl + C &= 0
\end{aligned} \qquad (c)$$

Investigating the possibility of curved forms of equilibrium, we observe that the only

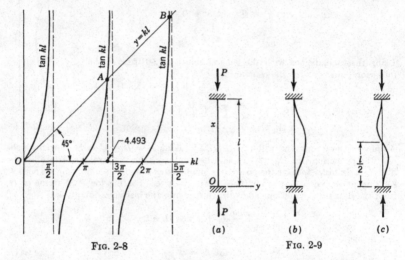

Fig. 2-8   Fig. 2-9

way to have a nontrivial solution of these four equations is to have the determinant of the coefficients equal to zero. This determinant is

$$\begin{vmatrix}
0 & 1 & 0 & 1 \\
k & 0 & 1 & 0 \\
\sin kl & \cos kl & l & 1 \\
k \cos kl & -k \sin kl & 1 & 0
\end{vmatrix}$$

and setting it equal to zero gives the equation

$$2(\cos kl - 1) + kl \sin kl = 0$$

Noting that $\sin kl = 2 \sin (kl/2) \cos (kl/2)$ and $\cos kl = 1 - 2 \sin^2(kl/2)$, we can write this equation in the form

$$\sin \frac{kl}{2} \left(\frac{kl}{2} \cos \frac{kl}{2} - \sin \frac{kl}{2} \right) = 0 \qquad (d)$$

One solution of this equation is

$$\sin \frac{kl}{2} = 0$$

and therefore $kl = 2n\pi$ and

$$P_{cr} = \frac{4n^2\pi^2 EI}{l^2} \tag{2-13}$$

Noting that $\sin kl = 0$ and $\cos kl = 1$ whenever $\sin kl/2 = 0$, we find from Eqs. (c) the following values of the constants:

$$A = C = 0 \qquad B = -D$$

and the equation for the deflection curve is

$$y = B \left(\cos \frac{2n\pi x}{l} - 1 \right) \tag{2-14}$$

If $n = 1$, we obtain the lowest critical load [see Eq. (2-6)] and the column assumes the symmetrical buckled shape shown in Fig. 2-9b.

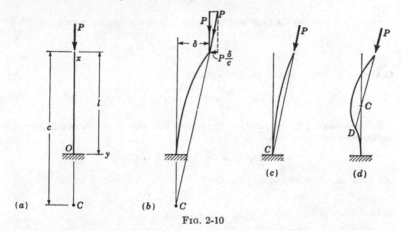

FIG. 2-10

A second solution of Eq. (d) is obtained by setting the term in parentheses equal to zero, giving the equation

$$\tan \frac{kl}{2} = \frac{kl}{2}$$

The lowest root of this equation is $kl/2 = 4.493$, and therefore

$$P_{cr} = \frac{8.18\pi^2 EI}{l^2} \tag{2-15}$$

which corresponds to the antisymmetric buckling pattern shown in Fig. 2-9c. However, since this critical value is larger than the previous value for symmetrical buckling, it is of practical interest only in the case of a column with lateral support at midheight.

Column with Load through a Fixed Point. In the preceding examples the direction of the compressive force P was assumed to remain constant during buckling of the bar. Let us consider now a case in which a change in direction of the force P

occurs.   Assume, for example, that the force P is produced by the tension of a cable which always passes through the fixed point C on the x axis, as shown in Fig. 2-10. The lower end of the column is built in, and the upper end is free to move laterally.

This problem differs from the usual Euler case (see Fig. 2-1) because during buckling there is a shearing force at the upper end of the bar.   This force is equal to the horizontal component of the tensile force P in the cable (Fig. 2-10b), and since for small deflections the vertical component of the force can be taken equal to P, we obtain

$$V = -\frac{P\delta}{c}$$

Substituting this expression for V into the general equation for the shearing force [Eq. (1-4)] gives the following condition at the upper end of the bar $(x = l)$:

$$EI \frac{d^3y}{dx^3} + P \frac{dy}{dx} = \frac{P\delta}{c}$$

or

$$\frac{d^3y}{dx^3} + k^2 \frac{dy}{dx} = \frac{k^2\delta}{c}$$

A second condition at the upper end of the bar is that the bending moment is zero, or

$$\frac{d^2y}{dx^2} = 0 \qquad \text{at } x = l$$

At the lower end of the bar the conditions are

$$y = \frac{dy}{dx} = 0 \qquad \text{at } x = 0$$

Again evaluating the constants in the general solution (2-10), we obtain from the conditions at the lower end

$$B + D = 0 \qquad Ak + C = 0$$

and from the conditions at the upper end

$$C = \frac{\delta}{c}$$
$$A \sin kl + B \cos kl = 0$$

Solving these equations for the constants and then substituting into Eq. (2-10), we obtain

$$y = \frac{\delta}{kc} [(\tan kl)(\cos kx - 1) + kx - \sin kx] \qquad (e)$$

As a final condition, the deflection at the upper end of the bar is δ, and hence, from Eq. (e), we conclude that

$$\tan kl = kl \left(1 - \frac{c}{l}\right) \qquad (2\text{-}16)$$

Equation (2-16) gives the value of the critical load for any particular value of the ratio c/l.   The solution of this equation is facilitated by the use of tables[1] of the function $(\tan x)/x$.   Table 2-1 gives values of kl and P_{cr} for various values of c/l as determined from Eq. (2-16).

[1] See *ibid.*, p. 32 of Addenda.

TABLE 2-1. CRITICAL LOADS FOR COLUMN WITH LOAD THROUGH A FIXED POINT
[From Eq. (2-16)]

$\dfrac{c}{l}$	0	0.2	0.4	0.6	0.8	1.0	1.2	1.5
kl	4.493	4.438	4.346	4.173	3.790	π	2.654	2.289
$\dfrac{P_{cr}}{\pi^2 EI/l^2}$	2.05	2.00	1.91	1.76	1.46	1	0.714	0.531
$\dfrac{c}{l}$	2.0	3.0	4.0	5.0	8.0	10	20	∞
kl	2.029	1.837	1.758	1.716	1.657	1.638	1.602	$\dfrac{\pi}{2}$
$\dfrac{P_{cr}}{\pi^2 EI/l^2}$	0.417	0.342	0.313	0.298	0.278	0.272	0.260	0.25

If c is greater than l, as assumed in Fig. 2-10b, the right-hand side of Eq. (2-16) is negative and the smallest value of kl which satisfies the equation[1] is between $\pi/2$ and π (see Fig. 2-8). This means that the critical load is greater than $\pi^2 EI/4l^2$, which was found previously for the case shown in Fig. 2-1. This can be explained by noting that the transverse force $P\delta/c$ counteracts the tendency for lateral buckling and hence a larger critical load is required. If c increases, the value of kl approaches the value $\pi/2$, and when c finally becomes infinitely large, we have

$$kl = \frac{\pi}{2} \qquad P_{cr} = \frac{\pi^2 EI}{4l^2}$$

which is the same result as for the previous case (Fig. 2-1) when the load remains always vertical.

When $c = l$, the fixed point C coincides with the lower end of the bar (Fig. 2-10c), the right-hand side of Eq. (2-16) vanishes, and we obtain

$$kl = \pi \qquad P_{cr} = \frac{\pi^2 EI}{l^2}$$

which is the same as for the fundamental case of buckling. This can be explained by noting that when the line of action of P passes through the base of the column, the bending moment at that point vanishes and the bar is in the same condition as a bar with hinged ends.

If the distance c is less than l, the right-hand side of Eq. (2-16) is positive and the smallest value of kl which satisfies the equation is somewhere between π and $3\pi/2$. The deflection curve then has an inflection point D as shown in Fig. 2-10d. Finally, when $c = 0$, Eq. (2-16) becomes the same as Eq. (2-11) and we have the case of a bar pinned at the top and fixed at the base (Fig. 2-7).

Bar with Rounded Ends. When a bar with rounded ends buckles laterally (Fig. 2-11), there will be a displacement b of the line of action of the compressive forces P.

[1] The trivial solution, $kl = 0$, is excluded as usual.

Since the angle of rotation θ of the end of the bar is small, the displacement (see Fig. 2-11b) will be

$$b = R\theta \tag{f}$$

where R is the radius of the hemispherical end of the bar.

Assuming a symmetrical shape of buckling (Fig. 2-11c) and taking the origin of coordinates at the center of the bar, we conclude at once that the constants A and C in the general solution (2-10) must be zero. This can be seen from the condition of symmetry which requires that the terms in Eq. (2-10) give a deflection curve which is symmetrical about the center of the bar.[1] From the condition that $y = 0$ when

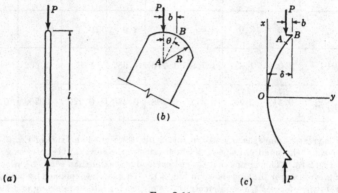

$$(a) \qquad \qquad (b) \qquad \qquad (c)$$

Fig. 2-11

$x = 0$, we get $B = -D$, so that the equation of the deflection curve becomes

$$y = D(1 - \cos kx) \tag{g}$$

The bending moment at any section of the column is equal to

$$M = -P(\delta - b - y) \tag{h}$$

and the bending moment at the end B is

$$(M)_{x=l/2} = Pb = PR\theta \approx PR\frac{dy}{dx}$$

Thus we have the following condition at the upper end of the bar:

$$EI\frac{d^2y}{dx^2} = -PR\frac{dy}{dx} \qquad \text{at } x = \frac{l}{2}$$

Using this condition with Eq. (g) gives

$$1 + kR\tan\frac{kl}{2} = 0$$

or

$$\frac{kl}{2}\tan\frac{kl}{2} = -\frac{l}{2R} \tag{2-17}$$

[1] The following end conditions can be used to establish the same result: $dy/dx = 0$ at $x = 0$; $y = \delta$ at $x = l/2$ and $x = -l/2$.

TABLE 2-2. CRITICAL LOADS FOR COLUMN WITH ROUNDED ENDS
[From Eq. (2-17)]

$\dfrac{l}{2R}$	0	1	2	4	6	8	10	15	20	30	40	50	∞
$\dfrac{kl}{2}$	π	2.798	2.459	2.043	1.874	1.791	1.743	1.682	1.653	1.625	1.611	1.603	$\dfrac{\pi}{2}$
$\dfrac{P_{cr}}{\pi^2 EI/l^2}$	4	3.17	2.45	1.69	1.42	1.30	1.23	1.15	1.11	1.07	1.05	1.04	1

Equation (2-17) can be used to determine the critical load for symmetrical buckling. If R is equal to zero, we obtain

$$\frac{kl}{2} = \frac{\pi}{2} \qquad P_{cr} = \frac{\pi^2 EI}{l^2} \tag{2-18}$$

which agrees with the usual result for a bar with pin ends. As R increases indefinitely, we approach the condition of a bar with flat ends and Eq. (2-17) gives

$$\frac{kl}{2} = \pi \qquad P_{cr} = \frac{4\pi^2 EI}{l^2}$$

which is the critical load for a fixed-end column. Table 2-2 gives values of $kl/2$ and P_{cr} from Eq. (2-17) for various values[1] of the ratio $l/2R$.

2.3. The Use of Beam-column Theory in Calculating Critical Loads.

Instead of the differential equation of the deflection curve being used for calculating critical loads, as was done in the two preceding articles, the problem can be solved in many cases by using results already obtained for beam-columns. It was shown in Chap. 1 that at certain values of the compressive force P the deflections and bending moments in a beam-column tend to increase indefinitely. Those values of the compressive force are evidently critical values.

As an example, let us consider the case of a beam-column AB pinned at one end and fixed at the other, as shown in Fig. 1-10. If the beam is subjected to a uniform lateral load q, the bending moment at the built-in end [see Eq. (1-37)] is

$$M_b = -\frac{ql^2}{8} \frac{\chi(u)}{\psi(u)} = -\frac{ql^2}{8} \frac{4(\tan 2u)(\tan u - u)}{u(\tan 2u - 2u)} \tag{a}$$

This moment increases indefinitely when the denominator of the expression approaches zero, provided the numerator does not also approach zero. This condition gives

$$\tan 2u = 2u$$

[1] Equation (2-17) is solved readily using tables of the function $x \tan x$; see *ibid.*, p. 32 of Addenda.

or, substituting $kl = 2u$ [see Eq. (1-13)],

$$\tan kl = kl$$

This result is the same as obtained previously by integration of the differential equation [see Eq. (2-11)]. Thus the critical value of the compressive force is that value for which the bending moment at the built-in end becomes infinitely large regardless of the magnitude of the lateral load.

The same procedure can be used to determine critical loads for a bar with elastically restrained ends (Fig. 1-13). When the bar is subjected to a lateral load, the moments acting at the ends are obtained from Eqs. (1-43):

$$-\frac{M_a}{\alpha} = \theta_{0a} + \frac{M_a l}{3EI} \psi(u) + \frac{M_b l}{6EI} \phi(u) \qquad (2\text{-}19)$$

$$-\frac{M_b}{\beta} = \theta_{0b} + \frac{M_b l}{3EI} \psi(u) + \frac{M_a l}{6EI} \phi(u) \qquad (2\text{-}20)$$

In these equations α and β are coefficients of end restraint [see Eqs. (1-41)], θ_{0a} and θ_{0b} represent the angles of rotation at the ends due to lateral load only, and the functions $\phi(u)$ and $\psi(u)$ are given by Eqs. (1-27) and (1-28). The moments M_a and M_b are end moments acting on the member AB (Fig. 1-13) and are positive in the directions shown. Solving Eqs. (2-19) and (2-20) for the moment M_a gives

$$M_a = \frac{-\theta_{0a}\left[\frac{1}{\beta} + \frac{l}{3EI}\psi(u)\right] + \theta_{0b}\left[\frac{l}{6EI}\phi(u)\right]}{\left[\frac{1}{\alpha} + \frac{l}{3EI}\psi(u)\right]\left[\frac{1}{\beta} + \frac{l}{3EI}\psi(u)\right] - \left[\frac{l}{6EI}\phi(u)\right]^2} \qquad (b)$$

The solution for M_b is obtained similarly and has the same expression in the denominator. Therefore the moments M_a and M_b become infinitely large when the denominator of Eq. (b) becomes zero. Thus the equation for determining the critical condition is

$$\left[\frac{1}{\alpha} + \frac{l}{3EI}\psi(u)\right]\left[\frac{1}{\beta} + \frac{l}{3EI}\psi(u)\right] - \left[\frac{l}{6EI}\phi(u)\right]^2 = 0 \qquad (2\text{-}21)$$

For particular values of α and β, this equation can be solved for u and the critical load determined. This method of calculation is useful in analyzing rigid frames and continuous beams, as discussed in the following articles.

In the particular case of symmetry (Fig. 2-12), we have

$$\alpha = \beta \qquad \theta_{0a} = \theta_{0b} \qquad M_a = M_b \qquad (c)$$

and Eqs. (2-1,9) and (2-20) are replaced by the single equation

$$- \frac{M_a}{\alpha} = \theta_{0a} + \frac{M_a l}{3EI} \psi(u) + \frac{M_a l}{6EI} \phi(u) \tag{2-22}$$

Solving this equation for M_a and setting the denominator of the resulting expression equal to zero, we obtain the equation for the critical load:

$$\frac{1}{\alpha} + \frac{l}{3EI} \psi(u) + \frac{l}{6EI} \phi(u) = 0$$

Now substituting the expressions for $\psi(u)$ and $\phi(u)$ from Eqs. (1-27) and (1-28) and also noting that $\tan u = (1 - \cos 2u)/\sin 2u$, we write this equation in the form

$$\frac{\tan u}{u} = - \frac{2EI}{\alpha l} \tag{2-23}$$

Values of u found from this equation lie between the limits $\pi/2$ and π. The value $\pi/2$ corresponds to $\alpha = 0$, which means that the ends of the bar are free to rotate, and the critical load is given by Eq. (2-5) for the

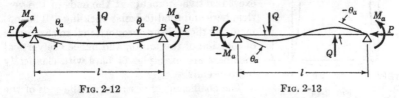

FIG. 2-12   FIG. 2-13

fundamental case. When the ends of the bar are rigidly built in, the coefficient α becomes infinite, the value of u is π, and the critical load is $P_{cr} = 4\pi^2 EI/l^2$. For intermediate values of α Eq. (2-23) can be solved readily using tables[1] of the function $(\tan x)/x$.

If the loading on the symmetrically supported bar is antisymmetrical (Fig. 2-13), we have

$$\alpha = \beta \qquad \theta_{0a} = - \theta_{0b} \qquad M_a = - M_b \tag{d}$$

and the critical load is determined by the equation

$$\frac{1}{\alpha} + \frac{l}{3EI} \psi(u) - \frac{l}{6EI} \phi(u) = 0$$

or

$$\frac{3}{u}\left(\frac{1}{u} - \frac{1}{\tan u}\right) = - \frac{6EI}{\alpha l} \tag{2-24}$$

Values of the critical load from this equation correspond to antisymmetric buckling patterns. The equation is solved easily for any value of α by using Table A-1, Appendix, since the expression on the left-hand side is

[1] *Ibid.*

the function ψ but with u in place of $2u$. As one of the limiting cases we have $\alpha = 0$ for pin ends; hence $u = \pi$ and $P_{cr} = 4\pi^2 EI/l^2$, which corresponds to the antisymmetric buckling shape shown in Fig. 2-6b. For fixed ends, α becomes infinite, $u = 4.493$, and the critical load is given by Eq. (2-15).

2.4. Buckling of Frames.[1] Since each member of a framework with rigid joints is in the condition of a bar with elastically restrained ends, the method described in the preceding article will be used for considering the buckling of frames. As a simple example, let us consider a frame $ABCD$ which is symmetrical with respect to horizontal and vertical axes (Fig. 2-14). The vertical members of the frame are compressed by axial forces P, and it is assumed that lateral movement of the joints is prevented by external constraints. When the load P reaches its critical value, the vertical bars begin to buckle as indicated by the dotted lines. This buckling is accompanied by bending of the two horizontal bars AB and CD. These bars exert reactive moments at the ends of the vertical bars and tend to resist buckling. The moments at the ends are proportional to the angles of rotation of the joints, and hence the vertical members are examples of bars with elastically built-in ends.

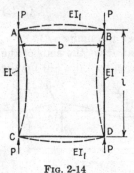

Fig. 2-14

The coefficient of restraint at the ends of the vertical bars is found from a consideration of the bending of the horizontal bars by couples at the ends. Denoting by EI_1 the flexural rigidity of the horizontal bars, the expression for the coefficient α is

$$\alpha = \frac{2EI_1}{b} \tag{a}$$

Since the vertical members buckle in a symmetrical shape, the critical load can be obtained from Eq. (2-23) of the preceding article. If the flexural rigidity of the vertical bars is denoted by EI, this equation becomes

$$\frac{\tan u}{u} = - \frac{Ib}{I_1 l} \tag{b}$$

and the critical load can be found from this equation in each particular case.

[1] The first discussion of problems of the stability of members of a rectangular frame was given by F. Engesser, "Die Zusatzkräfte und Nebenspannungen eiserner Fachwerkbrücken," Berlin, 1893. See also H. Zimmermann, "Knickfestigkeit der Stabverbindungen," Berlin, 1925.

If the frame in Fig. 2-14 consists of four identical bars, Eq. (b) becomes

$$\frac{\tan u}{u} = -1$$

and the lowest root of this equation is

$$u = \frac{kl}{2} = 2.029$$

and hence $\qquad P_{cr} = \frac{16.47EI}{l^2}$

If the horizontal bars are absolutely rigid, the right-hand side of Eq. (b) becomes zero and therefore $\tan u = 0$, $u = \pi$, and

$$P_{cr} = \frac{4\pi^2 EI}{l^2}$$

This is the case of a bar with fixed ends. Finally, if $I_1 = 0$ we obtain $u = \pi/2$ and

$$P_{cr} = \frac{\pi^2 EI}{l^2}$$

as for a bar with pinned ends.

If the horizontal members of the frame are subjected to the action of compressive forces Q, as shown in Fig. 2-15, the coefficients of end restraint α will be diminished. Instead of Eq. (a) we must use the expression

$$\alpha = \frac{2EI_1}{b} \frac{u_1}{\tan u_1} \qquad (c)$$

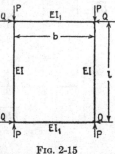

FIG. 2-15

which is obtained from Eq. (1-34). The quantity u_1 for the horizontal member is

$$u_1 = \frac{b}{2} \sqrt{\frac{Q}{EI_1}}$$

In calculating the critical value of the compressive force P, we again use Eq. (2-23) of the preceding article and obtain

$$\frac{\tan u}{u} = - \frac{Ib}{I_1 l} \frac{\tan u_1}{u_1} \qquad (2\text{-}25)$$

With the quantities on the right-hand side of this equation known, the critical force P can be found assuming that the horizontal bars do not buckle first.

If it is desired to determine critical values of the force Q (Fig. 2-15), the above procedure will give the equation

$$\frac{\tan u_1}{u_1} = -\frac{I_1 l}{Ib}\frac{\tan u}{u}$$

which is the same as Eq. (2-25). Thus Eq. (2-25) defines limiting values of the two axial forces P and Q. For example, if the frame is square with all members having the same flexural rigidity, Eq. (2-25) becomes

$$\frac{\tan u}{u} = -\frac{\tan u_1}{u_1} \tag{2-26}$$

This equation is plotted graphically in Fig. 2-16. If the values of P and Q locate a point on the curved line in the figure, buckling will occur. It is

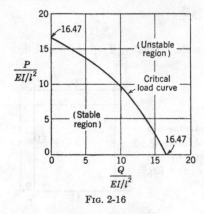

FIG. 2-16

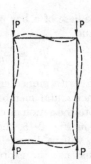

FIG. 2-17

seen that the critical value of P decreases as Q increases, and vice versa, as would be expected. If the values of P and Q locate a point below the curved line, no buckling will occur; hence the portion of the graph below the curve represents a stable region.

Other forms of buckling for the frame in Fig. 2-14 are possible. For example, in Fig. 2-17 is shown a buckled configuration in which the members have inflection points at their mid-points. This case corresponds to the antisymmetric buckling case of the preceding article, and the critical load is found from Eq. (2-24) by substituting

$$\alpha = \frac{6EI_1}{b} \tag{d}$$

which represents the coefficient of end restraint. Critical loads for the buckling mode of Fig. 2-17 are larger than those for the symmetrical mode of Fig. 2-14 and hence are not usually so important.

If the flexural rigidities of the two horizontal members of the frame in Fig. 2-14 are different, the end conditions for the compressed vertical members are no longer the same and the critical load will be obtained from Eq. (2-21) of the preceding article.

In the case shown in Fig. 2-18a the vertical bar is rigidly built in at the base and elastically supported at the top. Then β is infinite, α is finite,[1] and Eq. (2-21) becomes

$$4\psi(u) \left[\frac{3EI}{\alpha l} + \psi(u) \right] = [\phi(u)]^2 \qquad (2\text{-}27)$$

This equation can be solved for each particular case by trial and error, using Table A-1, Appendix, for values of the functions $\psi(u)$ and $\phi(u)$.

For the case shown in Fig. 2-18b the coefficient β is zero and α is finite. When β approaches zero in Eq. (2-21), the factor

$$\frac{1}{\alpha} + \frac{l}{3EI} \psi(u) \qquad (e)$$

must approach zero also and the equation for calculating the critical load is obtained by equating expression (e) to zero. This gives the relation

$$\psi(u) = -\frac{3EI}{\alpha l} \qquad (2\text{-}28)$$

Fɪɢ. 2-18

When the horizontal member is considered as a beam with one end built in and the other hinged and the flexural rigidity of this member is denoted by EI_1, the magnitude of α in Eq. (2-28) is $4EI_1/b$ and the equation for the critical load becomes

$$\psi(u) = -\frac{3Ib}{4I_1 l} \qquad (2\text{-}29)$$

Assuming, for instance, that $b = l$ and $I = I_1$, we find from Eq. (2-29) that

$$\psi(u) = -\frac{3}{4} \qquad 2u = kl = 3.83 \qquad P_{cr} = \frac{14.7EI}{l^2}$$

In the limiting case where α becomes infinite, Eq. (2-28) coincides with Eq. (2-11) and we have the solution for a column fixed at one end and pinned at the other (Fig. 2-7).

---

[1] In order for this discussion to be exact, it must be assumed that the compressive force P is applied to the vertical bar before the rigid connection is made. Then there will be no bending in the horizontal bar before buckling occurs. The small changes in the lengths of the bars at buckling are neglected also.

In the previous discussion it was assumed that the ends of the compressed members do not displace laterally. Let us consider now the case shown in Fig. 2-19 in which a frame with compressed vertical members is free to move laterally at the top. If the frame has a vertical axis of symmetry, each vertical member can be considered separately as a compressed bar free at the lower end and elastically built in at the upper end. Taking the coordinate axes as shown in the figure, the differential equation of the deflection curve of the bar AB is

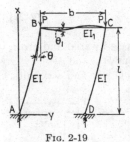

FIG. 2-19

$$EI \frac{d^2y}{dx^2} = - Py$$

The solution of this equation, satisfying conditions at the lower end, is

$$y = A \sin kx \qquad (f)$$

At the upper end the angles θ and θ_1 must be equal and since the horizontal bar BC is bent by two couples, each equal to $P(y)_{x=l}$, the condition at the upper end is

$$\left(\frac{dy}{dx}\right)_{x=l} = P \frac{b}{6EI_1} (y)_{x=l}$$

or, by using expression (f),

$$k \cos kl = \frac{Pb \sin kl}{6EI_1} \qquad (g)$$

If the horizontal bar is absolutely rigid, $EI_1 = \infty$, and we obtain

$$\cos kl = 0 \qquad kl = \frac{\pi}{2} \qquad P_{cr} = \frac{\pi^2 EI}{4l^2}$$

In the general case, Eq. (g) can be represented in the following form:

$$kl \tan kl = \frac{6I_1 l}{Ib} \qquad (h)$$

and the critical value of the load P can be found for any numerical value of the ratio $I_1 l/Ib$. Assuming that all three bars of the frame are identical, we obtain

$$kl \tan kl = 6$$

from which

$$kl = 1.35 \qquad \text{and} \qquad P_{cr} = \frac{1.82EI}{l^2}$$

2.5. Buckling of Continuous Beams. In calculating critical compressive forces for continuous beams, we can again make use of beam-column formulas. Previously, formulas were derived for the bending of continuous beams subjected to longitudinal compression in addition to transverse bending (see Art. 1.10). The critical values of the compressive forces will now be obtained as the values at which even a slight lateral load will produce infinite deflection. As in Art. 1.10, the compressive forces in the continuous beam are assumed to be constant within each span but may vary from one span to the next.

Considering two consecutive spans of a bar on several supports (Fig. 1-14) the relation between the three consecutive bending moments at the supports is given by Eq. (1-44):

$$M_{n-1}\phi(u_{n-1}) + 2M_n\left[\psi(u_{n-1}) + \frac{l_n I_{n-1}}{l_{n-1} I_n}\psi(u_n)\right]$$
$$+ M_{n+1}\frac{l_n}{l_{n-1}}\frac{I_{n-1}}{I_n}\phi(u_n) = -\frac{6EI_{n-1}}{l_{n-1}}(\theta_{0n} + \theta'_{0n}) \quad (a)$$

There will be as many equations of this type as there are statically indeterminate moments, provided the ends of the continuous beam are simply supported. If the ends are fixed, then two additional equations, expressing the condition of fixity, must be used in addition to Eqs. (a). The coefficients in these equations contain the functions

FIG. 2-20

$\phi(u)$ and $\psi(u)$ and depend on the magnitudes of the compressive force P. The critical values of these forces are those values for which the bending moments, as solved from Eq. (a), become infinitely large. This requires that the determinant of the left-hand side of Eqs. (a) be made equal to zero. In this way an equation for calculating the critical values of the compressive forces is obtained.[1]

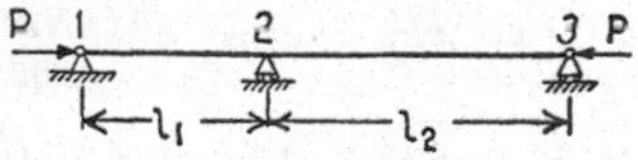

Let us consider a bar on three supports, with hinged ends, and compressed by forces P applied at the ends (Fig. 2-20). In this case there is only one unknown moment M_2 and Eq. (a) becomes

$$2M_2\left[\psi(u_1) + \frac{l_2 I_1}{l_1 I_2}\psi(u_2)\right] = -\frac{6EI_1}{l_1}(\theta_{02} + \theta'_{02})$$

The critical value of the compressive force is now obtained from the condition that M_2 becomes infinite, which means that

$$\psi(u_1) + \frac{l_2 I_1}{l_1 I_2}\psi(u_2) = 0 \qquad (b)$$

[1] There is an exceptional case in which buckling may occur with all bending moments at the supports equal to zero. This takes place when all spans of the bar have such proportions and such compressive forces that $u_1 = u_2 = u_3 = \cdots$. In this case buckling of each span is not influenced by the adjacent spans and the critical values of the forces are calculated for each span as for a bar with hinged ends.

Assuming that the cross section of the bar is the same for both spans, we have

$$u_1 = \frac{k_1 l_1}{2} = \frac{l_1}{2}\sqrt{\frac{P}{EI}} \qquad u_2 = \frac{k_2 l_2}{2} = \frac{l_2}{2}\sqrt{\frac{P}{EI}}$$

$$\frac{u_1}{u_2} = \frac{l_1}{l_2}$$

Equation (b) can now be put in the form

$$\frac{\psi(u_1)}{\psi(u_1 l_2/l_1)} = -\frac{l_2}{l_1} \qquad\qquad (c)$$

which can be solved by using values of the function $\psi(u)$ from Table A-1 in the Appendix. If we take $l_2 = 2l_1$, Eq. (c) becomes

$$\frac{\psi(u_1)}{\psi(2u_1)} = -2$$

and from Table A-1 we find $2u_1 = 1.93$, whence

$$P_{cr} = \frac{(1.93)^2 EI}{l_1^2} = \frac{3.72 EI}{l_1^2} = \frac{14.9 EI}{l_2^2}$$

It is seen that the value of the critical load lies between the two values $\pi^2 EI/l_1^2$ and $\pi^2 EI/l_2^2$, calculated for separate spans as if each were a bar with hinged ends. The stability of the shorter span is reduced, owing to the action of the longer span, while the stability of the longer span is increased.

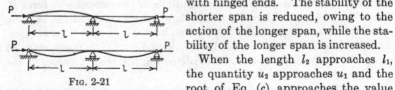

FIG. 2-21

When the length l_2 approaches l_1, the quantity u_2 approaches u_1 and the root of Eq. (c) approaches the value $2u_1 = 2u_2 = \pi$. In this case the bending moment at the middle support is zero and each span can be considered as a bar with hinged ends (see footnote on p. 67). The next root of Eq. (c), for u_2 approaching u_1, is $2u_1 = 2u_2 = 4.493$. Then

$$P_{cr} = \frac{20.19 EI}{l_1^2} \approx \frac{\pi^2 EI}{(0.7 l_1)^2}$$

The two forms of the buckled bar are shown in Fig. 2-21. Only the first form, corresponding to the smallest compressive force, is of practical significance.

As a second example, let us consider a bar on four supports (Fig. 2-22) and assume that $l_1 = l_3$ and $I_1 = I_3$. Spans 1 and 3 are compressed axially by forces P and may be considered as bars having one end hinged and the other end elastically built in. It can be seen that the shape of the buckled bar will be approximately as shown in Fig. 2-22b and, from symmetry, the bending moments at supports 2 and 3 will be equal. Since the compressive force in the second span is zero, we obtain

$$\phi(u_2) = \psi(u_2) = 1$$

and Eq. (a) gives

$$2\psi(u_1) + \frac{3l_2 I_1}{l_1 I_2} = 0 \tag{d}$$

This equation coincides with Eq. (2-28) of the previous article if we substitute $\alpha = 2EI_2/l_2$ in that equation.

As a last example, let us consider the case shown in Fig. 2-23. A compressed bar AB is rigidly connected to a column at C, so that any lateral buckling of the bar must be accompanied by bending of the column. In the solution of this problem Eqs. (1-48) and (1-49) of Art. 1.10 will be used. If M_c and M'_c are the values of the bending moments in

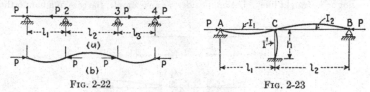

FIG. 2-22   FIG. 2-23

the two adjacent cross sections to the left and to the right of support C, respectively, and if there is no lateral load, these equations become

$$2M_c\psi(u_1) + 2M'_c \frac{l_2}{l_1}\frac{I_1}{I_2}\psi(u_2) = 0 \tag{e}$$

$$\frac{h}{3EI'}(M'_c - M_c) = \frac{M_c l_1}{3EI_1}\psi(u_1) \tag{f}$$

in which EI' and h are the flexural rigidity and the length of the column. The critical value of the compressive force is found by setting the determinant of Eqs. (e) and (f) equal to zero, which yields

$$\psi(u_1) + \psi(u_2)\frac{l_2}{l_1}\frac{I_1}{I_2}\left[1 + \frac{l_1}{h}\frac{I'}{I_1}\psi(u_1)\right] = 0 \tag{g}$$

This equation can be solved in each particular case by using Table A-1. In the particular case where $l_1 = l_2$ and $I_1 = I_2$, Eq. (g) becomes

$$\psi(u_1)\left[2 + \frac{l_1}{h}\frac{I'}{I_1}\psi(u_1)\right] = 0$$

from which

$$\psi(u_1) = 0 \qquad \text{or} \qquad \psi(u_1) = -\frac{2hI_1}{l_1 I'}$$

The first of these two solutions gives the same value for P_{cr} as that obtained for the bar shown in Fig. 2-7. It corresponds to a deflection curve symmetrical with respect to C, in which the column does not bend at all. The second solution, which gives a smaller value for P_{cr}, corresponds to a nonsymmetrical shape of the buckled bar as shown in Fig. 2-23. Only this second solution has any practical significance. For any particular case it is obtained readily from Table A-1. Taking, for example, $2hI_1/l_1I' = 1$, we find from the table, $2u_1 = kl_1 = 3.73$ and

$$P_{cr} = \frac{13.9EI}{l_1^2}$$

2.6. Buckling of Continuous Beams on Elastic Supports. A compressed member may be supported at several intermediate points by supports which are not completely rigid. For example, a compression member of a truss may be supported laterally at one or more intermediate points by other members of the truss.[1] The general method of solving such problems consists of using Eqs. (1-46) for a continuous bar with supports not on a straight line. Let us consider, as an example, the case of a bar on three

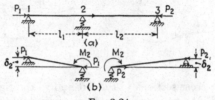

Fig. 2-24

supports (Fig. 2-24) in which the middle support is elastic. If the bar buckles under the action of the compressive forces, the center reaction R_2 will be proportional to the deflection δ_2. Let α_2 be the spring constant of the support, that is, the load which will produce unit deflection of the support, so that

$$R_2 = \alpha_2\delta_2 \qquad (a)$$

All the supports of the beam are such that the cross sections of the beam at the supports can rotate freely during bending. Considering the two adjacent spans as simply supported beams acted upon by forces P_1, P_2, and couples M_2 (Fig. 2-24b), the reaction R_2, from statics, is

$$R_2 = \frac{P_1\delta_2}{l_1} + \frac{P_2\delta_2}{l_2} - \frac{M_2}{l_1} - \frac{M_2}{l_2} \qquad (b)$$

[1] The first problems of the type discussed in this article were solved by Jasinsky, who considered the lateral buckling of compression diagonals of lattice trusses. See "Scientific Papers of F. S. Jasinsky," vol. 1, p. 145, St. Petersburg, 1902. This important paper by Jasinsky on the buckling of columns also appeared in a French translation; see *Ann. ponts chaussées*, 1894. See also H. Zimmermann, "Die Knickfestigkeit eines Stabes mit elastischer Querstützung," W. Ernst and Sohn, Berlin, 1906, and his papers in *Sitzb. Berlin Akad. Math. physik. Kl.*, 1905, p. 898; 1907, pp. 235 and 326; 1909, pp. 180 and 348.

From (a) and (b) we obtain

$$\alpha_2 \delta_2 = \frac{P_1 \delta_2}{l_1} + \frac{P_2 \delta_2}{l_2} - \frac{M_2}{l_1} - \frac{M_2}{l_2} \tag{c}$$

Another equation for support 2 is obtained from the general Eq. (1-46). Observing that the moments M_1 and M_3 for the supports 1 and 3 are zero and that the angle β_2 for the middle support (see Fig. 1-15) is

$$\beta_2 = \frac{\delta_2}{l_1} + \frac{\delta_2}{l_2}$$

Eq. (1-46) becomes

$$2M_2 \left[\psi(u_1) + \frac{l_2}{l_1} \frac{I_1}{I_2} \psi(u_2) \right] = \frac{6EI_1}{l_1} \left(\frac{\delta_2}{l_1} + \frac{\delta_2}{l_2} \right) \tag{d}$$

Buckling of the bar becomes possible when Eqs. (c) and (d) yield a solution for M_2 and δ_2 different from zero. Hence the critical values of P_1 and P_2 are found by setting the determinant of these two equations equal to zero, which gives

$$2 \left[\alpha_2 - \frac{P_1}{l_1} - \frac{P_2}{l_2} \right] \left[\psi(u_1) + \frac{l_2}{l_1} \frac{I_1}{I_2} \psi(u_2) \right] = - \frac{6EI_1}{l_1} \frac{(l_1 + l_2)^2}{l_1^2 l_2^2} \tag{e}$$

In this equation P_1 and P_2 are unknown; if the ratio between them is given, their values can be found by using Table A-1. In the particular case where $P_1 = P_2$ and $I_1 = I_2$, Eq. (e) can be simplified by using the expression for $\psi(u)$ [see Eq. (1-28)] and finally represented in the following form:

$$\sin 2u_1 \sin 2u_2 = 2(u_1 + u_2) \sin 2(u_1 + u_2) \left[\frac{l_1 l_2}{(l_1 + l_2)^2} - \frac{P}{\alpha_2(l_1 + l_2)} \right] \tag{f}$$

In a more general case where there are several intermediate elastic supports, we can write for each support two equations similar to (c) and (d); the critical values of the compressive forces are determined by equating the determinant of this system of equations to zero.

If the cross section is constant and the compressive force the same for all spans, it is advantageous to use Eqs. (1-50) instead of Eqs. (1-46) in calculating the critical value of the compressive force. To illustrate this method of solution, let us consider again the beam on three supports (Fig. 2-24). Equation (1-50) for this case becomes

$$\frac{q(l_1 + l_2)^4}{16EIu^4} \left[\frac{\cos \left(1 - \dfrac{2l_1}{l_1 + l_2} \right) u}{\cos u} - 1 \right] - \frac{q(l_1 + l_2)^2}{8EIu^2} l_1 l_2$$

$$- \frac{\sin kl_1}{Pk \sin k(l_1 + l_2)} R_2 \sin kl_2 + \frac{l_1 l_2}{P(l_1 + l_2)} R_2 = \frac{R_2}{\alpha_2} \tag{g}$$

The critical value of the compressive force is that value at which the deflections, and hence the reaction R_2, begin to increase indefinitely.[1] This requires that the coefficient of R_2 in Eq. (g) becomes zero. Thus, in calculating the critical load, we obtain the following equation:

$$- \frac{\sin kl_1 \sin kl_2}{Pk \sin k(l_1 + l_2)} + \frac{l_1 l_2}{P(l_1 + l_2)} - \frac{1}{\alpha_2} = 0 \tag{h}$$

Observing that $kl_1 = 2u_1$ and $kl_2 = 2u_2$, we find that Eq. (h) coincides with Eq. (f) obtained before in another way.

[1] If there are two equal spans, buckling may occur in such a way that $R_2 = 0$. In this case the buckling condition of each span is the same as that of a bar with hinged ends and it is not necessary to consider the continuous bar.

In discussing solutions of Eq. (f), let us begin with several simple cases. If $\alpha_2 = \infty$, Eq. (e), from which Eq. (f) is obtained as a particular case, coincides with Eq. (b) of the previous article (see p. 67), which was obtained for a bar on three rigid supports. Thus we can use the solution from that article.

If α_2 approaches zero, the second term in the parentheses on the right side of Eq. (f) approaches infinity, and the equation can be satisfied only if $\sin 2(u_1 + u_2)$ approaches zero simultaneously. The critical load is then obtained from the equation

$$\sin 2(u_1 + u_2) = \sin k(l_1 + l_2) = 0$$

from which

$$P_{cr} = \frac{\pi^2 EI}{(l_1 + l_2)^2}$$

This coincides with the critical load for a bar with hinged ends and of length $l = l_1 + l_2$.

If the two spans are equal, we have $u_1 = u_2$, $l_1 = l_2 = l/2$, and Eq. (f) can be put in a simpler form:

$$\sin 2u_1 \left[-\sin 2u_1 + 8u_1 \cos 2u_1 \left(\frac{1}{4} - \frac{P}{\alpha_2 l}\right) \right] = 0 \tag{i}$$

The upper limit for the critical value of the compressive force is obtained by assuming that the intermediate support is absolutely rigid ($\alpha_2 = \infty$). Then the shape of the buckled bar is that shown in Fig. 2-25a and the critical value of the compressive force is obtained from the equation

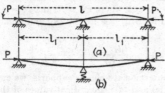

$$2u_1 = \pi$$

which gives

$$P_{cr} = \frac{\pi^2 EI}{l_1^2} = \frac{4\pi^2 EI}{l^2} \tag{j}$$

FIG. 2-25

The lower limit for the critical load is obtained by assuming that the intermediate support is absolutely flexible ($\alpha_2 = 0$). The shape of the deflection curve of the buckled bar is then as in Fig. 2-25b, and we have

$$2u_1 = \frac{\pi}{2} \qquad P_{cr} = \frac{\pi^2 EI}{l^2} \tag{k}$$

For any intermediate value of the rigidity of the elastic support we have

$$\frac{\pi}{2} < 2u_1 \leq \pi \tag{l}$$

There are two possibilities for the left side of (i) to become zero. Either $\sin 2u_1 = 0$, which gives for the critical load the value (j), or the expression in the brackets may become zero. From the inequality (l) it can be concluded that $\sin 2u_1$ is positive and $\cos 2u_1$ negative. Hence, the expression in brackets may become zero only if

$$\frac{P}{\alpha_2 l} \geq \frac{1}{4}$$

and the corresponding smallest value of P is

$$P = \frac{\alpha_2 l}{4} \tag{m}$$

If this value is larger than (j), the condition $\sin 2u_1 = 0$ determines the critical value of the load and the shape of the buckled bar is that shown in Fig. 2-25a. The limiting value of the rigidity of the support at which this shape of buckling occurs is obtained from Eq. (m) by substituting for P the value (j). Then

$$\frac{4\pi^2EI}{l^2} = \frac{\alpha_2 l}{4}$$

from which

$$\alpha_2 = \frac{16\pi^2EI}{l^3} \tag{n}$$

For smaller values of α_2, the flexibility of the intermediate support should be considered; the value of P_{cr} is obtained by determining the value of P at which the expression in the brackets of Eq. (i) becomes zero.

In Fig. 2-26 a curve is plotted which shows the variation of the critical load with the rigidity of the intermediate support. In this curve the ratios $P_{cr}:\pi^2EI/l^2 = P_{cr}:P_e$ are taken as ordinates and the ratios

$$\alpha_2 l:\pi^2EI/l^2 = \alpha_2 l/P_e$$

as abscissas. The curve deviates but very little from a straight line so that the critical load increases in approximately the same proportion as the rigidity of the support.

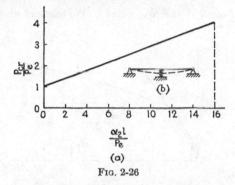

(a)

FIG. 2-26

If the spans are not equal, the general Eq. (f) should be considered in calculating the critical load. The lower limit for P_{cr} is given, as before, by Eq. (k). For determining the upper limit we assume $\alpha_2 = \infty$. Then the right side of Eq. (f) becomes zero when

$$2(u_1 + u_2) = 2\pi \tag{o}$$

At the same time the left side is negative, since one of the two angles $2u_1$ and $2u_2$ is larger and the other smaller than π. Both sides of Eq. (f) can be made equal (i.e., the upper limit for P will be found) by taking for $2(u_1 + u_2)$ a quantity somewhat smaller than 2π. This indicates that in the case of rigid supports any lateral displacement of the intermediate support from the middle position diminishes the value of the critical load.

After the upper and lower limits for P are determined by taking $\alpha_2 = \infty$ and $\alpha_2 = 0$, the critical value of the compressive force for any intermediate value of α_2 is obtained by solving Eq. (f) by the trial-and-error method. This solution is simplified if we note that for

$$\alpha_2 = \frac{P(l_1 + l_2)}{l_1 l_2} \tag{p}$$

the right side of Eq. (f) is zero. Assuming $l_1 > l_2$, the smallest value of P which makes the left side of the equation equal to zero is obtained from the equation

$$\sin 2u_1 = 0$$

from which $2u_1 = \pi$ and $(2u_1 + u_2) = \pi l/l_1$. If we take α_2 smaller than (p), the value of $2(u_1 + u_2)$ must be smaller than $\pi l/l_1$ found above; at the same time it must be larger than the value π found for $\alpha_2 = 0$. Hence, the root of Eq. (f) must lie within the limits

$$\pi < 2(u_1 + u_2) < \frac{\pi l}{l_1} \qquad (q)$$

For values of α_2 larger than that given by (p), the quantity $2(u_1 + u_2)$ will be larger than $\pi l/l_1$ and at the same time it must be smaller than 2π, as was explained before. Hence the limits for the root of Eq. (f) are

$$\frac{\pi l}{l_1} < 2(u_1 + u_2) < 2\pi \qquad (r)$$

By using (q) and (r), Eq. (f) can be solved by the trial-and-error method for any particular value of α_2.

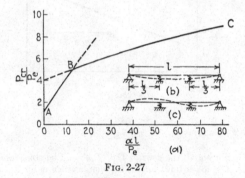

FIG. 2-27

As another problem of the stability of a bar on elastic supports, let us consider a continuous beam of constant cross section simply supported at the ends on rigid supports and having several equally spaced intermediate elastic supports of equal rigidity. It is sometimes necessary to select the common rigidity of the intermediate supports so that they will not deflect when the bar buckles, and will, therefore, be equivalent to absolutely rigid supports.[1] We have already discussed the case of one intermediate support and obtained Eq. (n) for determining the value of α_2 at which the support behaves as though it were absolutely rigid.

The same problem can be solved with little difficulty in the general case of any number of intermediate supports by using Eqs. (1-50). Take a continuous beam with two intermediate supports each having a spring constant of α (Fig. 2-27b). Equa-

[1] This problem was first discussed by J. G. Boobnov, "Theory of Structure of Ships," vol. 1, p. 259, St. Petersburg, 1913. See also W. B. Klemperer and H. B. Gibbons, *Z. angew. Math. u. Mech.*, vol. 13, p. 251, 1933 and M. A. Lazard, *Ann. inst. tech. bâtiment et trav. publ.*, no. 88, September, 1949.

tions (1-50) for three equal spans, each of length $l/3$, are

$$-\frac{\sin\dfrac{2kl}{3}}{Pk\sin kl}R_3\sin\frac{kl}{3}+\frac{2}{3P}R_3\frac{l}{3}-\frac{\sin\dfrac{kl}{3}}{Pk\sin kl}R_2\sin\frac{kl}{3}+\frac{1}{3P}R_2\frac{l}{3}=\frac{R_3}{\alpha}$$

$$-\frac{\sin\dfrac{kl}{3}}{Pk\sin kl}R_3\sin\frac{kl}{3}-\frac{\sin\dfrac{kl}{3}}{Pk\sin kl}R_2\sin\frac{2kl}{3}+\frac{1}{3P}\left(R_3\frac{l}{3}+R_2\frac{2l}{3}\right)=\frac{R_2}{\alpha}$$

$$(s)$$

If the supports are absolutely rigid, the bar buckles, so that there will be inflection points at the supports and each span is in the same condition as a bar with hinged ends of length $l/3$. The critical value of the compressive force is then obtained from the equation

$$\frac{kl}{3}=\pi$$

from which,

$$P_{cr}=\frac{9\pi^2EI}{l^2}$$

Assume now that the supports are elastic and that their rigidity approaches the limiting value at which the supports behave as though they were absolutely rigid. In this case the critical value of the compressive force approaches the value obtained above for absolutely rigid supports, and we can assume that

$$\frac{kl}{3}=\pi-\Delta \qquad (t)$$

where Δ is a small quantity. Substituting (t) in Eqs. (s) and neglecting small quantities, we finally obtain

$$\frac{l}{9P}R_2+\left(\frac{2l}{9P}-\frac{1}{\alpha}\right)R_3=0$$

$$\left(\frac{2l}{9P}-\frac{1}{\alpha}\right)R_2+\frac{l}{9P}R_3=0$$

$$(u)$$

The value of α at which the critical load approaches that for absolutely rigid supports is obtained by equating the determinant of Eqs. (u) to zero. Then

$$\left(\frac{2l}{9P}-\frac{1}{\alpha}\right)^2-\left(\frac{l}{9P}\right)^2=0$$

from which

$$\alpha=\frac{9P}{l} \qquad (v)$$

where

$$P=\frac{9\pi^2EI}{l^2}$$

The variation in the critical load due to an increase in the rigidity of the supports can be handled in the same manner as was explained for one elastic support. The results of such an investigation[1] are shown in Fig. 2-27a, in which the ratio P_{cr}/P_e is plotted against $\alpha l/P_e$ where $P_e=\pi^2EI/l^2$. When the rigidity of the supports is small, the deflection curve of the buckled bar has no inflection points (Fig. 2-27b). The curve AB in Fig. 2-27a corresponds to this condition. For greater rigidity of the supports an inflection point occurs at the middle (Fig. 2-27c). This condition is represented in Fig. 2-27a by the curve BC. When α approaches the magnitude given by Eq. (v), the critical load approaches the value $9\pi^2EI/l^2$ and the points of support become inflection points. A further increase in the rigidity of the supports has no effect on the buckling of the bar.

[1] See Klemperer and Gibbons, *loc. cit.*

Proceeding in the same way for m equal spans of length l/m, we obtain for the necessary rigidity of the elastic supports, at which they behave as if absolutely rigid, the expression

$$\alpha = \frac{mP}{\gamma l} \tag{2-30}$$

in which m is the number of spans, γ a numerical factor which depends on the number of spans, and $P = m^2\pi^2EI/l^2$ is the critical load calculated for one span as for a bar of length l/m with hinged ends. Several values of the factor γ are given in Table 2-3. It is seen that γ diminishes as the number of spans increases and approaches the value $\gamma = 0.250$.

TABLE 2-3. VALUES OF THE FACTOR γ IN EQ. (2-30)

m	2	3	4	5	6	7	9	11
γ	0.500	0.333	0.293	0.276	0.268	0.263	0.258	0.255

2.7. Large Deflections of Buckled Bars (The Elastica). In the discussions of the preceding articles of this chapter it was found that the deflection of a bar was indeterminate at the critical load. This indicated that at the critical load the bar could have any value of deflection, provided the deflection remained small. This conclusion is reached because of the nature of the differential equations which were used for calculating the critical loads. These equations were based upon the approximate expression d^2y/dx^2 for the curvature of the buckled bar. If the exact expression for curvature is used, there will be no indefiniteness in the value of the deflection. The shape of the elastic curve, when found from the exact differential equation, is called the *elastica*.[1]

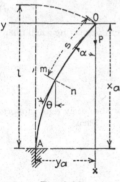

FIG. 2-28

Let us begin by considering the slender rod shown in Fig. 2-28, fixed at the base and free at the upper end. If the load P is taken somewhat larger than the critical value [Eq. (2-4)], a large deflection of the bar is produced. Taking the coordinate axes as shown in the figure and measuring the distance s along the axis of the bar from the origin O, we find that the exact expression for the curvature of the bar is $d\theta/ds$. Since the bending moment in the bar is equal to the flexural

---

[1] The problem of the elastica was first investigated by Euler, *loc. cit.*, and also by Lagrange, Sur la figure des colonnes, *Misc. Taurinensia*, vol. 5, 1770–1773. Lagrange's work is reprinted in "Oeuvres de Lagrange," vol. 2, pp. 125–170, Gauthier-Villars, Paris, 1868. For a historical discussion see Timoshenko, "History of Strength of Materials," pp. 30–40, McGraw-Hill Book Company, Inc., New York, 1953.

rigidity times the curvature, the exact differential equation of the deflection curve is

$$EI \frac{d\theta}{ds} = -Py \tag{a}$$

As in our previous discussions, the change in length of the column due to compression will be neglected.[1]
Differentiating Eq. (a) with respect to s and using the relation

$$\frac{dy}{ds} = \sin \theta$$

we obtain
$$EI \frac{d^2\theta}{ds^2} = -P \sin \theta \tag{b}$$

Thus the differential equation of the deflection curve is of the same form as the differential equation for the oscillations of a pendulum. In the equation for the pendulum the quantity EI is replaced by the moment of inertia of the pendulum with respect to the axis of rotation, s is replaced by the time, and P by the weight of the pendulum multiplied by the distance of the center of gravity from the axis of rotation. This analogy between the deformation of a slender rod loaded only at its ends and the rotation of a rigid body about a fixed point was discovered by Kirchhoff[2] and is known as *Kirchhoff's dynamical analogy*.

In solving Eq. (b), we begin by multiplying both sides by $d\theta$ and integrating, so that

$$\int \frac{d^2\theta}{ds^2} \frac{d\theta}{ds} ds = -k^2 \int \sin \theta \, d\theta$$

where $k^2 = P/EI$. This equation can be expressed in the form

$$\frac{1}{2} \int \frac{d}{ds} \left(\frac{d\theta}{ds} \right)^2 ds = -k^2 \int \sin \theta \, d\theta$$

and upon integrating we obtain

$$\frac{1}{2} \left(\frac{d\theta}{ds} \right)^2 = k^2 \cos \theta + C$$

where C is a constant of integration which is determined from the conditions at the upper end of the bar. At the upper end we have $d\theta/ds = 0$, since the bending moment is zero, and also $\theta = \alpha$. These conditions give

$$C = -k^2 \cos \alpha$$

[1] This assumption is justified for the usual structural materials.
[2] G. R. Kirchhoff, *J. Math. (Crelle)*, vol. 56, 1859. The analogy was further developed by W. Hess, *Math. Ann.*, vol. 25, 1885.

and therefore

$$\left(\frac{d\theta}{ds}\right)^2 = 2k^2(\cos\theta - \cos\alpha)$$

or

$$\frac{d\theta}{ds} = \pm k\sqrt{2}\sqrt{\cos\theta - \cos\alpha}$$

Since $d\theta/ds$ is always negative, as seen from Fig. 2-28, the positive sign will be dropped from this equation. Solving for ds gives

$$ds = -\frac{d\theta}{k\sqrt{2}\sqrt{\cos\theta - \cos\alpha}}$$

and the total length of the bar, after the limits of integration are interchanged, is

$$l = \int ds = \int_0^\alpha \frac{d\theta}{k\sqrt{2}\sqrt{\cos\theta - \cos\alpha}}$$
$$= \frac{1}{2k}\int_0^\alpha \frac{d\theta}{\sqrt{\sin^2\frac{\alpha}{2} - \sin^2\frac{\theta}{2}}} \quad (c)$$

This integral can be simplified by using the notation $p = \sin(\alpha/2)$ and by introducing a new variable ϕ in such a manner that

$$\sin\frac{\theta}{2} = p\sin\phi = \sin\frac{\alpha}{2}\sin\phi \quad (d)$$

It is seen from these relations that when θ varies from 0 to α, the quantity $\sin\phi$ varies from 0 to 1; hence ϕ varies from 0 to $\pi/2$. We also find from Eq. (d), by differentiation, that

$$d\theta = \frac{2p\cos\phi\,d\phi}{\cos(\theta/2)} = \frac{2p\cos\phi\,d\phi}{\sqrt{1 - p^2\sin^2\phi}} \quad (e)$$

Substituting in Eq. (c) and noting that

$$\sqrt{\sin^2\frac{\alpha}{2} - \sin^2\frac{\theta}{2}} = p\cos\phi \quad (f)$$

we obtain

$$l = \frac{1}{k}\int_0^{\pi/2} \frac{d\phi}{\sqrt{1 - p^2\sin^2\phi}} = \frac{1}{k}K(p) \quad (g)$$

The integral appearing in Eq. (g) is known as a *complete elliptic integral of the first kind* and is designated by $K(p)$. The value of the integral K depends only on p and is tabulated numerically, for various values of $p = \sin\alpha/2$, in many engineering handbooks. With such a table available, the value of p (and hence the value of the angle α at the top of the bar) can be found readily for any value of the load P.

When the deflection of the bar is very small, α and p will also be small and the term $p^2 \sin^2 \phi$ can be neglected in comparison with unity in Eq. (g). Then we have

$$l = \frac{1}{k} \int_0^{\pi/2} d\phi = \frac{\pi}{2k} = \frac{\pi}{2} \sqrt{\frac{EI}{P}}$$

and

$$P = P_{cr} = \frac{\pi^2 EI}{4l^2}$$

which is the value of the critical load given by Eq. (2-4).

As the value of α increases, the integral K and the load P increase also. For example, take the case where $\alpha = 60°$ and $p = \sin \alpha/2 = \frac{1}{2}$. From a table of elliptic integrals we find that $K(\frac{1}{2}) = 1.686$ and therefore

$$l = 1.686 \sqrt{\frac{EI}{P}} \qquad P = \frac{2.842 EI}{l^2}$$

Taking the ratio of P to the critical load gives

$$\frac{P}{P_{cr}} = \frac{4(2.842)}{\pi^2} = 1.152$$

Thus a load which is 15.2 per cent greater than the Euler load, at which buckling first begins, will produce a deflection such that the tangent at the top (Fig. 2-28) has an angle of 60° with the vertical.

In Table 2-4 are given values of the ratio P/P_{cr} for various values of the angle α.

TABLE 2-4. LOAD-DEFLECTION DATA FOR A BUCKLED BAR (FIG. 2-28)

α	0°	20°	40°	60°	80°	100°	120°	140°	160°	176°
P/P_{cr}	1	1.015	1.063	1.152	1.293	1.518	1.884	2.541	4 029	9.116
x_a/l	1	0.970	0.881	0.741	0.560	0.349	0.123	−0.107	−0.340	−0.577
y_a/l	0	0.220	0.422	0.593	0.719	0.792	0.803	0 750	0 625	0.421

In calculating deflections of the bar we note that

$$dy = \sin \theta \, ds = - \frac{\sin \theta \, d\theta}{k \sqrt{2} \sqrt{\cos \theta - \cos \alpha}}$$

Then the total deflection of the top of the bar in the horizontal direction (Fig. 2-28) is

$$y_a = \frac{1}{2k} \int_0^\alpha \frac{\sin \theta \, d\theta}{\sqrt{\sin^2 \alpha/2 - \sin^2 \theta/2}} \tag{h}$$

From Eq. (d) we have $\sin \theta/2 = p \sin \phi$ and therefore

$$\cos \frac{\theta}{2} = \sqrt{1 - p^2 \sin^2 \phi}$$

Using the relation $\sin \theta = 2(\sin \theta/2)(\cos \theta/2)$, we now find that

$$\sin \theta = 2p \sin \phi \sqrt{1 - p^2 \sin^2 \phi} \qquad (i)$$

Substituting expressions (e), (f), and (i) into Eq. (h) and changing the limits accordingly, we obtain

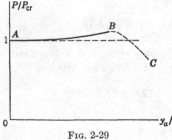

FIG. 2-29

$$y_a = \frac{2p}{k} \int_0^{\pi/2} \sin \phi \, d\phi = \frac{2p}{k} \qquad (j)$$

Thus the deflection of the bar can be calculated by first selecting a value of α (or p), then determining k (and hence P) from Eq. (g), and finally finding y_a from Eq. (j). For example, again taking $\alpha = 60°$ and $p = \frac{1}{2}$, we have $kl = K(\frac{1}{2}) = 1.686$ and

$$y_a = \frac{l}{1.686} = 0.593l$$

Numerical results obtained in this way for various values of α are given in the last line of Table 2-4.

The relation between the deflection y_a and the load P is shown graphically in Fig. 2-29 by the curve AB. The curve is tangent to the horizontal line $P = P_{cr}$ at point A where the deflection is zero. Thus the increase in the load P, corresponding to a small increment of deflection, is a small quantity of second order. This explains why the deflection was found to be indefinite in magnitude when the approximate expression for curvature was used. It should be noted that the curve AB can be used only up to the proportional limit of the material. Beyond this limit the resistance of the

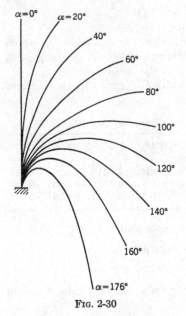

FIG. 2-30

bar to bending diminishes and a curve similar to that indicated by the dotted line BC is obtained.

The coordinate distance x_a (Fig. 2-28) can be calculated in a similar way, and the result is

$$x_a = \frac{2}{k} \int_0^{\pi/2} \sqrt{1 - p^2 \sin^2 \phi} \, d\phi - l$$

$$= \frac{2}{k} E(p) - l \qquad (k)$$

where $E(p)$ is the *complete elliptic integral of the second kind*. Numerical results obtained from this equation are given in Table 2-4.

The coordinates of intermediate points along the deflection curve can be calculated also, using elliptic integrals. The shapes of the deflection curves for various values of α are given in Fig. 2-30. It is apparent that a slight increase of the load above the critical value is sufficient to produce a large deflection of the bar.[1]

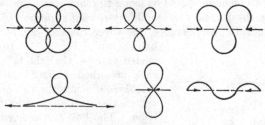

FIG. 2-31

In the above discussion it was assumed that the bar was built in at one end and free at the other. The results obtained can be used also in the case of a bar with hinged ends. In this case the curves of Fig. 2-30 represent only one-half of the length of the bar. The curves shown in Fig. 2-31, representing possible forms of equilibrium of a thin wire, can all be obtained by combining curves from Fig. 2-30. Thus, the forces necessary to produce such bending can be found from Table 2-4. If the deflections of a bar with hinged ends remain small, the relation between

---

[1] Numerous investigators, in addition to those already mentioned, have dealt with the problem of the elastica. A modification of Kirchhoff's theory to include bars with initial curvature was given by A. Clebsch in his book "Theorie der Elasticität fester Körper," Leipzig, 1862 (French translation by Saint-Venant, 1883). Numerous special cases of large deflections were discussed by L. Saalschütz, "Der belastete Stab," Leipzig, 1880. See also E. Collignon, *Ann. ponts et chaussées*, vol. 17, p. 98, 1889; C. Kriemler, "Labile und stabile Gleichgewichtsfiguren," Dissertation, Karlsruhe, 1902; M. Born, Dissertation, Göttingen, 1909; and A. N. Krylov, Sur les formes d'équilibre des pièces chargées debout, *Bull. acad. sci. USSR*, 1931, p. 963.

the deflection δ at the center and the load can be represented by the following approximate formula:[1]

$$\frac{\delta}{l} = \frac{\sqrt{8}}{\pi} \sqrt{\frac{P}{P_{cr}} - 1} \left[1 - \frac{1}{2}\left(\frac{P}{P_{cr}} - 1\right) \right] \qquad (l)$$

2.8. The Energy Method. From the previous discussions it is seen that if the centrally applied compressive force is smaller than its critical value, a compressed bar remains straight and this straight form of equilibrium is stable. If we increase the load slightly above its critical value, there are, theoretically, two possible forms of equilibrium. One possibility is for the bar to remain straight, and the other is for the bar to buckle sideways. Experiments show that the straight form is unstable and that a bar will always buckle sideways under the action of a load greater than the critical value.

The question of the stability of various forms of equilibrium of a compressed bar can be investigated by using the same methods as are used in investigating the stability of equilibrium configurations of rigid-body systems. Consider, for instance, the three cases of equilibrium of the

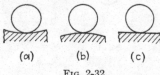

(a)   (b)   (c)

Fig. 2-32

ball shown in Fig. 2-32. It can be concluded at once that the ball on the concave spherical surface (a) is in *stable equilibrium*, while the ball on the convex spherical surface (b) is in *unstable equilibrium*. The ball on the horizontal plane (c) is said to be in *indifferent* or *neutral equilibrium*. The type of equilibrium can be ascertained by considering the energy of the system. In the first case (Fig. 2-32a) any displacement of the ball from its position of equilibrium will raise the center of gravity. A certain amount of work is required to produce such a displacement; thus the potential energy of the system increases for any small displacement from the position of equilibrium. In the second case (Fig. 2-32b), any displacement from the position of equilibrium will lower the center of gravity of the ball and will decrease the potential energy of the system. Thus in the case of stable equilibrium the energy of the system is a *minimum* and in the case of unstable equilibrium it is a *maximum*. If the equilibrium is indifferent (Fig. 2-32c), there is no change in energy during a displacement.

For each of the systems shown in Fig. 2-32 stability depends only on the shape of the supporting surface and does not depend on the weight of the ball. In the case of a compressed column (see Arts. 2.1 and 2.2) we find

[1] This formula can be obtained by representing the deflection at the center of the bar in series form and taking only the first few terms.

that the column may be stable or unstable, depending on the magnitude of the axial load. A model representing such a condition is shown in Fig. 2-33. A vertical bar AB, considered to be infinitely rigid, is hinged at the bottom and supported by a spring BC at the top. The bar carries a centrally applied load P. For small values of this load the vertical position of the bar is stable, and if a disturbing force produces lateral displacement at B, the bar will return to its vertical position under the action of the spring. The critical value of the load P can be found from a consideration of the energy of the system. Assume that a small lateral displacement occurs at B, so that the bar becomes inclined to the vertical by a small angle α, Fig. 2-33a. Owing to this displacement, the load P is lowered by the amount

$$l(1 - \cos \alpha) \approx \frac{l\alpha^2}{2} \qquad (a)$$

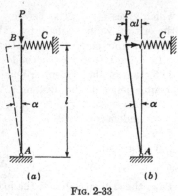

and the decrease in the potential energy of the load P, equal to the work done by P, is

$$\frac{Pl\alpha^2}{2}$$

At the same time the spring elongates by the amount αl and the increase in strain energy of the spring is

$$\frac{\beta(\alpha l)^2}{2}$$

(a)   (b)

FIG. 2-33

where β denotes the spring constant. The system will be stable if

$$\frac{\beta(\alpha l)^2}{2} > \frac{Pl\alpha^2}{2}$$

and will be unstable if

$$\frac{\beta(\alpha l)^2}{2} < \frac{Pl\alpha^2}{2}$$

Therefore the critical value of the load P is found from the condition that

$$\frac{\beta(\alpha l)^2}{2} = \frac{Pl\alpha^2}{2}$$

from which $P_{cr} = \beta l$.

The same conclusion can be reached in another way by considering the equilibrium of the forces acting on the bar. If by some disturbance the bar is brought to a slightly inclined position (Fig. 2-33b), there will be two forces acting on the upper end B of the bar, the vertical force P and the horizontal force $\beta \alpha l$ of the spring. If the moment of the force in the spring with respect to point A is larger than the moment of the force P, that is, if

$$\beta \alpha l^2 > P \alpha l$$

we shall evidently have stable equilibrium and the force in the spring will restore the bar to its initial vertical position. Conversely, if

$$\beta \alpha l^2 < P \alpha l$$

the vertical position will be unstable and the system will collapse after a slight disturbance. The critical value of P is found from the equation

$$\beta \alpha l^2 = P \alpha l$$

which gives $P_{cr} = \beta l$ as before. Thus we have two methods for finding the critical value of the force. We can obtain it from energy considerations or from equations of statics.

In using the energy method we begin by assuming some small lateral deflection of the system. This deflection means an increase ΔU in the strain energy of the system. At the same time the load P will move through a small distance and do work equal to ΔT. The system is stable in its undeflected form if

$$\Delta U > \Delta T$$

and unstable if

$$\Delta U < \Delta T$$

Thus the critical value of the load P is found from the equation

$$\Delta U = \Delta T \tag{2-31}$$

which represents the condition when the equilibrium configuration changes from stable to unstable.

As a second example of the use of the energy method, consider the system shown in Fig. 2-34a, consisting of three equal bars of length $l/3$. The bars are assumed to be rigid and are joined by pin connections at the supports. The supports at A and D are unyielding, while the supports at B and C consist of elastic springs defined by the spring constant β. If the axial compressive force is sufficiently large, the system may buckle, resulting in deflections δ_1 and δ_2 at supports B and C (Fig. 2-34a). For small deflections, the angle of inclination α_1 of the bar AB is $3\delta_1/l$, the angle $\alpha_2 = 3\delta_2/l$, and the angle between bar BC and the horizontal is $(3\delta_2 - 3\delta_1)/l$. The distance λ moved by the force P is found by using Eq. (a) for each of the three bars, giving

$$\lambda = \frac{1}{2}\frac{l}{3}\left[\left(\frac{3\delta_1}{l}\right)^2 + \left(\frac{3\delta_2 - 3\delta_1}{l}\right)^2 + \left(\frac{3\delta_2}{l}\right)^2\right]$$

$$= \frac{3}{l}(\delta_1{}^2 - \delta_1\delta_2 + \delta_2{}^2)$$

and the work done by the force P is $\Delta T = P\lambda$.

The strain energy stored in the elastic supports during buckling is

$$\Delta U = \frac{\beta}{2} (\delta_1{}^2 + \delta_2{}^2)$$

The critical condition is found from Eq. (2-31) as follows:

$$\frac{3P}{l} (\delta_1{}^2 - \delta_1\delta_2 + \delta_2{}^2) = \frac{\beta}{2} (\delta_1{}^2 + \delta_2{}^2)$$

and therefore

$$P = \frac{\beta l}{6} \frac{\delta_1{}^2 + \delta_2{}^2}{\delta_1{}^2 - \delta_1\delta_2 + \delta_2{}^2} = \frac{\beta l}{6} \frac{N}{D} \qquad (b)$$

where N and D represent the numerator and denominator of the fraction. To find the critical value of P from Eq. (b) we must adjust the deflections δ_1 and δ_2, which are

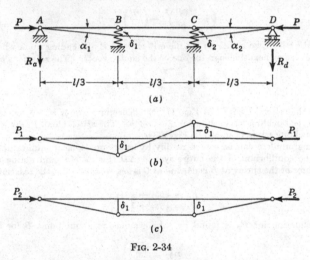

FIG. 2-34

unknown, so as to make P a minimum value. The minimum condition is expressed by the equations

$$\frac{\partial P}{\partial \delta_1} = 0 \qquad \frac{\partial P}{\partial \delta_2} = 0 \qquad (c)$$

and from the first of Eqs. (c) we obtain

$$\frac{\partial P}{\partial \delta_1} = \frac{\beta l}{6} \frac{D(\partial N/\partial \delta_1) - N(\partial D/\partial \delta_1)}{D^2} = 0$$

or

$$\frac{\partial N}{\partial \delta_1} - \frac{N}{D} \frac{\partial D}{\partial \delta_1} = 0$$

Combining this last equation with Eq. (b) we obtain

$$\frac{\partial N}{\partial \delta_1} - \frac{6P}{\beta l} \frac{\partial D}{\partial \delta_1} = 0 \qquad (d)$$

and in a similar manner, the second of Eqs. (c) gives

$$\frac{\partial N}{\partial \delta_2} - \frac{6P}{\beta l} \frac{\partial D}{\partial \delta_2} = 0 \tag{e}$$

Substituting the values of N and D from Eq. (b) into Eqs. (d) and (e), we obtain the following two equations for the deflections δ_1 and δ_2:

$$\delta_1 \left(1 - \frac{6P}{\beta l}\right) + \delta_2 \left(\frac{3P}{\beta l}\right) = 0$$
$$\delta_1 \left(\frac{3P}{\beta l}\right) + \delta_2 \left(1 - \frac{6P}{\beta l}\right) = 0 \tag{f}$$

Except for the trivial solution when $\delta_1 = \delta_2 = 0$, the only possibility for a solution of Eqs. (f) is if the determinant is zero, which gives

$$\left(1 - \frac{6P}{\beta l}\right)^2 - \left(\frac{3P}{\beta l}\right)^2 = 0$$

This equation has two solutions for the critical load P, depending upon whether we take a positive or negative sign for one of the square roots. The solutions are

$$P_1 = \frac{\beta l}{9} \quad \text{and} \quad P_2 = \frac{\beta l}{3} \tag{g}$$

If we use the critical load P_1 in Eqs. (f), the deflections are $\delta_1 = -\delta_2$, which corresponds to the buckling mode shown in Fig. 2-34b. The critical load P_2 gives $\delta_1 = \delta_2$, and the buckling mode is shown in Fig. 2-34c.

The same problem can be solved readily by using equations of equilibrium. Considering the equilibrium of the entire system AD (Fig. 2-34a) and noting that the reactive force of the spring at B is $\beta\delta_1$ and at C is $\beta\delta_2$, we obtain for the reactions at the ends

$$R_a = \tfrac{2}{3}\beta\delta_1 + \tfrac{1}{3}\beta\delta_2$$
$$R_d = \tfrac{1}{3}\beta\delta_1 + \tfrac{2}{3}\beta\delta_2$$

Another equation for R_a is found by taking moments about point B for bar AB, which gives

$$P\delta_1 = \frac{R_a l}{3}$$

and similarly, for CD,

$$P\delta_2 = \frac{R_d l}{3}$$

Combining these four equations we obtain

$$\delta_1 \left(2 - \frac{9P}{\beta l}\right) + \delta_2 = 0$$
$$\delta_1 + \delta_2 \left(2 - \frac{9P}{\beta l}\right) = 0$$

which have solutions $P_1 = \beta l/9$ and $P_2 = \beta l/3$ as before.

The same two general methods for solving problems of stability can be applied in the case of buckling of elastic bars. In the previous articles of this chapter the critical load was calculated from the differential equations which expressed the conditions for equilibrium of the buckled bar. Let us now apply the energy method to these cases and use Eq. (2-31). The quantity ΔU in that equation will represent the strain energy of bending added to the bar when it is deflected laterally,[1] and ΔT will represent the work done by the compressive force P.

Consider first the column shown in Fig. 2-1b fixed at the base and free at the upper end. The deflection curve of the slightly buckled bar is given by the equation [see Eq. (2-2)]

$$y = \delta\left(1 - \cos\frac{\pi x}{2l}\right) \qquad (h)$$

The bending moment at any cross section is

$$M = -P(\delta - y) = -P\delta\cos\frac{\pi x}{2l} \qquad (2\text{-}32)$$

and the corresponding strain energy of bending is[2]

$$\Delta U = \int_0^l \frac{M^2\,dx}{2EI} = \frac{P^2\delta^2 l}{4EI} \qquad (i)$$

The vertical movement of the load P during buckling is found from Eq. (1-55):

$$\lambda = \frac{1}{2}\int_0^l \left(\frac{dy}{dx}\right)^2 dx = \frac{\delta^2\pi^2}{16l}$$

and the corresponding work produced by P is

$$\Delta T = \frac{P\delta^2\pi^2}{16l} \qquad (j)$$

Substituting expressions (i) and (j) into Eq. (2-31), we obtain for the critical load the same value as given by Eq. (2-4). The exact value of the critical load is obtained, since the correct expression for the deflection curve [Eq. (h)] was used. This expression was obtained previously by integration of the differential equation.

In a similar way, the buckled shape of the deflection curve for the fundamental case (see Fig. 2-6) is

$$y = A\sin\frac{\pi x}{l}$$

---

[1] There is also a change in the strain energy of compression, but a more detailed investigation shows that this small change of energy can be neglected and the bar considered as inextensible; see Timoshenko, Z. Math. u. Physik, vol. 58, p. 337, 1910, and A. Pflüger, "Stabilitätsprobleme der Elastostatik," Springer-Verlag, Berlin, 1950.

[2] See Timoshenko, "Strength of Materials," 3d ed., part I, p. 317, D. Van Nostrand Company, Inc., Princeton, N.J., 1955.

and the bending moment is

$$M = Py = PA \sin \frac{\pi x}{l}$$

from which
$$\Delta U = \int_0^l \frac{M^2 \, dx}{2EI} = \frac{P^2 A^2 l}{4EI}$$

and also
$$\Delta T = P\lambda = \frac{P\pi^2 A^2}{4l}$$

Equating ΔU and ΔT gives

$$P_{cr} = \frac{\pi^2 EI}{l^2}$$

which is the Euler load for a pin end column. Again we see that the energy method gives the exact value of the critical load when the true shape of the deflection curve is known. In many cases the true shape of the buckled bar is not known, however, and then the energy method can be used to find an approximate value of the critical load, as discussed in the following article.

2.9. Approximate Calculation of Critical Loads by the Energy Method. The principal use of the energy method in problems of buckling is the determination of approximate values of critical loads for cases in which an exact solution of the differential equation of the deflection curve is either unknown or too complicated. In such cases we must begin by assuming a reasonable shape for the deflection curve. While it is not essential for an approximate solution that the assumed curve satisfy completely the end conditions of the bar, it should at least satisfy the conditions pertaining to deflections and slopes.

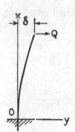

FIG. 2-35

As a first example, consider again the case of a bar fixed at the base and free at the top (Fig. 2-1b) and assume that we do not know the exact shape of the deflection curve. An approximation to the shape of the deflection curve is obtained from the deflection curve of a cantilever beam loaded at the free end, as shown in Fig. 2-35. The equation of this curve is

$$y = \frac{Qx^2}{6EI} (3l - x) = \frac{\delta x^2}{2l^3} (3l - x) \tag{a}$$

This curve is different from the true curve, given by Eq. (h), p. 87, but it satisfies the required end conditions, since it has a vertical tangent at the lower end and zero curvature at the top. The strain energy of bending of the buckled bar is [see Eq. (2-32)]

$$\Delta U = \int_0^l \frac{M^2 \, dx}{2EI} = \frac{P^2}{2EI} \int_0^l (\delta - y)^2 \, dx$$

Now substituting Eq. (a) we find

$$\Delta U = \frac{17}{35} \frac{P^2 \delta^2 l}{2EI}$$

The vertical displacement of the load P is

$$\lambda = \frac{1}{2} \int_0^l \left(\frac{dy}{dx} \right)^2 dx = \frac{3}{5} \frac{\delta^2}{l}$$

Equating ΔU and ΔT, we obtain

$$\frac{17}{35} \frac{P^2 \delta^2 l}{2EI} = \frac{3}{5} \frac{P \delta^2}{l}$$

from which $$P_{cr} = \frac{42}{17} \frac{EI}{l^2} = 2.4706 \frac{EI}{l^2}$$

The correct value of the critical load is

$$P_{cr} = \frac{\pi^2 EI}{4 l^2} = 2.4674 \frac{EI}{l^2}$$

and we see that the error of the approximate solution is only 0.13 per cent.
The energy method gives a very satisfactory approximation to the true critical load, provided the shape of the assumed curve is reasonably close to the exact curve. Usually it is especially important, for good results, that the end conditions of the bar be satisfied by the assumed curve. However, there are cases when even a very rough assumption will still give a satisfactory result. For instance, we might assume the deflec-tion curve in the above example to be a parabola given by the equation

$$y = \frac{\delta x^2}{l^2} \tag{b}$$

and then

$$\Delta U = \int_0^l \frac{M^2 \, dx}{2EI} = \frac{P^2 \delta^2}{2EI} \int_0^l \left(1 - \frac{x^2}{l^2} \right)^2 dx = \frac{8}{15} \frac{P^2 \delta^2 l}{2EI}$$

The vertical displacement of the upper end of the buckled bar in this case is

$$\lambda = \frac{1}{2} \int_0^l \left(\frac{dy}{dx} \right)^2 dx = \frac{2}{3} \frac{\delta^2}{l}$$

and Eq. (2-31) gives

$$\frac{8}{15} \frac{P^2 \delta^2 l}{2EI} = \frac{2}{3} \frac{P \delta^2}{l}$$

from which $$P_{cr} = 2.50 \frac{EI}{l^2}$$

The error of this approximate solution is about 1.3 per cent. A satisfactory approximation for the critical load is thus obtained, although the assumed parabolic curve is a very poor representation of the true curve. It does not satisfy the end conditions, since it has constant curvature along the length while in the actual curve the curvature is zero at the top of the bar and a maximum at the bottom.

In the calculation of the strain energy of bending, the expression

$$\Delta U = \int_0^l \frac{M^2\,dx}{2EI} = \int_0^l \frac{P^2(\delta - y)^2\,dx}{2EI} \tag{2-33}$$

was used in the above discussion. An alternate form of this expression is

$$\Delta U = \frac{EI}{2}\int_0^l \left(\frac{d^2y}{dx^2}\right)^2 dx \tag{2-34}$$

Equations (2-33) and (2-34) are both exact as long as the true expression for the deflection curve is being used. However, when an assumed curve is used, expression (2-33) is preferable, since then the accuracy of the approximate solution will depend on the accuracy of y, whereas if expression (2-34) is used, the accuracy of the solution will depend on the accuracy of d^2y/dx^2. In the selection of a deflection curve, y is usually obtained with much better accuracy than d^2y/dx^2.[1]

The energy method always gives values of the critical load which are larger than the true value unless the assumed deflection curve happens to be the correct one. This follows from the fact that the true shape is the only one which represents a deflection configuration for which each element of the bar is in equilibrium. To have the bar in equilibrium with an incorrect shape of buckling requires that additional constraints be introduced in order to maintain that shape. The additional constraints naturally can only increase the rigidity of the bar, and hence the critical load becomes larger than its true value. Thus, if several assumed deflection curves are used, the lowest critical load found from those assumed curves will be the most accurate.

A further improvement in the energy method can be made by taking, for the assumed deflection curve, an expression which involves several parameters. The shape of the deflection curve can then be altered by changing the values of the parameters. The most accurate result for the critical load P will correspond to values of the parameters such that P has a minimum value.[2]

---

[1] For a mathematical proof that the use of Eq. (2-33) results in better accuracy than is given by Eq. (2-34), see H. A. Lang, *Quart. Appl. Math.*, vol..5, p. 510, 1947.

[2] See Timoshenko, *Bull. Polytech. Inst., Kiev* (in Russian), 1910, and later, Sur la stabilité des systèmes élastiques, *Ann. ponts et chaussées*, Paris, 1913. This latter paper appears in "The Collected Papers of Stephen P. Timoshenko," pp. 92–224, McGraw-Hill Book Company, Inc., New York, 1954.

As an example, let us take a prismatic bar with hinged ends (Fig. 2-36) and assume that the bar has buckled under the action of the compressive forces P. Any shape of the deflection curve of such a bar can be represented by the trigonometric series (see Art. 1.11)

$$y = a_1 \sin \frac{\pi x}{l} + a_2 \sin \frac{2\pi x}{l} + a_3 \sin \frac{3\pi x}{l} + \cdots \qquad (c)$$

By varying the parameters $a_1, a_2, \ldots$, we can obtain various shapes of the deflection curve. The strain energy of bending of the buckled bar is

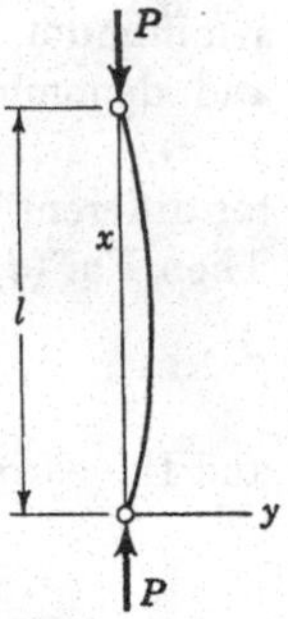

$$\Delta U = \int_0^l \frac{M^2 \, dx}{2EI} = \frac{P^2}{2EI} \int_0^l y^2 \, dx$$

or, by substituting the series (c) for y and performing the integration (see Art. 1.11), we obtain

$$\Delta U = \frac{P^2 l}{4EI} \sum_{n=1}^{n=\infty} a_n{}^2$$

During buckling the forces P produce the work [see Eq. (1-56)]

$$\Delta T = P\lambda = \frac{P}{2} \int_0^l \left(\frac{dy}{dx}\right)^2 dx = \frac{\pi^2 P}{4l} \sum_{n=1}^{n=\infty} n^2 a_n{}^2$$

Then Eq. (2-31) becomes

$$\frac{P^2 l}{4EI} \sum_{n=1}^{n=\infty} a_n{}^2 = \frac{\pi^2 P}{4l} \sum_{n=1}^{n=\infty} n^2 a_n{}^2$$

and we obtain

$$P = \frac{\pi^2 EI}{l^2} \frac{\displaystyle\sum_{n=1}^{n=\infty} n^2 a_n{}^2}{\displaystyle\sum_{n=1}^{n=\infty} a_n{}^2} \qquad (d)$$

FIG. 2-36

To find the critical value of the compressive force P we must adjust the parameters $a_1, a_2, \ldots$ in such a manner as to make expression (d) a minimum. This is readily accomplished in this case on the basis of the following considerations: Imagine a series of fractions of the type

$$\frac{a}{b}, \frac{c}{d}, \frac{e}{f} \cdots \qquad (e)$$

in which each of the numbers $a, b, c, \ldots$ is assumed positive. If we add the numerators and denominators together, we obtain the fraction

$$\frac{a + c + e + \cdots}{b + d + f + \cdots} \tag{f}$$

This fraction is evidently of some intermediate value between the largest and the smallest of the fractions (e). Expression (d) is analogous to the fraction (f). Therefore, it follows that if we wish to make expression (d) a minimum, we must take only one term of the series in the numerator and denominator. In other words, we must take all parameters a_1, $a_2, a_3, \ldots$, except one, equal to zero. It is seen also that the parameter different from zero must be a_1, since it has the smallest coefficient n^2. Then Eq. (d) gives

$$P_{cr} = \frac{\pi^2 EI}{l^2}$$

and the corresponding deflection curve is

$$y = a_1 \sin \frac{\pi x}{l}$$

In this way we arrive at the same result as was obtained before by integration (see Art. 2.1).

Expression (c) for the deflection curve of a buckled bar with hinged ends can be used in more complicated cases when the cross section of the bar is variable or when the compressive forces are distributed along its length. Then the true deflection curve is no longer a sine curve, and by taking the first few terms in the series (c) we can obtain a satisfactory approximate solution. Assume, for example, that the first two terms of the series (c) are to be used. Then the quantities ΔU and ΔT can be calculated readily in terms of a_1 and a_2. Equating ΔU and ΔT then gives an equation from which P is found in terms of a_1 and a_2. Finally, the values of a_1 and a_2 must be determined in such a way that P has a minimum value, and therefore we have the equations

$$\frac{\partial P}{\partial a_1} = 0 \qquad \frac{\partial P}{\partial a_2} = 0 \tag{g}$$

Equations (g) determine the critical value of the load P and also give the ratio a_1/a_2, thereby defining the shape of the approximate deflection curve. This method will be illustrated in subsequent articles.

In a more general case we can assume that the buckling curve is represented by the equation

$$y = a_1 f_1(x) + a_2 f_2(x) + a_3 f_3(x) + \cdots \tag{h}$$

in which $f_1(x), f_2(x), \ldots$ are functions of x satisfying the conditions at the ends of the bar and $a_1, a_2, \ldots$ are constants defining the amplitudes of the terms. For example, in the case of a bar with hinged ends (Fig. 2-36) the end conditions for each function $f_i(x)$ are[1]

$$f_i(x) = f_i''(x) = 0 \qquad \text{at } x = 0 \text{ and } x = l \tag{i}$$

Assuming that the work of the external forces and the change in strain energy during buckling are given by the expressions

$$\Delta T = \frac{P}{2} \int_0^l (y')^2 \, dx \qquad \Delta U = \frac{P^2}{2EI} \int_0^l y^2 \, dx$$

we obtain from Eq. (2-31)

$$P = \frac{EI \int_0^l (y')^2 \, dx}{\int_0^l y^2 \, dx} \tag{2-35}$$

To find the value of P_{cr} we have to select the coefficients $a_1, a_2, \ldots$ in expression (h) so as to make expression (2-35) a minimum. This requires that the derivatives of expression (2-35) with respect to each coefficient a_i must vanish, and we obtain equations of the form

$$\int_0^l y^2 \, dx \frac{\partial}{\partial a_i} \int_0^l (y')^2 \, dx - \int_0^l (y')^2 \, dx \frac{\partial}{\partial a_i} \int_0^l y^2 \, dx = 0$$

or, using Eq. (2-35),

$$\frac{\partial}{\partial a_i} \int_0^l (y')^2 \, dx - \frac{P}{EI} \frac{\partial}{\partial a_i} \int_0^l y^2 \, dx = 0 \tag{j}$$

Observing that, from Eq. (h),

$$\frac{\partial y}{\partial a_i} = f_i(x) \qquad \frac{\partial y'}{\partial a_i} = f_i'(x)$$

we obtain from Eq. (j)

$$\int_0^l y' f_i'(x) \, dx - \frac{P}{EI} \int_0^l y f_i(x) \, dx = 0 \tag{k}$$

If we substitute expression (h) for y into this equation and perform the indicated integrations, we obtain a homogeneous linear equation in $a_1, a_2, \ldots$. The number of equations (k) will be equal to the number of coefficients $a_1, a_2, \ldots$. We can satisfy these equations by putting $a_1 = a_2 = a_3 = \cdots = 0$, but then the deflection (h) vanishes and there is no buckling. To have buckling, at least some of the quantities $a_1, a_2, \ldots$ must be different from zero, and this is possible only if the determinant of Eq. (k) vanishes. Thus, setting this determinant equal to zero gives the equation for calculating P_{cr}.

Integrating by parts, we can transform the first term of Eq. (k) as follows:

$$\int_0^l y' f_i'(x) \, dx = [y' f_i(x)]_0^l - \int_0^l y'' f_i(x) \, dx \tag{l}$$

Observing that $f_i(x)$ vanishes at both ends of the bar, as follows from the conditions (i) for a bar with hinged ends, we conclude that the first term on the right-hand side of

---

[1] The primes denote differentiation with respect to x, that is, $f_i'(x) = df_i(x)/dx$, $f_i''(x) = d^2 f_i(x)/dx^2, \ldots, y' = dy/dx, y'' = d^2 y/dx^2, \ldots$

Eq. (l) vanishes.   Equation (k) will then be represented in the form

$$\int_0^l \left(y'' + \frac{P}{EI} y \right) f_i(x) \, dx = 0 \qquad (m)$$

The differential equation of the deflection curve of the buckled bar (Fig. 2-36) is

$$y'' + \frac{P}{EI} y = 0 \qquad (n)$$

The exact expression for y must satisfy this equation, but if we wish to find an approximate solution for y having the form (h), the coefficients a_1, a_2, . . . must be selected so that the left-hand side of Eq. (m) vanishes.   Substituting the approximate expression (h) into Eq. (m) and using the notation

$$k^2 = \frac{P}{EI}$$

we find that the linear equations for determining a_1, a_2, . . . take the form

$$a_1 \int_0^l [f_1''(x) + k^2 f_1(x)] f_i(x) \, dx + a_2 \int_0^l [f_2''(x) + k^2 f_2(x)] f_i(x) \, dx + \cdots = 0 \qquad (o)$$

The determinant of this system of equations, when equated to zero, gives the equation for determining the approximate value of P_{cr}.

Equation (n) for the buckling of a prismatic bar with hinged ends is very simple, and we can obtain easily the exact value of P_{cr}.   Often, however, the differential equation will be more complicated.   For example, if the cross section of the bar in Fig. 2-36 is variable, then I in Eq. (n) is no longer constant and the rigorous solution of the equation becomes complicated. We can then use for y the approximate expression (h) and calculate for P_{cr} an approximate value, proceeding as explained above.[1]

FIG. 2-37

2.10. Buckling of a Bar on an Elastic Foundation.   If there are many equally spaced elastic supports of equal rigidity, their action on the buckled bar (Fig. 2-37) can be replaced by the action of a continuous elastic medium.   The reaction of the medium at any cross section of the bar is then proportional to the deflection at that section.   If α is the spring constant of the individual supports and a is the distance between them, the rigidity of the equivalent elastic medium is expressed by the quantity

$$\beta = \frac{\alpha}{a} \qquad (a)$$

The quantity β is called the *modulus of the foundation* and has the dimensions of a force divided by the square of a length.   It represents the

---

[1] This method of obtaining approximate solutions of differential equations was developed by Swiss scientist Walter Ritz; see his famous papers in *Z. reine u. angew. Math.*, vol. 135, pp. 1–61, 1908, and *Ann. Physik*, vol. 28, p. 737, 1909.   See also his "Gesammelte Werke," Paris, 1911.

magnitude of the reaction of the foundation per unit length of the bar if the deflection is equal to unity.

In calculating the critical value of the compressive force, we can use the energy method.[1] The general expression for the deflection curve of a bar with hinged ends can be represented by the series

$$y = a_1 \sin \frac{\pi x}{l} + a_2 \sin \frac{2\pi x}{l} + a_3 \sin \frac{3\pi x}{l} + \cdots \tag{b}$$

The strain energy of bending[2] of the bar is [Eq. (1-53), p. 25]

$$\Delta U_1 = \frac{EI}{2} \int_0^l \left(\frac{d^2 y}{dx^2}\right)^2 dx = \frac{\pi^4 EI}{4l^3} \sum_{n=1}^{n=\infty} n^4 a_n^2 \tag{c}$$

In calculating the strain energy of the elastic foundation, we note that the lateral reaction on an element dx of the bar is $\beta y\, dx$ and that the corresponding energy is $(\beta y^2/2)\, dx$. Then the total energy of deformation of the elastic medium is

$$\Delta U_2 = \frac{\beta}{2} \int_0^l y^2\, dx$$

or, substituting series (b) for y,

$$\Delta U_2 = \frac{\beta l}{4} \sum_{n=1}^{n=\infty} a_n^2 \tag{d}$$

The work done by the compressive forces P, from Eq. (1-56), is

$$\Delta T = \frac{P\pi^2}{4l} \sum_{n=1}^{n=\infty} n^2 a_n^2 \tag{e}$$

Substituting (c), (d), and (e) in Eq. (2-31), we obtain

$$\frac{\pi^4 EI}{4l^3} \sum_{n=1}^{n=\infty} n^4 a_n^2 + \frac{\beta l}{4} \sum_{n=1}^{n=\infty} a_n^2 = \frac{P\pi^2}{4l} \sum_{n=1}^{n=\infty} n^2 a_n^2 \tag{f}$$

---

[1] See Timoshenko's paper in *Bull. Polytech. Inst.*, St. Petersburg, 1907. Another method of solving the problem is given by H. Zimmermann, *Zentr. Bauverwaltung*, 1906. The case in which a bar is elastically supported along only a portion of its length has been discussed by Hjalmar Granholm, "On the Elastic Stability of Piles Surrounded by a Supporting Medium," Stockholm, 1929.

[2] For convenience, we use here Eq. (2-34) instead of Eq. (2-33). The results of this analysis are exact and hence there is no question of relative accuracy involved.

from which

$$P = \frac{\pi^2 EI}{l^2} \frac{\sum\limits_{n=1}^{n=\infty} n^2 a_n^2 + \frac{\beta l^4}{\pi^4 EI} \sum\limits_{n=1}^{n=\infty} a_n^2}{\sum\limits_{n=1}^{n=\infty} n^2 a_n^2} \tag{2-36}$$

To determine the critical value of the load P, it is necessary to find a relation between the coefficients a_1, a_2, . . . which will make expression (2-36) a minimum. This result is accomplished by making all the coefficients except one equal to zero, as explained in the preceding article. This means that the deflection curve of the bar is a simple sine curve, and if we let a_m be the coefficient different from zero, we obtain

$$y = a_m \sin \frac{m\pi x}{l} \tag{g}$$

and the critical load is[1]

$$P = \frac{\pi^2 EI}{l^2} \left(m^2 + \frac{\beta l^4}{m^2 \pi^4 EI} \right) \tag{2-37}$$

where m is an integer. Equation (2-37) gives the critical load P as a function of m, which represents the number of half sine waves in which the bar subdivides at buckling and the properties of the beam and of the foundation. Thus the lowest critical load may occur with $m = 1, 2, 3, . . .$, depending on the values of the other constants.

In order to determine the value of m which makes Eq. (2-37) a minimum, we begin by considering the special case when β equals zero. Then there is no resisting foundation, and from Eq. (2-37), we see that m must be taken equal to 1. This is the familiar case of buckling of a bar with hinged ends. If β is very small, but greater than zero, we must again take $m = 1$ in Eq. (2-37). Thus, for a very flexible elastic medium, the bar buckles without an intermediate inflection point. By gradually increasing β, we finally arrive at a condition where P in Eq. (2-37) is smaller for $m = 2$ than for $m = 1$. At this value of the modulus of the elastic foundation the buckled bar will have an inflection point at the middle. The limiting value of the modulus β at which the transition from $m = 1$ to $m = 2$ occurs is found from the condition that at this limiting value of β expression (2-37) should give the same value for P independently of whether $m = 1$ or $m = 2$. Thus we obtain

$$1 + \frac{\beta l^4}{\pi^4 EI} = 4 + \frac{\beta l^4}{4\pi^4 EI}$$

[1] Note that in this case the energy method gives exact results.

from which $$\frac{\beta l^4}{\pi^4 EI} = 4 \tag{h}$$

For values of β smaller than that given by Eq. (h), the deflection curve of the buckled bar has no inflection point and $m = 1$. For β somewhat larger than that given by Eq. (h), there will be an inflection point at the middle and the bar will be subdivided into two half-waves ($m = 2$).

By increasing β, we obtain conditions in which the number of half-waves is $m = 3, 4, \ldots$. To find the value of β at which the number of half-waves changes from m to $m + 1$, we proceed as above for $m = 1$ and $m = 2$. In this way we obtain the equation

$$m^2 + \frac{\beta l^4}{m^2 \pi^4 EI} = (m + 1)^2 + \frac{\beta l^4}{(m + 1)^2 \pi^4 EI}$$

from which $$\frac{\beta l^4}{\pi^4 EI} = m^2(m + 1)^2 \tag{2-38}$$

For given dimensions of the bar and for a given value of β, this equation can be used for determining m, the number of half-waves. Substituting m in Eq. (2-37), the value of the critical load is obtained. It is seen that in all cases formula (2-37) can be represented in the form

$$P_{cr} = \frac{\pi^2 EI}{L^2} \tag{2-39}$$

where a reduced length L is substituted for the actual length l of the bar. A series of values of L/l, calculated from Eqs. (2-37) and (2-38), are given in Table 2-5 for various values of $\beta l^4/16EI$.

TABLE 2-5. REDUCED LENGTH L FOR A BAR ON AN ELASTIC FOUNDATION[1]

$\beta l^4/(16EI)$	0	1	3	5	10	15	20	30	40	50	75	100
L/l	1	0.927	0.819	0.741	0.615	0.537	0.483	0.437	0.421	0.406	0.376	0.351
$\beta l^4/(16EI)$	200	300	500	700	1,000	1,500	2,000	3,000	4,000	5,000	8,000	10,000
L/l	0.286	0.263	0.235	0.214	0.195	0.179	0.165	0.149	0.140	0.132	0.117	0.110

[1] Note that the table is calculated for values of $\beta l^4/16EI$ rather than for $\beta l^4/\pi^4 EI$.

As β increases, the number of half-waves also increases. Then, when 1 is neglected in comparison with m, Eq. (2-38) becomes

$$\frac{\beta l^4}{\pi^4 EI} = m^4 \qquad \text{or} \qquad \frac{l}{m} = \pi \sqrt[4]{\frac{EI}{\beta}} \tag{2-40}$$

By substituting this value of the wave length l/m in Eq. (2-37), we obtain

$$P_{cr} = \frac{2m^2 \pi^2 EI}{l^2} \tag{2-41}$$

and P_{cr} is two times larger than for a hinged bar of length l/m. An example of a long bar elastically supported is found, for instance, in the case of welded railway rails. At high temperatures considerable compressive stress may be produced in the rails, and lateral buckling may occur if the resistance of the foundation to lateral movement is not sufficient.

In the above discussion a continuous distribution of lateral reactions was assumed. The above formulas can be used also with reasonable accuracy in the case of isolated elastic supports, provided the proportions of the bar and the lateral rigidity of the supports are such that not less than three supports correspond to one half-wave of the buckled bar. If there are fewer than three supports per half-wave length, the critical value of the load should be calculated as explained in Art. 2.6.

2.11. Buckling of a Bar with Intermediate Compressive Forces. In the design of compression members, the case of a bar compressed by intermediate axial forces is sometimes encountered. A simple example of this type is shown in Fig. 2-38, in which a bar with hinged ends is compressed by forces applied at the ends and also by a force P_2 applied at an intermediate cross section.[1] If the compressive forces exceed their critical value slightly, the bar buckles as shown by the dashed line in Fig. 2-38. Let δ be the deflection at C where the load P_2 is applied; then the lateral reactions Q at the supports become

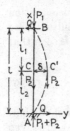

FIG. 2-38

$$Q = \frac{\delta P_2}{l}$$

Assuming, for generality, that the upper and lower parts of the bar have two different cross-sectional moments of inertia I_1 and I_2, the differential equations of the deflection curve for the two portions of the bar are

$$EI_1 \frac{d^2 y_1}{dx^2} = -P_1 y_1 - Q(l - x) = -P_1 y_1 - \frac{\delta P_2}{l}(l - x) \qquad (a)$$

$$EI_2 \frac{d^2 y_2}{dx^2} = -P_1 y_2 - Q(l - x) + P_2(\delta - y_2)$$

$$= -P_1 y_2 - \frac{\delta P_2}{l}(l - x) + P_2(\delta - y_2) \qquad (b)$$

Using the following notations

$$k_1{}^2 = \frac{P_1}{EI_1} \qquad k_2{}^2 = \frac{P_2}{EI_2} \qquad k_3{}^2 = \frac{P_1 + P_2}{EI_2} \qquad k_4{}^2 = \frac{P_2}{EI_1}$$

---

[1] This problem was discussed by Jasinsky, *loc. cit.* Additional cases were considered by W. J. Duncan, *Engineering*, p. 180, 1952. The case where the bar has an initial curvature was discussed by S. I. Sergev, *Univ. Wash. Eng. Expt. Sta. Bull.* 113, 1945.

the general solutions of Eqs. (a) and (b) are

$$y_1 = C_1 \sin k_1 x + C_2 \cos k_1 x - \frac{\delta}{l} \frac{k_4{}^2}{k_1{}^2} (l - x) \qquad l_2 \leq x \leq l$$

$$y_2 = C_3 \sin k_3 x + C_4 \cos k_3 x + \frac{\delta}{l} \frac{k_2{}^2}{k_3{}^2} x \qquad 0 \leq x \leq l_2$$

The constants of integration C_1, C_2, C_3, and C_4 are obtained from the following end conditions of the two portions of the buckled bar:

$$(y_1)_{x=l} = 0 \qquad (y_1)_{x=l_2} = \delta \qquad (y_2)_{x=l_2} = \delta \qquad (y_2)_{x=0} = 0$$

From these conditions we find

$$C_1 = - \frac{\delta(k_1{}^2 l + k_4{}^2 l_1) \cos k_1 l}{k_1{}^2 l \sin k_1 l_1}$$

$$C_2 = -C_1 \tan k_1 l$$

$$C_3 = \frac{\delta(k_3{}^2 l - k_2{}^2 l_2)}{k_3{}^2 l \sin k_3 l_2} \qquad C_4 = 0$$

Substituting these values in the condition of continuity at C

$$\left(\frac{dy_1}{dx}\right)_{x=l_2} = \left(\frac{dy_2}{dx}\right)_{x=l_2}$$

we obtain the following equation for calculating the critical loads:

$$\frac{k_4{}^2}{k_1{}^2} - \frac{k_1{}^2 l + k_4{}^2 l_1}{k_1 \tan k_1 l_1} = \frac{k_2{}^2}{k_3{}^2} + \frac{k_3{}^2 l - k_2{}^2 l_2}{k_3 \tan k_3 l_2} \tag{2-42}$$

In each particular case, knowing the ratios

$$\frac{P_1 + P_2}{P_1} = m \qquad \frac{I_2}{I_1} = n \qquad \frac{l_2}{l_1} = p \tag{c}$$

we can find, by trial and error, the smallest value of the load $P_1 + P_2$ at which Eq. (2-42) is satisfied. This is the critical value of the compressive force. It can always be represented by the formula

$$(P_1 + P_2)_{cr} = \frac{\pi^2 E I_2}{L^2} \tag{2-43}$$

in which L is the *reduced length* of the bar. In the particular case where $l_1 = l_2$, which is commonly encountered in design, the values of L can be obtained from Table 2-6.

The above method of calculating critical values of compressive loads can be used also where axial forces are applied at several intermediate cross sections. The differential equation for each portion of the deflection curve between any two consecutive forces can be easily set up, but the amount of calculation necessary to obtain the final criterion for deter-

TABLE 2-6. VALUES OF L/l FOR COLUMN IN FIG. 2-38, WITH $l_1 = l_2$

m \ n	1.00	1.25	1.50	1.75	2.00	3.00
1.00	1.00	0.95	0.91	0.89	0.87	0.82
1.25	1.06	1.005	0.97	0.94	0.915	
1.50	1.12	1.06	1.02	0.99	0.96	
1.75	1.18	1.11	1.07	1.04	1.005	
2.00	1.24	1.16	1.12	1.08	1.05	

mining critical loads, analogous to Eq. (2-42), increases rapidly with an increase in the number of intermediate loads. It is advantageous in such cases to use one of the approximate methods.

Applying the energy method to the above example (Fig. 2-38) and assuming as a first approximation that the deflection curve is a sine curve,

$$y = \delta \sin \frac{\pi x}{l}$$

we find that the bending moments for the two portions of the curve are, for the case when $l_1 = l_2 = l/2$,

$$M_1 = P_1 y + \frac{\delta P_2}{l}(l - x)$$

$$M_2 = (P_1 + P_2)y - \frac{\delta P_2 x}{l}$$

Substituting in the expression for the strain energy of bending we obtain

$$\Delta U = \int_{l/2}^{l} \frac{M_1{}^2\, dx}{2EI_1} + \int_0^{l/2} \frac{M_2{}^2\, dx}{2EI_2} = \frac{\delta^2}{2EI_1}\left(P_1{}^2\frac{l}{4} + P_2{}^2\frac{l}{24} + P_1 P_2\frac{2l}{\pi^2}\right)$$
$$+ \frac{\delta^2}{2EI_2}\left[(P_1 + P_2)^2\frac{l}{4} + P_2{}^2\frac{l}{24} - P_2(P_1 + P_2)\frac{2l}{\pi^2}\right] \quad (d)$$

The work done by the forces P_1 and P_2 during buckling is

$$\Delta T = \frac{P_1}{2}\int_0^l \left(\frac{dy}{dx}\right)^2 dx + \frac{P_2}{2}\int_0^{l/2}\left(\frac{dy}{dx}\right)^2 dx = \frac{\delta^2\pi^2}{4l}\left(P_1 + \frac{1}{2}P_2\right) \quad (e)$$

Substituting (d) and (e) in Eq. (2-31) (see p. 84) and using notations (c), we obtain

$(P_1 + P_2)_{cr}$

$$= \frac{(\pi^2 EI_2/l^2)(m + 1)}{m + \dfrac{m}{6}\left(\dfrac{m-1}{m}\right)^2 - \dfrac{8}{\pi^2}(m - 1) + n\left[\dfrac{1}{m} + \dfrac{m}{6}\left(\dfrac{m-1}{m}\right)^2 + \dfrac{8}{\pi^2}\dfrac{m-1}{m}\right]} \quad (2\text{-}44)$$

Substituting in this formula various values for m and n and comparing the results with those given in Table 2-6, we find that in all cases the formula gives errors less than 1 per cent, which is sufficient accuracy for all practical purposes.

2.12. Buckling of a Bar under Distributed Axial Loads. If longitudinal compressive forces are continuously distributed along a bar, the differential equation of the deflection curve of the buckled bar is no longer an equation with constant coefficients. The solution of the equation usually requires the application of infinite series or recourse to one of the approximate methods, such as the energy method. Both of these methods will now be illustrated by an example.

Consider the problem of buckling of a prismatic bar (Fig. 2-39) due to its own weight.[1] The lower end of the bar is vertically built in, the upper end is free, and the weight is uniformly distributed along the length. If the bar buckles as shown by the dotted line, the differential equation of the deflection curve is

$$EI \frac{d^2y}{dx^2} = \int_x^l q(\eta - y) \, d\xi \qquad (a)$$

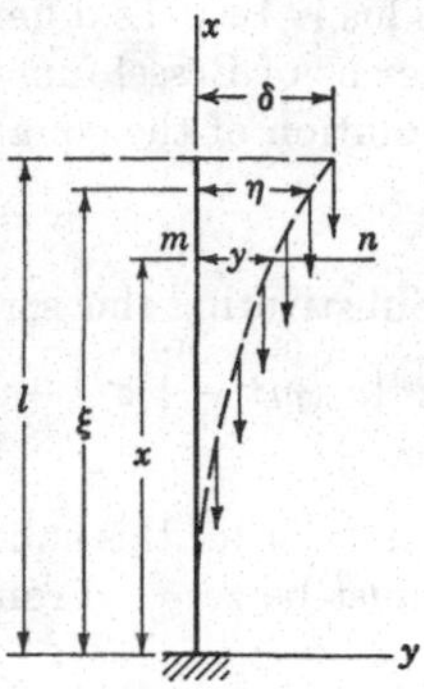

where the integral on the right-hand side of the equation represents the bending moment at any cross section mn produced by the uniformly distributed load of intensity q. Differentiating Eq. (a) with respect to x, we obtain the equation

$$EI \frac{d^3y}{dx^3} = -q(l - x) \frac{dy}{dx} \qquad (b)$$

Fig. 2-39

To simplify the discussion, let us now introduce, instead of x, a new independent variable z by taking

$$z = \frac{2}{3} \sqrt{\frac{q}{EI} (l - x)^3} \qquad (c)$$

Then, by differentiation, we obtain

$$\frac{dy}{dx} = -\frac{dy}{dz} \sqrt[3]{\frac{3}{2} \frac{qz}{EI}} \qquad (d)$$

$$\frac{d^2y}{dx^2} = \left(\frac{3}{2} \frac{q}{EI}\right)^{\frac{2}{3}} \left(\frac{1}{3} z^{-\frac{1}{3}} \frac{dy}{dz} + z^{\frac{1}{3}} \frac{d^2y}{dz^2}\right) \qquad (e)$$

$$\frac{d^3y}{dx^3} = \frac{3}{2} \frac{q}{EI} \left(\frac{1}{9} z^{-1} \frac{dy}{dz} - \frac{d^2y}{dz^2} - z \frac{d^3y}{dz^3}\right) \qquad (f)$$

---

[1] This problem was discussed first by Euler, but he did not succeed in obtaining a satisfactory solution; see I. Todhunter and K. Pearson, "History of the Theory of Elasticity," vol. 1, pp. 39–50, Cambridge, 1886. The problem was solved by A. G. Greenhill, *Proc. Cambridge Phil. Soc.*, vol. 4, 1881. In his paper Greenhill indicates a variety of buckling problems which can be solved by using Bessel functions. Independently, the same problem was discussed in a very complete manner by F. S. Jasinsky, *loc. cit.* See also J. Dondorff, "Die Knickfestigkeit des geraden Stabes mit veränderlichem Querschnitt und veränderlichem Druck, ohne und mit Querstützen," Dissertation, Düsseldorf, 1907, and N. Grishcoff, *Bull. acad. sci., Kiev*, 1930.

Substituting in Eq. (b) and letting

$$\frac{dy}{dz} = u \tag{g}$$

we obtain

$$\frac{d^2u}{dz^2} + \frac{1}{z}\frac{du}{dz} + \left(1 - \frac{1}{9z^2}\right)u = 0 \tag{h}$$

This is Bessel's differential equation, and its solution can be expressed in terms of Bessel functions.[1] These functions are evaluated by taking the solution of the equation in the form of an infinite series:

$$u = z^m(a_0 + a_1z + a_2z^2 + a_3z^3 + \cdots) \tag{i}$$

Substituting this series and its derivatives in Eq. (h), we obtain

$$z^m\{a_0(m^2 - \tfrac{1}{9})z^{-2} + a_1[(m + 1)^2 - \tfrac{1}{9}]z^{-1} + a_0 + a_2[(m + 2)^2 - \tfrac{1}{9}]$$
$$+ (a_1 + a_3[(m + 3)^2 - \tfrac{1}{9}])z + \cdots\} = 0 \tag{j}$$

In order for this equation to be satisfied, the coefficient of each power of z must be zero.   From the first term we obtain

$$m^2 - \tfrac{1}{9} = 0 \qquad m = \pm\tfrac{1}{3}$$

The second term then gives $a_1 = 0$, the fourth term $a_3 = 0$, etc., and it is seen that the terms in series (i) involving odd powers of z must vanish. The third term in Eq. (j) gives

$$a_2 = -\frac{a_0}{(m + 2)^2 - \tfrac{1}{9}}$$

and by considering additional terms we arrive at the general relation

$$a_n = -\frac{a_{n-2}}{(m + n)^2 - \tfrac{1}{9}} \tag{k}$$

Corresponding to the two values of m, there are two series which satisfy Eq. (h).   By evaluating the coefficients in those series from Eq. (k), we can write the general solution of Eq. (h) in the form

$$u = C_1z^{-\frac{1}{3}}\left(1 - \frac{3}{8}z^2 + \frac{9}{320}z^4 - \cdots\right)$$
$$+ C_2z^{\frac{1}{3}}\left(1 - \frac{3}{16}z^2 + \frac{9}{896}z^4 - \cdots\right) \tag{l}$$

In this equation C_1 and C_2 are constants of integration, and the series, except for constant factors, represent Bessel functions of the first kind, of orders $-\tfrac{1}{3}$ and $+\tfrac{1}{3}$, respectively.

[1] See, for example, T. V. Kármán and M. A. Biot, "Mathematical Methods in Engineering," chap. 2, McGraw-Hill Book Company, Inc., New York, 1940.

The constants C_1 and C_2 must be determined from the end conditions of the bar. Since the upper end of the bar is free, we have the condition

$$\left(\frac{d^2y}{dx^2}\right)_{x=l} = 0$$

Observing that $z = 0$ when $x = l$ and also using Eqs. (e) and (g), we can express this condition as

$$\left(\frac{1}{3} z^{-\frac{1}{3}} u + z^{\frac{1}{3}} \frac{du}{dz}\right)_{z=0} = 0$$

Substituting Eq. (l) for u into this equation, we obtain $C_2 = 0$ and hence

$$u = C_1 z^{-\frac{1}{3}}\left(1 - \frac{3}{8} z^2 + \frac{9}{320} z^4 - \cdots\right) \tag{m}$$

At the lower end of the bar the condition is

$$\left(\frac{dy}{dx}\right)_{x=0} = 0$$

With the use of Eqs. (c), (d), and (g), this condition is expressed in the form

$$u = 0 \qquad \text{when } z = \frac{2}{3}\sqrt{\frac{ql^3}{EI}}$$

The value of z which makes $u = 0$ can be found from Eq. (m) by trial and error or from a table of values of the Bessel function of order $-\frac{1}{3}$. Tables giving the zeros of the Bessel functions are also available.[1] The lowest value of z which makes $u = 0$, corresponding to the lowest buckling load, is found to be $z = 1.866$, and hence

$$\frac{2}{3}\sqrt{\frac{ql^3}{EI}} = 1.866$$

or
$$(ql)_{cr} = \frac{7.837 EI}{l^2} \tag{n}$$

This is the critical value of the uniform load for the bar shown in Fig. 2-39.

When the same method is used, the case of the combined action of a uniform compressive load ql and a compressive force P applied at the ends can be studied. If the conditions at the ends are like those shown in Fig. 2-39, and if the uniformly distributed load q is absent, the critical value of the load P applied at the top is

$$P_{cr} = \frac{\pi^2 EI}{4l^2}$$

[1] See, for example, Jahnke and Emde, op. cit., p. 167.

The uniform load ql reduces the critical value of the load P and we can put

$$P_{cr} = \frac{mEI}{l^2} \tag{2-45}$$

where the factor m, smaller than $\pi^2/4$, gradually diminishes when the load ql increases and approaches zero when ql approaches the value given by Eq. (n). Using the notation

$$n = ql \div \frac{\pi^2 EI}{4l^2}$$

the values of the coefficient m in Eq. (2-45) for various values of n can be computed and are given in Table 2-7.[1]

TABLE 2-7. VALUES OF m IN EQ. (2-45)

n	0	0.25	0.50	0.75	1.0	2.0	3.0	3.18	4.0	5.0	10.0
m	$\pi^2/4$	2.28	2.08	1.91	1.72	0.96	0.15	0	-0.69	-1.56	-6.95

In calculating the effect of the uniform load ql on the magnitude of P_{cr}, we can obtain a reasonably good approximation by assuming that the effect of ql is equivalent to a load $0.3ql$ applied at the top of the bar. (The value $0.3ql$ was calculated from the data in Table 2-7.) Thus the critical load is

$$P_{cr} \approx \frac{\pi^2 EI}{4l^2} - 0.3ql$$

When the uniform load is larger than that given by Eq. (n), P_{cr} becomes negative and a tensile force P must be applied in order to prevent the bar from buckling.

In the calculation of the critical value of the distributed compressive loads, the energy method can also be used to advantage. In the case represented in Fig. 2-39 for instance, we can take as a first approximation for the deflection curve

$$y = \delta \left(1 - \cos \frac{\pi x}{2l}\right) \tag{o}$$

This is the true curve for the case where buckling occurs under the action of a compressive load applied at the end [see Eq. (2-2)]. In the case of a uniformly distributed axial load, the true curve is more complicated, as was shown in the previous discussion. Nevertheless, the curve given by Eq. (o) satisfies the geometrical end conditions and can be taken as a suitable curve for an approximate calculation.

The bending moment at any cross section mn (Fig. 2-39) is

$$M = \int_x^l q(\eta - y)\, d\xi \tag{p}$$

[1] See Grishcoff, *loc. cit.*

Substituting Eq. (*o*) for *y* in this expression and also observing that

$$\eta = \delta \left(1 - \cos \frac{\pi \xi}{2l} \right)$$

we obtain

$$M = q\delta \left[(l - x) \cos \frac{\pi x}{2l} - \frac{2l}{\pi} \left(1 - \sin \frac{\pi x}{2l} \right) \right]$$

Substituting this equation in the expression for the strain energy of bending, we obtain

$$\Delta U = \int_0^l \frac{M^2 \, dx}{2EI} = \frac{\delta^2 q^2 l^3}{2EI} \left(\frac{1}{6} + \frac{9}{\pi^2} - \frac{32}{\pi^3} \right) \qquad (q)$$

The work done by the distributed axial load during lateral buckling will be calculated next. Owing to the inclination of an element *ds* of the deflection curve at the cross section *mn* (Fig. 2-39), the upper part of the load undergoes a downward displacement equal to

$$ds - dx \approx \frac{1}{2} \left(\frac{dy}{dx} \right)^2 dx$$

and the corresponding work done by this load is

$$\frac{1}{2} q(l - x) \left(\frac{dy}{dx} \right)^2 dx$$

Therefore, the total work produced by the load during buckling, by using Eq. (*o*), is

$$\Delta T = \frac{1}{2} q \int_0^l (l - x) \left(\frac{dy}{dx} \right)^2 dx = \frac{\pi^2 \delta^2 q}{8} \left(\frac{1}{4} - \frac{1}{\pi^2} \right) \qquad (r)$$

Substituting (*q*) and (*r*) in Eq. (2-31), we obtain as a first approximation for the critical value of the weight

$$(ql)_{cr} = \frac{7.89EI}{l^2}$$

Comparing this result with Eq. (*n*), obtained by integration of the differential equation, it is seen that the error of the first approximation is less than 1 per cent, and thus it is accurate enough for any practical application.

An even better approximation can be obtained by taking *y* as a function of several parameters and then adjusting the parameters so as to make $(ql)_{cr}$ a minimum. To illustrate this method, let us again consider the column in Fig. 2-39 and assume that

$$y = \delta_1 \left(1 - \cos \frac{\pi x}{2l} \right) + \delta_2 \left(1 - \cos \frac{3\pi x}{2l} \right) \qquad (s)$$

This equation satisfies the geometrical conditions at the ends of the bar and contains two parameters δ_1 and δ_2. Substituting in expression (*p*) for the bending moment, we obtain

$$M = q\delta_1 \left[(l - x) \cos \frac{\pi x}{2l} - \frac{2l}{\pi} \left(1 - \sin \frac{\pi x}{2l} \right) \right]$$
$$+ q\delta_2 \left[(l - x) \cos \frac{3\pi x}{2l} + \frac{2l}{3\pi} \left(1 + \sin \frac{3\pi x}{2l} \right) \right]$$

Now substituting in Eq. (q) for the strain energy of bending and performing the integration, we find that

$$\Delta U = \frac{q^2 l^3}{2EI} (\delta_1{}^2\alpha + 2\delta_1\delta_2\beta + \delta_2{}^2\gamma) \tag{t}$$

in which

$$\alpha = \frac{1}{6} + \frac{9}{\pi^2} - \frac{32}{\pi^3} = 0.04650 \qquad \beta = \frac{32}{9\pi^3} - \frac{1}{12\pi^2} = 0.10622$$
$$\gamma = \frac{1}{6} + \frac{1}{\pi^2} + \frac{32}{27\pi^3} = 0.30621$$

Substituting expression (s) into expression (r), we obtain

$$\Delta T = \frac{q\pi^2}{8} (\delta_1{}^2\alpha' + 2\delta_1\delta_2\beta' + \delta_2{}^2\gamma') \tag{u}$$

where

$$\alpha' = \frac{1}{4} - \frac{1}{\pi^2} = 0.14868 \qquad \beta' = \frac{3}{\pi^2} = 0.30396$$
$$\gamma' = \frac{9}{4} - \frac{1}{\pi^2} = 2.14868$$

Substituting (t) and (u) in Eq. (2-31), we obtain

$$(ql)_{\mathrm{cr}} = \frac{\pi^2 EI}{4l^2} \frac{\delta_1{}^2\alpha' + 2\delta_1\delta_2\beta' + \delta_2{}^2\gamma'}{\delta_1{}^2\alpha + 2\delta_1\delta_2\beta + \delta_2{}^2\gamma} \tag{v}$$

The conditions for $(ql)_{\mathrm{cr}}$ to be a minimum are

$$\frac{\partial(ql)_{\mathrm{cr}}}{\partial\delta_1} = 0 \qquad \frac{\partial(ql)_{\mathrm{cr}}}{\partial\delta_2} = 0$$

or

$$\frac{\pi^2 EI}{4l^2} \frac{\partial}{\partial\delta_1} (\delta_1{}^2\alpha' + 2\delta_1\delta_2\beta' + \delta_2{}^2\gamma') - (ql)_{\mathrm{cr}} \frac{\partial}{\partial\delta_1} (\delta_1{}^2\alpha + 2\delta_1\delta_2\beta + \delta_2{}^2\gamma) = 0$$
$$\frac{\pi^2 EI}{4l^2} \frac{\partial}{\partial\delta_2} (\delta_1{}^2\alpha' + 2\delta_1\delta_2\beta' + \delta_2{}^2\gamma') - (ql)_{\mathrm{cr}} \frac{\partial}{\partial\delta_2} (\delta_1{}^2\alpha + 2\delta_1\delta_2\beta + \delta_2{}^2\gamma) = 0$$

After differentiation we obtain

$$\delta_1 \left[\frac{\pi^2 EI}{2l^2} \alpha' - 2(ql)_{\mathrm{cr}}\alpha \right] + \delta_2 \left[\frac{\pi^2 EI}{2l^2} \beta' - 2(ql)_{\mathrm{cr}}\beta \right] = 0$$
$$\delta_1 \left[\frac{\pi^2 EI}{2l^2} \beta' - 2(ql)_{\mathrm{cr}}\beta \right] + \delta_2 \left[\frac{\pi^2 EI}{2l^2} \gamma' - 2(ql)_{\mathrm{cr}}\gamma \right] = 0$$

The possibility of buckling occurs when these equations give for δ_1 and δ_2 solutions different from zero. This requires that the determinant of the equations must be equal to zero; i.e.,

$$\left[\frac{\pi^2 EI}{2l^2} \alpha' - 2(ql)_{\mathrm{cr}}\alpha \right] \left[\frac{\pi^2 EI}{2l^2} \gamma' - 2(ql)_{\mathrm{cr}}\gamma \right] - \left[\frac{\pi^2 EI}{2l^2} \beta' - 2(ql)_{\mathrm{cr}}\beta \right]^2 = 0$$

or

$$4(ql)_{\mathrm{cr}}{}^2(\alpha\gamma - \beta^2) - 2(ql)_{\mathrm{cr}} \frac{\pi^2 EI}{2l^2} (\alpha\gamma' + \alpha'\gamma - 2\beta\beta') + (\alpha'\gamma' - \beta'^2) \left(\frac{\pi^2 EI}{2l^2} \right)^2 = 0$$

Solving this quadratic equation for $(ql)_{cr}$ and substituting numerical values for the constants, we obtain

$$(ql)_{cr} = \frac{7.84EI}{l^2}$$

This value practically coincides with that given in Eq. (n).

TABLE 2-8. VALUES OF m IN EQ. (2-46)

n	0	0.25	0.50	0.75	1.0	2.0	3.0
m	π^2	8.63	7.36	6.08	4.77	$-.657$	-4.94

By using the energy method we can also consider a vertical bar hinged at the ends and submitted to the action of its own weight ql in addition to compressive forces P applied at the ends (Fig. 2-40). The critical values of P can be represented by the equation

$$P_{cr} = \frac{mEI}{l^2} \qquad (2\text{-}46)$$

in which the numerical factor m depends on the value of the ratio

$$n = ql \div \frac{\pi^2 EI}{l^2}$$

Several values of the factor m are given in Table 2-8.

An approximation for the critical load P is obtained by assuming that one-half of the weight ql of the bar is applied at the top, i.e., by taking

$$P_{cr} = \frac{\pi^2 EI}{l^2} - \frac{ql}{2}$$

FIG. 2-40

For large values of n, P_{cr} is negative, which indicates that in such cases tensile forces P should be applied at the ends to prevent the bar from lateral buckling.

The energy method can be applied advantageously in various cases of distributed compressive loads acting on a bar. With the use of this method the integration of equations with variable coefficients, requiring the use of infinite series, is replaced by the simple problem of finding the minimum of a certain expression, such as the right-hand side of Eq. (v) above. When the number of terms in the expression for the deflection curve is increased, as in Eq. (s) above, the accuracy of the solution can be increased, although the first approximation is usually sufficient for practical applications.

2.13. Buckling of a Bar on an Elastic Foundation under Distributed Axial Loads. In this article we shall consider the buckling of the bar shown in Fig. 2-41a. The bar is subjected to a distributed axial load q and supported by a continuous elastic foundation. The axial load q will be assumed to have the distribution shown in Fig. 2-41a and b; that is, the intensity of distributed load at the ends is q_0 and the load is directed toward the center of the bar. The load q decreases linearly to the center, where it has zero value. This distribution of load represents approximately the variation in compressive stress in the top chord of a bridge truss, as will be shown later. The modulus of the elastic foundation is denoted as β (see p. 94) and, when multiplied by the deflection y, gives the reaction of the foundation per unit length of bar.

The beam of Fig. 2-41a can be analyzed by solving the differential equation of the deflection curve of the buckled bar. The equation can be integrated by the use of infinite series, as explained in the previous article. The same result, however, can be obtained more easily by using the energy method. The deflection curve of the buckled bar in the case of hinged ends can be represented by the series

$$y = a_1 \sin \frac{\pi x}{l} + a_2 \sin \frac{2\pi x}{l} + a_3 \sin \frac{3\pi x}{l} + \cdots \tag{a}$$

Assuming that the cross section of the bar is constant along its length, the strain

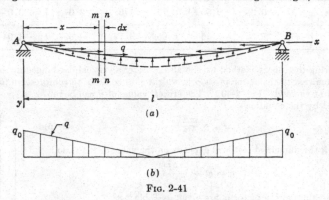

(a)

(b)

Fig. 2-41

energy of bending[1] of the bar, together with the strain energy of the foundation, is (see Art. 2.10)

$$\Delta U = \frac{\pi^4 EI}{4l^3} \sum_{n=1}^{n=\infty} n^4 a_n{}^2 + \frac{\beta l}{4} \sum_{n=1}^{n=\infty} a_n{}^2 \tag{b}$$

In calculating the work produced by the distributed compressive load during bending, we note that the intensity of this load at any cross section, distance x from the left support (Fig. 2-41b), is

$$q = q_0 \left(1 - \frac{2x}{l}\right) \tag{c}$$

where q_0 is the intensity of load at the ends. Considering an element of the bar between two consecutive cross sections mm and nn, the axial load to the right of the cross section mm will be displaced toward the immovable support A, owing to the small inclination of this element during buckling, by the amount $\frac{1}{2}(dy/dx)^2 \, dx$ and will produce the work

$$-\frac{1}{2}\left(\frac{dy}{dx}\right)^2 dx \int_x^l q_0\left(1 - \frac{2x}{l}\right) dx = \frac{q_0}{2l} x(l - x) \left(\frac{dy}{dx}\right)^2 dx$$

The total work produced by the compressive load during bending is

$$\Delta T = \frac{q_0}{2l} \int_0^l x(l - x) \left(\frac{dy}{dx}\right)^2 dx$$

[1] For convenience, Eq. (2-34) is used instead of Eq. (2-33). Since this may mean poor accuracy if only one term of series (a) is used, it is necessary to consider two- and three-term approximations also.

Substituting in this expression the series (a) for y and using the formulas

$$\int_0^l x \cos^2 \frac{m\pi x}{l}\, dx = \frac{l^2}{4} \qquad \int_0^l x^2 \cos^2 \frac{m\pi x}{l}\, dx = \frac{l^3}{6} + \frac{l^3}{4m^2\pi^2}$$

$$\int_0^l x \cos \frac{n\pi x}{l} \cos \frac{m\pi x}{l}\, dx = 0 \qquad \text{when } m + n \text{ is an even number}$$

$$\int_0^l x \cos \frac{n\pi x}{l} \cos \frac{m\pi x}{l}\, dx = -\frac{2l^2}{\pi^2}\frac{m^2 + n^2}{(m^2 - n^2)^2} \qquad \text{when } m + n \text{ is an odd number}$$

$$\int_0^l x^2 \cos \frac{n\pi x}{l} \cos \frac{m\pi x}{l}\, dx = \frac{2l^3}{\pi^2}\frac{m^2 + n^2}{(m^2 - n^2)^2}(-1)^{m+n}$$

we finally obtain

$$\Delta T = \frac{q_0}{2}\left[\sum_{n=1}^{\infty} a_n^2\left(\frac{n^2\pi^2}{12} - \frac{1}{4}\right) - 4\sum_n\sum_m a_n a_m \frac{nm(m^2 + n^2)}{(m^2 - n^2)^2}\right] \qquad (d)$$

where the double series in the brackets contains only terms in which the sum $(m + n)$ is even and m is not equal to n. Substituting (b) and (d) in Eq. (2-31), we obtain for the compressive force the expression

$$\frac{q_0 l}{4} = \frac{\dfrac{\pi^4 EI}{8l^2}\displaystyle\sum_{n=1}^{\infty} n^4 a_n^2 + \dfrac{\beta l^2}{8}\displaystyle\sum_{n=1}^{\infty} a_n^2}{\displaystyle\sum_{n=1}^{\infty} a_n^2\left(\dfrac{n^2\pi^2}{12} - \dfrac{1}{4}\right) - 4\displaystyle\sum_n\sum_m a_n a_m \dfrac{nm(m^2 + n^2)}{(m^2 - n^2)^2}} \qquad (e)$$

The quantity $q_0 l/4$ represents the compressive force at the center of the bar. Next, the problem is to find such relations between the coefficients $a_1, a_2, a_3, \ldots$ as to make expression (e) a minimum. Proceeding as before and equating to zero the derivatives of this expression with respect to $a_1, a_2, \ldots$, we finally arrive at a system of homogeneous linear equations in $a_1, a_2, \ldots$ of the following type:

$$\left[(n^4 + \gamma)\pi^2 - 2\alpha\left(\frac{n^2\pi^2}{3} - 1\right)\right]a_n + 16\alpha\sum_m a_m \frac{nm(m^2 + n^2)}{(m^2 - n^2)^2} = 0 \qquad (f)$$

in which, for simplification, we use the notation

$$\alpha = \frac{q_0 l}{4} \div \frac{\pi^2 EI}{l^2} \qquad \gamma = \frac{\beta l^4}{\pi^4 EI} \qquad (g)$$

The summation in the second term of Eq. (f) is extended over all values of m different from n such that $(m + n)$ is an even number. Thus, Eq. (f) can be subdivided into two groups, one containing the coefficients a_m with all values of m taken odd and the second with all values of m taken even.

The equations of the first group are

$$\left[(1 + \gamma)\pi^2 - 2\alpha\left(\frac{\pi^2}{3} - 1\right)\right]a_1 + \alpha\left(\frac{15}{2}a_3 + \frac{65}{18}a_5 + \frac{175}{72}a_7 + \cdots\right) = 0$$

$$\frac{15}{2}\alpha a_1 + [(3^4 + \gamma)\pi^2 - 2\alpha(3\pi^2 - 1)]a_3 + \alpha\left(\frac{255}{8}a_5 + \frac{609}{50}a_7 + \cdots\right) = 0$$

$$\frac{65}{18}\alpha a_1 + \frac{255}{8}\alpha a_3 + \left[(5^4 + \gamma)\pi^2 - 2\alpha\left(\frac{25}{3}\pi^2 - 1\right)\right]a_5 + \alpha\left(\frac{1,295}{18}a_7 + \cdots\right) = 0$$

$$\frac{175}{72}\alpha a_1 + \frac{609}{50}\alpha a_3 + \frac{1,295}{18}\alpha a_5 + \left[(7^4 + \gamma)\pi^2 - 2\alpha\left(\frac{49}{3}\pi^2 - 1\right)\right]a_7 + \cdots = 0$$

$$\cdots\cdots\cdots\cdots\cdots\cdots\cdots\cdots\cdots\cdots\cdots\cdots\cdots\cdots\cdots \qquad (h)$$

The equations of the second group are:

$$\left[(2^4 + \gamma)\pi^2 - 2\alpha\left(\frac{4}{3}\pi^2 - 1\right)\right]a_2 + \alpha\left(\frac{160}{9}a_4 + \frac{15}{2}a_6 + \cdots\right) = 0$$

$$\frac{160}{9}\alpha a_2 + \left[(4^4 + \gamma)\pi^2 - 2\alpha\left(\frac{16}{3}\pi^2 - 1\right)\right]a_4 + \alpha\left(\frac{1{,}248}{25}a_6 + \cdots\right) = 0$$

$$\frac{15}{2}\alpha a_2 + \frac{1{,}248}{25}\alpha a_4 + \left[(6^4 + \gamma)\pi^2 - 2\alpha\left(\frac{36}{3}\pi^2 - 1\right)\right]a_6 + \cdots = 0$$

. . . . (i)

Buckling of the bar becomes possible when one of the above two systems of equations gives for coefficients a_m a solution different from zero, i.e., when the determinant of system (h) or of system (i) becomes equal to zero. The system (h) corresponds to a symmetrical shape of the buckled bar, while system (i) corresponds to an antisymmetrical shape of the buckled bar.

Let us begin with the case where the rigidity of the elastic medium is very small. In this case the deflection curve of the buckled bar has only one half-wave (see Art. 2.10) and is symmetrical with respect to the middle. Therefore Eqs. (h) should be used. The first approximation is obtained by taking only the first term in the series (a) and putting $a_3 = a_5 = \cdots = 0$. Then the first equation of (h) will give for a_1 a solution different from zero only if

$$(1 + \gamma)\pi^2 - 2\alpha\left(\frac{\pi^2}{3} - 1\right) = 0$$

from which

$$\alpha = \frac{\pi^2(1 + \gamma)}{2(\frac{1}{3}\pi^2 - 1)}$$

Using notations (g), we finally obtain

$$\left(\frac{q_0 l}{4}\right)_{cr} = \frac{\pi^2 EI}{l^2}\frac{\pi^2(1 + \gamma)}{2(\frac{1}{3}\pi^2 - 1)} \tag{j}$$

If there is no lateral elastic resistance and if the bar is compressed by axial load distributed as shown in Fig. 2-41b, the quantity γ in Eq. (j) becomes zero [see notations (g)] and we obtain

$$\left(\frac{q_0 l}{4}\right)_{cr} = 2.15\frac{\pi^2 EI}{l^2} \tag{k}$$

Thus the critical load is more than twice as large as in the case where the bar is compressed by loads applied only at the ends.

To obtain a better approximation for the critical compressive force, we take the two terms in expression (a) with coefficients a_1 and a_3 The corresponding two equations, from system (h), are

$$\left[(1 + \gamma)\pi^2 - 2\alpha\left(\frac{\pi^2}{3} - 1\right)\right]a_1 + \frac{15}{2}\alpha a_3 = 0$$

$$\frac{15}{2}\alpha a_1 + [(3^4 + \gamma)\pi^2 - 2\alpha(3\pi^2 - 1)]a_3 = 0$$

Taking γ equal to zero and equating to zero the determinant of the above two equations, we obtain

$$\left[\pi^2 - 2\alpha\left(\frac{\pi^2}{3} - 1\right)\right]\left[81\pi^2 - 2\alpha(3\pi^2 - 1)\right] - \left(\frac{15}{2}\right)^2\alpha^2 = 0$$

Solving this equation for α, we obtain

$$\alpha = 2.06 \qquad \left(\frac{q_0l}{4}\right)_{cr} = 2.06\frac{\pi^2EI}{l^2} \qquad (l)$$

With the use of three terms of the series (a) with the coefficients a_1, a_3, and a_5 and the three equations of system (h), a third approximation can be calculated. Such calculations show that the error of the second approximation, given by Eq. (l), is less than 1 per cent, so that further approximations are of no practical importance and we can put

$$\left(\frac{q_0l}{4}\right)_{cr} = 2.06\frac{\pi^2EI}{l^2} = \frac{\pi^2EI}{(0.696l)^2}$$

Thus the reduced length in this case is

$$L = 0.696l$$

When a greater restraint is supplied by the elastic foundation, the buckled form of the bar may have two half-waves, and we obtain an inflection point at the middle of the bar. To calculate the critical load in such a case, the system (i) should be used.

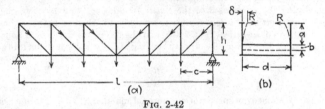

Fig. 2-42

With a further increase of β, the buckled bar has three half-waves, and we must again use the system of equations (h) in calculating the critical value of the compressive load. In all these cases the critical load can be represented by the equation

$$\left(\frac{q_0l}{4}\right)_{cr} = \frac{\pi^2EI}{L^2} \qquad (2\text{-}47)$$

in which the reduced length L depends on the rigidity of the elastic foundation. Several values of the ratio L/l are given in Table 2-9. It is seen from the table that,

TABLE 2-9. REDUCED LENGTH L IN EQ. (2-47)

$\beta l^4/16EI$	0	5	10	15	22 8	56.5	100	162.8	200	300	500	1,000
L/l	0.696	0 524	0.443	0.396	0.363	0.324	0.290	0.259	0.246	0.225	0.204	0.174

when the rigidity of the elastic foundation increases, the ratio L/l approaches the values obtained before for a uniformly compressed bar (see Table 2-5).

The results derived above for the bar in Fig. 2-41a can be applied to the problem of the stability of the upper chord of a low-truss bridge, or pony truss (Fig. 2-42a and b). In the absence of upper chord bracing, the lateral buckling (see Fig. 2-42b) of the top chord is resisted by the elastic reactions of the vertical and diagonal members of the truss. At the supports there are usually frames or bracing members of considerable rigidity, so that the ends of the chord can be considered as immovable in the lateral direction. Thus the upper chord can be considered as a bar with hinged ends compressed by forces distributed along its length and elastically supported at intermediate points. A general method of solving problems of this type is discussed in Art. 2.6.

However, the amount of work necessary to obtain the critical value of the compressive force increases rapidly with the number of elastic supports.[1] The stability of the compressed chord can be increased by increasing the rigidity of the lateral supports. For a constant cross section of the chord and a constant compressive force, the minimum rigidity, at which the supports begin to behave as though they were absolutely rigid, is found from Eq. (2-30). If the proportions of the compressed chord and verticals of the bridge (Fig. 2-42) are such that the half-wave length of the buckled chord is large in comparison with one panel length of the bridge (say the half-wave length is not less than three panels), a great simplification of the problem can be obtained by replacing the elastic supports by an equivalent elastic foundation and replacing the concentrated compressive forces, applied at the joints, by a continuously distributed load. Assuming that the bridge is uniformly loaded, the compressive forces transmitted to the chord by the diagonals are proportional to the distances from the middle of the span, and the equivalent compressive load distribution is as shown in Fig. 2-41b by the shaded areas.

In calculating the modulus β of the elastic foundation, equivalent to the elastic resistance of the verticals,[2] it is necessary to establish the relation between the force R applied at the top of a vertical (Fig. 2-42b) and the deflection that would be produced if the upper chord were removed. If only bending of the vertical is taken into account, then

$$\delta = \frac{Ra^3}{3EI_1}$$

where I_1 is the moment of inertia of one vertical member. Taking into account the bending of the floor beam and using notations indicated in the figure, we obtain

$$\delta = \frac{Ra^3}{3EI_1} + \frac{R(a+b)^2d}{2EI_2}$$

where I_2 is the moment of inertia of the cross section of the floor beam. The force necessary to produce a deflection δ equal to unity is then

$$R_0 = \frac{1}{\dfrac{a^3}{3EI_1} + \dfrac{(a+b)^2d}{2EI_2}}$$

and the modulus of the equivalent elastic foundation is

$$\beta = \frac{R_0}{c}$$

where c is the distance between verticals.

[1] Several numerical examples of calculations of the stability of a compressed chord as a bar on elastic supports can be found in the book by H. Müller-Breslau, "Graphische Statik," vol. 2, part 2, 1908. See also paper by A. Ostenfeld, *Beton u. Eisen*, vol. 15, 1916. An analysis using the moment-distribution method was given by F. Kerekes and C. L. Hulsbos, Elastic Stability of the Top Chord of a Three-span Continuous Pony Truss Bridge, *Iowa Eng. Expt. Sta. Bull.* 177, 1954.

[2] Since the diagonals are tension members, their rigidity is small in comparison with that of the struts and can be neglected.

If the truss has parallel chords and a large number of panels, the maximum intensity q_0 of the axial load is, from statics,

$$q_0 = \frac{Q}{2h}$$

where Q is the total load on one truss and h is the depth of the truss.

With β and q_0 thus determined, the critical load can be calculated for the upper chord in the manner described above.[1]

The method developed above for the case of a bar of uniform cross section supported by an elastic medium of constant modulus along the length of the bar can be extended to include cases of chords of variable cross section and cases where the rigidities of the elastic supports vary along the length.[2]

2.14. Buckling of Bars with Changes in Cross Section.

An examination of the bending-moment diagram for a buckled bar indicates that a bar of uniform cross section is not the most economical form to carry compressive loads. It is evident in the case of a compressed bar with hinged ends, for example, that the stability can be increased by removing a portion of the material from the ends and increasing the cross section over the middle portion (see Fig. 2-43b). In steel structures such bars are very often used. The cross section usually changes abruptly, since the increase in section is accomplished by riveting or welding additional plates or angles along portions of the column. A simple case of such a column was discussed

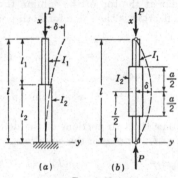

(a)   (b)

FIG. 2-43

previously in Art. 2.11. Another example is shown in Fig. 2-43a. To determine the critical value of the load P in this case, it is necessary to write separately the differential equation of the deflection curve for each portion of the column. If I_1 and I_2 are the moments of inertia of the cross sections for the upper and lower portions of the column, respectively, these equations are

---

[1] In this form the problem of the stability of low-truss bridges was first discussed by Jasinsky in "Scientific Papers of F. S. Jasinsky," vol. 1, p. 145, St. Petersburg, 1902. Some corrections of Jasinsky's results were discussed by Timoshenko by using the energy method, in *Bull. Polytech. Inst., Kiev* (in Russian), 1910, and Sur la stabilité des systèmes élastiques, *Ann. ponts et chaussées*, Paris, 1913. See also Timoshenko, Problems Concerning Elastic Stability in Structures, *Trans. ASCE*, vol. 94, 1930.

[2] Several applications of the energy method in design of through bridges are given in the paper by S. Kasarnowsky and D. Zetterholm, *Der Bauingenieur*, vol. 8, p. 760, 1927. See also papers by A. Hrennikoff and by K. Kriso in *Publ. Intern. Assoc. Bridge Structural Eng.*, vol. 3, 1935.

$$EI_1 \frac{d^2y_1}{dx^2} = P(\delta - y_1)$$
$$EI_2 \frac{d^2y_2}{dx^2} = P(\delta - y_2)$$
(a)

Using the notations,

$$k_1{}^2 = \frac{P}{EI_1} \qquad k_2{}^2 = \frac{P}{EI_2}$$

and taking into account the conditions at the built-in end of the column, we find that the solutions of Eqs. (a) are

$$y_1 = \delta + C \cos k_1x + D \sin k_1x$$
$$y_2 = \delta(1 - \cos k_2x)$$

The constants of integration C and D are obtained from the conditions that at the top of the column the deflection is δ and that at $x = l_2$ the deflection is the same for both portions. Hence,

$$\delta + C \cos k_1l + D \sin k_1l = \delta$$
$$\delta + C \cos k_1l_2 + D \sin k_1l_2 = \delta(1 - \cos k_2l_2)$$

from which

$$C = -D \tan k_1l \qquad D = \frac{\delta \cos k_2l_2 \cos k_1l}{\sin k_1l_1}$$

Since the two portions of the deflection curve have the same tangent at $x = l_2$, we obtain the equation

$$\delta k_2 \sin k_2l_2 = -Ck_1 \sin k_1l_2 + Dk_1 \cos k_1l_2$$

Substituting for C and D the above values, we finally obtain the transcendental equation

$$\tan k_1l_1 \tan k_2l_2 = \frac{k_1}{k_2}$$
(b)

for calculating the critical load. Knowing the ratios I_1/I_2 and l_1/l_2, the solution of this equation can be found in each particular case by a trial-and-error method.

By the substitution of $a/2$ for l_2 and $l/2$ for l, the results obtained from Eq. (b) can be applied also for a column with hinged ends and symmetrical with respect to the middle cross section (Fig. 2-43b). The critical value of the load in this case can be represented by the formula

$$P_{cr} = \frac{mEI_2}{l^2}$$
(2-48)

in which m is a numerical factor depending on the ratios a/l and I_1/I_2.

TABLE 2-10. VALUES OF THE FACTOR m IN EQ. (2-48)

I_1/I_2 \ a/l	0.2	0.4	0.6	0.8
0.01	0.15	0.27	0.60	2.26
0.1	1.47	2.40	4.50	8.59
0.2	2.80	4.22	6.69	9.33
0.4	5.09	6.68	8.51	9.67
0.6	6.98	8.19	9.24	9.78
0.8	8.55	9.18	9.63	9.84

Several values of this factor, calculated from Eq. (b), are given[1] in Table 2-10.

This same method can be used also if the number of changes in the cross section of the bar is greater than that considered above. Naturally, with an increase in the number of changes, the derivation of the equation for calculating the critical load and the solution of this equation become more complicated,[2] so that it is advisable to use one of the approximate methods.

Solution by Energy Method. Considering again the case represented in Fig. 2-43a and using the energy method, we can take as a first approximation for the deflection curve

$$y = \delta \left(1 - \cos \frac{\pi x}{2l} \right) \qquad (c)$$

Proceeding as before, we find the following expressions for the strain energy of bending and for the work done by the compressive forces P during buckling:

$$\Delta U = \int_0^{l_2} \frac{M^2\,dx}{2EI_2} + \int_{l_2}^{l} \frac{M^2\,dx}{2EI_1} = \frac{P^2\delta^2}{2EI_2} \left(\int_0^{l_2} \cos^2 \frac{\pi x}{2l}\,dx + \frac{I_2}{I_1} \int_{l_2}^{l} \cos^2 \frac{\pi x}{2l}\,dx \right)$$
$$= \frac{P^2\delta^2}{2EI_2} \left[\frac{l_2}{2} + \frac{I_2}{I_1}\frac{l_1}{2} + \frac{l}{2\pi}\left(1 - \frac{I_2}{I_1}\right) \sin \frac{\pi l_2}{l} \right] \qquad (d)$$

$$\Delta T = \frac{P}{2} \int_0^l \left(\frac{dy}{dx} \right)^2 dx = \frac{\pi^2 P \delta^2}{16l} \qquad (e)$$

Substituting (d) and (e) in Eq. (2-31) gives

$$P_{cr} = \frac{\pi^2 E I_2}{4l^2} \frac{1}{\dfrac{l_2}{l} + \dfrac{l_1}{l}\dfrac{I_2}{I_1} - \dfrac{1}{\pi}\left(\dfrac{I_2}{I_1} - 1 \right) \sin \dfrac{\pi l_2}{l}} \qquad (2-49)$$

[1] Table 2-10 was calculated by A. N. Dinnik, "Design of Columns of Varying Cross Section," translated from the Russian by M. Maletz, *Trans. ASME*, vol 54, 1932. A similar table for bars with fixed ends is also given in this paper.

[2] Several examples of this type of problem have been discussed by A. Franke, *Z. Math. u. Physik*, vol. 49, 1901. See also Timoshenko, Buckling of Bars of Variable Cross-Section, *Bull. Polytech. Inst., Kiev*, 1908, and S. Falk, *Ingr.-Arch.*, vol. 24, p. 85, 1956.

For a bar with hinged ends (Fig. 2-43b), by substituting $a/2$ for l_2 and $l/2$ for l, Eq. (2-49) becomes

$$P_{cr} = \frac{\pi^2 E I_2}{l^2} \frac{1}{\dfrac{a}{l} + \dfrac{l-a}{l}\dfrac{I_2}{I_1} - \dfrac{1}{\pi}\left(\dfrac{I_2}{I_1} - 1\right)\sin\dfrac{\pi a}{l}} \tag{2-50}$$

Comparison of the results obtained from Eq. (2-50) with values of m from Table 2-10 shows that this approximate solution gives very satisfactory results if the ratio I_2/I_1 is not very large. Taking, for instance, $I_1/I_2 = 0.4$ and $a/l = 0.2$ and 0.6, we obtain from Eq. (2-50) $m = 5.14$ and 8.61, respectively, instead of the numbers 5.09 and 8.51, as given in Table 2-10, which is sufficiently accurate for all practical purposes.[1]

This same procedure can be used for a bar consisting of several portions of different cross sections. In such cases additional integrals appear in Eq. (d), one for each portion of the bar, but these integrals are readily evaluated numerically.

The use of the method of successive approximations for buckling problems of this type is described in the next article.

2.15. The Determination of Critical Loads by Successive Approximations.

The method of successive approximations is used to determine critical loads in cases where the exact solution is unknown or very complicated. Whereas the energy method always gives a value for the critical load which is higher than the true value (see p. 90), the method of successive approximations provides a means of obtaining both lower and upper bounds to the critical load. Thus the accuracy of the approximate solution is known, and the successive approximation procedure can be continued until the desired accuracy is obtained.

In the determination of critical buckling loads by this method, a deflection curve for the buckled bar is first assumed. Based upon these assumed deflections, the bending moments in the bar are calculated in terms of the axial force P. Then, knowing the bending moments, we can determine the deflections of the bar by any of the standard methods of strength of materials, such as the conjugate-beam method or double-integration method. Equating the originally assumed deflections to the latter values gives an equation from which the critical load is calculated. This process is now repeated, using the final set of deflections from the first calculations as a new approximation to the true values. The result of this second approximation will be another equation for the critical load, giving a more accurate value than the first equation. The process is continued until there is very little difference between the assumed and calculated deflections, in which case the critical load is nearly exact.

The assumed deflections and the corresponding calculated values can be equated at any point along the axis of the bar in obtaining the equation for the critical load. The lowest value of the critical load found in this

[1] Solution of several examples of this kind can be found in the book by E. Elwitz, "Die Lehre von der Knickfestigkeit," vol. 1, p. 222, Düsseldorf, 1918.

way represents a lower limit, and the highest value represents an upper limit. Thus, at each step of the calculations the critical load is known to be within certain limits. A more accurate value of the critical load is obtained by using average values of the deflections, as will be shown in the examples to follow.[1]

In order to illustrate the method of successive approximations, we shall begin with the simple case of a bar with hinged ends (Fig. 2-44a) for

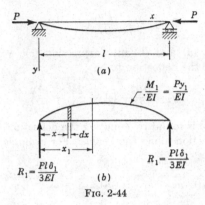

Fig. 2-44

which the exact solution is known. As a first approximation, assume that the deflection curve for the buckled bar is a parabola:

$$y_1 = \frac{4\delta_1 x(l - x)}{l^2} \tag{a}$$

This equation gives an assumed deflection curve which has zero value at the ends and maximum deflection at the center equal to δ_1. The bending moment at any section of the bar is $M_1 = Py_1$, and the deflection caused by these moments can be found readily by the conjugate-beam method.[2] The conjugate beam is shown in Fig. 2-44b and is considered to be

[1] The procedure for determining critical loads described in this article is equivalent to an integration by successive approximations of the differential equation for a buckled bar. This method of solving differential equations has been used widely. It was originated by H. A. Schwarz, "Gesammelte Werke," vol. 1, pp. 241–265. See also P. Funk, Mitt. Hauptvereines deut. Ingr. Tschechoslowaki, Nos. 21 and 22, Brünn, 1931. The application of the method to buckling problems is due to F. Engesser, Z. Österr. Ingr. u. Architek. Vereines, 1893. The graphical method is due to L. Vianello, Z. Ver. deut. Ingr., vol. 42, p. 1436, 1898. A mathematical proof of the convergence of the method was given by E. Trefftz, Z. Angew. Math. u. Mech., vol. 3, p. 272, 1923; see also the book by A. Schleusner, "Zur Konvergenz des Engesser-Vianello-Verfahrens," Berlin, 1938.

[2] See, for example, Timoshenko, "Strength of Materials," 3d ed., part I, p. 155, D. Van Nostrand Company, Inc., Princeton, N.J., 1955.

loaded with the fictitious load M_1/EI. The desired deflections are numerically equal to the bending moments in the conjugate beam. The fictitious reactions of the conjugate beam are

$$R_1 = \frac{Pl\delta_1}{3EI} \tag{b}$$

and the bending moment at any section x_1 is

$$R_1 x_1 - \int_0^{x_1} \frac{Py_1}{EI} (x_1 - x) \, dx \tag{c}$$

Substituting Eqs. (a) and (b) into expression (c), we find the second approximation for the deflection to be

$$y_2 = \frac{Pl\delta_1 x}{3EI} \left(1 - \frac{2x^2}{l^2} + \frac{x^3}{l^3} \right) \tag{d}$$

The critical load is found by equating the deflection y_2 and the deflection y_1 at some section along the beam. For example, at the center of the beam we have

$$(y_1)_{x=l/2} = \delta_1 \qquad (y_2)_{x=l/2} = \delta_2 = \frac{5Pl^2\delta_1}{48EI} \tag{e}$$

and equating these expressions gives

$$P_{cr} = \frac{48EI}{5l^2} = \frac{9.6EI}{l^2}$$

which is about 2.7 per cent smaller than the true critical load. To obtain a more accurate result, we can calculate average values of the deflections y_1 and y_2 as follows:

$$(y_1)_{av} = \frac{1}{l} \int_0^l y_1 \, dx = \frac{2}{3} \delta_1 \tag{2-51}$$

$$(y_2)_{av} = \frac{1}{l} \int_0^l y_2 \, dx = \frac{Pl^2\delta_1}{15EI} \tag{2-52}$$

Equating the average values of y_1 and y_2 gives

$$P_{cr} = \frac{10EI}{l^2}$$

which is about 1.3 per cent higher than the correct value. Finally, if it is desired to determine upper and lower bounds on P_{cr}, we need to find the maximum and minimum values of the ratio y_1/y_2. From Eqs. (a) and (d) we obtain

$$\frac{y_1}{y_2} = \frac{12EI}{Pl^2} \frac{l^2(l - x)}{l^3 - 2x^2l + x^3}$$

which has a maximum value at $x = 0$ and a minimum value at $x = l/2$. These values are

$$\left(\frac{y_1}{y_2}\right)_{max} = \frac{12EI}{Pl^2} \qquad \left(\frac{y_1}{y_2}\right)_{min} = \frac{9.6EI}{Pl^2}$$

and therefore the critical load is between the values

$$\frac{9.6EI}{l^2} < P_{cr} < \frac{12EI}{l^2}$$

The successive approximation cycle can now be repeated, using y_2 from Eq. (d) as the assumed deflection. This expression can be written in the form

$$y_2 = \frac{16\delta_2 x}{5l}\left(1 - \frac{2x^2}{l^2} + \frac{x^3}{l^3}\right)$$

where δ_2 equals the deflection at the center of the bar [see Eq. (e)]. The bending moment in the bar (Fig. 2-44a) is then Py_2, and the load on the conjugate beam is $M_2/EI = Py_2/EI$. Calculating the fictitious bending moments in the conjugate beam gives the third approximation for the deflection as

$$y_3 = \frac{8Pl^2\delta_2}{75EI}\left(3\frac{x}{l} - 5\frac{x^3}{l^3} + 3\frac{x^5}{l^5} - \frac{x^6}{l^6}\right)$$

Equating the deflections y_2 and y_3 at the center of the beam gives

$$\delta_2 = \frac{61Pl^2\delta_2}{600EI}$$

from which

$$P_{cr} = \frac{9.836EI}{l^2}$$

which is about 0.35 per cent below the correct value. If the average values of y_2 and y_3 are equated, we find that

$$P_{cr} = \frac{9.882EI}{l^2}$$

which is approximately 0.12 per cent above the true value. The ratio of the deflections is

$$\frac{y_2}{y_3} = \frac{30EI}{Pl^2}\frac{l^2(l^3 - 2x^2l + x^3)}{3l^5 - 5x^2l^3 + 3x^4l - x^5}$$

Determining maximum and minimum values of this ratio leads to the result

$$\frac{9.836EI}{l^2} < P_{cr} < \frac{10EI}{l^2}$$

Thus, by the method of successive approximations we can obtain upper and lower limits to the critical load, and the method can be continued until the results are as accurate as desired.   It is seen that values of the critical load obtained by using average values of the deflections are usually more accurate than those obtained by selecting at random the deflection at a particular section of the bar, such as the center.

Numerical Procedure.   When the bar has a cross section which varies along the span, a numerical procedure of successive approximations is useful. Instead of assuming the deflection y as some function of x, the beam is divided into segments

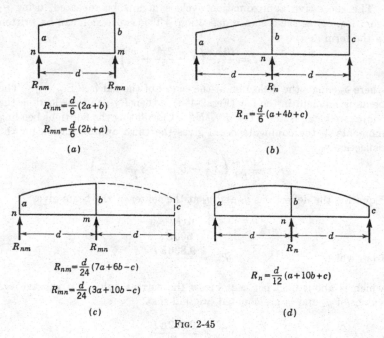

$$R_{nm}=\frac{d}{6}(2a+b)$$

$$R_{mn}=\frac{d}{6}(2b+a)$$

(a)

$$R_n=\frac{d}{6}(a+4b+c)$$

(b)

$$R_{nm}=\frac{d}{24}(7a+6b-c)$$

$$R_{mn}=\frac{d}{24}(3a+10b-c)$$

(c)

$$R_n=\frac{d}{12}(a+10b+c)$$

(d)

FIG. 2-45

and a numerical value of deflection assumed for each division point, or station, along the beam.   Then subsequent calculations are made in tabular form, calculating ordinates to the M/EI diagram and deflections in the conjugate beam at each station. Comparing the final deflections with the initially assumed values determines the critical load, as explained above.

This method will be illustrated by determining the critical load for the hinged end column[1] shown in Fig. 2-46.   Only the left-hand half of the bar is shown, since the

[1] This numerical procedure was presented in very complete form by N. M. Newmark, Numerical Procedure for Computing Deflections, Moments and Buckling Loads, *Trans. ASCE*, vol. 108, p. 1161, 1943.   This paper also gives examples of bars with other end conditions.   The method is applicable to bars with any variation in cross section and to bars with varying axial load.

column is symmetrical about the center. The ratio $I_1/I_2 = 0.4$, and the ratio $a/l = 0.6$, where a equals the length of the enlarged central portion of the bar (see Fig. 2-43b). The bar is divided into a total of 10 segments, each of length $l/10$, and the division points are designated by station numbers.

The first step is to assume a set of deflections y_1 representing a first approximation. The values selected in Fig. 2-46 are ordinates to a sine curve. For convenience in the calculations the values are multiplied by 100 and divided by δ_1, which is the deflection at the center of the bar. The common factors in each case are shown in the right-hand column. On the next line, values of M_1/EI are tabulated and represent the intensities of load on the conjugate beam at the station points. These values are equal to Py_1/EI and are expressed in terms of the common factor listed at the right.

The fictitious load on the conjugate beam is represented by an irregular load diagram and it is convenient, therefore, to replace the actual load by a series of concentrated loads acting at the station points. The values of the concentrated loads, denoted by R in the table, are computed from the formulas[1] in Fig. 2-45. If the fictitious loading (M/EI diagram) between two stations varies linearly or is assumed linear, then the formulas in Fig. 2-45a and b can be used. In these figures, d represents the distance between stations while a and b are the ordinates to the M/EI diagram. Figure 2-45b is used when the fictitious load is continuous over the station point. If the M/EI diagram has an abrupt change at the station, then the formulas of Fig. 2-45a must be used separately for the loads on either side of the station.

If the M/EI diagram is represented by a smooth curve, as is usually the case, a suitable approximation is obtained by calculating the fictitious concentrated loads on the basis of a second-degree parabola. The parabola is determined so as to pass through three consecutive points on the M/EI fictitious loading curve (see Fig. 2-45c and d) and gives a good approximation to the true curve. The formulas in Fig. 2-45c give the equivalent concentrated loads due to a distributed load between stations n and m only. Thus, these formulas are used when the load changes abruptly at the station point. The ordinate to the M/EI diagram labeled c may be an extrapolated value if, for some reason, no actual value exists. The formula in Fig. 2-45d is used when the curve is continuous over the station point.

Returning now to Fig. 2-46, the value of the concentrated load R_1 at station 1 is determined from Fig. 2-45d, and we have

$$R_1 = \frac{d}{12}(a + 10b + c)$$

$$= \frac{0.1l}{12}[0 + 10(78) + 148]\frac{P\delta_1}{100EI_2} = 7.7\frac{P\delta_1 l}{100EI_2}$$

At station 2 there is an abrupt change in the M/EI diagram and therefore Fig. 2-45c must be used for the segments on each side of station 2. The computation is as follows:

$$R_{21} = \frac{d}{24}(7a + 6b - c)$$

$$= \frac{0.1l}{24}[7(148) + 6(78) - 0]\frac{P\delta_1}{100EI_2} = 6.3\frac{P\delta_1 l}{100EI}$$

$$R_{23} = \frac{d}{24}(7a + 6b - c)$$

$$= \frac{0.1l}{24}[7(59) + 6(81) - 95]\frac{P\delta_1}{100EI_2} = 3.3\frac{P\delta_1 l}{100EI_2}$$

$$R_2 = R_{21} + R_{23} = 9.6\frac{P\delta_1 l}{100EI_2}$$

[1] *Ibid.*

At stations 3, 4, and 5 the formula of Fig. 2-45d is used again and the results are shown in the table. The value of the concentrated load at the end of the beam is not calculated, since it will have no effect on the determination of fictitious moments in the conjugate beam.

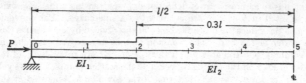

Station number	0	1	2	3	4	5	Common factor
y_1	0	31	59	81	95	100	$\dfrac{\delta_1}{100}$
M_1/EI	0	78	148 59	81	95	100	$\dfrac{P\delta_1}{100EI_2}$
R		7.7	9.6	8.0	9.4	9.9	$\dfrac{P\delta_1 l}{100EI_2}$
Average slope		39.6	31.9	22.3	14.3	4.9		$\dfrac{P\delta_1 l}{100EI_2}$
y_2	0	3.96	7.15	9.38	10.81	11.30	$\dfrac{P\delta_1 l^2}{100EI_2}$
y_1/y_2		7.83	8.25	8.64	8.79	8.85	$\dfrac{EI_2}{Pl^2}$
y_2	0	35.0	63.3	83.0	95.7	100	$\dfrac{\delta_2}{100}$
M_2/EI	0	87.5	158.2 63.3	83.0	95.7	100	$\dfrac{P\delta_2}{100EI_2}$
R		8.61	10.32	8.24	9.50	9.93	$\dfrac{P\delta_2 l}{100EI_2}$
Average slope		41.63	33.02	22.70	14.46	4.96		$\dfrac{P\delta_2 l}{100EI_2}$
y_3		4.163	7.465	9.735	11.181	11.677	$\dfrac{P\delta_2 l^2}{100EI_2}$
y_2/y_3		8.407	8.480	8.526	8.559	8.564	$\dfrac{EI_2}{Pl^2}$

FIG. 2-46

The next step is to calculate the fictitious shearing forces in the conjugate beam. These forces are constant between the stations and represent values of average slope in the actual bar. The first value given in the table represents the fictitious reaction of the conjugate beam and is obtained by the following evident calculation:

$$7.7 + 9.6 + 8.0 + 9.4 + \tfrac{1}{2}(9.9) = 39.65$$

This value is recorded in the table as the fictitious shearing force, or average slope, in the first segment of the beam. The shear in the next segment is obtained by subtracting the value of the concentrated load from the shear in the previous segment, and this process is continued until the middle of the bar is reached.

The deflections in the bar are calculated directly from the values of average slope, noting that the deflection at station 1 is equal to the value of average slope in the first segment times the distance between stations; the deflection at station 2 is equal to the deflection at 1 plus the next value of average slope times the distance between stations; etc. Finally, the ratios of the assumed deflections y_1 to the new values y_2 are determined. Considering the maximum and minimum values of the ratios, we see that the upper and lower limits for P_{cr} are

$$\frac{7.83EI_2}{l^2} < P_{cr} < \frac{8.85EI_2}{l^2}$$

To calculate an approximate value of the critical load, we replace Eqs. (2-51) and (2-52) by summations and obtain

$$(y_1)_{av} = \frac{1}{l} \sum y_1 \, \Delta x \qquad (f)$$

$$(y_2)_{av} = \frac{1}{l} \sum y_2 \, \Delta x \qquad (g)$$

The ratio of $(y_1)_{av}$ to $(y_2)_{av}$ is equal to the ratio of the sums of the deflections y_1 and y_2, since the segment length Δx is constant. In summing up the values of y_1 and y_2 we must take into account both halves of the beam. Thus for the ratio of the sums of the deflections we have

$$\frac{(y_1)_{av}}{(y_2)_{av}} = \frac{2(31 + 59 + 81 + 95) + 100}{2(3.96 + 7.15 + 9.38 + 10.81) + 11.30} \frac{EI_2}{Pl^2}$$

$$= 8.55 \frac{EI_2}{Pl^2}$$

and the critical load is approximately.

$$P_{cr} = \frac{8.55EI_2}{l^2}$$

The exact value (see Table 2-10) is $P_{cr} = 8.51EI_2/l^2$, and thus it is seen that very accurate results are obtained with only one cycle of successive approximation computations.

The results can be improved by repeating the cycle of calculations, as shown in Fig. 2-46. The second cycle begins with deflections y_2, which are proportional to the deflections y_2 found from the first set of computations. These values can be multiplied by any constant factor in order to adjust the order of magnitude of the figures. In this case they are multiplied by 100/11.3 in order to give δ_2 as the deflection at the center. The results of the second cycle show that the load P_{cr} is between the values

$$\frac{8.407EI_2}{l^2} < P_{cr} < \frac{8.564EI_2}{l^2}$$

and the value obtained by taking the ratio of the sums of the deflections is

$$P_{cr} = \frac{8.52EI_2}{l^2}$$

which is practically the same as the exact value.

Graphical Method. A graphical method of successive approximations can also be used for the calculation of critical loads. In this method the first step, as previously, is to assume a shape for the deflection curve of the buckled bar. This curve will also represent the bending-moment diagram for the bar, but to a different scale, since $M = Py$. Now considering the bending-moment diagram divided by EI as a fictitious lateral load and constructing the corresponding funicular curve, we obtain the new deflection curve. If, by adjusting the value of P, the new curve can be brought into complete coincidence with the assumed curve, this will indicate that the assumed curve is the true deflection curve and that the corresponding P is the correct value of the critical load. Usually the two curves will be different, but by adjusting the value of P we can make the deflections equal at any one point, such as at the middle of the span. In this way we obtain an approximate value for the critical load. To get a better approximation, we take the constructed funicular curve as a second approximation for the deflection curve and repeat again the same construction as above.

Instead of calculating the critical load from the condition that the deflections of the two consecutive curves at a certain point are equal, we can use average values of

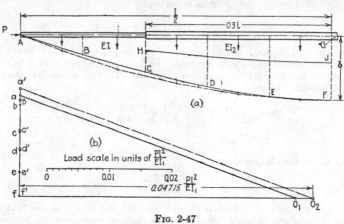

FIG. 2-47

the deflections, as before, and take the ratio of the areas under the two deflection curves. The critical load is calculated by equating this ratio to unity. Proceeding with the construction of consecutive funicular curves in the same way and calculating the critical load after each cycle, we can approximate the critical load more and more closely.[1]

To illustrate the graphical method, let us consider again the column shown in Fig. 2-46, with $I_1/I_2 = 0.4$ and $a/l = 0.6$. The graphical solution for the critical load is shown in Fig. 2-47. Since the bar is symmetrical with respect to the center G, only half of the construction is given.

A portion of a sine curve $ABCDEF$ is selected as the trial deflection curve. The bending-moment diagram for any compressive force P is the area $AB \ldots FGA$ with ordinates multiplied by P. The load for the conjugate beam is the bending-moment

[1] The mathematical proof of this statement is discussed by R. von Mises, *Monatsschr. Math. Physik*, vol. 22, p. 33, 1911, and by E. Trefftz, *Z. angew. Math. u. Mech.*, vol. 3, p. 272, 1923. See also A. Pflüger, "Stabilitäts-probleme der Elastostatik," p. 200, Springer-Verlag, Berlin, 1950.

diagram divided by EI. Therefore, the area $ACHJGA$ with ordinates multiplied by P/EI_1 is this load when the bending-moment ordinates on the middle portion of the column have been reduced by the ratio I_1/I_2.

This load is divided into sections, as shown by the dotted lines. Each section is replaced by an equal load acting at its centroid, as indicated by the arrows.[1] These loads are plotted on the load diagram $abcdef$ (Fig. 2-47b).

Point O_1 is the position of the pole of the force polygon for which the corresponding funicular polygon passes through A and horizontally through F. The curve tangent to this latter polygon is the deflection curve for the assumed bending-moment diagram. Since the two curves do not check very closely, the new curve is used for a second trial. The new load diagram is $a'b'c'd'e'f'$ and the new pole is found to be O_2. Its corresponding funicular polygon is found practically to coincide with the one previously drawn, showing that the second trial curve was very close to the actual curve of buckling.

To find the value of P necessary to keep the column in this deflected position, the deflections at the center are equated. The assumed deflection was δ, and the deflection obtained by construction is the product of the pole distance of the force polygon

$$O_2f' = 0.04715 \frac{Pl^2}{EI_1}$$

and the ordinate δ in the equilibrium polygon. Then

$$0.04715 \frac{Pl^2}{EI_1} \delta = \delta$$

and
$$P_{cr} = 21.2 \frac{EI_1}{l^2} = 8.48 \frac{EI_2}{l^2}$$

which is very close to the exact value.

In this problem a sine curve was used for the first trial curve, although it can easily be seen that, since it is the true curve of buckling for a uniform bar, it will not have sufficient curvature along the portion AC of the curve. Had the sine curve been deliberately altered to give more curvature along this portion, a satisfactory value for P_{cr} would have been obtained with only one approximation. For example, using a parabola as a trial curve, we find the first pole distance to be

$$O_1f = 0.0472 \frac{Pl^2}{EI_1}$$

from which the critical load is

$$P_{cr} = 8.48 \frac{EI_2}{l^2}$$

Thus the accuracy of this first approximation (for an assumed parabolic curve) is equal to that of the second approximation when a sine curve is used. The fact that the two curves checked very closely when starting with a parabola indicated that a second trial was unnecessary.

2.16. Bars with Continuously Varying Cross Section.

In order to decrease the weight of compression members, columns with gradually changing cross section are sometimes used. The differential equation of the deflection curve for these cases was derived by Euler, who discussed

[1] Alternatively, loads may be placed at points A, B, C, D, E, and F, as was done in the numerical solution, provided they are evaluated accordingly.

columns of various shapes, including a truncated cone and pyramid.[1] The stability of bars bounded by a surface of revolution of the second degree was discussed by Lagrange.[2]

A case of considerable practical importance, in which the moment of inertia of the cross section varies according to a power of the distance along the bar, has also been investigated.[3] Let us begin by considering a bar with the lower end built in and the upper end free (Fig. 2-48a). If the moment of inertia of the cross section varies as a power of the distance

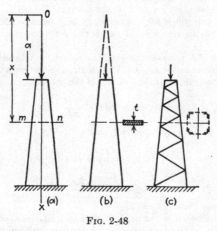

Fig. 2-48

from the fixed point O, we can express the moment of inertia at any cross section mn in the form

$$I_x = I_1 \left(\frac{x}{a}\right)^n \tag{a}$$

where I_1 is the moment of inertia at the top of the bar. By taking various values for n, we obtain various shapes of the column. When $n = 1$, we obtain the case of a column in the form of a plate of constant thickness

---

[1] A German translation of this work can be found in Ostwald's "Klassiker der exakten Wissenschaften," no. 175, Leipzig, 1910.

[2] *Loc. cit.* Other problems of this same type were discussed by A. N. Dinnik, *Phil. Mag.*, vol. 10, p. 785, 1930.

[3] See A. N. Dinnik, *Isvest. Gornogo Inst.*, Ekaterinoslav, 1914, and *Vestnik Ingenerov*, Moscow, 1916. The principal results of these papers have been translated into English: Design of Columns of Varying Cross Section, by A. N. Dinnik (translated by M. Maletz), *Trans. ASME*, vol. 51, 1929, and vol. 54, 1932. Independently the same problem was discussed by A. Ono, *Mem. Coll. Eng., Kyushu Imp. Univ.*, Fukuoka, Japan, vol. 1, 1919. See also L. Bairstow and E. W. Stedman, *Engineering*, vol. 98, p. 403, 1914, and A. Morley, *Engineering*, vol. 97, p. 566, 1914, and vol. 104, p. 295, 1917.

t and of varying width (Fig. 2-48b).   The assumption $n = 2$ represents, with sufficient accuracy, the case of a built-up column consisting of four angles connected by diagonals (Fig. 2-48c).   In this case the cross-sectional area of the column remains constant and the moment of inertia is approximately proportional to the square of the distance of the centroids of the angles from the axes of symmetry of the cross section. Finally, by taking $n = 4$, we obtain such cases as a solid truncated cone or a pyramid.

In discussing the deflection curve of the buckled bar, we shall take the coordinate axes as shown in Fig. 2-49.   Then the differential equation of the deflection curve is

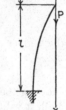

$$EI_1\left(\frac{x}{a}\right)^n \frac{d^2 y}{dx^2} = -Py \qquad (b)$$

This equation can be solved by means of Bessel functions for any value of n.   In the particular case of $n = 2$, however, the solution can be obtained in a very simple manner. For $n = 2$, Eq. (b) has the form

$$\frac{EI_1}{a^2} x^2 \frac{d^2 y}{dx^2} = -Py \qquad (c)$$

FIG. 2-49

which can be reduced to an equation with constant coefficients by the substitution

$$\frac{x}{a} = e^z \qquad (d)$$

From Eq. (d) we obtain $dz/dx = 1/x$ and therefore

$$\frac{dy}{dx} = \frac{dy}{dz}\frac{dz}{dx} = \frac{1}{x}\frac{dy}{dz} \qquad (e)$$

$$\frac{d^2 y}{dx^2} = \frac{d}{dx}\left(\frac{1}{x}\frac{dy}{dz}\right) = \frac{1}{x}\frac{d}{dx}\frac{dy}{dz} + \frac{dy}{dz}\frac{d}{dx}\frac{1}{x} = \frac{1}{x^2}\frac{d^2 y}{dz^2} - \frac{1}{x^2}\frac{dy}{dz} \qquad (f)$$

Rearranging expression (f) then gives

$$x^2 \frac{d^2 y}{dx^2} = \frac{d^2 y}{dz^2} - \frac{dy}{dz} \qquad (g)$$

which, when substituted into Eq. (c), gives the following differential equation with constant coefficients:

$$\frac{d^2 y}{dz^2} - \frac{dy}{dz} + \frac{Pa^2}{EI_1} y = 0 \qquad (h)$$

The general solution of Eq. (h) is

$$y = \sqrt{e^z}\,(A \sin \beta z + B \cos \beta z) \qquad (i)$$

where A and B are constants of integration and the quantity

$$\beta = \sqrt{\frac{Pa^2}{EI_1} - \frac{1}{4}} \qquad (j)$$

is assumed to be real and positive. Using Eq. (d), the solution (i) is expressed in the form

$$y = \sqrt{\frac{x}{a}}\left[A \sin\left(\beta \ln \frac{x}{a}\right) + B \cos\left(\beta \ln \frac{x}{a}\right)\right] \qquad (k)$$

From the condition at the upper end of the bar $(y = 0$ at $x = a)$, we find $B = 0$. At the lower end of the bar we have the condition

$$\frac{dy}{dx} = 0 \qquad \text{at } x = a + l$$

which gives

$$\tan\left(\beta \ln \frac{a+l}{a}\right) + 2\beta = 0 \qquad (l)$$

Knowing the dimensions a and l in each particular case, we can find the smallest value of β which satisfies Eq. (l) by trial and error. Then, from Eq. (j), the lowest critical load can be obtained. This value can be represented by the general formula

$$P_{cr} = \frac{mEI_2}{l^2} \qquad (2\text{-}53)$$

where I_2 is the moment of inertia at the lower end of the column $(x = a + l)$. The factor m depends on the ratio a/l only, and values are given in Table 2-11. Note that as I_1/I_2 approaches unity, the factor m approaches $\pi^2/4$.

TABLE 2-11. VALUES OF THE FACTOR m IN EQ. (2-53) FOR $n = 2$

I_1/I_2	0	0.1	0.2	0.3	0.4	0.5	0.6	0.7	0.8	0.9	1.0
m	0.250	1.350	1.593	1 763	1.904	2.023	2.128	2.223	2.311	2.392	$\pi^2/4$

In the case of a solid conical bar, we put $n = 4$ in Eq. (b), and the differential equation for buckling becomes

$$\frac{EI_1}{a^4} x^4 \frac{d^2y}{dx^2} = -Py \qquad (m)$$

If the substitution $x = 1/t$ is made, this equation can be brought to the form

$$\frac{d^2y}{dt^2} + \frac{2}{t}\frac{dy}{dt} + \frac{Pa^4}{EI_1} y = 0 \qquad (n)$$

which is a form of Bessel's differential equation and has the solution

$$y = t^{-\frac{1}{2}}[A_1 J_{-\frac{1}{2}}(\alpha t) + B_1 Y_{-\frac{1}{2}}(\alpha t)] \qquad (o)$$

where A_1 and B_1 are constants of integration, $J_{-\frac{1}{2}}(\alpha t)$ and $Y_{-\frac{1}{2}}(\alpha t)$ represent Bessel functions of the order $-\frac{1}{2}$ of the first and second kind, respectively, and

$$\alpha = \sqrt{\frac{Pa^4}{EI_1}} \qquad (p)$$

The Bessel functions of order $-\frac{1}{2}$ are expressible in the form

$$J_{-\frac{1}{2}}(\alpha t) = \frac{\cos(\alpha t)}{\sqrt{(\pi/2)\alpha t}}$$

$$Y_{-\frac{1}{2}}(\alpha t) = \frac{\sin(\alpha t)}{\sqrt{(\pi/2)\alpha t}}$$

and therefore the solution of Eq. (n) is

$$y = \frac{1}{t}\sqrt{\frac{2}{\pi\alpha}}\left[A_1 \cos(\alpha t) + B_1 \sin(\alpha t)\right]$$

and the general solution of Eq. (m) is

$$y = x\left[A\cos\frac{\alpha}{x} + B\sin\frac{\alpha}{x}\right] \qquad (q)$$

where A and B are constants of integration.

Using the conditions at the two ends of the bar gives the equation

$$\tan\frac{\alpha l}{a(a+l)} = -\frac{\alpha}{a+l}$$

or

$$\frac{\tan\gamma}{\gamma} = -\frac{a}{l} \qquad (2\text{-}54)$$

where

$$\gamma = \frac{\alpha l}{a(a+l)} = \frac{l}{a+l}\sqrt{\frac{Pa^2}{EI_1}} \qquad (2\text{-}55)$$

Equation (2-54) can be solved readily for γ, for any particular value of a/l, by using tables[1] of $(\tan x)/x$. Knowing γ, the critical load can be found from Eq. (2-55) and expressed in the form given by Eq. (2-53). Several values of m for this case are given in Table 2-12.

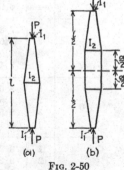

FIG. 2-50

By substituting $l/2$ for l in Eq. (2-53), we can use all the preceding results to obtain the critical load for a bar with hinged ends, symmetrical with respect to the middle cross section (Fig. 2-50a).

A more general case is obtained by combining the solution of Eq. (b) with that of the differential equation for a prismatic bar. In this way we can obtain the critical load for a bar with hinged ends having a central prismatic portion (Fig. 2-50b). The end regions may be of various shapes, corresponding to various values of the exponent n in Eq. (a). The critical load can be represented again by Eq. (2-53). Values of the factor m are given for this case in Table 2-13.

[1] See p. 32 of Addenda to Jahnke and Emde, *op. cit.*

TABLE 2-12. VALUES OF THE FACTOR m IN EQ. (2-53) FOR $n = 4$

I_1/I_2	0.1	0.2	0.3	0.4	0.5	0.6	0.7	0.8	0.9	1.0
m	1.202	1.505	1.710	1.870	2.002	2.116	2.217	2.308	2.391	$\pi^2/4$

TABLE 2-13. VALUES OF THE FACTOR m IN EQ. (2-53) FOR THE BAR IN FIG. 2-50b

I_1/I_2	n	a/l					
		0	0.2	0.4	0.6	0.8	1
0.1	1	6.48	7.58	8.63	9.46	9.82	π^2
	2	5.40	6.67	8.08	9.25	9.79	
	3	5.01	6.32	7.84	9.14	9.77	
	4	4.81	6.11	7.68	9.08	9.77	
0.2	1	7.01	7.99	8.91	9.63	9.82	π^2
	2	6.37	7.49	8.61	9.44	9.81	
	3	6.14	7.31	8.49	9.39	9.81	
	4	6.02	7.20	8.42	9.38	9.80	
0.4	1	7.87	8.59	9.19	9.70	9.84	π^2
	2	7.61	8.42	9.15	9.63	9.84	
	3	7.52	8.38	9.12	9.62	9.84	
	4	7.48	8.33	9.10	9.62	9.84	
0.6	1	8.61	9.12	9.55	9.76	9.85	π^2
	2	8.51	9.04	9.48	9.74	9.85	
	3	8.50	9.02	9.46	9.74	9.85	
	4	8.47	9.01	9.45	9.74	9.85	
0.8	1	9.27	9.54	9.69	9.83	9.86	π^2
	2	9.24	9.50	9.69	9.82	9.86	
	3	9.23	9.50	9.69	9.81	9.86	
	4	9.23	9.49	9.69	9.81	9.86	
1.0		π^2	π^2	π^2	π^2	π^2	π^2

When various values are taken for the ratios I_1/I_2 and a/l and various values of the number n, a variety of cases of practical importance can be solved by using this table.[1]

Column with Distributed Axial Load. If a column of variable cross section is submitted to the action of a distributed axial load, the differential equation of the deflec-

[1] This table was calculated by Dinnik, *loc. cit.* Examples illustrating the use of the table are given in that paper and also in Timoshenko, "Strength of Materials," 3d ed., part II, p. 169, D. Van Nostrand Company, Inc., Princeton, N.J., 1956.

tion curve of the buckled column can always be integrated by using Bessel functions, provided the flexural rigidity and the intensity of the distributed load can be represented by the equations

$$EI = EI_2 \left(\frac{x}{l}\right)^n \qquad q = q_2 \left(\frac{x}{l}\right)^p \tag{r}$$

where I_2 and q_2 are the moment of inertia and the intensity of load at the lower built-in

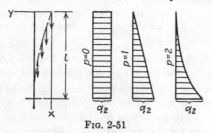

FIG. 2-51

end of the column (see Fig. 2-51). The critical value of the compressive force will always be given by the equation

$$P_{cr} = \int_0^l q_2 \left(\frac{x}{l}\right)^p dx = \frac{mEI_2}{l^2} \tag{2-56}$$

Several values of the factor m for various values of n and p in Eqs. (r) are given[1] in Table 2-14.

TABLE 2-14. VALUES OF THE FACTOR m IN EQ. (2-56) FOR COLUMN
IN FIG. 2-51

n \ p	0	1	2	3	4	5
0	7.84	16.1	27.3	41.3		
1	5.78	13.0	23.1	36.1	52.1	
2	3.67	9.87	18.9	30.9	45.8	63.6
3		6.59	14.7	25.7	39.5	

By substituting $l/2$ for l in Eq. (2-56), we obtain the critical compressive force for a bar with hinged ends which is symmetrical (and symmetrically loaded) with respect to the middle cross section.

Column of Minimum Weight. It is sometimes of practical interest to find the shape of a solid column such that its weight, for a given value of the critical load, will be a minimum. Lagrange was the first to undertake the solution to this problem.[2] He stated that the problem was to find the curve which, by its revolution about an axis in its plane, determines the column of greatest efficiency. His conclusion (incorrect) was that the most efficient column is a column of constant circular cross section. A further investigation of the same problem was made by Clausen.[3] He left the form

---

[1] This table was calculated by Dinnik, *loc. cit.*

[2] Lagrange, *loc. cit.*

[3] *Bull. phys.-math. acad.*, St. Petersburg, vol. 9, pp. 368–379, 1851. See also E. L. Nicolai, *Bull. Polytech. Inst., St. Petersburg*, vol. 8, p. 255, 1907, and H. Blasius, *Z. Math. u. Physik*, vol. 62, pp. 182–197, 1914.

of the cross sections undetermined and assumed only that they were similar and similarly placed. The result of his investigation was that the most efficient column has a volume $\sqrt{3}/2$ times the volume of the cylindrical column of the same strength. A. Ono[1] also arrived at approximately the same result by using various values of the exponent n in Eq. (b). He found that the value of n for a minimum volume of the bar was $n = 0.93$ and that the corresponding volume is 87 per cent of the volume of a prismatic bar of the same strength.

When exact results are not available, the method of successive approximations described in the preceding article can be used to determine critical loads for columns of varying cross section.

2.17. The Effect of Shearing Force on the Critical Load. In the preceding derivations of the equations for the critical loads, we used the differential equation of the deflection curve in which the effect of shearing force on the deflection was neglected. When buckling occurs, however, there will be shearing forces acting on the cross sections of the bar. The effect of these forces on the critical load will now be discussed for the

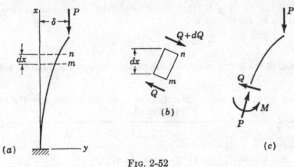

Fig. 2-52

column shown in Fig. 2-52a. The shearing forces Q acting on an element of length dx, between two cross sections m and n, are shown in Fig. 2-52b. The magnitude of this shearing force[2] (see Fig. 2-52c) is

$$Q = P \frac{dy}{dx} \qquad (a)$$

The change in slope of the deflection curve produced by the shearing force is nQ/AG, where A is the total cross-sectional area of the column, G the modulus in shear, and n a numerical factor depending on the shape of the cross section. For a rectangular cross section the factor $n = 1.2$, and for a circular cross section $n = 1.11$. For an I beam bent about the minor axis of the cross section (that is, bent in the plane of the flanges) the factor $n \approx 1.2A/A_f$, where A_f is the area of the two flanges. This

[1] Ono, *loc. cit.*

[2] Note that the shearing forces Q act on cross sections which are normal to the axis of the bar.

value lies within the range 1.4 to 2.8 for the usual I beam and plate girder sections. If an I beam bends in the plane of the web (about the major axis) the factor $n \approx A/A_w$, where A_w is the area of the web. For this case, values of n from 2 to 6 are typical for rolled steel sections.

The rate of change of slope produced by the shearing force Q represents the additional curvature due to shear and is equal to

$$\frac{n}{AG}\frac{dQ}{dx} = \frac{nP}{AG}\frac{d^2y}{dx^2}$$

The total curvature of the deflection curve is now obtained by adding the curvature produced by the shearing force to the curvature produced by the bending moment. Then, for the column in Fig. 2-52, the differential equation of the deflection curve becomes

$$\frac{d^2y}{dx^2} = \frac{P(\delta - y)}{EI} + \frac{nP}{AG}\frac{d^2y}{dx^2}$$

or
$$\frac{d^2y}{dx^2} = \frac{P}{EI(1 - nP/AG)}(\delta - y) \tag{b}$$

This equation differs from Eq. (2-1) only by the factor $(1 - nP/AG)$ in the denominator on the right-hand side. Proceeding as in Art. 2.1, we obtain for the critical value of the load P the equation

$$\frac{P}{EI(1 - nP/AG)} = \frac{\pi^2}{4l^2}$$

from which
$$P_{cr} = \frac{P_e}{1 + nP_e/AG} \tag{2-57}$$

where $P_e = \pi^2 EI/4l^2$ represents the Euler critical load for this case. Thus, owing to the action of shearing forces, the critical load is diminished in the ratio[1]

$$\frac{1}{1 + nP_e/AG} \tag{c}$$

This ratio is very nearly equal to unity for solid columns, such as a column of rectangular cross section or a column with I cross section. Hence in these cases the effect of shearing force can usually be neglected. For built-up columns consisting of struts connected by lacing bars or batten plates, the shear effect may become of practical importance and is considered further in Art. 2.18. A graph of Eq. (2-57) is given later in Fig. 2-54.

[1] This result was obtained first by F. Engesser, *Zentr. Bauverwaltung*, vol. 11, p. 483, 1891. See also F. Nussbaum, *Z. Math. u. Physik*, vol. 55, p. 134, 1907. A refinement to expression (c), taking into account Poisson's ratio, was given by R. Gran Olsson, Det Kongelige Norske Videnskabers Selskab, *Forhandlinger*, vol. 10, no. 21, p. 79, 1937 (in German).

By using the appropriate differential equation of the deflection curve, analogous to Eq. (b), the effect of shear can be investigated for compressed bars with any other end conditions.

Energy Method. When the effect of shearing forces on the critical load is considered, the energy method can also be used. Take, as an example, a bar with hinged ends (Fig. 2-36). The general expression for the deflection curve is

$$y = a_1 \sin \frac{\pi x}{l} + a_2 \sin \frac{2\pi x}{l} + a_3 \sin \frac{3\pi x}{l} + \cdots$$

Taking into account the strain energy of shear, we find

$$\Delta U = \int_0^l \frac{M^2 \, dx}{2EI} + \int_0^l \frac{nQ^2}{2AG} \, dx = \frac{P^2}{2EI} \int_0^l y^2 \, dx + \frac{nP^2}{2AG} \int_0^l \left(\frac{dy}{dx}\right)^2 \, dx$$

$$= \frac{P^2 l}{4EI} \sum_{m=1}^{m=\infty} a_m{}^2 + \frac{n\pi^2 P^2}{4AGl} \sum_{m=1}^{m=\infty} m^2 a_m{}^2 \quad (d)$$

The work done by the forces P during buckling is

$$\Delta T = \frac{\pi^2 P}{4l} \sum_{m=1}^{m=\infty} m^2 a_m{}^2 \tag{e}$$

Substituting in Eq. (2-31), we obtain

$$P = \frac{\pi^2 EI}{l^2} \frac{\Sigma m^2 a_m{}^2}{\sum \left(1 + \dfrac{n\pi^2 EI m^2}{AGl^2}\right) a_m{}^2} \tag{f}$$

The smallest value of P is obtained by taking only the first term in the series of expression (f). Then

$$P_{cr} = \frac{\pi^2 EI}{l^2} \frac{1}{1 + nP_e/AG} \tag{2-58}$$

in which

$$P_e = \frac{\pi^2 EI}{l^2}$$

Thus, owing to shear, the critical load is diminished in the ratio (c), provided the Euler load P_e is taken for a bar with hinged ends.

Modified Shear Equation. In the preceding discussion of the effect of shear, Eq. (a) was used for the evaluation of the shearing force Q. Another expression can be obtained by considering the deformation of an element mn (Fig. 2-53) cut out from the column shown in Fig. 2-52. The angle $d\theta$ in the figure represents the change in slope due to the bending moment $M = P(\delta - y)$, with θ measured from the direction of the x axis (vertical) to the normal N to the cross section. Due to shearing strains γ, there is an additional slope measured from the normal N to the tangent T to the axis of the deflected column. Thus the slope of the deflected curve is

$$\frac{dy}{dx} = \theta + \gamma = \theta + \frac{nQ}{AG} \tag{g}$$

The axial force P has a component in the direction N equal to $P \cos \theta \approx P$ and a component $Q = P \sin \theta \approx P\theta$. Substituting in Eq. (g) the slope becomes

$$\frac{dy}{dx} = \theta + \frac{nP\theta}{AG} = \theta\left(1 + \frac{nP}{AG}\right) \tag{h}$$

Observing that $d\theta/dx = M/EI = P(\delta - y)/EI$ we obtain from Eq. (h) the following expression for curvature:

$$\frac{d^2y}{dx^2} = \frac{P(\delta - y)}{EI}\left(1 + \frac{nP}{AG}\right) \tag{i}$$

The difference between Eq. (i) and the previous Eq. (b) is due to the fact that in the derivation of Eq. (b) the shear force is calculated from the total slope dy/dx of the deflection curve [see Eq. (a)] whereas in the derivation of Eq. (i) only the angle of rotation of the cross section is used. Solving this equation in the same manner as before (see Art. 2.1) we find that the critical load is

$$P_{cr} = \frac{\sqrt{1 + 4nP_e/AG} - 1}{2n/AG} \tag{2-59}$$

where $P_e = \pi^2 EI/4l^2$. The differences in the results obtained from Eqs. (2-57) and (2-59) are negligible for solid columns. However, for cases in which the effect of

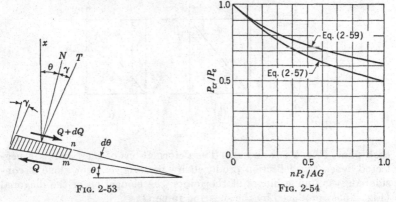

FIG. 2-53   FIG. 2-54

shear is unusually large, as in the buckling of helical springs (see Art. 2.19), Eq. (2-59) may be more accurate, although Eq. (2-57) is more on the safe side. For a column with hinged ends Eq. (2-59) can also be used, provided P_e is taken as the Euler critical load for a bar with hinged ends. A graph of Eqs. (2-57) and (2-59) is given in Fig. 2-54 for comparison.

2.18. Buckling of Built-up Columns. *Laced Column.* The critical load for a laced column is always less than for a solid column having the same cross-sectional area and the same slenderness ratio l/r. This decrease in the critical load is due primarily to the fact that the effect of shear on deflections is much greater for a laced column than for a solid bar. The actual value of the critical load depends upon the detailed arrangement and dimensions of the lacing bars.

If the laced column (Fig. 2-55a) has a large number of panels, Eq. (2-57), derived for a solid bar, can be adapted to the calculation of the critical load.   We can write Eq. (2-57) in the form

$$P_{or} = \frac{P_e}{1 + P_c/P_d} \tag{2-60}$$

where P_e is the Euler critical load and the quantity $1/P_d$ for a laced
column corresponds to n/AG for a solid bar.   Thus the factor $1/P_d$ is
the quantity by which the shearing force Q is multiplied in order to obtain
the additional slope γ of the deflection curve due to shear.   Thus we have

$$\gamma = \frac{Q}{P_d} \tag{a}$$

and to determine the quantity $1/P_d$ in any particular case we must
investigate the lateral displacements produced by the shearing force.

Consider first the laced column shown in Fig. 2-55a.   The shear displacement is due to the lengthening and shortening of the lacing bars in

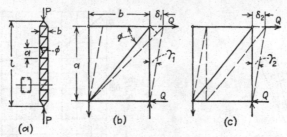

Fig. 2-55

each panel (Fig. 2-55b and c).   The deformations of these bars are neglected when the deflection produced by bending moment alone is considered.   Assuming hinges at the joints, the elongation of the diagonal
(Fig. 2-55b) produced by the shearing force Q is

$$\frac{Qa}{A_d E \sin \phi \cos \phi}$$

in which ϕ is the angle between the diagonal and horizontal bars, $Q/\cos \phi$
is the tensile force in the diagonal, $a/\sin \phi$ is the length of the diagonal,
and A_d is the cross-sectional area of two diagonals, one on each side of
the column (Fig. 2-55a).   The corresponding lateral displacement is

$$\delta_1 = \frac{Qa}{A_d E \sin \phi \cos^2 \phi} \tag{b}$$

Considering next the shortening of the battens, or horizontal lacing bars
(Fig. 2-55c), we find that the corresponding lateral displacement is

$$\delta_2 = \frac{Qb}{A_b E} \tag{c}$$

where b is the length of the battens between hinges and A_b is the cross-sectional area of two battens, one on each side of the column.

The total angular displacement produced by the shearing force Q is obtained from Eqs. (b) and (c) as follows:

$$\gamma = \frac{\delta_1 + \delta_2}{a} = \frac{Q}{A_d E \sin \phi \cos^2 \phi} + \frac{Qb}{aA_b E} \tag{d}$$

From Eq. (a) we then obtain

$$\frac{1}{P_d} = \frac{1}{A_d E \sin \phi \cos^2 \phi} + \frac{b}{aA_b E} \tag{e}$$

When this value is substituted in Eq. (2-60), the critical load for a strut with hinged ends (Fig. 2-55a) is[1]

$$P_{cr} = \frac{\pi^2 E I}{l^2} \frac{1}{1 + \dfrac{\pi^2 E I}{l^2} \left(\dfrac{1}{A_d E \sin \phi \cos^2 \phi} + \dfrac{b}{aA_b E} \right)} \tag{2-61}$$

In this expression I is the moment of inertia of the cross section of the strut. If the cross-sectional areas A_b and A_d are small in comparison with the area of the channels (Fig. 2-55a) or other main members, the critical load from Eq. (2-61) may be considerably lower than the Euler value. Thus the laced column may be considerably weaker than a solid strut with the same EI, but since the amount of material used is less, the laced column may be more economical.

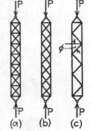

Fig. 2-56

An equation similar to Eq. (2-61) can be obtained when there are two diagonal lacing bars in each panel (Fig. 2-56a). Under the action of a shearing force, one diagonal is stressed in tension and the other in compression. The battens do not take part in the transmission of shearing force, and the system is equivalent to that shown in Fig. 2-56b. The critical load is obtained in this case from Eq. (2-61) by omitting the term containing A_b and doubling the cross-sectional area A_d. Thus we have

$$P_{cr} = \frac{\pi^2 E I}{l^2} \frac{1}{1 + \dfrac{\pi^2 E I}{l^2} \dfrac{1}{A_d E \sin \phi \cos^2 \phi}} \tag{2-62}$$

[1] Equation (2-61) was derived first by F. Engesser, *Zentr. Bauverwaltung*, vol. 11, p. 483, 1891. For a further discussion of the same problem, in connection with the collapse of the Quebec Bridge, see the following papers: F. Engesser, *ibid.*, vol. 27, p. 609, 1907, and *Z. Ver. deut. Ingr.*, p. 359, 1908; L. Prandtl, *Z. Ver. deut. Ingr.*, 1907; and Timoshenko, Buckling of Bars of Variable Cross Section, *Bull. Polytech. Inst., Kiev*, 1908.

in which A_d denotes the cross-sectional area of four diagonals, two on each side of the column in the same panel. Equation (2-62) can be used also in the case of a single system of diagonal bars (Fig. 2-56c), provided A_d is the area of two diagonals and ϕ is measured as shown.

Columns with Batten Plates. In the case of a strut made with battens only, as in Fig. 2-57, to obtain the lateral displacement produced by the shearing force Q, we must consider the deformation of an element of the strut between the sections mn and m_1n_1. Assuming that the deflection curves of the channels have points of inflection at these sections, the bending of the element will be as shown in Fig. 2-57b. The lateral

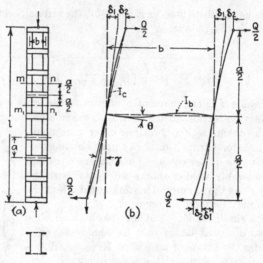

Fig. 2-57

deflection consists of the sum of the displacements δ_1, due to bending of the batten, and δ_2, due to bending of the channels. There are couples $Qa/2$ acting at the ends of the batten, and the angle of rotation at each end of the batten is

$$\theta = \frac{Qab}{12EI_b}$$

where b is the length of the batten and EI_b is its flexural rigidity. The lateral displacement δ_1 produced by this bending of the batten is

$$\delta_1 = \frac{\theta a}{2} = \frac{Qa^2b}{24EI_b} \tag{f}$$

The displacement δ_2 can be found from the expression for the deflection of a cantilever beam. Thus we obtain

$$\delta_2 = \frac{Q}{2}\left(\frac{a}{2}\right)^3 \frac{1}{3EI_c} = \frac{Qa^3}{48EI_c} \tag{g}$$

in which EI_c is the flexural rigidity of one of the vertical channels. The total angular displacement produced by the shearing force Q is

$$\gamma = \frac{\delta_1 + \delta_2}{\frac{1}{2}a} = \frac{Qab}{12EI_b} + \frac{Qa^2}{24EI_c} \tag{h}$$

and from Eq. (a), we obtain

$$\frac{1}{P_d} = \frac{ab}{12EI_b} + \frac{a^2}{24EI_c}$$

Substituting $1/P_d$ into Eq. (2-60), we obtain

$$P_{cr} = \frac{\pi^2 EI}{l^2} \frac{1}{1 + \frac{\pi^2 EI}{l^2}\left(\frac{ab}{12EI_b} + \frac{a^2}{24EI_c}\right)} \tag{2-63}$$

in which the factor $\pi^2 EI/l^2$ represents the critical load for the entire column calculated as for a solid column. It is seen that when the flexural rigidity of the battens is small, the critical load is much lower than that given by Euler's formula.[1]

In the calculation of the angular displacement γ, the shear in the batten can be taken into consideration also. From Fig. 2-57b it can be seen that the shearing force in the batten is

$$\frac{Qa}{b}$$

and the corresponding shearing strain is

$$\frac{nQa}{bA_bG} \tag{i}$$

where A_b is the cross-sectional area of two battens and n equals 1.2, since the battens have a rectangular cross section. Adding expression (i) to the previous Eq. (h), we obtain

$$P_{cr} = \frac{\pi^2 EI}{l^2} \frac{1}{1 + \frac{\pi^2 EI}{l^2}\left(\frac{ab}{12EI_b} + \frac{a^2}{24EI_c} + \frac{na}{bA_bG}\right)} \tag{2-64}$$

instead of Eq. (2-63).

[1] Another equation for the critical load of a column with batten plates is given on p. 151.

If the vertical channels of the built-up column represented in Fig. 2-57 are very flexible, or if the distance between battens is large, collapse of the column may occur as a result of local buckling of the channels between two consecutive battens. To take this possibility of buckling into account, let us consider an element of the column between two battens and assume that it is buckled as shown in Fig. 2-58. Assuming that the rigidity of the battens is very large, the critical value of the compressive force at which the assumed buckling will occur (see Fig. 2-4) is

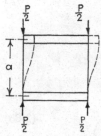

FIG. 2-58

$$P = \frac{2\pi^2 E I_c}{a^2} \qquad (j)$$

The effect of the axial load $P/2$ on the bending of the vertical channels can now be taken into account by writing Eq. (g) in the form [see Eq. (a), Art. 1.11]

$$\delta_2 = \frac{Q a^3}{48 E I_c} \frac{1}{1 - \alpha} \qquad (k)$$

where

$$\alpha = \frac{P_{cr}}{2\pi^2 E I_c / a^2} \qquad (l)$$

Using expression (k) for δ_2, we express the critical load P_{cr} for the strut in Fig. 2-57 in the form

$$P_{cr} = \frac{\pi^2 E I}{l^2} \frac{1}{1 + \frac{\pi^2 E I}{l^2}\left[\frac{ab}{12 E I_b} + \frac{a^2}{24 E I_c}\frac{1}{(1 - \alpha)} + \frac{na}{b A_b G} \right]} \qquad (2\text{-}65)$$

Since α depends upon P_{cr}, this equation can be solved only by trial and error. It should be noticed also that the critical load for the column between battens (Fig. 2-58) is always less than given by Eq. (j), inasmuch as the battens are not rigid. This means that the true value of α is larger than given by Eq. (l), and hence the true critical load is less than obtained from Eq. (2-65). However, these differences are not of practical significance, since the term in the denominator of Eq. (2-65) containing I_c is usually small compared with the term containing I_b.

Experiments have been made[1] which show satisfactory agreement with Eq. (2-65) in all cases in which the number of panels (Fig. 2-57a) is larger than six. The design of built-up columns is discussed in Art. 4.6.

Columns with Perforated Cover Plates.[2] The cross section of a typical column with perforated cover plates is shown in Fig. 2-59a. In the calculation of the cross-sectional area and moment of inertia of the col-

---

[1] See Timoshenko, *Ann. ponts et chaussées*, series 9, vol. 3, p. 551, 1913.

[2] A very complete report concerning columns of this type was made by M. W. White and B. Thürlimann, Study of Columns with Perforated Cover Plates, *AREA Bull.* 531, 1956.

umn, the properties of the net area (section nn) can be used with sufficient accuracy for most practical purposes. In determining the lateral displacement due to shearing force Q, we again consider an element from the column (Fig. 2-59b). This element is similar to the element from the column with batten plates (Fig. 2-57b), except that instead of a narrow batten plate we have the portion of the cover plate between perforations. Thus we finally obtain the idealized element of Fig. 2-59c, where the horizontal cross member can be considered as infinitely rigid. The lengths of the vertical projections, which are treated as cantilever beams,

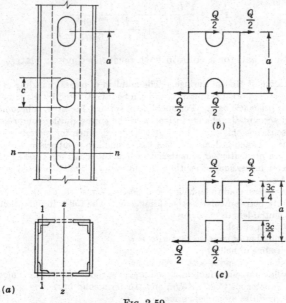

FIG. 2-59

will be somewhere between $c/2$ and $a/2$, where c is the length of a perforation. The value $3c/4$ is reasonable and gives results which agree with experiments.[1]

The equations for a column with batten plates can now be modified for this case. Since the cross member (analogous to a batten) is infinitely rigid, we can substitute $I_b = \infty$ in Eq. (f) and obtain $\delta_1 = 0$. The displacement δ_2 is determined as the deflection of a cantilever [Eq. (g)], and we obtain

$$\delta_2 = \frac{Q}{2}\left(\frac{3c}{4}\right)^3 \frac{1}{3EI_f} = \frac{9Qc^3}{128EI_f} \qquad (m)$$

[1] *Ibid.*

where I_f represents the moment of inertia of the "flange" of the column, that is, the entire effective area of the column on one side of the z axis, taken about the centroid of the flange (axis 1-1). The angular displacement due to Q is

$$\gamma = \frac{\delta_1 + \delta_2}{\frac{1}{2}a} = \frac{9Qc^3}{64aEI_f}$$

and therefore

$$\frac{1}{P_d} = \frac{9c^3}{64aEI_f} \tag{n}$$

from which we obtain

$$P_{cr} = \frac{\pi^2 EI}{l^2} \frac{1}{1 + \dfrac{\pi^2 EI}{l^2}\left(\dfrac{9c^3}{64aEI_f}\right)} \tag{2-66}$$

as the critical load for a column with perforated cover plates.

2.19. Buckling of Helical Springs. The problem of lateral buckling of compressed helical springs[1] is of practical interest and can be investigated by the same methods as were used for prismatic bars. For buckling of springs, however, it becomes necessary to take into account the change in length of the spring during compression. In all previous discussions this effect was not considered, since for materials such as steel and wood the change in length of the bar is small in comparison with the original length. For a prismatic bar of material such as rubber or for a spring, the length of the bar after compression rather than the initial length must be used in determining the critical load.

The spring is assumed to be helical and close coiled, so that each coil lies in a plane which is nearly perpendicular to the spring axis. The following notation is used:

l_0 = initial length of spring
h_0 = pitch of helix
n_0 = number of coils, so that $n_0 h_0 = l_0$
R = radius of helix
l = length of spring after compression
$\alpha_0, \beta_0, \gamma_0$ = flexural, shearing, and compressive rigidities of the unloaded spring (analogous to EI, AG/n, and AE for a solid bar)
α, β, γ = rigidities of loaded spring

The number of coils per unit length of spring increases by the factor l_0/l when the spring is compressed. Hence the rigidities decrease by the factor l/l_0, and we have

$$\alpha = \alpha_0 \frac{l}{l_0} \qquad \beta = \beta_0 \frac{l}{l_0} \qquad \gamma = \gamma_0 \frac{l}{l_0} \tag{a}$$

[1] The first investigation in this field was made by E. Hurlbrink, *Z. Ver. deut. Ingr.*, vol. 54, p. 138, 1910. The problem was further discussed by R. Grammel, *Z. angew. Math. u. Mech.*, vol. 4, p. 384, 1924; C. B. Biezeno and J. J. Koch, *ibid.*, vol. 5, p. 279, 1925; and H. Ziegler, *Ingr.-Arch.*, vol. 10, p. 227, 1939. The final form of the buckling equations was obtained by J. A. Haringx, *Proc. Konink. Ned. Akad. Wetenschap.*, vol. 45, p. 533, 1942 (in English). A very complete presentation of the buckling of springs was given by Haringx, *Philips Research Repts.*, vol. 3, 1948, and vol. 4, 1949, published by Philips Research Laboratories, Eindhoven, Netherlands.

The effect of shearing force is important in determining the critical load for a spring, and Eq. (2-59) will be used.   For a spring with hinged ends we have

$$P_e = \frac{\pi^2 \alpha}{l^2} \tag{b}$$

and substitution into Eq. (2-59) gives

$$P_{cr} = \frac{\sqrt{1 + (4\pi^2 \alpha/l^2\beta)} - 1}{2/\beta}$$

Next, substituting Eqs. (a), we obtain

$$P_{cr} = \frac{\sqrt{1 + (4\pi^2\alpha_0/l^2\beta_0)} - 1}{2l_0/\beta_0 l} \tag{c}$$

From a consideration of the compression of the spring, we have

$$\frac{l_0 - l}{l_0} = \frac{P_{cr}}{\gamma_0} \quad \text{or} \quad l = l_0 \left(1 - \frac{P_{cr}}{\gamma_0} \right) \tag{d}$$

Substituting Eq. (d) for l into Eq. (c), we obtain the following equation for the critical load:

$$\left(\frac{P_{cr}}{\gamma_0} \right)^2 \left(\frac{\gamma_0}{\beta_0} - 1 \right) + \frac{P_{cr}}{\gamma_0} - \frac{\pi^2 \alpha_0}{l_0^2 \gamma_0} = 0$$

from which

$$\frac{P_{cr}}{\gamma_0} = \frac{1 \pm \sqrt{1 - \dfrac{4\pi^2 \alpha_0}{l_0^2 \gamma_0} \left(1 - \dfrac{\gamma_0}{\beta_0} \right)}}{2 \left(1 - \dfrac{\gamma_0}{\beta_0} \right)} \tag{2-67}$$

Since only the lower value of P_{cr}/γ_0 is of practical interest, the minus sign should be used in Eq. (2-67).   The compressed length of the buckled spring can now be found from Eq. (d).

For a spring with wire of circular cross section the flexural and shearing rigidities are[1]

$$\alpha_0 = \frac{EIl_0}{\pi R n_0} \frac{1}{1 + E/2G} \tag{e}$$

$$\beta_0 = \frac{EIl_0}{\pi R^3 n_0} \tag{f}$$

where I is the moment of inertia of the cross section of the circular wire about a diameter.   The compressive rigidity is[2]

$$\gamma_0 = \frac{GIl_0}{\pi R^3 n_0} \tag{g}$$

Substituting Eqs. (e) to (g) in Eq. (2-67) and also taking $E/G = 2.6$, which corresponds to Poisson's ratio of 0.3, we obtain the following expression for the critical load:

$$\frac{P_{cr}}{\gamma_0} = 0.8125 \left[1 \pm \sqrt{1 - 27.46 \left(\frac{R}{l_0} \right)^2} \right] \tag{2-68}$$

[1] See Timoshenko, "Strength of Materials," 3d ed., part II, Eq. (264), p. 297, and Eq. (o), p. 298, D. Van Nostrand Company, Inc., Princeton, N.J., 1956.
[2] Ibid., part I, Eq. (162), p. 293, 1955.

Figure 2-60 shows a graph of P_{cr}/γ_0 as a function of l_0/R. The solid-line curve was obtained using the minus sign in Eq. (2-68), and the dotted curve was obtained using the plus sign. The graph shows that there is a critical value $(l_0/R = 5.24)$ below which the spring will not buckle. These calculations are in good agreement with experiments, providing the number of coils is not small and the coils do not touch before buckling occurs.[1]

2.20. Stability of a System of Bars. Several problems dealing with buckling of built-up columns were discussed in Article 2.18 on the basis of certain simplifying assumptions. To obtain a more satisfactory solution of these problems, an application of the general theory of stability of a system of elastic bars is necessary.[2] Let us begin with a consideration of trusses which have hinged joints and consider as a first problem the simple case of a system consisting of only two bars (Fig. 2-61), such that under the action of a vertical load P the vertical bar of the system is compressed and there is no force in the inclined bar.

Assuming that the vertical bar is absolutely rigid but the inclined bar remains elastic, the critical value of the compressive force P can be obtained easily by using the

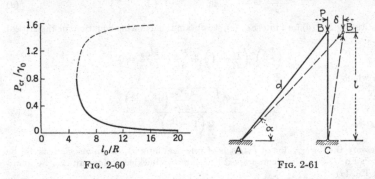

FIG. 2-60   FIG. 2-61

energy method or the equilibrium method (see p. 84). Using the second method and assuming that under the action of the vertical load the system may have a form of equilibrium as indicated by the dashed lines, we must determine the magnitude of the load necessary to keep the system in equilibrium in such a displaced position. If δ is the small displacement of joint B, the tensile force in the inclined bar is $A_d E \delta \cos \alpha/d$, where d is the length of the inclined bar, α its angle of inclination, and A_d its cross-sectional area. The equation of static equilibrium of joint B_1 (Fig. 2-61) in the hori-

---

[1] For the results of experiments, as well as an analysis considering other end conditions of the spring and other shapes of cross section of the spring wire, see Haringx, *loc. cit.*

[2] Such a theory was developed by R. von Mises *Z. angew. Math. u. Mech.*, vol. 3, p. 407, 1923. It was applied to various cases of laced columns by R. von Mises and J. Ratzersdorfer, *ibid.*, vol. 5, p. 218, 1925, and vol. 6, p. 181, 1926. Other cases were discussed by H. Müller-Breslau, "Die Neueren Methoden der Festigkeitslehre und der Statik der Baukonstruktionen," 4th ed., p. 398, Leipzig, 1913, and 5th ed., p. 380, Leipzig, 1924; L. Mann, *Z. Bauwesen*, vol. 59, p. 539, 1909; K. Ljungberg, *Der Eisenbau*, p. 100, 1922; M. Grüning, "Die Statik des ebenen Tragwerkes," Berlin, 1925; Wilhelm Wenzel, "Über die Stabilität des Gleichgewichtes ebener elastischer Stabwerke," Dissertation, University of Berlin, 1929.

zontal direction is

$$A_d E \frac{\delta \cos^2 \alpha}{d} = P \frac{\delta}{l}$$

from which $\qquad\qquad P_{cr} = A_d E \sin \alpha \cos^2 \alpha \qquad\qquad\qquad$ (a)

To obtain a more accurate solution of the problem, the elasticity of the vertical bar should be considered also. Under the action of the load P this bar will be compressed. There is no force in the inclined bar if we base our calculations on the initial unde-formed configuration of the system, as is usually done in calculating forces in bars. Therefore any compression of the vertical bar produces lateral displacement of the system, and joint B begins to move laterally at the beginning of loading. In such a case there does not exist a definite critical value of the load P at which lateral dis-placement becomes possible. To eliminate the necessity of considering the above lateral displacement, let us assume the vertical bar to be compressed by the load P first; only after this deformation is the inclined bar attached to joint B. Thus we finally have the bar BC in a vertical position with the compressive force P acting on it and the inclined bar free from stresses. In calculating the critical value of P, we proceed as before and assume a small lateral displacement of joint B. Owing to this displacement, a force in the inclined bar and a change in compressive force in the vertical bar will be produced. Hence, any lateral displacement δ of joint B will be accompanied by a vertical displacement δ_1 of the same joint, due to change in com-pression of the vertical bar. If X denotes the tensile force produced in the inclined bar, the corresponding increase in the compressive force in the vertical bar is $X \sin \alpha$ and its shortening is

$$\delta_1 = \frac{Xl \sin \alpha}{A_v E}$$

in which A_v is the cross-sectional area of the vertical bar. The total elongation of the inclined bar is equal to $\delta \cos \alpha - \delta_1 \sin \alpha$. Then the force X is found from the equation

$$\frac{Xd}{A_d E} = \delta \cos \alpha - \frac{Xl \sin^2 \alpha}{A_v E}$$

from which $\qquad\qquad X = \dfrac{A_d E \delta \cos \alpha}{d[1 + (A_d/A_v) \sin^3 \alpha]}$

Writing the equation of equilibrium of joint B as

$$X \cos \alpha = \frac{P\delta}{l}$$

and substituting the above expression for X, we obtain

$$P_{cr} = \frac{A_d E \sin \alpha \cos^2 \alpha}{1 + (A_d/A_v) \sin^3 \alpha} \qquad\qquad (b)$$

Comparing this with expression (a), obtained before, it is seen that the effect of com-pression of the vertical bar is given by the second term in the denominator of formula (b).

The critical load calculated from Eq. (b) will be of practical interest only when it is smaller than the critical load for the vertical member considered as a bar with hinged ends, since otherwise the system will fail owing to buckling of the vertical bar, and not as a result of the lateral displacement shown in Fig. 2-61. Thus we can write the equation

$$\frac{A_d E \sin \alpha \cos^2 \alpha}{1 + (A_d/A_v) \sin^3 \alpha} \lessgtr \frac{\pi^2 E A_v r_v^2}{l^2}$$

in which l/r_v is the slenderness ratio of the vertical bar.   Assuming that E is the same for both bars, the above equation can be written in the following form:

$$A_d \sin \alpha \cos^2 \alpha \lessgtr \frac{\pi^2 A_v r_v^2}{l^2} + \frac{A_d r_v^2 \pi^2 \sin^3 \alpha}{l^2}$$

Remembering that $(r_v/l)^2$ is usually a very small quantity, it can be concluded that the kind of instability represented in Fig. 2-61 can occur only if A_d is very small in comparison with A_v or if the angle α is very small.   In both cases the second term in the denominator of formula (b) is very small and can be neglected; thus formula (a) is sufficiently accurate for practical purposes.

In the foregoing discussion it was assumed that the vertical bar was compressed first by the load P and that only after this was the inclined bar attached.   If the system is assembled in the unstressed condition, then, as was mentioned before, the application of any vertical load P will produce some lateral displacement.   This condition will be analogous to that of a bar compressed by forces applied with a small eccentricity.   From the very beginning of compression such a bar starts to bend, but this bending is very small and begins to increase rapidly only as the load approaches the critical value calculated for a centrally compressed bar.   The same action occurs in the above system, and the lateral displacement begins to increase rapidly only as the load P approaches the value given by formula (b).

Let us consider now the general case of a truss with hinged joints, and for simplification, let us assume that all joints are in the same plane.   If we denote the initial length of a member between any two joints i and k by l_{ik} and the length of the same member after loading by a_{ik}, then the force in the member produced by the loading is $A_{ik}E(a_{ik} - l_{ik})/l_{ik}$, where A_{ik} is the cross-sectional area of the member.   Equations of equilibrium for any joint k can be written in the usual form:

$$\sum_i \frac{A_{ik}E(a_{ik} - l_{ik}) \cos \alpha_{ik}}{l_{ik}} - X_k = 0$$

$$\sum_i \frac{A_{ik}E(a_{ik} - l_{ik}) \sin \alpha_{ik}}{l_{ik}} - Y_k = 0$$

$$(c)$$

where α_{ik} denotes the angle between the member ik and the x axis after the deformation of the truss; X_k and Y_k are the components of any external load applied at joint k, and the summation is extended over all members meeting at joint k.   In calculating the critical value of the load, we proceed as before and assume an infinitely small displacement of the system from the position of equilibrium and find the value of the load necessary to keep the system in equilibrium in this displaced position.   This value will be the critical load.   Let δx_k and δy_k denote the components of the small displacement of a joint k and δx_i and δy_i the same for a joint i; then from a simple geometrical consideration (Fig. 2-62) it can be seen that the small change δa_{ik} in the length of a member ik and the small change $\delta \alpha_{ik}$ in the angle α_{ik} corresponding to the above small displacements are

$$\delta a_{ik} = (\delta x_k - \delta x_i) \cos \alpha_{ik} + (\delta y_k - \delta y_i) \sin \alpha_{ik}$$

$$\delta \alpha_{ik} = \frac{1}{a_{ik}} [-(\delta x_k - \delta x_i) \sin \alpha_{ik} + (\delta y_k - \delta y_i) \cos \alpha_{ik}]$$

$$(d)$$

Substituting $a_{ik} + \delta a_{ik}$ for a_{ik}, and $\alpha_{ik} + \delta \alpha_{ik}$ for α_{ik} in Eqs. (c), the equations of equilibrium for the new configuration of the system are obtained.   Remembering that the displacements δx and δy are infinitely small, and using Eqs. (c), the above

equations become

$$\sum_i \cos \alpha_{ik} \frac{A_{ik}E}{l_{ik}} \delta a_{ik} - \sum_i \sin \alpha_{ik} \frac{A_{ik}E(a_{ik} - l_{ik})}{l_{ik}} \delta \alpha_{ik} = 0$$

$$\sum_i \sin \alpha_{ik} \frac{A_{ik}E}{l_{ik}} \delta a_{ik} + \sum_i \cos \alpha_{ik} \frac{A_{ik}E(a_{ik} - l_{ik})}{l_{ik}} \delta \alpha_{ik} = 0$$

(e)

Writing equations of this form for all joints[1] and substituting for δa_{ik} and $\delta \alpha_{ik}$ their values from Eqs. (d), we obtain as many homogeneous linear equations for determining δx and δy as the number of independent displacements δx and δy. The assumed displaced form of equilibrium becomes possible when these equations may yield, for the displacements δx and δy, solutions different from zero. Thus we arrive at the conclusion that the critical value of the load is obtained by equating to zero the determinant of those equations.

Let us apply this method in the case discussed previously (Fig. 2-63). Since hinges 1 and 3 are fixed, there are only two independent displacements δx_2 and δy_2. The

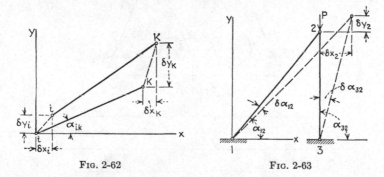

FIG. 2-62   FIG. 2-63

changes of the lengths a_{12} and a_{32}, and of the angles α_{12} and α_{32}, from Eqs. (d) are

$$\delta a_{12} = \delta x_2 \cos \alpha_{12} + \delta y_2 \sin \alpha_{12} \qquad \delta a_{32} = \delta y_2$$

$$\delta \alpha_{12} = \frac{1}{a_{12}} (-\delta x_2 \sin \alpha_{12} + \delta y_2 \cos \alpha_{12}) \qquad \delta \alpha_{32} = -\frac{\delta x_2}{a_{32}}$$

Substituting in Eqs. (e), we obtain

$$\delta x_2 \left(-\frac{P}{a_{32}} + \frac{A_{12}E}{a_{12}} \cos^2 \alpha_{12} \right) + \delta y_2 \frac{A_{12}E \sin \alpha_{12} \cos \alpha_{12}}{a_{12}} = 0$$

$$\delta x_2 \frac{A_{12}E \sin \alpha_{12} \cos \alpha_{12}}{a_{12}} + \delta y_2 \left(\frac{A_{12}E \sin^2 \alpha_{12}}{a_{12}} + \frac{A_{32}E}{a_{32}} \right) = 0$$

Equating to zero the determinant of these equations, the critical value of the load P is obtained, which agrees with the value (b) given before.

Applying the same method to the case shown in Fig. 2-64 and denoting by l and A the length and the cross-sectional area of the horizontal bars and by l_1 and A_1 the cor-

[1] In the case of a joint at a fixed support, δx and δy are zero. In the case of a joint on rollers, only one of the displacements δx and δy is independent.

responding values for the inclined bars, we find that the critical value of the load P is[1]

$$P_{cr} = \frac{AE}{\cot^2 \alpha \left(3 + \dfrac{2l_1 A}{lA_1 \cos^2 \alpha} \right)} \qquad (f)$$

In the case of very rigid inclined bars it becomes

$$P_{cr} = \frac{AE}{3 \cot^2 \alpha}. \qquad (g)$$

This latter result can be obtained from a simple consideration of the displaced form of equilibrium shown in Fig. 2-64 by dotted lines. If X is the decrease in the compressive force in the upper horizontal bar due to the assumed displacement of the system, the increase in the compressive force of the lower horizontal bars is $\frac{1}{2}X$ and the additional

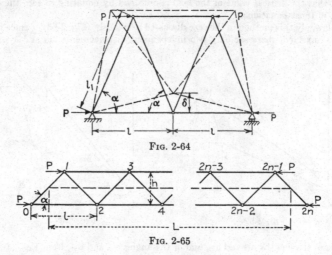

Fig. 2-64

Fig. 2-65

shortening of these bars is $Xl/(2EA)$. Taking into account this shortening and also the deflection δ of the middle joint of the lower chord, we note that the elongation of the upper horizontal bar is equal to $2\left(\dfrac{l_1 \delta \sin \alpha}{l} - \dfrac{Xl}{4AE} \right)$ and the force X can be found from the equation

$$\frac{l_1 \delta \sin \alpha}{l} - \frac{Xl}{4AE} = \frac{Xl}{2AE}$$

which gives

$$X = \frac{4AEl_1 \delta \sin \alpha}{3l^2}$$

Substituting this in the equation of equilibrium for the displaced system

$$P\delta = Xl_1 \sin \alpha$$

which is obtained by passing a section through the truss and by taking moments about the displaced position of the middle joint of the lower chord, we obtain the above value (g) for the critical load.

[1] See R. von Mises and J. Ratzersdorfer, *Z. Math. u. Physik*, vol. 5, p. 227, 1925.

In the case of a truss with many panels (Fig. 2-65), assuming that the diagonals are very rigid, we obtain for the critical value of the compressive force[1]

$$(2P)_{cr} = m \frac{\pi^2 EI}{L^2} \tag{2-69}$$

where $\qquad m = \frac{(4n-2)^2}{\pi^2} \tan^2 \frac{\pi}{4n-2} \qquad n = \text{number of panels}$

$$L = \left(n - \frac{1}{2}\right) l \qquad I = \frac{Ah^2}{2} = \frac{A}{2}\left(\frac{l \tan \alpha}{2}\right)^2$$

A = cross-sectional area of a chord member

It is seen that, when n increases, the factor m approaches unity and the critical value of the axial compressive load $2P$ approaches Euler's value for a solid bar of length L and with $I = Ah^2/2$ (Fig. 2-65).

In another extreme case, where the chords are very rigid in comparison with the diagonals and where n is large, the result of rigorous calculations coincides completely with formula (2-62), obtained before by an approximate method. The same result is obtained also when the diagonals and the chords have rigidities of the same order of magnitude.[2]

The above results were obtained by assuming ideal hinges at all joints. If we assume that the chords are continuous bars and that only the diagonals are attached by hinges, rigorous calculations show[3] that the additional rigidity can be calculated approximately by the method used in deriving Eq. (2-63). Hence the approximate Eqs. (2-62) and (2-63) can always be used in practical calculations provided the number of panels is not small, say not less than six.

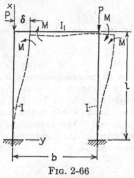

In discussing frame structures, we begin with the single frame shown in Fig. 2-66. Let I and l be the cross-sectional moment of inertia and the length of verticals and I_1 and b the corresponding quantities for the horizontal bar. Assuming that under the

Fig. 2-66

action of the vertical loads P the frame buckles sideways as shown in the figure and denoting by M the moments at the rigid joints, we find the differential equation for the deflection curve of a vertical is

$$EI \frac{d^2 y}{dx^2} = P(\delta - y) - M \tag{h}$$

the solution of which is

$$y = A \cos kx + B \sin kx + \delta - \frac{M}{P}$$

where $\qquad k^2 = \frac{P}{EI}$

Determining the constants A and B from the conditions at the built-in end, we obtain

$$y = \left(\delta - \frac{M}{P}\right)(1 - \cos kx) \tag{i}$$

[1] See R. von Mises and J. Ratzersdorfer, Z. angew. Math. Mech., vol. 5, p. 218, 1925. and vol. 6, p. 181, 1926.

[2] Ibid.

[3] See Wenzel, loc. cit.

The conditions at the upper end of the vertical are

$$(y)_{z=-l} = \delta \qquad \left(\frac{dy}{dx}\right)_{z=-l} = \frac{Mb}{6EI_1} \qquad (j)$$

Substituting Eq. (i) for y, we obtain

$$\delta \cos kl + \frac{M}{P}(1 - \cos kl) = 0$$

$$\delta k \sin kl - \frac{M}{P}\left(k \sin kl + \frac{bP}{6EI_1}\right) = 0 \qquad (k)$$

When the determinant of these equations is equated to zero, the following equation for calculating the critical value of the load P is obtained:

$$\frac{kl}{\tan kl} = -\frac{6lI_1}{bI} \qquad (l)$$

If the horizontal bar is absolutely rigid, then $I_1 = \infty$ and we obtain

$$kl = \pi \qquad P_{cr} = \frac{\pi^2 EI}{l^2} \qquad (m)$$

In another extreme case, where $I_1 = 0$, we obtain from Eq. (l)

$$kl = \frac{\pi}{2} \qquad P_{cr} = \frac{\pi^2 EI}{4l^2} \qquad (n)$$

For all intermediate cases the value of P_{cr} is obtained by solving Eq. (l).

In the above discussion the changes in lengths of the verticals at buckling have been neglected. From Fig. 2-66 it is seen that, owing to the action of the moments M, tension in the left vertical and compression in the right vertical equal to $2M/b$ will be produced. The corresponding change in length is $2Ml/(AEb)$ and the resulting rotation of the horizontal bar is $4Ml/(AEb^2)$. Hence the second condition of (j) becomes

$$\left(\frac{dy}{dx}\right)_{z=-l} = \frac{Mb}{6EI_1} + \frac{4Ml}{AEb^2}$$

Making the corresponding changes in Eqs. (k), we finally obtain for determining the critical value of the load P

$$\frac{kl}{\tan kl} = -\frac{6lI_1}{bI}\frac{1}{1 + 24lI_1/Ab^3} \qquad (o)$$

The last factor on the right side of this equation represents the effect of the axial deformation of the verticals on the magnitude of the critical load. Since I_1 is usually very small in comparison with Ab^2, this effect is small and usually can be neglected.

In the case where the frame (Fig. 2-66) is not symmetrical or is not symmetrically loaded, the problem of determining the critical value of the load becomes more complicated, since it is necessary to consider displacements and rotations of both upper joints.[1]

In the case of a square frame with all members equal and equally compressed (Fig. 2-67), buckling will occur as shown in the figure and each bar is in the condition of a bar with hinged ends, so that the critical compressive force is given by Eq. (2-5). For any regular polygon with n equal sides and with each member compressed by the same axial force (Fig. 2-68), the critical force in the bars is given by the following equations:[2]

[1] Several examples of this kind are discussed by Mises and Ratzersdorfer, *loc. cit.*
[2] *Ibid.*

$$\text{For } n > 3 \qquad P_{cr} = \left(\frac{4\pi}{n}\right)^2 \frac{EI}{l^2}$$

$$\text{For } n = 3 \qquad P_{cr} = (1.23\pi)^2 \frac{EI}{l^2}$$

(2-70)

For more complicated framed structures, a general theory of stability has been developed.[1] Using a method analogous to that applied above for trusses with hinged joints, the critical values of the loads are found by equating to zero the determinant of the system of homogeneous linear equations representing the conditions of equilibrium of the joints of the system in a slightly deflected condition.

Applying this method to the case of columns with batten plates (Fig. 2-57) and using the same notations as in Art. 2.18, we obtain the following expression for the critical load in the case of very rigid battens:

$$P_{cr} = \frac{2z^2 EI_c}{a^2}$$

(2-71)

The numerical factor z is obtained from the transcendental equation

$$\frac{4I_c}{A_c b^2} = \frac{1 - \cos (\pi/n)}{\cos (\pi/n) - \cos z} \frac{\sin z}{z}$$

(p)

where $n = l/a$ denotes the number of panels in the column, A_c is the cross-sectional area of one channel, b is the length of a batten, equal to the distance between the center

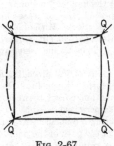

Fig. 2-67

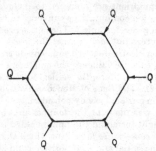

Fig. 2-68

lines of the channels, and I_c is the moment of inertia of the cross-sectional area of one channel. Calculations show that the approximate equation (2-64) is in satisfactory agreement with the more rigorous solution Eq. (2-71).

Take, as a numerical example,[2]

$$\frac{A_c b^2}{4I_c} = 180 \qquad n = 10$$

$$I = 2I_c + \frac{A_c b^2}{2} = 362I_c$$

Substituting these data into Eq. (p), we obtain

$$z = 2.583$$

[1] *Ibid.*

[2] These numerical values are taken from the dimensions of the specimens used by Timoshenko in experiments with models of batten-plate columns, see *Ann. ponts et chaussées*, series 9, vol. 3, p. 551, 1913.

and the critical load, from Eq. (2-71), becomes

$$P_{cr} = 0.369 \frac{\pi^2 EI}{l^2}$$

The approximate Eq. (2-64) gives, for $I_b = \infty$ and $A_b = \infty$,

$$P_{cr} = \frac{\pi^2 EI}{l^2} \frac{1}{1 + \pi^2 I a^2 / 24 I_c l^2} = 0.403 \frac{\pi^2 EI}{l^2}$$

Thus in this extreme case, where the critical load for the batten-plate column is only about 40 per cent of that calculated for the corresponding solid column, the error of the approximate equation is about 9 per cent. For cases where the critical load does not differ much from the value calculated for a solid column, Eq. (2-64) represents a better approximation and can be used with sufficient accuracy in practical design and also in cases where the flexural rigidity of the batten plates is not large.

The stability of a framework with rigid joints can be investigated by using the moment-distribution method. In the use of this method, a particular set of values of the external loads is assumed and the corresponding axial forces in the bars are determined, assuming that the truss has pin joints. Then an arbitrary moment is applied to one of the joints of the frame and the moments in the frame are distributed in the usual way. However, stiffness and carry-over factors used in the moment-distribution computations are modified to include the effect of axial load.[1] If the moment-distribution computations converge to finite values for the final end moments, the frame is, in general, stable. The entire process is then repeated using increased loads on the structure but maintaining the loads in the same proportion. If the loads are above the critical value, the moment-distribution computations will not converge, in general, to finite values of the end moments in the bars. Thus by successive applications of this procedure, the critical load is determined.[2]

2.21. The Case of Nonconservative Forces. In many of the preceding articles we began our analysis by considering a slightly buckled form of the column, but no mention was made of the manner in which the deformation varied between the initial straight form and the final buckled form. This was permissible, since we dealt only with *conservative forces*, for which the work done during a displacement depends only on the initial and final positions and is independent of the path of the point of application of the force. For example, the work done by a gravity or weight force depends only on the lowering of the center of gravity of the object. We made use of this fact in calculating critical loads by the energy method, since the work done by the external forces was taken as $\Delta T = P\lambda$, where λ represented the distance between the initial and final locations of the point of application of the force P, measured in the direction of the force. Likewise, when the differential equation of equilibrium was used, the existence of a buckled form was assumed and the smallest force required to maintain the bar in that form was determined from the end conditions. This critical force was independent of the path of the bar in reaching the assumed buckled shape.

In the case of nonconservative forces both of the above methods, called *static methods*, may prove inadequate, and we must use instead a *dynamic criterion* of

---

[1] Modified values of these factors were determined by B. W. James, *NACA Tech. Note* 534, 1935.

[2] The validity of this method was established by N. J. Hoff, *J. Aeronaut. Sci.*, vol. 8, no. 3, pp. 115-119, 1941. See also N. J. Hoff, B. A. Boley, S. V. Nardo, and S. Kaufman, *Trans. ASCE*, vol. 116, p. 958, 1951.

stability.[1] An example of a nonconservative force acting on a column is shown in Fig. 2-69a. It is assumed that a constant compressive force P is applied to the column and always acts during buckling in the direction of the tangent to the deflection curve at the top of the column. In this case it is not possible to calculate the work done by the force P during buckling on the basis of the initial position A and the final position A' only. It is necessary also to know the direction of the tangent at every intermediate position. To show this, let us resolve the force P into a vertical component P_1, which can be taken equal to P if the deflections remain small, and a horizontal component H (Fig. 2-69b). If the tangent at A remains vertical during buckling, as shown by the dotted line, and then is allowed to rotate only after reaching point A', there will be no work produced by the horizontal component H. Thus the total work done would be the same as in the usual case of a column with vertical load. A different result is obtained if the tangent at A and the force P rotate continuously during buckling, since then the horizontal component H also produces work. It is apparent that a definite value of the work done by the force P can be obtained only if additional information regarding the deflection curve of the column during buckling is available. This conclusion shows that the energy method cannot be used in this case for calculating P_{cr}.

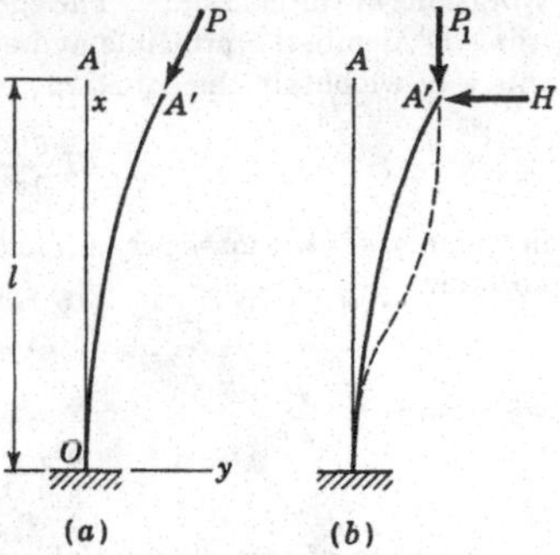

If we attempt to use the differential equation (2-9) for determining P_{cr}, we observe that the lower end of the column is fixed (Fig. 2-69a), and therefore the conditions are

$$y = \frac{dy}{dx} = 0 \qquad \text{at } x = 0 \qquad (a)$$

At the upper end of the column the bending moment is zero and the shearing force V is $-H$, which is equal to $-P\,dy/dx$. Thus from Eqs. (1-3) and (1-4), Art. 1.2, we have the conditions

$$\frac{d^2y}{dx^2} = \frac{d^3y}{dx^3} = 0 \qquad \text{at } x = l \qquad (b)$$

FIG. 2-69

Using these conditions for determining the four constants of integration in the general solution [Eq. (2-10)], we find that the conditions can be satisfied only by taking all the constants equal to zero. From this we would conclude that only the straight form of equilibrium is possible within the elastic range. This is the conclusion from a static point of view. Let us next investigate the stability of the column from a dynamic point of view.

In stating the dynamic criterion of stability, we begin by assuming that the loaded column is subjected to an initial disturbance which produces small vibrations. If these vibrations decrease with time, we can conclude that the straight form of equilibrium of the column is stable. The initial vibrations gradually die out owing to damping action, which is always present, and after an interval of time the column returns to its initial straight form. On the other hand, if the external forces acting on the column are such that the amplitude of the vibrations begins to grow without limit, the straight form of equilibrium of the column is unstable.

[1] See the papers by H. Ziegler in *Advances in Appl. Mech.*, vol. 4, pp. 357–403, 1956, and *Ingr.-Arch.*, vol. 20, no. 1, p. 49, 1952.

In the previous cases dealing with conservative forces, it was shown (see p. 84) that if the load was below the critical value, any deflection of the column from its initial straight form corresponded to an increase of the total potential energy of the system (strain energy of the bar plus potential energy of the load). If a small amount of kinetic energy is imparted to the column by an initial impulse and vibrations ensue, it can be concluded from the principle of conservation of energy that there will be no tendency for the vibrations to grow. Hence the critical loads determined by the static methods will satisfy the dynamic criterion of stability also. In the case of nonconservative forces, the conditions are different. If small vibrations of the bar are initiated, the forces may produce positive work, which will result in a progressive growing of the amplitude of vibrations. Such a condition indicates that we have the case of instability.

Returning to our previous example of a bar subjected to a nonconservative force (Fig. 2-69) and using the dynamic criterion of stability, we have to investigate small vibrations of the column.[1] The equation for this vibration is found from Eq. (1-5) by using D'Alembert's principle and substituting inertia forces for the lateral load. In this way we obtain the equation

$$EI \frac{\partial^4 y}{\partial x^4} + P \frac{\partial^2 y}{\partial x^2} + \frac{q}{g} \frac{\partial^2 y}{\partial t^2} = 0 \qquad (c)$$

in which q/g is the mass per unit length of the column. Dividing by EI and using the notation

$$k^2 = \frac{P}{EI} \qquad a = \frac{q}{gEI} \qquad (d)$$

we obtain

$$\frac{\partial^4 y}{\partial x^4} + k^2 \frac{\partial^2 y}{\partial x^2} + a \frac{\partial^2 y}{\partial t^2} = 0 \qquad (e)$$

The solution of this differential equation with constant coefficients can be taken in the form

$$y = A_0 f(x) e^{i\omega t} \qquad (f)$$

where $i = \sqrt{-1}$. Substituting into Eq. (e), we obtain for $f(x)$ the ordinary differential equation

$$\frac{d^4 f(x)}{dx^4} + k^2 \frac{d^2 f(x)}{dx^2} - a\omega^2 f(x) = 0$$

for which the general solution is

$$f(x) = A \cosh \lambda_1 x + B \sinh \lambda_1 x + C \cos \lambda_2 x + D \sin \lambda_2 x \qquad (g)$$

where

$$\lambda_1 = \left[\sqrt{a\omega^2 + \frac{k^4}{4}} - \frac{k^2}{2} \right]^{\frac{1}{2}}$$

$$\lambda_2 = \left[\sqrt{a\omega^2 + \frac{k^4}{4}} + \frac{k^2}{2} \right]^{\frac{1}{2}} \qquad (h)$$

For the determination of the constants in the solution (g) we have the four end conditions given by (a) and (b). Substituting Eq. (g) in those conditions we obtain four homogeneous linear equations for A, B, C, and D. To satisfy these equations we can put $A = B = C = D = 0$, which gives the straight form of equilibrium of the

[1] This investigation was made by Max Beck, Z. angew. Math. u. Physik, vol. 3, p. 225, 1952. Independently the same problem was discussed by K. S. Dejneko and M. J. Leonov, Appl. Math. Mech. (Russian), vol. 19, p. 738, 1955.

column. To obtain another solution, the determinant of the equations must vanish. This gives the frequency equation

$$2a\omega^2 + k^4 + 2a\omega^2 \cosh \lambda_1 l \cos \lambda_2 l + k^2 \sqrt{a\omega^2} \sinh \lambda_1 l \sin \lambda_2 l = 0 \qquad (2\text{-}72)$$

from which the radian frequency ω can be calculated for any value of the force P. It is seen that for $\omega = 0$, that is, for the static condition of the column, there is no value of k^2, different from zero, which can satisfy Eq. (2-72). This means that there is no value of P at which the column can stay in a slightly buckled condition, and we thus reach the same conclusion as in the previous calculations using the static method.

Now considering ω different from zero and also taking $P = 0$, we obtain

$$k^2 = 0 \qquad \lambda_1 = \lambda_2 = \sqrt[4]{a\omega^2}$$

and Eq. (2-72) becomes

$$\cosh (l \sqrt[4]{a\omega^2}) \cos (l \sqrt[4]{a\omega^2}) = -1 \qquad (i)$$

This is the known frequency equation for a prismatic cantilever bar[1] and the first two frequencies ω_1 and ω_2 are given in the second column of Table 2-15.

TABLE 2-15. FREQUENCIES OF COMPRESSED COLUMN (FROM EQ. 2-72)

$\dfrac{Pl^2}{\pi^2 EI}$	0	0.5	1.0	1.5	2.0	2.001
$\omega_1^2 \dfrac{al^4}{\pi^4}$	0.125	0.26	0.30	0.46	0.96	0.98
$\omega_2^2 \dfrac{al^4}{\pi^4}$	4.86	4.2	3.3	2.6	1.02	0.99

If we now consider increasing values of P and take continually larger values of k^2 in Eq. (2-72), we obtain the squares of the first two frequencies ω_1 and ω_2, as given in the second and third lines of Table 2-15. The values of ω^2 are positive, and when the corresponding values of ω are substituted into Eqs. (g) and (f), we obtain solutions of Eq. (e) in the form

$$y = f(x) \sin \omega t \qquad \text{and} \qquad y = f(x) \cos \omega t$$

These equations represent simple harmonic vibrations of constant amplitude, and thus we conclude that the column is stable. There is, however, a definite value of P at which the character of the vibration of the column changes. We see from Table 2-15 that the values of ω_1^2 and ω_2^2 approach each other as P increases, and a more refined calculation shows that for

$$\frac{Pl^2}{\pi^2 EI} = 2.008 \qquad (j)$$

these values coincide; that is, the frequency Eq. (2-72) has a double root. With a further increase of P the roots become complex and of the form $\omega = m + in$, where m and n are real numbers. Substituting in Eq. (f), we obtain solutions of the form

$$y = f(x)e^{(n+im)t} \qquad \text{and} \qquad y = f(x)e^{(-n+im)t}$$

[1] See Timoshenko, "Vibration Problems in Engineering," 3d ed. in collaboration with D. H. Young, p. 338, D. Van Nostrand Company, Inc., Princeton, N.J., 1955.

Since either n or $-n$ is a positive number, the corresponding value of y increases indefinitely with time, indicating that the column is unstable. Thus from Eq. (j) we obtain the critical value of the force P as

$$P_{cr} = \frac{2.008\pi^2 EI}{l^2} \tag{2-73}$$

Thus by applying the dynamic concept of stability we obtained for the tangential force P a definite critical value which we were unable to find by static considerations. No definite conclusion can be made (as yet) regarding the practical value of this result, since no method has been devised for applying a tangential force to a column during bending.

A similar problem arises if we assume that compressive forces are distributed along the axis of the column and always act in the direction of the tangent to the deflection curve.

Compressed and Twisted Shaft. As another example of buckling under the action of nonconservative forces, let us consider the case of a shaft subjected to the action of an

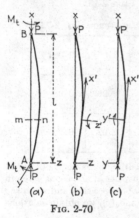

FIG. 2-70

axial compressive force P and a twisting couple M_t, Fig. 2-70. The ends of the shaft are assumed to be attached to the supports by ideal spherical hinges, or universal joints, and are free to rotate in any direction. Let us assume also that during buckling the forces P and the couples M_t retain their initial directions. Under such conditions the couple M_t is not conservative, since its work depends on the manner in which the tangent at the end of the bar moves during buckling. To show this, let us consider the lower end A of the shaft. The tangent to the slightly buckled shaft at A can be brought to its inclined position by rotation about the y and z axes through angles dz/dx and dy/dx. During such rotations the couple M_t, which always lies in a horizontal plane, does not produce any work. However, the tangent can be brought to the inclined position in another way. We can, for example, rotate the tangent about the y or z axis by the angle $(\sqrt{dy^2 + dz^2})/dx$ and then bring it to the final position by rotation about the x axis. During the latter rotation the couple produces work, and thus it is evident that we have a nonconservative system of loading. Assuming that the behavior of the twisting couples is ascertained in some way, let us determine the critical loads from the equations of equilibrium of the buckled shaft, using the usual static approach.

The deflection curve in this case will not be a plane curve, and we must consider the two projections of the curve as shown in Fig. 2-70b and c. We assume also that the principal moments of inertia of the cross section of the bar are equal, so that any two central perpendicular axes in the plane of the cross section can be taken as the principal axes.[1] Considering any cross section mn of the bar and taking the principal axes parallel to y and z, we find that the directions of these axes after buckling will be y' and z'. In deriving the differential equations of the deflection curve, we consider the upper portion of the bar and calculate the moments of the forces applied to this part of the bar with respect to the y' and z' axes. Taking the moments positive as

[1] The case of a shaft with two different flexural rigidities was discussed by R. Grammel, *Z. angew. Math. u. Mech.*, vol. 3, p. 262, 1923.

indicated in the figures, we find that the moments, with respect to the y' and z' axes, of the compressive force P are Pz and $-Py$, respectively. The moments of the twisting couple M_t with respect to the same axes are $-M_t\, dy/dx$ and $-M_t\, dz/dx$. The differential equations of the deflection curve in each plane now become

$$EI \frac{d^2z}{dx^2} = -Pz + M_t \frac{dy}{dx}$$

$$EI \frac{d^2y}{dx^2} = -Py - M_t \frac{dz}{dx}$$

(k)

The general solutions of these equations are

$$y = A \sin (m_1 x + \alpha_1) + B \sin (m_2 x + \alpha_2)$$

$$z = A \cos (m_1 x + \alpha_1) + B \cos (m_2 x + \alpha_2)$$

(l)

in which A, B, α_1, and α_2 are constants of integration and m_1 and m_2 are the two roots of the quadratic equation

$$EIm^2 + M_t m - P = 0$$

(m)

Substituting (l) into Eqs. (k), it can be shown that these equations are satisfied.

For determining the constants A, B, α_1, and α_2 we have the following conditions at the ends:

$$(y)_{x=0} = 0 \qquad (y)_{x=l} = 0 \qquad (z)_{x=0} = 0 \qquad (z)_{x=l} = 0 \qquad (n)$$

Substituting expressions (l) for y and z, we obtain:

$$A \sin \alpha_1 + B \sin \alpha_2 = 0 \qquad A \cos \alpha_1 + B \cos \alpha_2 = 0$$
$$A \sin (m_1 l + \alpha_1) + B \sin (m_2 l + \alpha_2) = 0$$
$$A \cos (m_1 l + \alpha_1) + B \cos (m_2 l + \alpha_2) = 0$$

On substituting for $B \sin \alpha_2$ and $B \cos \alpha_2$ from the first two equations in the second two, we find that

$$A[\sin (m_1 l + \alpha_1) - \sin (m_2 l + \alpha_1)] = 0$$
$$A[\cos (m_1 l + \alpha_1) - \cos (m_2 l + \alpha_1)] = 0$$

from which it follows that $m_1 l$ and $m_2 l$ differ by a multiple of 2π. The smallest values for M_t and P at which buckling will occur are obtained from the condition

$$m_1 l - m_2 l = \pm 2\pi$$

or, by using Eq. (m),

$$\frac{M_t^2}{4EI} + P = \frac{\pi^2 EI}{l^2}$$

(2-74)

When M_t is zero, this equation gives the known Euler formula for the critical load. When P is equal to zero, we obtain the value of the torque, which, acting alone, will produce buckling of the shaft.[1] If the shaft is in tension, the sign of P in Eq. (2-74) must be changed. Thus the stability of the shaft against buckling produced by a torque is increased if a tensile force is applied.

This same problem was discussed by the use of the dynamic concept of stability, and from a consideration of the lateral vibration of the shaft the same expression (2-74)

---

[1] Formula (2-74) was obtained by A. G. Greenhill, *Proc. Inst. Mech. Engrs.* (*London*), 1883. Further discussion of the problem is given in the papers by E. L. Nicolai, Dissertation, St. Petersburg, 1916 (Russian), and *Z. angew. Math. u. Mech.*, vol. 6, p. 30, 1926.

was obtained.[1] The problem of buckling of shafts with other end conditions was discussed by several authors,[2] but again, in the case of nonconservative forces, the question of how to apply the forces mechanically remains unanswered. Thus at present there is no experimental verification of the results. In the theory of structures in the elastic range we usually encounter stationary conservative loads and a static approach to stability problems is satisfactory. On the other hand, the dynamic concept of stability is essential in the analysis of problems where the forces vary with time. Some examples of this type will be discussed in the next article.

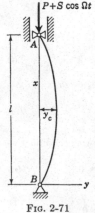

FIG. 2-71

2.22. Stability of Prismatic Bars under Varying Axial Forces. Let us begin the discussion with the simple case of a uniform bar with hinged ends (Fig. 2-71) subjected to the action of the axial compressive force

$$P + S \cos \Omega t \qquad (a)$$

This force consists of a stationary part P and a periodically varying part $S \cos \Omega t$, having amplitude S and radian frequency Ω. The total axial compressive force therefore varies between $P + S$ and $P - S$. Experience has shown that a slender bar can withstand without buckling a maximum force $P + S$ which is larger than the Euler load $P_e = \pi^2 EI/l^2$. Also, at certain values of the frequency Ω of the pulsating force, violent lateral vibrations of the bar are produced, so that the bar is unstable at these frequencies.[3] In studying this problem the dynamic concept of stability will be used. We shall assume that the bar is straight and perfectly elastic and that small lateral vibrations are produced by some impulse. During these vibrations the upper end of the bar moves slightly up and down and, at certain frequencies, the external pulsating force may produce positive work, resulting in increasing amplitude of the vibrations. This indicates the condition of instability.[4]

In studying the lateral vibrations we shall use the differential equation (c) of the preceding article which in the case of the pulsating load (a) becomes[5]

$$EI \frac{\partial^4 y}{\partial x^4} + (P + S \cos \Omega t) \frac{\partial^2 y}{\partial x^2} + \frac{q}{g} \frac{\partial^2 y}{\partial t^2} = 0 \qquad (b)$$

[1] See H. Ziegler, *Z. angew. Math. u. Phys.*, vol. 2, p. 265, 1951, and A. Troesch, *Ingr.-Arch.*, vol. 20, p. 258, 1952.

[2] A review of these cases is given in the paper by H. Ziegler in *Advances in Appl. Mech.*, vol. 4, pp. 357–403, 1956.

[3] A similar phenomenon can be demonstrated easily by applying to a string a pulsating tensile force. At certain frequencies of the force violent lateral vibrations of the string can be produced.

[4] This problem was first solved by N. M. Belajev, Engineering Structures and Structural Mechanics, "Collection of Papers," pp. 149–167, Leningrad, 1924. See also E. Mettler, *Mitt. Forsch. Anst. GHH-Konz.*, vol. 8, p. 1, 1940, and K. Klotter, *Forsch. Ing.-Wesen*, vol. 12, p. 209, 1941. A similar problem in the case of a bar with built-in ends was discussed by F. Weidenhammer, *Ingr.-Arch.*, vol. 19, p. 162, 1951. Numerous problems on stability of structures under pulsating loads are given in the book by B. B. Bolotin, "Dynamic Stability of Elastic Systems," Moscow, 1956 (in Russian).

[5] It is assumed here that the period of the pulsating force is very large in comparison with the period of the fundamental longitudinal vibration of the bar and therefore we can take the axial force as constant along the length of the bar.

Let us take the solution of this equation in the form

$$y = Af(t) \sin \frac{\pi x}{l} \qquad (c)$$

which satisfies the conditions at the hinged ends of the bar and represents the deflection in the form of a half sine wave. Substituting Eq. (c) into Eq. (b), we obtain the following equation for determining the function $f(t)$:

$$\frac{d^2f(t)}{dt^2} + \frac{g\pi^2}{ql^2}\left(\frac{\pi^2 EI}{l^2} - P\right)f(t) - \left(\frac{g\pi^2}{ql^2}S\cos\Omega t\right)f(t) = 0 \qquad (d)$$

In the particular case when $S = 0$ and $P < \pi^2 EI/l^2$, we have a static load smaller than the Euler load and Eq. (d) gives a simple harmonic vibration with frequency

$$\omega^2 = \frac{g\pi^2}{ql^2}\left(\frac{\pi^2 EI}{l^2} - P\right) \qquad (e)$$

For $P = 0$ we have

$$\omega_0^2 = \frac{g\pi^4 EI}{ql^4} \qquad (f)$$

This is the square of the radian frequency of lateral vibration for a pin end prismatic bar without axial load.[1] From Eq. (e) we see that with an increase in the load P the frequency ω decreases and when the load reaches the Euler value, the frequency becomes zero. Under this load there will no longer be any vibration, and the bar is in equilibrium in a slightly deflected form. Thus for $S = 0$, Eq. (d) gives the same value for the critical load as found previously from static considerations.

Let us consider now the case when S is not zero and the bar is submitted to the action of a pulsating load. To simplify the writing of Eq. (d) we introduce the following notations:

$$p = \frac{P}{P_e} \qquad s = \frac{S}{P_e} \qquad (g)$$

Then Eq. (d) becomes

$$\frac{d^2f(t)}{dt^2} + \omega_0^2(1 - p)f(t) - \omega_0^2(s\cos\Omega t)f(t) = 0 \qquad (h)$$

Now let us replace t by a new variable τ defined by the relation

$$\tau = \Omega t \qquad (i)$$

Since

$$\frac{d^2f(t)}{dt^2} = \Omega^2 \frac{d^2f(\tau)}{d\tau^2} \qquad (j)$$

Eq. (h) becomes

$$\frac{d^2f(\tau)}{d\tau^2} + (a + b\cos\tau)f(\tau) = 0 \qquad (k)$$

where

$$a = \frac{\omega_0^2}{\Omega^2}(1 - p) \qquad b = -\frac{\omega_0^2}{\Omega^2}s \qquad (l)$$

In each particular case the quantities a and b can be calculated readily by using Eqs. (f) and (g). Investigation shows[2] that the character of the solution of Eq. (k) depends on the numerical values of a and b. At certain values of these quantities the solution

[1] See Timoshenko, "Vibration Problems in Engineering," 3d ed. in collaboration with D. H. Young, p. 332, D. Van Nostrand Company, Inc., Princeton, N.J., 1955.

[2] Such an investigation was made by M. J. O. Strutt, Z. Physik, vol. 69, p. 597. 1931, and Ergeb. Math., vol. 1, p. 24, 1932.

gives a vibration which grows with time and thereby indicates an unstable condition. This is shown in Fig. 2-72, where values of a and b given by coordinates of ·points in the unshaded areas represent an unstable condition. The shaded areas indicate regions of stability.[1]

Starting with the case of small values of s we have to consider points in the vicinity of the horizontal axis, and we see that instability occurs when $a = \frac{1}{4}, 1, 2\frac{1}{4}, \ldots$. If we take $p = 0$, we reach the first critical value of a when $\Omega = 2\omega_0$, which shows that a small periodically varying axial force may produce violent lateral vibration of the bar if its frequency is twice the fundamental frequency of lateral vibration of the bar.[2] The next critical condition corresponds to $a = 1$, or $\Omega = \omega_0$. When p is gradually increased from 0 to 1, the critical values of Ω decrease. Taking, for example, $p = \frac{1}{2}$, we get for the first critical frequency $\Omega = \sqrt{2}\,\omega_0$. When the magnitude of the variable portion of the load increases, b will increase also; then the regions of instability

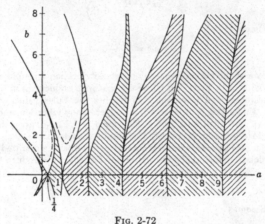

FIG. 2-72

in Fig. 2-72 become wider and, instead of critical points, we get for Ω gradually widening critical ranges.

In the preceding discussion the effect of damping was not considered. If it is taken into consideration, the curves in Fig. 2-72 must be somewhat altered, as shown by the dotted lines. The regions of instability are reduced, and finite values of S are required to produce lateral vibrations in the region of the critical points. Furthermore the required values of S are seen to increase with the order of the critical frequencies, and thus frequencies of higher orders become of no practical importance.

We see also from Fig. 2-72 that the maximum compressive force $P + S$, consistent with stability, may become much higher than the Euler load. Let us take, as an example, the values $p = 0$ and $\frac{1}{4} < a < 1$. We see that between these values of a

---

[1] Equation (k) is known as the Mathieu equation, and additional references on the theory of this equation are given in the paper by S. Lubkin and J. J. Stoker, *Quart. Appl. Math.*, vol. 1, no. 3, p. 215, 1943.

[2] This is readily apparent if we note that the pulsating load produces positive work when it is in compression and when the vibrating bar is moving from the middle position to the extreme position. Likewise, it produces positive work when it is in tension and the bar is moving from the extreme to the middle position.

there are values of b greater than 2 in the region of stability.   This means that the maximum compressive force can be more than twice the Euler load without producing lateral buckling of the bar.   It is assumed that the bar is sufficiently slender that the maximum stress always remains below the elastic limit of the material.

As another example of the action of a force which varies with time, consider again the bar in Fig. 2-71 and imagine that the lower end of the bar is stationary while the upper end A moves downward[1] with a constant velocity c.   Assuming an initial crookedness of the bar

$$y_0 = a \sin \frac{\pi x}{l} \qquad (m)$$

the differential equation for the lateral displacement y_c of the center of the bar during loading was derived and integrated.   The results of this integration for one particular case are represented in Fig. 2-73.   The vertical axis of the graph gives values of the

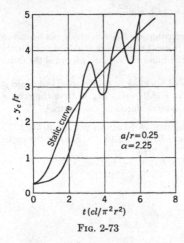

FIG. 2-73

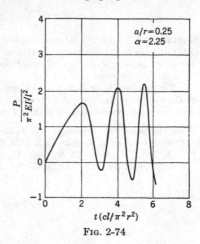

FIG. 2-74

dimensionless quantity y_c/r, where y_c is the deflection at the center and r is the radius of gyration of the cross section.   The horizontal axis represents time and is expressed in dimensionless form.   The graph is plotted for an initial crookedness given by the ratio $a/r = 0.25$ and also for the value of the parameter

$$\alpha = \frac{\pi^8 E I r^4 g}{c^2 l^6 q} = 2.25$$

For comparison, the static deflection of the bar for the same displacement of the upper end A is also plotted in the figure.   Owing to the inertia forces, the dynamic deflections of the bar at the beginning of loading lag behind the values for static loading.   With a further increase in time the bar is accelerated sufficiently to cause the deflections to increase rapidly.   Finally the deflections exceed the static values, and lateral vibrations of the bar ensue.   The corresponding values of the axial compressive

[1] This problem was discussed by N. J. Hoff, *J. Appl. Mech.*, vol. 18, p. 68, 1951. See also N. J. Hoff, S. V. Nardo, and B. Erickson, *Proc. 1st U.S. Natl. Congr. Appl. Mech.*, ASME, New York, 1952.

force are shown in Fig. 2-74.   In this example the velocity c of the upper end of the bar was about 4 in. per sec.

The problem of lateral buckling of an initially curved slender bar was studied also in the case in which a constant compressive force P acts for a short interval of time.[1] Assuming again that the time interval during which the force P acts is large in comparison with the period of longitudinal vibration of the bar, we may consider the compressive force as constant along the bar.   The equation for lateral motion of the bar then becomes

$$EI \frac{\partial^4 y}{\partial^4 x} + P \frac{\partial^2}{\partial x^2} (y + y_0) + \frac{q}{g} \frac{\partial^2 y}{\partial t^2} = 0$$

in which y_0 denotes the initial deflection of the axis of the bar from the straight form and $y + y_0$ represents the total deflection.   Again, assuming the deflection for a bar with hinged ends in the form

$$y_0 = a \sin \frac{\pi x}{l} \qquad y = Af(t) \sin \frac{\pi x}{l}$$

we obtain for $f(t)$ a differential equation with constant coefficients which can be solved readily in each particular case.   In this way it can be shown that the bar can withstand safely compressive forces P larger than the Euler critical load provided the duration is sufficiently short.

Similar problems were discussed in the case of buckling of thin plates, and again it was shown that plates can carry stresses in the middle plane higher than the critical values provided the duration of action of the external forces is very short.[2]

[1] See the paper by C. Koning and J. Taub, *Luftfahrt-Forsch.*, vol. 10, p. 55, 1933. An English translation appears in *NACA Tech. Mem.* 748, 1934.   See also the paper by J. H. Meier, *J. Aeronaut. Sci.*, vol. 12, p. 433, 1945.

[2] See the paper by G. A. Zizicas, *Trans. ASME*, vol. 74, p. 1257, 1952.

CHAPTER 3

INELASTIC BUCKLING OF BARS

3.1. Inelastic Bending. Before beginning the discussion of inelastic buckling, let us review the theory of bending of beams when the material is stressed beyond the proportional limit.[1] The theory is based upon the assumption that cross sections of the beam remain plane during bending, and hence longitudinal strains are proportional to their distances from the neutral surface. Let us assume also that the same relation exists between stress and strain as in the case of simple tension and compression, represented by the stress-strain diagram in Fig. 3-1.

Let us begin with a beam of rectangular cross section, Fig. 3-2, and assume that the radius of curvature of the neutral surface produced by the bending moments M is equal to ρ. In such a case the unit elongation of a fiber at distance y from the neutral surface is

$$\epsilon = \frac{y}{\rho} \qquad (a)$$

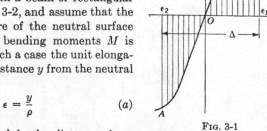

FIG. 3-1

Denoting by h_1 and h_2 the distances from the neutral axis to the lower and upper surfaces of the beam, respectively, we find that the elongations in the extreme fibers are

$$\epsilon_1 = \frac{h_1}{\rho} \qquad \epsilon_2 = -\frac{h_2}{\rho} \qquad (b)$$

It is seen that the elongation or contraction of any fiber is readily obtained provided we know the position of the neutral axis and the radius of curvature ρ. These two quantities can be found from the two equations of

[1] For a more complete discussion of inelastic bending, see Timoshenko, "Strength of Materials," 3d ed., part II, chap. 9, D. Van Nostrand Company, Inc., Princeton, N.J., 1956.

statics:

$$\int_A \sigma \, dA = b \int_{-h_2}^{h_1} \sigma \, dy = 0 \tag{c}$$

$$\int_A \sigma y \, dA = b \int_{-h_2}^{h_1} \sigma y \, dy = M \tag{d}$$

The first of these equations states that the sum of the normal forces acting on any cross section of the beam vanishes, since these forces represent a couple. The second equation states that the moment of the same forces with respect to the neutral axis is equal to the bending moment M.

Equation (c) is now used for determining the position of the neutral axis. From Eq. (a) we have

$$y = \rho\epsilon \qquad dy = \rho \, d\epsilon \tag{e}$$

Substituting into Eq. (c), we obtain

$$\int_{-h_2}^{h_1} \sigma \, dy = \rho \int_{\epsilon_2}^{\epsilon_1} \sigma \, d\epsilon = 0 \tag{f}$$

Hence the position of the neutral axis is such that the integral $\int_{\epsilon_2}^{\epsilon_1} \sigma \, d\epsilon$ vanishes. To determine this position we use the curve AOB in Fig. 3-1,

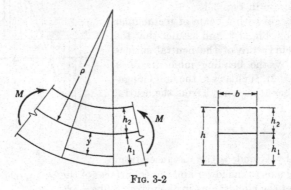

Fig. 3-2

which represents the tension-compression test diagram for the material of the beam, and we denote by Δ the sum of the absolute values of the maximum elongation and the maximum contraction, which is

$$\Delta = \epsilon_1 - \epsilon_2 = \frac{h_1}{\rho} + \frac{h_2}{\rho} = \frac{h}{\rho} \tag{g}$$

To solve Eq. (f), we have only to mark the length Δ on the horizontal axis in Fig. 3-1 in such a way as to make the two areas, shaded in the figure, equal. In this manner we obtain the strains ϵ_1 and ϵ_2 in the extreme fibers. Equations (b) then give

$$\frac{h_1}{h_2} = \left| \frac{\epsilon_1}{\epsilon_2} \right| \tag{h}$$

which determines the position of the neutral axis. Observing that the elongations ϵ are proportional to the distance from the neutral axis, we conclude that the curve AOB also represents the distribution of bending stresses along the depth of the beam if h is substituted for Δ.

In calculating the radius of curvature ρ we use Eq. (d). Substituting for y and dy their values from Eqs. (e), we represent Eq. (d) in the following form:

$$b\rho^2 \int_{\epsilon_2}^{\epsilon_1} \sigma\epsilon \, d\epsilon = M \tag{i}$$

By observing that $\rho = h/\Delta$ from Eq. (g), we can write Eq. (i), after a simple transformation, as follows:

$$\frac{bh^3}{12} \frac{1}{\rho} \frac{12}{\Delta^3} \int_{\epsilon_2}^{\epsilon_1} \sigma\epsilon \, d\epsilon = M \tag{j}$$

Comparing this result with the known equation

$$\frac{EI}{\rho} = M \tag{k}$$

for bending of beams following Hooke's law, we conclude that beyond the proportional limit the curvature produced by a moment M can be calculated from the equation

$$\frac{E'I}{\rho} = M \tag{3-1}$$

in which E' is defined by the expression

$$E' = \frac{12}{\Delta^3} \int_{\epsilon_2}^{\epsilon_1} \sigma\epsilon \, d\epsilon \tag{3-2}$$

The integral in this expression represents the moment with respect to the vertical axis through the origin O of the shaded area shown in Fig. 3-1. Since the ordinates of the curve in the figure represent stresses and the abscissas represent strains, the integral and also E' have the dimensions of pounds per square inch, which are the same dimensions as the modulus E.

The magnitude of E' for a given material, corresponding to a given curve in Fig. 3-1, is a function of Δ or of h/ρ. Taking several values of Δ and using the curve in Fig. 3-1 as previously explained, we determine for each value of Δ the corresponding extreme elongations ϵ_1 and ϵ_2 and from Eq. (3-2) determine the corresponding values of E'. In this way a curve representing E' as a function of $\Delta = h/\rho$ is obtained. In Fig. 3-3 such a curve is shown for structural steel with $E = 30 \times 10^6$ psi and the proportional limit equal to 30,000 psi.   In this case, for $\Delta < 0.002$, E' remains constant and equal to E.   With such a curve the moment corresponding to any assumed curvature can be readily calculated from Eq. (3-1), and we can plot a curve, Fig. 3-4, giving the moment M as a function of Δ. For small values of Δ the material follows Hooke's law, and the curvature is proportional to the bending moment M, as shown in Fig. 3-4 by the straight line OC.   Beyond the proportional limit the rate of change of the curvature increases as the moment increases.

If the tension and compression portions of the stress-strain diagram are the same, the neutral axis passes through the centroid of the cross section

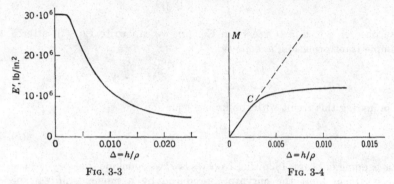

FIG. 3-3   FIG. 3-4

and we obtain the following simplified expressions:

$$h_1 = h_2 = \frac{h}{2}$$

$$\epsilon_1 = -\epsilon_2 = \frac{\Delta}{2}$$

$$E' = \frac{24}{\Delta^3} \int_0^{\Delta/2} \sigma\epsilon \, d\epsilon$$

Also, within the elastic limit we have $\sigma = E\epsilon$ and

$$E' = \frac{24}{\Delta^3} E \int_0^{\Delta/2} \epsilon^2 \, d\epsilon = E$$

so that Eq. (3-1) reduces to the usual equation for elastic bending.

If instead of a rectangle we have any other symmetrical shape of cross section, the width b of the cross section is variable and Eqs. (c) and (d) must be written in the following form:

$$\int_{-h_2}^{h_1} b\sigma \, dy = \rho \int_{\epsilon_2}^{\epsilon_1} b\sigma \, d\epsilon = 0 \qquad (l)$$

$$\int_{-h_2}^{h_1} b\sigma y \, dy = \rho^2 \int_{\epsilon_2}^{\epsilon_1} b\sigma\epsilon \, d\epsilon = M \qquad (m)$$

Take as an example the case of a T section, Fig. 3-5. If we denote by ϵ' the longitudinal strain at the junction of the web and flange, Eqs. (l) and (m) can be written in the following form:

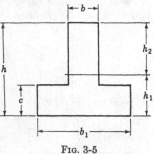

FIG. 3-5

$$\int_{\epsilon_2}^{\epsilon'} \sigma\, d\epsilon + \int_{\epsilon'}^{\epsilon_1} \frac{b_1}{b}\, \sigma\, d\epsilon = 0 \quad (n)$$

$$b\rho^2 \left(\int_{\epsilon_2}^{\epsilon'} \sigma\epsilon\, d\epsilon + \int_{\epsilon'}^{\epsilon_1} \frac{b_1}{b}\, \sigma\epsilon\, d\epsilon \right) = M \quad (o)$$

We see that in this case the ordinates of the tensile test curve AOB, Fig. 3-6, in the region corresponding to the flange of the cross section must be magnified in the ratio b_1/b. In determining the position of the neutral axis we proceed as in the preceding case and use the tension-compression test diagram, Fig. 3-6, and mark on the horizontal axis the position of the assumed length $\Delta = h/\rho$ such that the two shaded areas become numerically equal. In this manner the strains ϵ_1 and ϵ_2 in the extreme fibers are obtained. The strain ϵ' at the junction of the web and flange is obtained from the equation

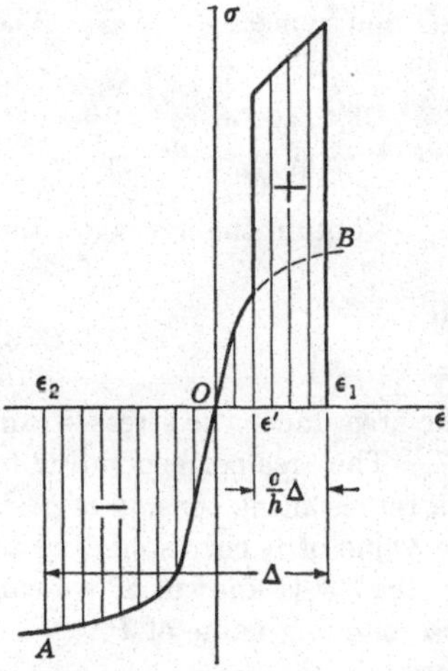

$$\frac{\epsilon_1 - \epsilon'}{\Delta} = \frac{c}{h}$$

in which c is the thickness of the flange (Fig. 3-5). Having determined the position of the neutral axis and observing that the expression in the parentheses of Eq. (o) represents the moment of the shaded areas in Fig. 3-6 with respect to the vertical axis through the origin O, we can readily calculate from Eq. (o) the moment M corresponding to the assumed value of $\Delta = h/\rho$. In this manner a curve similar to that shown in Fig. 3-4 can be constructed for a beam of T section. An I beam can be treated in a similar manner.

FIG. 3-6

3.2. Inelastic Bending Combined with Axial Load. In the case of simultaneous bending and compression of a beam, such as produced by an eccentrically applied compressive force, we can analyze the bending of the beam by the same method as described in the preceding article. Let us consider a rectanglar beam and again denote by ϵ_1 and ϵ_2 the strains in the extreme fibers on the convex and concave sides of the beam, respectively. Also, using the notation $\Delta = \epsilon_1 - \epsilon_2$, we can determine the radius of curvature ρ from Eq. (g) of the preceding article. The position of the neutral axis is determined by the values of ϵ_1 and ϵ_2 and is shifted from its position for pure bending by an amount defined by the strain ϵ_0 caused by the centrally applied load P (see Fig. 3-7).

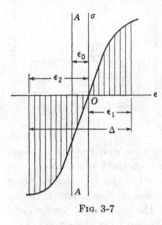

FIG. 3-7

The force acting at any cross section of the beam can be reduced to a compressive force P applied at the centroid of the cross section and a bending couple M. The values of P and M can be calculated in each particular case from statics by using the stress-strain diagram (Fig. 3-7). If y represents the distance from the neutral axis to any fiber of the beam (Fig. 3-2), the strain at any point is

$$\epsilon = \epsilon_0 + \frac{y}{\rho} \tag{a}$$

Rearranging Eq. (a) we obtain $y = \rho(\epsilon - \epsilon_0)$ and hence $dy = \rho\, d\epsilon$. The magnitude of the compressive force P is then

$$P = -b \int_{-h_2}^{h_1} \sigma \, dy = -b\rho \int_{\epsilon_2}^{\epsilon_1} \sigma \, d\epsilon = -\frac{bh}{\Delta} \int_{\epsilon_2}^{\epsilon_1} \sigma \, d\epsilon$$

Dividing this by the cross-sectional area bh, we obtain the average compressive stress

$$\sigma_c = \frac{P}{bh} = -\frac{1}{\Delta} \int_{\epsilon_2}^{\epsilon_1} \sigma \, d\epsilon \tag{3-3}$$

The integral in this expression represents the area under the stress-strain diagram, that is, the shaded area of Fig. 3-7. The area corresponding to compression is taken negative, and the area representing tension is positive. From Eq. (3-3) we can calculate the value of ϵ_2 corresponding to any assumed value of ϵ_1, provided the axial load P is known, or we can assume both ϵ_1 and ϵ_2 and calculate the corresponding value of P.

The bending moment is given by the expression

$$M = b \int_{-h_2}^{h_1} \sigma y \, dy = b\rho^2 \int_{\epsilon_2}^{\epsilon_1} \sigma(\epsilon - \epsilon_0) \, d\epsilon$$

or, since $\Delta = h/\rho$ and $I = bh^3/12$,

$$M = \frac{12I}{\rho\Delta^3} \int_{\epsilon_2}^{\epsilon_1} \sigma(\epsilon - \epsilon_0) \, d\epsilon \tag{b}$$

The integral in this expression represents the static moment of the shaded area of the stress-strain diagram (Fig. 3-7) with respect to the vertical axis AA. Thus the value of M can be calculated for any assumed values of ϵ_1 and ϵ_2. Equation (b) can be represented in the form

$$M = \frac{E''I}{\rho} \tag{3-4}$$

where

$$E'' = \frac{12}{\Delta^3} \int_{\epsilon_2}^{\epsilon_1} \sigma(\epsilon - \epsilon_0)\, d\epsilon \tag{3-5}$$

These equations have the same form as Eqs. (3-1) and (3-2) and reduce to those equations if the compressive load vanishes (and hence $\epsilon_0 = 0$).

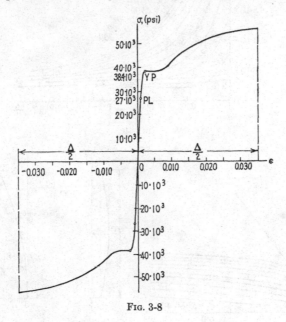

FIG. 3-8

By varying ϵ_1 and ϵ_2 in such a manner that σ_c remains constant, we obtain E'' as a function of $\Delta = \epsilon_1 - \epsilon_2 = h/\rho$ for any given value of σ_c. The resulting relation can be expressed graphically as shown in Fig. 3-9, which was plotted[1] for a structural steel having the stress-strain diagram shown in Fig. 3-8. When these curves are used with Eq. (3-4), the bending moment M can be represented as a function of Δ for each value of

[1] The curves in Fig. 3-9 are taken from the paper by M. Roš, *Proc. 2d Intern. Congr. Appl. Mech.*, Zürich, p. 368, 1926. The curves with $\sigma_c > 37,000$ psi are obtained assuming that bending occurs after yielding is produced by the compressive force.

σ_c as shown in Fig. 3-10. (In Fig. 3-10, the intermediate curves are calculated for the same values of σ_c as shown in Fig. 3-9.)

The shape of the deflection curve for an eccentrically loaded bar can be obtained by using the curves of Fig. 3-10 and applying approximate

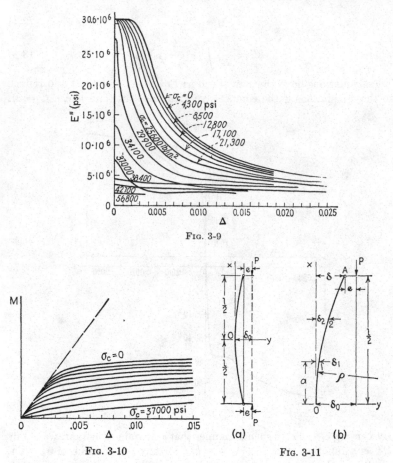

FIG. 3-9

FIG. 3-10

FIG. 3-11

methods of graphical or numerical integration.[1] For example, consider the bar shown in Fig. 3-11, symmetrical about the middle O, and of steel having the stress-strain curve shown in Fig. 3-8. We begin by assuming definite values of ϵ_1 and ϵ_2 for the center cross section and then calculate the corresponding values of P and σ_c from Eq. (3-3). Next, using the

---

[1] Several methods of integration are discussed in the paper by T. V. Kármán, *Forschungsarb.*, no. 81, Berlin, 1910.

curves of Fig. 3-10 we find the moment M which corresponds to this value of σ_c. Thus the bending moment and the compressive force for the cross section at the center of the beam are determined, and the distance $\delta_0 = M/P$ locates the line of action of the compressive force (Fig. 3-11b). Next we construct an element 0–1 of the deflection curve of small length a by using the radius $\rho = h/\Delta$ calculated for the middle of the bar. The deflection at the cross section 1 is approximately the same as for a flat circular arc. Thus we have $\delta_1 = a^2/2\rho$, and the bending moment is $M_1 = P(\delta_0 - \delta_1)$. With this moment, M_1, we find from Fig. 3-10 the corresponding value of Δ, denoted by Δ_1, and also calculate $\rho_1 = h/\Delta_1$.[1] Using this new radius, we construct the second portion 1–2 of the curve and calculate the deflection δ_2. Continuing these calculations, we arrive finally at the end A of the compressed bar and determine the deflection δ at this end and the eccentricity e of the load P corresponding to the assumed values of ϵ_1 and ϵ_2. Making such calculations for several values of ϵ_1 and ϵ_2 and selecting these values in each case so as to make P always the same, we finally obtain for P the deflection δ as a function of the eccentricity e.

The above calculations for the deflection curve can be generalized and expressed in dimensionless form, independent of any particular cross-sectional dimensions b and h of the rectangular column. First, it is apparent that the force P, producing certain deflections in a column, is proportional to the width b of the cross section. Thus if this width is changed, the force P must be changed in the same proportion in order to have the deflection curve unchanged. Second, to make the results independent of the depth h, we can make the calculations in terms of the dimensionless ratios ρ/h, δ/h, and l/h instead of the quantities ρ, δ, and l.

Following this procedure, we begin by assuming certain values for ϵ_1 and ϵ_2 and then determine the quantity

$$\Delta = \epsilon_1 - \epsilon_2 = \frac{h}{\rho}$$

Next, the average compressive stress σ_c is found from Eq. (3-3) and the bending moment from Eq. (3-4); that is,

$$M = \frac{E''}{12} \frac{h}{\rho} bh^2 \qquad (c)$$

The position of the line of action of the force P (Fig. 3-11b) is defined by the distance δ_0, for which we have $\delta_0 = M/P = M/\sigma_c bh$ or, using Eq. (c),

$$\frac{\delta_0}{h} = \frac{E''}{12\sigma_c} \frac{h}{\rho}$$

[1] A better approximation is obtained if we repeat the calculation for the first interval by taking the radius $(\rho + \rho_1)/2$ before going to the second interval.

For the deflection δ_1 at the end of the small interval a (Fig. 3-11b), we have $\delta_1 = a^2/2\rho$ or

$$\frac{\delta_1}{h} = \frac{a^2/h^2}{2(\rho/h)}$$

Thus, when we continue in this manner, the calculations are carried out in terms of dimensionless ratios and we finally obtain, for particular values of σ_c and eccentricity ratio e/h, the deflection δ/h as a function of the length l/h of the column. This function can be represented graphically by a curve. Several curves of this kind, calculated[1] for various values of σ_c and for $e/h = 0.005$, are shown in Fig. 3-12. Instead of values of l/h, values of the slenderness ratio l/r are taken as ordinates in this figure. It is seen that each of the curves has a certain maximum value of

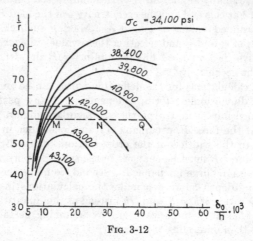

FIG. 3-12

l/r. For values of l/r below this maximum, we obtain two different values for the deflection, such as shown in the figure by the points M and N on the curve for $\sigma_c = 42,000$ psi. The point M, corresponding to the smaller deflection, represents the deflection which actually will be reached by a gradual increase of the load P from zero up to the final value equal to $\sigma_c bh$.

To obtain the deflection corresponding to the point N, we must apply some lateral forces to the column. If by using such forces we bring the bent column to the shape defined by the point N, we arrive again at a position of equilibrium. This equilibrium is unstable, however, since any further increase in the deflection does not require an increase of the

---

[1] These curves were calculated by T. V. Kármán, *op. cit.*, for steel having a yield-point stress equal to about 45,000 psi.

load but, instead, deflection proceeds with a diminishing of the load. To obtain, for instance, the deflection corresponding to the point Q, only 40,900 psi average compressive stress is required instead of 42,000 psi. For each of the maximum points, such as point K, the two possible forms of equilibrium coincide and the corresponding value of l/r is the maximum slenderness ratio at which the column can carry the compressive load $P = \sigma_c bh$, with an eccentricity ratio equal to 0.005. Thus, by using curves similar to those in Fig. 3-12, the relation between the slenderness of the column and the maximum load which it can carry can be established for any value of eccentricity.

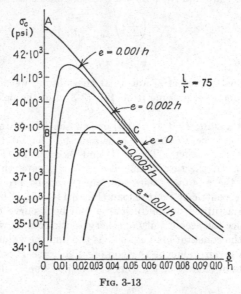

FIG. 3-13

The same limiting values of the compressive loads can be obtained in another way. It is seen in Fig. 3-12 that the points of intersection of horizontal lines and the curves, such as points M, N, and Q, give the relation between the direct compressive stress σ_c and the deflection δ for a given slenderness ratio l/r and an assumed eccentricity e. This relation can be represented by another curve. Several curves of this kind, calculated[1] for various values of the initial eccentricity and for a slenderness ratio $l/r = 75$, are given in Fig. 3-13. It is seen that for any initial eccentricity it is necessary to increase the load at the beginning in order to produce an increase of deflection, while beyond a certain limit, given by the maximum point of the curve, further deflection may proceed with

[1] *Ibid.*

a diminishing of the load. Thus the maximum points of the curves in Fig. 3-13 represent, for the given slenderness ratio and assumed eccentricity, the limiting value of the load which the column can carry.

The determination of the limiting values of the load can be simplified by assuming[1] that the deflection curve for the column in Fig. 3-11 is given by the equation

$$y = \delta \left(1 - \cos \frac{\pi x}{l} \right)$$

Then, from the usual approximate expression for the curvature of the deflection curve,[2] we find the curvature at the middle of the column to be

$$\frac{1}{\rho} = \left(\frac{d^2 y}{dx^2} \right)_{x=0} = \frac{\pi^2 \delta}{l^2}$$

from which

$$\delta = \frac{l^2}{\pi^2 \rho} = \frac{l^2 \Delta}{\pi^2 h} \tag{3-6}$$

For any assumed values of ϵ_1 and ϵ_2 at the middle, we next obtain the value of the compressive force from Eq. (3-3) and the deflection δ from Eq. (3-6).

Next, let us assume that ϵ_1 and ϵ_2 are chosen in such a manner that the compressive force P remains constant and that at the same time $\Delta = \epsilon_1 - \epsilon_2$ is increasing. Then M and $\delta_0 = M/P$ (Fig. 3-11) are increasing also. If the rate of increase of δ_0 is greater than the rate of increase of δ, the assumed deflection curve can be produced only by increasing the eccentricity e of the load (Fig. 3-11). If we have a reversed condition, the assumed deflection curve will be a curve of equilibrium only if we reduce the eccentricity e; otherwise the column will continue to deflect under the constant load P. Thus the limiting value of the eccentricity e for a given value of the compressive force P is that value at which the rates of change of δ and of δ_0 are the same. This means that

$$\frac{d\delta_0}{d\Delta} = \frac{d\delta}{d\Delta}$$

Substituting $\delta_0 = M/P$ and using formula (3-6), we obtain

$$\frac{dM}{d\Delta} = \frac{l^2 P}{\pi^2 h} \tag{3-7}$$

---

[1] Such an approximate solution was proposed by M. Roš and J. Brunner. See M. Roš, *loc. cit.*

[2] More accurate calculations, made by E. Chwalla, *Sitzber. Akad. Wiss., Wien*, vol. 137, IIa, p. 469, 1928, show that in the case of comparatively small eccentricities ($e/h < \frac{1}{6}$), the limiting compressive force producing failure is attained at a maximum deflection of less than $0.5h$. At such a deflection the elastic line is a flat curve, and the usual approximate expression for the curvature can be applied with sufficient accuracy.

Thus, to determine the limiting value of e for the assumed value of P, we need only to find on the corresponding curve in Fig. 3-10 a point for which the slope is given by the right-hand side of Eq. (3-7). Knowing the abscissa Δ and the moment M of this point, we easily obtain the values of $\delta_0 = M/P$, δ given by Eq. (3-6), and $e = \delta_0 - \delta$. When such calculations are repeated for several values of P, the carrying capacity of the column for a given value of e can be established.

The approximate method for determining the compressive load producing failure can be used also in the case of a column having a certain initial curvature. Assume, for instance, that the initial shape of the center line of the column (Fig. 3-14) is given by the equation

$$y_1 = a \cos \frac{\pi x}{l} \qquad (d)$$

and that, under the action of the compressive forces P, an additional deflection

$$y_2 = \delta \cos \frac{\pi x}{l} \qquad (e)$$

is produced. Then the change of curvature at the middle of the beam is

$$\frac{1}{\rho} - \frac{1}{\rho_0} = -\left(\frac{d^2 y_2}{dx^2}\right)_{x=0} = \delta \frac{\pi^2}{l^2} \qquad (f)$$

Assuming that the strains in the outermost fibers at the middle of the column are ϵ_1 and ϵ_2, we obtain $\Delta = \epsilon_1 - \epsilon_2$ and the corresponding change in curvature is equal to

$$\frac{1}{\rho} - \frac{1}{\rho_0} = \frac{\Delta}{h} \qquad (g)$$

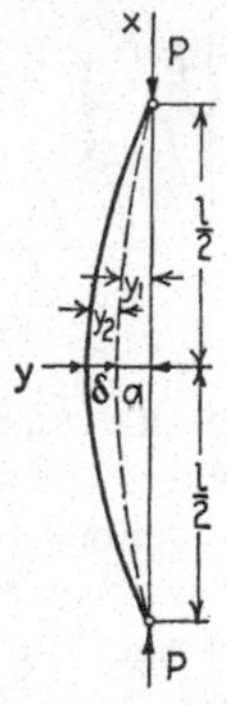

FIG. 3-14

From Eqs. (f) and (g) an equation, equivalent to Eq. (3-6), for calculation of δ can be obtained. The corresponding compressive force is obtained from Eq. (3-3), and the bending moment M at the middle from the curves in Fig. 3-10. Then the initial deflection a, which the column should have in order that the assumed bending can be actually produced by forces P, is obtained from the equation

$$P(a + \delta) = M$$

In order to obtain the limiting condition at which the load P brings the column to failure, we must proceed in exactly the same manner as in the case of an eccentrically loaded column and use Eq. (3-7). The results obtained in this way can be presented in the form of curves, each of which corresponds to a given initial deflection a and gives the values of the direct compressive stress σ_c, producing failure, as functions of the slenderness ratio l/r of the column.

3.3. Inelastic Buckling of Bars. Fundamental Case.

From the discussion of the preceding article (see Fig. 3-13) we see that the maximum load which a column can carry up to complete failure increases as the eccentricity of the applied axial load decreases. By gradually decreasing the eccentricity we finally arrive at the case of inelastic buckling of a perfectly straight, centrally loaded column. To obtain the corresponding critical stress, represented in Fig. 3-13 by point A, the Engesser-Kármán theory[1] has been applied.

It is assumed in this theory that up to the critical condition the column remains straight, and the critical load P_{cr} is calculated as the force required to maintain the column in a shape which is slightly deflected from the straight form of equilibrium. In discussing bending stresses corresponding to this small deflection, we observe that because of bending, there will be a small increase in the total compressive stress on the concave side of the column and a decrease in stress on the convex side. If the curve OBC in Fig. 3-15 represents the compression-test diagram for the material of the column and point C corresponds to the critical condition, then the stress-strain relation on the concave side of the column during small deflections is determined by the slope

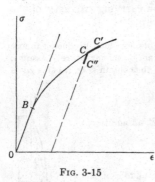

Fig. 3-15

of the tangent CC', called the *tangent modulus* E_t. On the convex side, where the stress diminishes because of bending, the stress-strain relation is defined by the slope of the line CC'', that is, by the initial modulus of elasticity E of the material. Then assuming that plane cross sections of the bar remain plane during bending, we find that the small bending

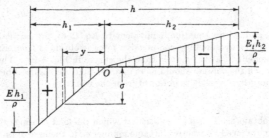

Fig. 3-16

stresses, superposed on the direct compressive stresses, will be distributed along the depth of the cross section as shown in Fig. 3-16. If ρ denotes the radius of curvature of the deflection curve, the maximum tensile and compressive stresses are Eh_1/ρ and E_th_2/ρ, respectively, and the position of the neutral axis O is found from the condition that the resultants of the tensile and compressive forces must be equal. In the case of a rectangular cross section of depth h this condition requires that

$$Eh_1{}^2 = E_th_2{}^2 \tag{a}$$

[1] F. Engesser, *Z. Ver. deut. Ingr.*, vol. 42, p. 927, 1898; T. V. Kármán, *loc. cit.*

Noting also that $h_1 + h_2 = h$, we obtain

$$h_1 = \frac{h \sqrt{E_t}}{\sqrt{E} + \sqrt{E_t}} \qquad h_2 = \frac{h \sqrt{E}}{\sqrt{E} + \sqrt{E_t}} \qquad (b)$$

If b is the width of the rectangular cross section, the bending moment represented by the stresses shown in Fig. 3-16 is

$$M = \frac{Eh_1}{\rho} \frac{bh_1}{2} \frac{2}{3} h = \frac{bh^3}{12\rho} \frac{4EE_t}{(\sqrt{E} + \sqrt{E_t})^2} \qquad (c)$$

This equation can be made to coincide with Eq. (1-3) for the deflection curve by introducing the quantity

$$E_r = \frac{4EE_t}{(\sqrt{E} + \sqrt{E_t})^2} \qquad (3\text{-}8)$$

which is called the *reduced modulus of elasticity*. With this notation we then obtain

$$M = \frac{E_r I}{\rho} \qquad (3\text{-}9)$$

or

$$Py = -E_r I \frac{d^2 y}{dx^2} \qquad (3\text{-}10)$$

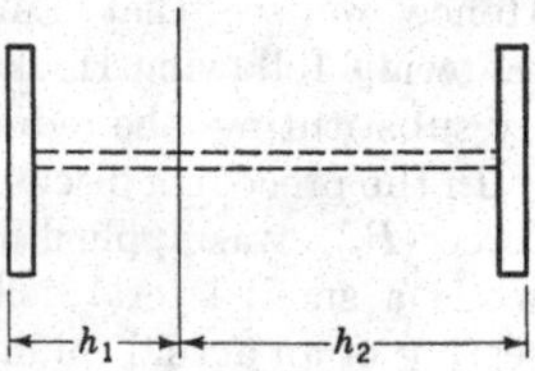

From the derivation of expression (3-8) for the reduced modulus, it is apparent that the magnitude of E_r depends not only on the mechanical properties of the material of the column but also on the shape of the cross section.

FIG. 3-17

Let us take as a second example an idealized I section, in which it is assumed that one-half of the cross-sectional area is concentrated in each flange and the area of the web is disregarded (Fig. 3-17). The forces in the flanges due to bending are

$$\frac{Eh_1}{\rho} \frac{A}{2} \qquad \text{and} \qquad \frac{E_t h_2}{\rho} \frac{A}{2}$$

so that

$$\frac{h_1}{h_2} = \frac{E_t}{E}$$

Again using the relation $h_1 + h_2 = h$, we obtain

$$h_1 = \frac{hE_t}{E + E_t} \qquad h_2 = \frac{hE}{E + E_t}$$

Since the moment of the internal forces is

$$M = \frac{Ah^2}{4\rho} \frac{2EE_t}{E + E_t}$$

we obtain finally

$$E_r = \frac{2EE_t}{E + E_t} \tag{3-11}$$

In a similar way, the expression for E_r can be found for any other shape of column cross section. The values of E_r from Eqs. (3-8) and (3-11) reduce to the modulus of elasticity E when E_t is constant and equal to E.

Returning now to Eq. (3-10) we see that it is of the same form as Eq. (2-1) for elastic buckling, except that E_r takes the place of E. Integrating this equation, we obtain for the critical load for a bar with hinged ends

$$(P_r)_{cr} = \frac{\pi^2 E_r I}{l^2} \tag{3-12}$$

and the corresponding critical stress is

$$(\sigma_r)_{cr} = \frac{\pi^2 E_r}{(l/r)^2} \tag{3-13}$$

Hence we see that Euler's column formula, derived previously for materials following Hooke's law, can also be used for inelastic materials by substituting[1] the reduced modulus E_r for the modulus of elasticity E.

In the preceding discussion it was assumed that the central compressive force $(P_r)_{cr}$ was applied first and then maintained at this constant value while a small lateral deflection was given to the column. During the testing of an actual column, the axial force increases simultaneously with lateral deflection. In such a case, the decrease of stress on the convex side of the column during the initial stages of bending may be compensated by the increase of direct compressive stress due to the continually increasing axial force. Thus the actual deformation may proceed without any release of stress in the fibers on the convex side, as was assumed in Fig. 3-16, and the stress-strain relation for the entire column is defined by the tangent modulus E_t. The differential equation of the deflection curve then becomes

$$Py = -E_t I \frac{d^2 y}{dx^2} \tag{3-14}$$

and for a column with hinged ends the critical load is

$$(P_t)_{cr} = \frac{\pi^2 E_t I}{l^2} \tag{3-15}$$

and the critical stress is

$$(\sigma_t)_{cr} = \frac{\pi^2 E_t}{(l/r)^2} \tag{3-16}$$

These latter expressions for critical load and critical stress differ from

[1] This theory of buckling is called the reduced modulus theory.

Eqs. (3-12) and (3-13), since they contain the tangent modulus E_t, which is somewhat smaller than the reduced modulus E_r and independent of the shape of the cross section. From this discussion[1] it follows that under a continuously increasing load the column begins to buckle as soon as the load reaches the value (3-15). Experimental results[2] obtained in testing solid circular rods of aluminum alloy are in satisfactory agreement with the tangent modulus theory. These results, represented in Fig. 3-18, show that for larger values of slenderness ratio the experimental points lie on Euler's curve and for shorter rods they agree with the tangent modulus curve. To construct such a curve we take several values of σ_{cr} and determine for each value the corresponding values of E_t from the compression-test diagram. Then, substituting the values of σ_{cr} and E_t

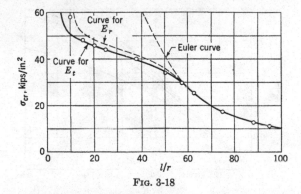

FIG. 3-18

into Eq. (3-16), we find the corresponding values of the slenderness ratio l/r.

Formulas (3-14) to (3-16) evidently can be applied in calculating critical loads in the case of perfectly elastic materials which do not follow Hooke's law. In such a case, if a small amount of bending occurs, the bending stresses are defined by the magnitude of the tangent modulus E_t.

It was already stated that the assumption that there is no release of stress on the convex side of the buckled column is correct only at the very beginning of buckling. If we wish to investigate deflections of a column with hinged ends beyond the value of the load given by Eq. (3-15), we have to take into account the release of stress on the convex side. This release of stress takes place primarily in the middle portion of the column where the bending stresses are greatest and occurs when the deflection of

---

[1] This theory, called the tangent modulus theory, was developed by F. R. Shanley, *J. Aeronaut. Sci.*, vol. 14, p. 261, 1947.

[2] See the paper by R. L. Templin, R. G. Sturm, E. C. Hartmann, and M. Holt, Aluminum Research Laboratories, Aluminum Company of America, Pittsburgh, 1938.

the column increases beyond the small initial deflection at the onset of buckling. The calculation of column deflections under this condition represents an involved problem,[1] and we shall give here only some final results of such an investigation. The solution can be simplified if the compression-test diagram is given by an analytic expression so that a formula for the tangent modulus can be obtained by differentiation. In the case of materials such as structural steel with a well-defined yield-point stress σ_{YP}, we shall use for the tangent modulus the expression[2]

$$\frac{d\sigma}{d\epsilon} = E_t = E \frac{\sigma_{YP} - \sigma}{\sigma_{YP} - c\sigma} \tag{3-17}$$

which gives $E_t = E$ for $\sigma = 0$ and $E_t = 0$ for $\sigma = \sigma_{YP}$. The corresponding compression-test diagrams for several values of the parameter c are shown in Fig. 3-19. Note that $c = 1$ corresponds to Hooke's law; that is, $\sigma = E\epsilon$. For structural steel the values $c = 0.96$ to 0.99 can be taken with good accuracy. With the use of Eq. (3-17)

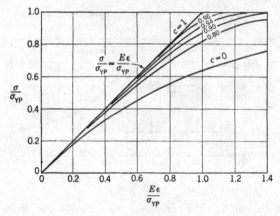

FIG. 3-19

for the tangent modulus, the maximum load P_{max} at which the column fails was calculated[3] for an idealized I section. The results of these calculations, for various values of the parameter c, are given in Fig. 3-20. The dotted line of this figure was used for constructing the dotted-line curve in Fig. 3-21. In this latter figure the curves for $(\sigma_r)_{cr}$ and $(\sigma_t)_{cr}$ from Eqs. (3-13) and (3-16) for two kinds of steel ($\sigma_{YP} = 51,200$ psi and $\sigma_{YP} = 34,100$ psi) are given also. It is seen that the stresses calculated on the basis of P_{max} are very close to $(\sigma_t)_{cr}$, and therefore the latter can be recommended for practical application. Similar results were obtained also for columns of rectangular

[1] See the papers by J. E. Duberg and T. W. Wilder, *NACA Tech. Note* 2267, 1951; U. Müllersdorf, *Der Bauingenieur*, vol. 27, p. 57, 1952; A. Pflüger, *Ingr.-Arch.*, vol. 20, p. 291, 1952; L. Hannes Larsson, *J. Aeronaut. Sci.*, vol. 23, p. 867, 1956. The results given in the following discussion are taken from the last paper.

[2] This expression for tangent modulus was suggested by Arvo Ylinen, *Teknillinen Aikakauslehti*, vol. 38, p. 9, 1948 (Finland). See also his paper in *Publ. Intern. Assoc. Bridge Structural Eng.*, vol. 16, p. 529, 1956.

[3] See L. Hannes Larsson, *loc. cit.*

cross section, and it can be concluded that for materials with a well-defined yield point, such as structural steel, the change from E_r to E_t does not result in a substantial change in the magnitude of σ_{cr}.

When Eq. (3-17) is used for materials which do not have a definite yield point, the stress σ_{YP} should be replaced by σ_{ult}. Suggested values of the parameter c are $c = 0.875$ for pine wood and $c = 0$ for concrete.

In conclusion, it is important to observe the differences in the typical load-deflection curves for elastic and inelastic (or plastic) buckling. In the case of slender bars which buckle elastically, the ideal load-deflection curve has the shape shown in Fig. 2-29. If there are inaccuracies present, the

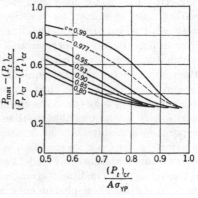

FIG. 3-20

curves have the forms shown in Fig. 4-3. However, in all cases, an increase in the deflection requires an increase in the load. The phenomenon of buckling is not sudden in character, and if the inaccuracies are reduced, it is possible to determine P_{cr} experimentally with good accuracy.

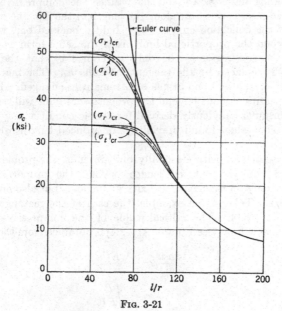

FIG. 3-21

In the case of inelastic buckling the phenomenon is quite different. From the curves in Fig. 3-13, it is apparent that when the load reaches the critical or maximum value, there is a rapid diminishing of the load with respect to an increase in deflection. This accounts for the suddenness with which inelastic buckling occurs. It is seen also that small inaccuracies have considerable influence on the value of the load P_{max} which the column can carry. Furthermore, at each value of the load there are two positions of equilibrium corresponding to points such as B and C in Fig. 3-13. Thus, as we approach the maximum load, an accidental force may cause the column to jump suddenly from the stable position B to the unstable position C. All these factors explain why experimental results become scattered in the plastic region. In later portions of this book we shall find in some cases of buckling of shells that curves similar to those in Fig. 3-13 are obtained also in the elastic region. As soon as such conditions occur, we can expect a character of suddenness in the buckling phenomenon and a wide scatter of experimental results.

3.4. Inelastic Buckling of Bars with Other End Conditions. The methods which were applied in the preceding article to the fundamental case of a bar with hinged ends can be applied also to bars with other end conditions. Columns with various end conditions were discussed for elastic buckling in Arts. 2.1 and 2.2. Since the compressive stress in each of these cases is constant along the length of the bar, the differential equation of the deflection curve of the slightly buckled bar, when compressed beyond the proportional limit, is of the same form as that used within the elastic region. The only difference is that the constant modulus E is replaced by the tangent modulus E_t. The mathematical expressions for the end conditions also remain unchanged. Hence the formulas for critical loads beyond the proportional limit will be obtained from the formulas previously derived for elastic conditions by replacing E by E_t. The values found previously for reduced lengths also remain unchanged.

In the case of bars with elastically built-in ends, the problem is more complicated and the reduced length beyond the proportional limit depends not only on the degree of fixity at the ends but also on the magnitude of l/r. Take, as an example, the case of the rectangular frame shown in Fig. 2-14. The critical value of the compressive forces P within the elastic limit [see Eq. (b), Art. 2.4] is obtained from the equation

$$\frac{\tan u}{u} = -\frac{EIb}{EI_1 l} \tag{a}$$

in which

$$u = \frac{l}{2}\sqrt{\frac{P}{EI}}$$

Beyond the proportional limit the flexural rigidity of the vertical compressed bars becomes $E_t I$ while the flexural rigidity of the horizontal bars remains unchanged. Hence the factor on the right-hand side of Eq. (a) changes with the value of σ_{cr}, and the root of this equation, which defines the reduced length, changes also.

As an example, let us consider a square frame with $I = I_1$. Within the elastic limit Eq. (a) for this case is

$$\frac{\tan u}{u} = -1$$

and gives

$$u = 2.029$$

so that

$$P_{cr} = \frac{\pi^2 E I}{(0.774l)^2}$$

and the reduced length is $L = 0.774l$. Beyond the proportional limit, Eq. (a) for the square frame becomes

$$\frac{\tan u}{u} = -\frac{E_t}{E} \tag{b}$$

where

$$u = \frac{l}{2}\sqrt{\frac{P}{E_t I}} = \frac{l}{2r}\sqrt{\frac{\sigma}{E_t}}$$

Since E_t is smaller than E, the root of this equation determining the critical load is larger and the corresponding reduced length is smaller than that obtained above for elastic conditions. Taking a series of values for σ_{cr} and the corresponding values of E_t, we can calculate from Eq. (b) the corresponding values of l/r and can represent by a curve the relation between the value of σ_{cr} and the slenderness ratio l/r. In this way it can be shown that the reduced length in this case varies from $L = 0.774l$ when $E_t = E$ to $L = 0.5l$ when $E_t = 0$. This result should be expected, since, with an increase of σ_{cr}, E_t and the flexural rigidity of the vertical bars decrease while the flexural rigidity of the horizontal bars remains unchanged. Thus the relative fixity at the ends of the vertical bars is increasing, and when σ_{cr} approaches the yield-point stress, the end conditions approach those of a bar with rigidly built-in ends, for which case $L = 0.5l$.

We have similar conditions in the case of continuous compressed bars on elastic supports and bars on elastic foundations. Beyond the proportional limit the relative rigidity of the supports and foundation is larger than within the elastic limit and the reduced length is smaller than that calculated before (see Arts. 2.5, 2.6, 2.10) on the basis of Hooke's law.

In the case of prismatic bars compressed by forces distributed along the length, the compressive stress is not constant along the length of the bar. Therefore in the case of buckling beyond the proportional limit, the tangent modulus E_t also is not constant for the entire length, so that a bar with variable flexural rigidity is obtained. As an example, consider the column shown in Fig. 2-38. If a compressive stress beyond the proportional limit is produced at the lower end of the bar, the stress in the upper portion of the bar continues to be within the elastic limit. For calculating the critical load in this case, it will be necessary to consider the lower portion of the bar as a portion with a variable flexural rigidity $E_t I$ while the upper portion has a constant flexural rigidity EI. Thus, the problem of calculating the critical load beyond the proportional limit becomes very complicated in this case. For an approximate calculation of the critical load, we can use the same formula as was obtained for elastic conditions [see Eq. (2-43)] and substitute for E the tangent modulus E_t, calculated for the lower end of the bar. This will be equivalent to the assumption that beyond the proportional limit the bar continues to possess a constant flexural rigidity and that this rigidity is the same as that at the lower end of the bar which is subjected to the highest compressive stress. Quite naturally, such an assumption results in too low a value for the critical load, and in using it we shall always be on the safe side.

A similar problem is also found in the case of bars of variable cross section. If the compressive stress varies along the length of the bar, an accurate calculation of the critical load beyond the proportional limit requires the introduction of the tangent modulus E_t for the inelastically compressed portions of the bar. If E_t is constant along these portions, as in the case shown in Fig. 2-38, the accurate calculation of the critical load can be accomplished without much difficulty. But the problem becomes more complicated if E_t is variable. We shall always be on the safe side if in such cases we use formulas derived for elastic conditions and substitute in them for E the tangent modulus E_t calculated for the cross section with the maximum compressive stress.

In the case of built-up columns the critical load beyond the proportional limit can be calculated also by introducing E_t instead of E. Take the case shown in Fig. 2-57. Under the action of an axial load the chords of the column will be uniformly compressed while the battens are unstressed. Hence in calculating the critical load beyond the proportional limit, we can use formula (2-64). It will be necessary to substitute E_t for E only in those terms relating to the chords and to keep E and G in those terms relating to the battens. Thus, if compression of the column exceeds the proportional limit, the battens become relatively more rigid and the properties of the built-up column approach those of a solid column.

CHAPTER 4

EXPERIMENTS AND DESIGN FORMULAS

4.1. Column Tests. The first experiments with buckling of centrally compressed prismatic bars were made by Musschenbroek.[1] As a result of his tests, he concluded that the buckling load was inversely proportional to the square of the length of the column, a result which was obtained by Euler 30 years later from mathematical analysis. At first, engineers did not accept the results of Musschenbroek's experiments and of Euler's theory. For instance, even Coulomb[2] continued to assume that the strength of a column was directly proportional to the cross-sectional area and independent of the length. These views were supported by experiments made on wooden and cast-iron columns of comparatively short length. Struts of this type usually fail under loads much less than Euler's critical load, and failure is due principally to crushing of the material and not to lateral buckling. E. Lamarle[3] was the first to give a satisfactory explanation of the discrepancy between theoretical and experimental results. He showed that Euler's theory is in agreement with experiments provided the fundamental assumptions of the theory regarding perfect elasticity of the material and ideal conditions at the ends are fulfilled.

Later experimenters[4] established definitely the validity of Euler's formula. In these experiments great care was taken to fulfill the end conditions assumed in the theory and to secure central application of the compressive load.[5] The tests showed that experimental values of σ_{cr}

[1] P. van Musschenbroek, "Introductio ad cohaerentiam corporum firmorum," Lugduni, 1729. See also a French translation by P. Massuet, "Essai de physique," Leyden, 1739.

[2] See Coulomb's memoir in the "Mémoires . . . par divers savans," Paris, 1776.

[3] E. Lamarle, *Ann. trav. publics de Belg.*, vol. 3, pp. 1–64, 1845; vol. 4, pp. 1–36, 1846.

[4] I. Bauschinger, *Mitt. mech.-tech. Lab. tech. Hochschule, München*, no. 15, 1889; A. Considère, *Congr. intern. des procédés de construct.*, Paris, vol. 3, p. 371, 1889; L. Tetmajer, "Die Gesetze der Knickung- und der zusammengesetzten Druckfestigkeit der technisch wichtigsten Baustoffe," 3d ed., Leipzig and Vienna, 1903.

[5] Considère was the first to introduce an adjustable arrangement at the ends so that the point of application of the load could be shifted slightly with the column under the load.

fall on Euler's curve provided the slenderness ratio of the column is
such that buckling occurs at a compressive stress below the proportional
limit of the material.   Figure 4-1 represents some test results[1] obtained
with mild structural steel specimens of various shapes of cross section.
It is seen that for $l/r > 105$ the results obtained follow Euler's curve
satisfactorily.   The value of l/r above which Euler's formula can be
applied depends on the proportional limit of the material and for high-
strength steels such as used in bridges is about 75.

Further progress in the experimental study of buckling problems was
accomplished by Kármán.[2]   In his experiments rectangular steel bars

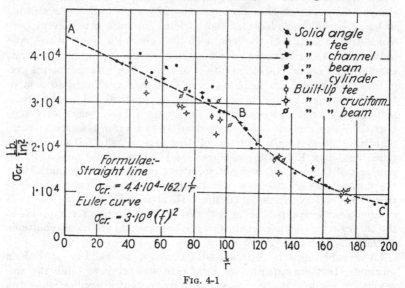

FIG. 4-1

with a proportional limit of 35,000 psi and a yield point of 46,000 psi
were tested.   Freedom of rotation of the ends of the columns was assured
by the use of knife-edges for applying the load, and the experimental
results checked Euler's formula with an accuracy of $1\frac{1}{2}$ per cent.   Kármán extended his experiments into the region of plastic deformation also.
By calculating the reduced modulus from the compression-test diagram,
as explained in Art. 3.3, he showed that Euler's formula can be applied
also in the case of shorter bars in which the critical stress exceeds the proportional limit of the material.   The upper curve in Fig. 4-2 gives values
of σ_{cr} as a function of the slenderness ratio calculated from Euler's formula

[1] From Tetmajer, *op. cit.*
[2] T. V. Kármán, *Forschungsarb.*, no. 81, 1910, Berlin.

by using the reduced modulus.   The lower curve in the same figure gives
the ultimate[1] stress when the load is applied with an eccentricity equal to
0.005h (see Fig. 3-13), where h is the depth of the cross section.   It is
seen that most of the experimental results, shown by the small circles,
are within the region between the two curves.   For $l/r > 90$, the experimental results follow Euler's curve very closely.   Between the proportional limit and the yield point the experimental results are in good
agreement with the results obtained theoretically by using the reduced
modulus.   For $l/r < 40$, the critical stress is above the yield point of the
material and the curve of σ_{cr} turns sharply upward.   Such values for
critical stresses can be obtained experimentally only if special precautions

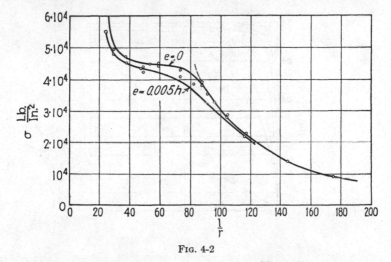

FIG. 4-2

are taken against buckling at the yield-point stress; thus they have no
practical significance in the design of columns.

In experimenting with the buckling of columns, it is usual practice to
represent the deflections of a column as a function of the centrally applied
load.   In an ideal case there will be no deflection up to the critical value
of the load, and above this point the load-deflection curve is as shown in
Fig. 2-29.   Owing to various kinds of imperfections, such as some
unavoidable initial curvature of the column, eccentricity in application of
the load, or nonhomogeneity of the material, the column begins to deflect
with the beginning of loading and usually fails before Euler's load is
reached.   The shapes of the load-deflection curves depend on the
accuracy with which the theoretical assumptions are fulfilled.   Several

---

[1] The stress corresponding to the maximum load.

curves of this kind, as obtained by various experimenters working within the elastic range, are shown in Fig. 4-3.   As the experimental techniques improved and the end conditions approached more closely the theoretical assumptions, the load-deflection curves approached more and more closely the horizontal line corresponding to the critical load.

Very accurate experiments were made in the Berlin-Dahlem materials testing laboratory by using a special construction for the end supports of the columns.[1]   In Fig. 4-4 some experimental results obtained in this laboratory are shown.   The material tested was a structural steel with a pronounced yield point at about 45,000 psi.   It is seen that for $l/r > 80$ the results follow Euler's curve very satisfactorily.   For shorter bars the

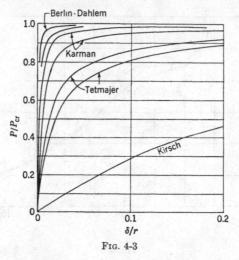

FIG. 4-3

yield-point stress should be considered as the critical stress.   Hence, for short columns it may be economical to use materials with a high yield point, whereas for long, slender columns this offers no advantage, since, for steel, the modulus of elasticity is practically unchanged.

The significance of the yield-point stress in column tests was indicated also by the experiments of the ASCE Special Committee on Steel Column Research.[2]   In this investigation special roller-bearing blocks were used to obtain hinged-end conditions.   This arrangement proved very useful for large columns requiring the application of considerable axial load,

[1] See paper by K. Memmler, *Proc. 2d Intern. Congr. Appl. Mech.*, Zürich, p. 357, 1926, and the book by W. Rein, "Versuche zur Ermittlung der Knickspannungen für verschiedene Baustähle," Springer-Verlag, Berlin, 1930.

[2] See *Trans. ASCE*, vol. 89, p. 1485, 1926; vol. 95, p. 1152, 1931; and vol. 98, p. 1376, 1933.

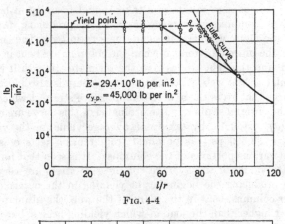

$$E = 29.4 \cdot 10^6 \text{ lb per in.}^2$$
$$\sigma_{y.p.} = 45{,}000 \text{ lb per in.}^2$$

FIG. 4-4

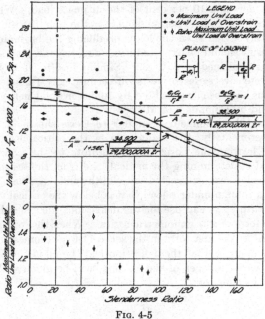

FIG. 4-5

which cannot be transmitted easily through knife edges. Figure 4-5 represents some of the results obtained by the committee in testing H-section columns under eccentrically applied loads. In one series of tests the eccentricities were in the plane of the web, and in the other series they were in the plane perpendicular to the web. The ratio e/s of the eccentricity to the core radius was taken equal to unity in both cases. The average compressive stresses $\sigma_c = P/A$, producing failure, are plotted in the figure against the slenderness ratio l/r. The small black circles give the ultimate strength when the columns are bent in the plane of the web, and the light circles give the ultimate strength if bending is perpendicular to the web. For comparison, there is also plotted in the figure a curve that gives the average compressive stress at which yielding begins in the outermost fiber. The ordinates of this curve are calculated from Eq. (d), Art. 1.13, by taking an average value for the yield-point stress ($\sigma_{YP} = 38,500$ psi) as obtained from tension tests of specimens taken from various portions of the columns. It is seen that for $l/r > 60$ the ultimate values of the average compressive stress are very close to the values producing the beginning of yielding in the outermost fibers. For shorter columns, bent in the plane of the web, the ultimate strength is somewhat higher than the load at which yielding begins, but the difference is not larger than 10 per cent of the ultimate load. When a short column is bent in the plane perpendicular to the web, the ultimate loads are considerably higher than the loads producing the beginning of yielding. This result should be expected if bending beyond the yield point is considered, as explained in Art. 3.2. Similar results were obtained by M. Rŏs.[1] Extensive series of tests on eccentrically loaded steel wide-flange columns have been carried out by Johnston and Cheney[2] at Lehigh University and by Campus and Massonnet.[3]

When a load-deflection curve similar to those shown in Fig. 4-3 is obtained experimentally, the magnitude of the critical load is usually obtained by drawing the horizontal asymptote to the curve. A very useful method of determining the critical load from the test data within the elastic region was suggested by Southwell.[4] Assuming that the deflection of a column under a load that is below the critical value is due to initial curvature, we can use for the deflections the general expression (1-61) in the form of a trigonometric series. When the load approaches the critical value, the first term in the series (1-61) becomes predominant and it can be assumed that the deflection δ at the middle of the column, measured at various stages of the loading, will be given with sufficient accuracy by the equation

$$\delta = \frac{a_1}{P_{cr}/P - 1} \tag{a}$$

in which a_1 is the initial deflection corresponding to the first term in

[1] M. Rŏs, *Proc. 2d Intern. Congr. Appl. Mech.*, Zürich, p. 368, 1926.

[2] B. Johnston and L. Cheney, Steel Columns of Rolled Wide Flange Section, *AISC Progr. Repts.* 1 and 2, 1942.

[3] F. Campus and C. Massonnet, *Compt. rend. recherches*, I.R.S.I.A., no. 17, Brussels, April, 1956.

[4] R. V. Southwell, *Proc. Roy. Soc., London*, series A, vol. 135, p. 601, 1932.

series (1-61). From this equation we obtain

$$\frac{\delta}{P} P_{cr} - \delta = a_1 \tag{b}$$

which shows that, if we plot the ratio δ/P against the measured deflection δ, the points will fall on a straight line (Fig. 4-6).[1] This line will cut the horizontal axis ($\delta/P = 0$) at the distance a_1 from the origin, and the inverse slope of the line gives the critical load.

If, instead of assuming an initial curvature, we assume that bending of the column is due to eccentric application of the load P, it will be found that the deflection at the middle can be represented with sufficient accuracy by expression (e), Art. 1.11; that is,

$$\delta = \frac{4e}{\pi} \frac{1}{P_{cr}/P - 1} \tag{c}$$

Considering a general case of the combined effect of an initial curvature and some eccentricity in application of the load, we find that the deflection at the middle of the column is

$$\delta = \left(a_1 + \frac{4e}{\pi}\right) \frac{1}{P_{cr}/P - 1} \tag{d}$$

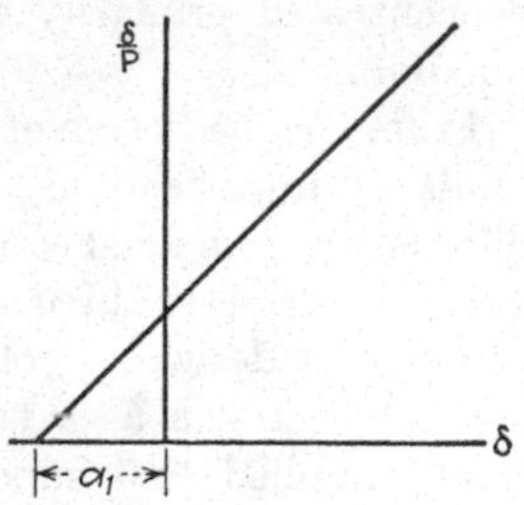

FIG. 4-6

and an equation, analogous to Eq. (b), again holds. Thus for any combination of initial curvature and eccentricity in application of the load, the critical load can be obtained as the inverse slope of a straight line such as shown in Fig. 4-6.

It is seen from Eq. (d) that, by taking $e = -\pi a_1/4$, we can eliminate the deflections at the middle of the column produced by a load below the critical value. This explains why a very accurate value for P_{cr} can be obtained by using an adjustable support for the ends of the column. In such experiments the magnitude of the eccentricities is determined by adjusting the point of application of the load in such a way as to compensate for the initial curvature.

It should be noted that an approximate value for the deflection due to eccentricity has been used in the derivation of Eq. (d). If the exact value of that deflection is used [see Eq. (1-33)], it can be shown that at the beginning of loading the column may deflect in one direction and later on suddenly buckle in the opposite direction.[2]

---

[1] This graph is usually called the Southwell plot. It has also been used for measured strains; see M. S. Gregory, *Civil Eng. (London)*, vol. 55, no. 642, 1960, and *Australian J. Appl. Sci.*, vol. 10, pp. 371–376, 1959.

[2] This phenomenon of a reversal in the direction of deflection has been investigated by H. Zimmermann, *Sitzber. Akad. Wiss., Berlin*, vol. 25, p. 262, 1923; see also his book, "Lehre vom Knicken auf neuer Grundlage," Berlin, 1930. Experiments are in satisfactory agreement with this theory; see K. Memmler, *loc. cit.*

In the case of shorter bars for which the critical load is above the proportional limit of the material, the load-deflection curves have the form indicated in Fig. 3-13.   It is seen that at a very small deflection the load reaches its maximum and then the column buckles suddenly, since the load necessary to maintain any further deflection falls off rapidly as the deflection is increased.   The maximum load reached in the experiment is usually taken as the critical load.   It can be seen that this maximum approaches the critical value calculated from Euler's equation by using the tangent modulus, when the eccentricity in the application of the load approaches zero.   The shape of the deflection curve of the buckled bar in this case is no longer sinusoidal, the permanent deformation being concentrated primarily at the middle, where the bending moment is a maximum.

From the discussion of this article it can be seen that, owing to various kinds of imperfections, actual columns will behave under load quite differently from ideal columns.   To take this into account in determining working stresses, three rather distinct approaches can be made to the problem of designing columns in structures: (1) the ideal-column formulas can be taken as a basis of column design, and a suitable factor of safety applied to compensate for the effect of various imperfections; (2) a factor of safety can be applied to an empirical formula, certain constants of which have been adjusted to make the formula fit the results of tests; and (3) the column can be assumed from the very beginning to have certain amounts of imperfection, and the safe load can be determined as a certain portion of the load at which yielding of the material begins.   Each of these methods of designing columns will be discussed in the following articles.

4.2. Ideal-column Formulas as a Basis of Column Design.   The experiments discussed in the previous article indicate that, in the case of a straight column with a centrally applied compressive force, the critical value of the compressive stress can be calculated with sufficient accuracy if the compression-test diagram for the material of the column is known. Euler's formula must be used for the calculations within the elastic range, while beyond the proportional limit the modified Euler's formula must be applied, using the tangent modulus E_t instead of E.   As a result of such calculations a diagram representing σ_{cr} as a function of the slenderness ratio can be obtained.   In Fig. 4-7 are shown two diagrams of this kind, calculated for two different kinds of structural steel (steel No. 54 with $\sigma_{YP} \approx 50,000$ psi and steel No. 37 with $\sigma_{YP} \approx 34,000$ psi).[1]   Up to

[1] These curves are taken from the paper by W. Gehler, *Proc. 2d Intern. Congr. Appl. Mech.*, Zürich, p. 364, 1926. E_r instead of E_t was used in these calculations, and there are given also the necessary portions of the compression-test diagrams and the curves representing the reduced moduli as functions of the direct compressive stress.

the proportional limit Euler's curve is used in each case, and above that limit curves based on the reduced moduli are used. At $l/r \approx 50$ the critical-stress diagram turns upward and σ_{cr} begins to increase with a further decrease in the slenderness ratio. This increase of σ_{cr} above the yield-point stress should not be considered in practical design, since it can be obtained only if special precautions are taken to prevent the column from buckling at the yield-point stress.

For practical application, each of the above diagrams can be replaced in the inelastic region by two straight lines, a horizontal line for the portion above the yield-point stress and an inclined line for the portion

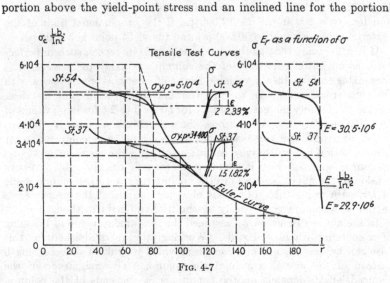

FIG. 4-7

between the yield point and the proportional limit. In this way a complete critical-stress diagram is obtained provided the yield point and the proportional limit of the material are determined from experiments. Such approximate diagrams are shown in Fig. 4-7 by dotted lines.

When the diagram of critical stress is known, the allowable stress is obtained for any value of the slenderness ratio by dividing σ_{cr} by the factor of safety. The selection of a proper factor of safety presents considerable difficulty in the design of columns. The principal cause of this difficulty lies in the fact that the behavior of columns under compression is different from that assumed in Euler's theory and is determined primarily by the magnitudes of the various imperfections, such as initial curvature of the column, eccentricity in application of the load, and nonhomogeneity of the material. The same types of imperfections are encountered also in other structures, such as beams subjected to the action of lateral loading, but in those cases the effect of the imperfections is negligible, while in the case of columns imperfections have a great effect on the deformation. There is little known about the magnitude of imperfections encountered in actual construction, and this fact is frequently compensated for by choosing a larger factor of safety.

The simplest method of choosing the factor of safety for columns is to assume that the effect of the various imperfections on the deformation and maximum fiber stress is independent of the slenderness ratio. Then the factor of safety will be constant for all values of l/r. For example, using a safety factor of 2.5 for the steel represented by the lower curve in Fig. 4-7, we find that the allowable stress for $l/r < 60$ will be 13,700 psi and for $l/r = 100$ it will be 12,000 psi if the proportional limit of the material is taken equal to 30,000 psi and the yield point is 34,000 psi.

If it is assumed that as the slenderness ratio increases, such imperfections as an initial curvature of the column are likely to increase, it is reasonable to introduce a variable factor of safety which increases with the slenderness ratio. In present German specifications, for instance, the factor of safety increases from 1.7 for $l/r = 0$ to 2.5 for large values of l/r. To simplify the calculations, these specifications give a constant allowable stress, the same as for simple tension, and compensate for column action by multiplying the force acting in the member by a *magnification factor* ω, which is equal to the ratio of the allowable stress in tension to the allowable stress in the column for the corresponding value of l/r. Having a table of values of ω, we can select the proper cross section of a column very readily by the trial-and-error method.

In the above discussion we assumed the fundamental case of buckling of a column with hinged ends. Working stresses established for this case can be used also in other cases provided we take a reduced length instead of the actual length of the column. The magnitude of the reduced length depends on the conditions at the ends of the column, on the manner of distribution of the compressive loads along the length of the column, and on the shape of the column if the cross section is not constant. Numerical data necessary for the calculation of the reduced length in the elastic range of the material were given for various cases in Chap. 2. In discussing buckling of columns beyond the proportional limit (see Art. 3.3), it was shown that the design will always be on the safe side if the reduced length is taken the same as in the elastic range and if, in calculating the reduced length, the cross section with the maximum compressive stress is taken as a basis.

The preceding discussion applies to solid columns and to columns built up of rolled sections, provided the sections are properly riveted or welded together.[1] The presence of rivets does not reduce appreciably the load capacity or flexural rigidity of the column, and in the calculation

---

[1] The question of required distances between rivets will be discussed in Art. 4.6.

of the slenderness ratio and cross-sectional area, it is usual practice to neglect the rivet holes.[1]

4.3. Empirical Formulas for Column Design. Instead of the critical-stress diagrams discussed in the previous article, empirical formulas are sometimes used in column design. One of the oldest formulas of this kind was originated by Tredgold[2] and later adapted by Gordon to represent the results of Hodgkinson's experiments. The final form of the formula was given by Rankine, and so it is called the *Rankine formula* or the *Rankine-Gordon* formula. The allowable average compressive stress by this formula is

$$(\sigma_c)_W = \frac{a}{1 + b(l/r)^2} \tag{a}$$

in which a is a stress and b is a numerical factor, both of which are constant for a given material. Tetmajer[3] showed that, to bring this equation into agreement with experiments, the factor b cannot be constant and must diminish as l/r increases. This fact is usually disregarded, and Eq. (a) still is used often in column design. By a proper selection of constants, it can be made to agree satisfactorily with the results of experiments within certain limits. For instance, the American Institute of Steel Construction (AISC) specifications of 1949 give for the safe stress (psi) on the gross section of a secondary compression member

$$(\sigma_c)_W = \frac{18,000}{1 + l^2/18,000r^2}$$

for l/r between 120 and 200. This same formula is given in the Building Code of New York City (1945) for main members with l/r between 60 and 120 and for secondary members with l/r between 60 and 200. For $l/r < 60$, an allowable stress of 15,000 psi is specified.

The *straight-line formula* gives the allowable average compressive stress in the form

$$(\sigma_c)_W = a - b\frac{l}{r} \tag{b}$$

in which the constants a and b depend upon the mechanical properties of the material and the safety factor. One such formula is used in the

---

[1] The question of the effect of rivet holes on the magnitude of the buckling load has been investigated by A. Föppl, *Mitt. Mech.-Tech. Lab. Tech. Hochschule, München*, no. 25, 1897. See also paper by Timoshenko in *Bull. Polytech. Inst., Kiev*, 1908.

[2] Regarding the history of the formula, see E. H. Salmon, "Columns," London, 1921; Todhunter and Pearson, "History of the Theory of Elasticity," vol. 1, p. 105, Cambridge, 1886; Timoshenko, "History of Strength of Materials," p. 208, McGraw-Hill Book Company, Inc., New York, 1953.

[3] Tetmajer, *loc. cit.*

Chicago Building Code and is given as

$$(\sigma_c)_W = 16,000 - 70\,\frac{l}{r} \qquad\qquad (c)$$

for $30 < l/r < 120$ for main members and for $30 < l/r < 150$ for secondary members. For values of $l/r < 30$, $(\sigma_c)_W = 14,000$ psi is used.

The above formula was obtained as a result of experiments by Tetmajer and Bauschinger[1] on structural steel columns with hinged ends. The experiments suggested for the critical value of the average compressive stress the formula

$$\sigma_{cr} = 48,000 - 210\,\frac{l}{r} \qquad\qquad (d)$$

Tetmajer recommended this formula for $l/r < 110$. Formula (c) is obtained from (d) by using a safety factor of 3. The Aluminum Company of America (ALCOA)[2] specifies a straight-line column formula for values of l/r below a certain limiting value and the Euler formula for l/r above that limit.

The *parabolic formula* proposed by J. B. Johnson[3] is also in common use. It gives for the allowable value of the average compressive stress

$$(\sigma_{cr})_W = a - b\left(\frac{l}{r}\right)^2 \qquad\qquad (e)$$

in which the constants a and b depend upon the mechanical properties of the material and the safety factor. For example, the AISC specifies the formula

$$(\sigma_{cr})_W = 17,000 - 0.485\left(\frac{l}{r}\right)^2$$

for $l/r < 120$. The American Railway Engineering Association (AREA) and the American Association of State Highway Officials (AASHO) specify

---

[1] See F. S. Jasinsky, "Scientific Papers," vol. 1, St. Petersburg, 1902, and *Ann. ponts et chaussées*, 7 série, vol. 8, p. 256, 1894. See also an extensive analysis of experimental results made by J. M. Moncrieff, *Proc. ASCE*, vol. 45, 1900, and his book "The Practical Column under Central or Eccentric Loading," New York, 1901.

[2] "Alcoa Structural Handbook," Aluminum Company of America, Pittsburgh, Pa., 1956. For the results of tests on aluminum alloy columns and comparison with the formulas see R. L. Templin, R. G. Sturm, E. C. Hartmann, and M. Holt, Column Strength of Various Aluminum Alloys, *Tech. Paper* 1, Aluminum Research Laboratories, Aluminum Company of America, 1938; H. N. Hill and J. W. Clark, Straightline Column Formulas for Aluminum Alloys, *Tech. Paper* 12, Aluminum Company of America, 1955.

[3] See C. E. Fuller and W. A. Johnston, "Applied Mechanics," vol. 2, p. 359, 1919. See also the paper by A. Ostenfeld, *Z. Ver. deut. Ingr.*, vol. 42, p. 1462, 1898.

$$(\sigma_{cr})_W = 15,000 - \frac{1}{4}\left(\frac{l}{r}\right)^2$$

for $l/r < 140$.

A variety of other column formulas have been derived by using the modified Euler formula [Eq. (3-13)] and taking the reduced modulus E_r as a function of the critical compressive stress. Formulas of this type were derived by Strand[1] and Frandsen.[2]

4.4. Assumed Inaccuracies as a Basis of Column Design. In the discussion of the application of Euler's formula in column design (Art. 4.2), it was indicated that the principal difficulty lies in the selection of a proper factor of safety to compensate for the various imperfections in a column. Under such circumstances it is logical to assume from the beginning that certain inaccuracies exist in the column rather than to assume an ideal case. Then a formula can be derived that contains not only the dimensions of the column and the quantities defining the mechanical properties of the material but also the values of the assumed inaccuracies. When these inaccuracies explicitly appear in a design formula, the selection of a safety factor can be put on a more reliable basis.

The principal imperfections that make the behavior of an actual column different from an ideal column are (1) unavoidable eccentricity in the application of the compressive load, (2) initial curvature of the column, and (3) nonhomogeneity of the material. In the discussion of load-deflection curves obtained from experiments with columns, it was shown (see p. 191) that the effect on the deflection of eccentricity in load application can be compensated for by assuming a properly chosen initial curvature of the column. Lack of homogeneity of the material can also be compensated for in an analogous manner. Assume for simplicity that a column consists of two parallel bars of different moduli joined together. To have uniform compression without lateral bending in such a column, the load must be applied at some point other than the centroid of the cross section. The position of this point depends not only on the shape of the cross section but also on the ratio of the two moduli. Hence, the effect of the nonhomogeneity of the material in this case is equivalent to the effect of a certain eccentricity and can also be compensated for by a properly chosen initial curvature of the column.

Many investigators have tried to establish the amount of eccentricity in load application by analyzing available experimental data on deflections of compressed columns. These calculations usually assume a

[1] Strand, Torbjörn, *Zentr. Bauverwaltung*, p. 88, 1914. See also R. Mayer, "Die Knickfestigkeit," p. 74, Springer-Verlag, Berlin, 1921.

[2] P. M. Frandsen, *Pub. Intern. Assoc. Bridge Structural Eng.*, Zürich, vol. 1, p. 195, 1932. See also the paper by W. R. Osgood, *Natl. Bur. Standards Research Paper* **492**, 1932.

constant eccentricity at both ends of the column and a certain value for the yield-point stress; then, from the magnitude of the load producing failure, the value of the eccentricity can be calculated. Analyzing Tetmajer's experiments in this way, Marston[1] and Jensen[2] found as average values of the ratio of the eccentricity to the core radius the values $e/s = 0.06$ and $e/s = 0.07$. The same values have been obtained also from the analysis of Lilly's experiments.[3] Instead of assuming that the eccentricity is proportional to the core radius, it seems more logical to

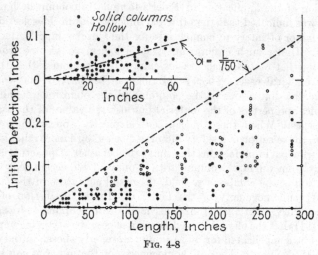

Fig. 4-8

assume that it depends on the length of the column.[4] On the basis of a comparative study of experimental data, Salmon recommends, for instance,

$$e = 0.001l \qquad (a)$$

The question of initial curvature also was studied by various experimenters. The results of these investigations were collected by Salmon[5] and are given in Fig. 4-8, in which the initial deflection a (the maximum distance of any point on the center line from the straight line joining the centroids of the end cross sections) is plotted against the length of the

[1] A. Marston, *Trans. ASCE*, vol. 39, p. 108, 1897.

[2] C. Jensen, *Engineering*, London, vol. 85, p. 433, 1908.

[3] W. E. Lilly, *Trans. ASCE*, vol. 76, p. 258, New York. Considerable data regarding inaccuracies in columns was collected by E. H. Salmon, "Columns," London, 1921. See also his discussion in *Trans. ASCE*, vol. 95, p. 1258, 1931.

[4] Several empirical formulas expressing the ratio e/s as a function of slenderness ratio are given in Salmon, *op. cit.*

[5] *Ibid.*

column. It is seen that practically all the points are below the straight line

$$a = \frac{l}{750} \qquad (b)$$

which is proposed for calculating the probable initial deflection in actual columns.

In addition to the effect of eccentricity in load application and initial curvature, the effect of nonhomogeneity of the material and unavoidable variation in the cross-sectional area of the column should be considered.

All the preceding imperfections can be replaced by an equivalent initial deflection of the column. To obtain this deflection from experiments, the load-deflection curves should be studied. From the discussion of the compression of initially curved bars (Art. 1.12) we know that for small loads an irregular behavior in the lateral deflection of the column can be expected. This conclusion was verified by a number of experimenters. When the load approaches the critical value, the first term in the series representing the deflection curve [see Eq. (1-61)] becomes predominant and we can find the equivalent initial deflection by plotting straight lines as shown in Fig. 4-6. There is little experimental data of this kind, and in choosing the value of an equivalent initial deflection, it is necessary to rely on the experimental data previously discussed. Assuming that all inaccuracies increase in proportion to the length of the column and considering load eccentricities as given by Eq. (a) and an initial deflection as given by Eq. (b), we can finally take as an initial deflection

$$a = \frac{l}{400} \qquad (c)$$

which will be sufficient to compensate for all probable imperfections in a column.[1]

If this method of design is adopted, the design of a column is reduced to the problem of compression of an initially curved bar, such as shown in Fig. 3-14. A solution for this case has been developed[2] by using Fourier series to represent the initial and final deflections. The results of the analysis are shown in Fig. 4-9 for steel having a yield point of 36,000 psi and proportional limit 30,000 psi. The curves give the limiting values of the average compressive stress σ_c producing failure, for several values of the initial curvature a/s. The curves show that, in the case of combined bending and compression and especially for small eccentricities, the yield

---

[1] H. Kayser, in a paper in *Bautechnik*, Berlin, vol. 8, 1930, by working backward from test results to find the amount of initial deflection that must have been present, found values of a ranging from $l/400$ to $l/1,000$. He recommended the use of $a = l/400$.

[2] H. M. Westergaard and W. R. Osgood, *Trans. ASME*, vol. 50, 1928.

load and the failure load are much nearer to each other than in the case of bending by transverse loading. Thus it seems logical to take as a basis for determining the allowable stresses the load that first produces the yield point stress in the column (see Art. 1.13). This method of procedure will eliminate the necessity of making the laborious computations required in the investigation of bending beyond the proportional limit.

If the initial deflection of the bar is given, the value of the average stress $(\sigma_c)_{YP}$ at which the yield-point stress is reached at the outermost fiber can be obtained with sufficient accuracy from curves similar to those in Fig. 4-9. From these curves it is possible to derive also a set of curves for various ratios of the initial deflection a to the length[1] of the column l. Since the ratio of the core radius s to the radius of gyration

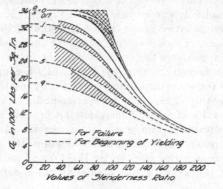

FIG. 4-9

depends on the shape of the cross section, the above curve will also depend on the shape of the cross section. In Figs. 4-10 and 4-11 two series of such curves[2] are shown calculated for a solid rectangular cross section and for a theoretical cross section in which all the material is assumed concentrated in the flanges. It is assumed that the structural steel has a yield-point stress of 36,000 psi and that the imperfections in the column are equivalent to an initial deflection a at the middle such that a/l has the values shown in the figures. These curves also indicate how $(\sigma_c)_{YP}$ varies when the inaccuracies in a column increase.

Having such curves, we can obtain the value of the average compressive

[1] In such case the initial deflection a increases with the slenderness ratio, and as a result of this $(\sigma_c)_{YP}$ decreases more rapidly than that given by the curves in Fig. 4-9.

[2] These curves are taken from the paper by D. H. Young, *Trans. ASCE*, vol. 101, 1936. Analogous curves, calculated on the assumption that the imperfections of the column are compensated for by a certain eccentricity, proportional to the length of the column, are given in the paper by Timoshenko, *Trans. ASME*, Applied Mechanics Division, vol. 1, p. 173, 1933.

stress $(\sigma_c)_{YP}$ at which yielding in the outermost fiber begins for any slenderness ratio l/r. It was noted above that the loads producing the beginning of yielding do not differ much from the loads producing complete failure and that they approach these loads as the slenderness ratio increases (see Fig. 4-9). By taking $(\sigma_c)_{YP}$ as a basis for calculating the allowable compressive stress $(\sigma_c)_W$ and using a constant factor of safety, we shall be on the safe side in all practical cases.[1] The margin of safety with respect to complete failure will be somewhat larger for smaller values of the slenderness ratio.

In the previous discussion it was assumed that the imperfections in a column are proportional to the length of the column and become very small for a column with a small slenderness ratio. Several authors have

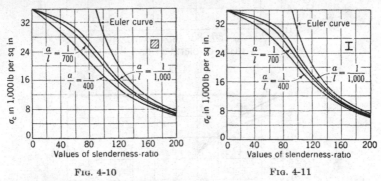

Fig. 4-10   Fig. 4-11

proposed dividing all imperfections into two kinds, one of which is independent of the length of the column and can be compensated for by an initial deflection proportional to the core radius s and another which can be compensated for by an initial deflection proportional to the length of the column.[2] Through the use of the curves shown in Fig. 4-9, a curve for $(\sigma_c)_{YP}$ for any values of the above two types of imperfection can be readily obtained. In Figs. 4-12 and 4-13 are shown two curves of this kind calculated on the assumption that the initial deflection compensating for all imperfections is given by the formula

$$a = 0.1s + \frac{l}{750}$$

[1] Actually, for very small values of a/s the load obtained on the basis of the yield-point stress may be somewhat greater than the failure load. This discrepancy is on the unsafe side but is not of consequence, since in practical design of columns larger values of a/s are assumed.

[2] Such a proposition was made first by F. S. Jasinsky, *loc. cit.* See also H. S. Prichard, *Proc. Eng. Soc. Western Penn.*, Pittsburgh, Pa., vol. 23, p. 325, 1908, and O. H. Basquin, *J. Western Soc. Eng.*, Chicago, vol. 18, p. 457, 1913.

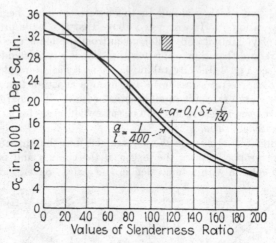

FIG. 4-12

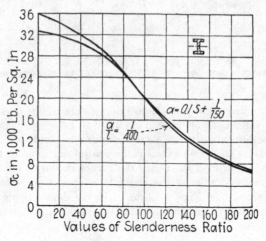

FIG. 4-13

One of these curves assumes a solid rectangular cross section, and the other a theoretical cross section in which all the material is concentrated in the flanges. For comparison, the previous two curves (from Figs. 4-10 and 4-11), calculated for $a = l/400$, are also shown in the figures.

4.5. Various End Conditions. In the discussion in the preceding article it was assumed that the ends of the compressed column were free to rotate. There are cases where actual conditions approach this assumption, but the ends of columns, as encountered in practice, usually are restrained to some degree and cannot rotate freely. The critical value of the load in such cases depends on the magnitude of the coefficients of restraint and becomes a maximum when the ends are completely fixed. Let us begin with a discussion of this extreme case. If the ends of the column are not free to rotate during compression, any eccentricities in the application of the compressive forces do not result in bending of the column and a straight column will undergo only a uniform compression. Hence, in the discussion of imperfections in columns with built-in ends, only initial curvature need be considered. The bending stresses due to this factor naturally will depend on the shape of the initial curvature of the column. Assuming, for example, an initial curvature (Fig. 4-14) as given by the sine curve $y = a \sin \pi x/l$, we conclude that for small loads P, where the deflections due to bending can be neglected in comparison with the initial deflections, there must be an eccentricity equal to $2a/\pi$ at the ends. This value of eccentricity is found by considering the total bending moment area, shown shaded in the figure. The total area must vanish in order to give zero rotation of one end of the column with respect to the other.

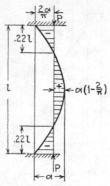

Fig. 4-14

With an increase of the compressive load, the additional deflection due to bending should be considered. By using Eqs. (l) and (m) of Art. 1.12, we find that the bending moments M_1 at the ends and the bending moment M_2 at the middle are

$$M_1 = -\frac{aP}{(1-\alpha)}\frac{\pi\alpha}{2u\tan u} \qquad M_2 = \frac{aP}{1-\alpha}\left(1 - \frac{\pi\alpha}{2u\sin u}\right) \qquad (a)$$

in which
$$\alpha = \frac{Pl^2}{\pi^2 EI} \qquad u = \frac{l}{2}\sqrt{\frac{P}{EI}} \qquad (b)$$

For small values of the load P we can assume that $u \approx \sin u \approx \tan u$ and neglect α in comparison with unity. Then

$$M_1 = -\frac{2aP}{\pi} \qquad M_2 = aP\left(1 - \frac{2}{\pi}\right) \qquad (c)$$

which requires an eccentricity of $2a/\pi$ in the application of the load, as mentioned above. It is seen that the moments at the ends are larger than the moment at the middle, and the ratio[1] of these moments is

[1] It should be noted that this ratio depends on the shape of the initial curvature. Assuming, for instance, that the initial curve is a parabola symmetrical about the middle of the bar, we find that $M_1/M_2 = -2$.

$$\frac{M_1}{M_2} \approx -1.75$$

This result is valid in the case of short columns for which the ultimate load is usually small in comparison with the Euler load.

With an increase of the load P the bending moments M_1 and M_2 increase also, and at the same time their ratio diminishes. Take, for instance, $P = \pi^2EI/l^2$, i.e., a compressive load equal to one-quarter of the critical load for a column with built-in ends. Then from Eqs. (a) we find that

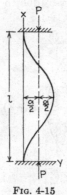

$$M_1 = -\frac{\pi aP}{4} \qquad M_2 = \frac{aP}{2}$$

and

$$\frac{M_1}{M_2} \approx -1.57$$

In the case of slender columns, the ultimate load, for a small initial curvature, approaches the value $4\pi^2EI/l^2$ and the quantity u in Eqs. (a) approaches the value π. The moments M_1 and M_2 at such values of u increase indefinitely, and their ratio approaches unity. The conditions of bending of the column approach those in which the initial curvature of the column is given by the equation

FIG. 4-15

$$y = \frac{a}{2}\left(1 - \cos\frac{2\pi x}{l}\right)$$

In this case (see Fig. 4-15) the moments at the ends and at the middle are always numerically equal and the maximum stress is the same as for a compressed column with hinged ends having a length equal to $l/2$ and an initial deflection represented by a sine curve with the deflection $a/2$ at the middle. The expression for the maximum bending moment in this case is [see Eq. (1-60)]

$$M = \frac{aP}{2(1 - \alpha/4)}$$

The ratio of the absolute value of the moment M_1, from Eqs. (a), to the moment M is

$$\frac{M_1}{M} = \frac{2\sqrt{\alpha}\left(1 - \dfrac{\alpha}{4}\right)}{(1 - \alpha)\tan\left(\pi\sqrt{\alpha}/2\right)}$$

For small values of α this ratio approaches the value $4/\pi$. With an increase of the load P, α increases and the ratio diminishes. For $\alpha = 1$ the ratio is equal to $3\pi/8$. When α approaches the value 4, corresponding to Euler's load for a column with built-in ends, the ratio of the moments approaches the value $8/3\pi$.

From this discussion it can be concluded that if the column represented in Fig. 4-14 is a slender column which fails principally because of bending stresses, we shall be on the safe side in taking the reduced length of the column equal to half of the actual length and in using the curves calculated for columns with hinged ends (Figs. 4-10 and 4-11). In the case of short columns the maximum bending stress may be somewhat larger than that obtained by using the above procedure, but in such columns the bending stresses at failure are usually small in comparison with the direct stress; hence, it seems satisfactory to use also in this case the reduced length $l/2$ and the curves in Figs. 4-10 and 4-11. Such procedure is further justified if we note that, in the discussion of columns with hinged ends, the magnitude of the initial curvature was chosen so as to compensate not only for the crookedness of the column but also for eccentricity in application of the load and that in the case of built-in ends this eccentricity is absent. Thus, discussing the design of columns with built-in ends on the basis of assumed inaccuracies, we arrive at the same conclusion regarding reduced length as we did when the diagram of critical stresses was used for determining working stresses.

Up to this point we have considered only two extreme cases, columns with hinged ends and columns with built-in ends. Compressed members in structures usually have intermediate end conditions, the restraint at the ends being dependent on the rigidity of adjacent members of the structure. The degree of fixity of the ends can be obtained only on the basis of investigation of the stability of the entire structure. Several examples of such investigations were discussed in Chap. 2, and it was shown that in each particular case the critical load for a compressed member of a structure can be calculated as for a column with hinged ends having a certain reduced length. When this reduced length is known, the design of the compressed member can be made by using curves such as those in Figs. 4-12 and 4-13.

Only in the case of the simplest structures can the stability conditions be established without much difficulty. Generally, in the design of compression members of a structure, the reduced length of these members is taken on the basis of some approximate consideration. For instance, in discussing the stability of the compressed top chord of a truss, it can be seen that the wind bracing and the members in the plane of the truss do not provide a large amount of resistance to lateral buckling of the chord members in alternate directions in successive panels. It is common practice to consider these members in lateral buckling as pin-ended columns, so that the actual length between the theoretical hinges should be used in this case. The same conclusion can be made also regarding the lateral buckling of compressed diagonals and verticals of a truss.

In considering the buckling of compressed members in the plane of a truss, it should be noted that, owing to rigidity of the joints, certain bending moments will be produced at the ends of these members. The magnitude of these moments can be calculated by the methods used in analyzing secondary stresses in trusses. In this analysis the effect of axial forces on the bending of truss members is usually neglected,[1] so that the moments are proportional to the loads. Under such a condition each compression member can be considered as an eccentrically loaded column with known eccentricities at the ends, and the allowable average compressive stress can be obtained by interpolation from the curves given in Figs. 1-28 to 1-31. The presence of initial curvature in a compressed member will add to the bending stresses and should be allowed for by superposing at each end an equivalent eccentricity e on the actual eccentricities mentioned above and calculated from the secondary stress analysis. These modified values of the eccentricities will then be used in calculating working stresses from the curves in Figs. 1-28 to 1-31. In a similar manner, the deflection of a compressed member due to its own weight can be compensated for by introducing certain additional eccentricity.

4.6. The Design of Built-up Columns. In discussing the buckling of built-up columns in Art. 2.18, we obtained Eq. (2-61) for calculating critical loads for laced columns and Eq. (2-64) for batten-plate columns. When these equations are used, the actual built-up column is replaced by an equivalent column of a reduced length, which is to be determined, in the case of a laced column (Fig. 2-55a), from the equation

$$L = l \sqrt{1 + \frac{\pi^2 EI}{l^2} \left(\frac{1}{A_d E \sin \phi \cos^2 \phi} + \frac{b}{A_b E a} \right)} \qquad (4\text{-}1)$$

and, in the case of a batten-plate column (Fig. 2-57), from the equation

$$L = l \sqrt{1 + \frac{\pi^2 EI}{l^2} \left(\frac{ab}{12 EI_b} + \frac{a^2}{24 EI_c} + \frac{na}{b A_b G} \right)} \qquad (4\text{-}2)$$

These formulas, derived for the case of buckling in the elastic range, can be used also beyond the elastic limit by replacing the modulus of elasticity E with the tangent modulus E_t (see Art. 3.3) in the expressions for the flexural rigidity EI of the column and flexural rigidity EI_c of one chord.

When the reduced length of a built-up column is determined, the allowable stress for the corresponding value of L/r is found from curves

---

[1] This can be justified if we note that the slenderness ratio of the chord members is usually small, so that the acting compressive forces are small in comparison with the Euler loads. The slenderness ratio of diagonals and verticals may be larger, but they are very often bent in an S shape. Under such conditions the effect of the axial forces on deflections, as can be concluded from the general discussion of Art. 1.11, is small.

such as are shown in Figs. 4-10 and 4-11 with a proper factor of safety. The use of these curves for built-up columns amounts to assuming that the imperfections are a function of the reduced length L rather than the true length. This means that a slightly higher factor of safety will be used in the case of built-up columns, which seems satisfactory.

A considerable number of experiments have been made with built-up columns,[1] but only in a few cases were the experiments made with the intention of verifying any theory. Of particular importance are experiments made by Petermann[2] and by J. Kayser,[3] dealing with batten-plate columns. The flexibility of such columns in the plane parallel to the batten plates depends very much on the dimensions of the battens and the distance between them. Experimental values of critical loads are in satisfactory agreement with those calculated by using Eq. (4-2).

In the design of built-up columns the proper dimensioning of the lacing bars and batten plates is of great practical importance. In these calculations we shall proceed as before and, as a basis for determining stresses in these parts, assume some imperfections in the column, such as initial curvature or eccentricity in load application. When this has been done, it will be possible to evaluate the maximum shearing force Q_{max} that arises for any value of the compressive load P. This maximum shearing force will then be calculated for the value of the load P at which yielding in the extreme fibers of the column begins. It is logical to design the lacing bars and batten plates on the basis of this maximum shearing force, so that they will yield simultaneously with the extreme fibers of the column.

The imperfections in a column should be so assumed that we have the most unfavorable condition as far as shearing forces are concerned. Possible types of imperfection, consisting of an initial curvature or an initial eccentricity in the application of the load, are shown in Fig. 4-16. The value of the initial deflection a or initial eccentricity e will be taken proportional to the length, which in this case is the reduced length of the column, calculated from Eq. (4-1) or (4-2). Considering the case in which the initial curvature is represented by one half-wave of a sine curve (Fig. 4-16a), we find that the maximum shearing force occurs at the ends of the column. For the small deflections which occur in practice we can take $Q_{max} = P\theta$. In the case of an S-shaped initial curvature (Fig. 4-16b), each half of the column can be considered as a column of the previous type but of length $l/2$ and deflection $\delta/2$. The initial shearing

---

[1] A discussion of some of these experiments is given in the report of the ASCE Special Committee on Steel Column Research, *loc. cit.* See also R. Mayer, "Die Knick-festigkeit," p. 387, Berlin, 1921, and D. Rühl, "Berechnung gegliederter Knickstäbe," Berlin, 1932.

[2] *Bauingenieur*, vol. 4, p. 1009, Berlin, 1926, and vol. 9, p. 509, 1931.

[3] *Bauingenieur*, vol. 8, p. 200, 1930.

forces at the ends will be the same as in the previous case, but the value of the shearing force at failure will be smaller, since, regardless of its initial shape, the column will buckle in one half-wave at the critical load for the length l, which is smaller than the critical load for the length $l/2$.

The case of two equal eccentricities in the same direction (Fig. 4-16c) is also more favorable than the case in Fig. 4-16a. Assuming that the eccentricity e is such that both columns fail at the same load P, we find that the bending moments at the middle will be equal at failure for both cases; hence the corresponding values of δ, θ, and Q_{max} will be smaller for case (c) than for case (a). In the case of equal eccentricities in opposite directions (Fig. 4-16d) there will be horizontal reactions at the ends equal to $2Pe/l$ and the maximum shearing force, equal to $2Pe/l + P\theta$, occurs

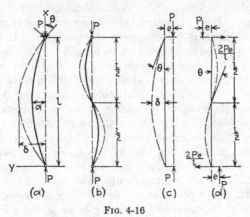

Fig. 4-16

at the middle. It is possible that under certain conditions this case will be more unfavorable than case (a). Thus, finally, we conclude that cases (a) and (d) should be considered in detail.[1]

Beginning with case (a) and using Eq. (1-60) for the deflection curve, we find that the angle θ at the ends is

$$\theta = \left(\frac{dy}{dx}\right)_{x=0} = \frac{\pi a}{l(1 - \alpha)} \tag{a}$$

where $\alpha = Pl^2/\pi^2 EI$, and we obtain for the maximum shearing force

$$Q_{max} = P \frac{\pi a}{l(1 - \alpha)}$$

Dividing both sides of this equation by the cross-sectional area A and

[1] This question is discussed by D. H. Young, *Proc. ASCE*, December, 1934, and *Publ. Intern. Assoc. Bridge Structural Eng.*, Zürich, vol. 2, p. 480, 1934.

using the notation $(\sigma_c)_{\text{YP}}$ for the average compressive stress when yielding in the extreme fiber of the column begins, we find that the value of the maximum shearing force per unit of cross-sectional area at which yielding begins is

$$\frac{Q_{\max}}{A} = (\sigma_c)_{\text{YP}} \frac{\pi a}{l(1 - \alpha)} \tag{4-3}$$

For any slenderness ratio l/r and a given initial deflection a, we find $(\sigma_c)_{\text{YP}}$ from curves such as shown in Figs. 4-10 and 4-11, with a proper factor of safety. Then $Q_{\max}$ is calculated from Eq. (4-3). This calculation can be simplified if we solve Eq. (i), Art. 1.13, for a and substitute its value in Eq. (4-3). We obtain, then,

$$\frac{Q_{\max}}{A} = \frac{\pi s}{l} [\sigma_{\text{YP}} - (\sigma_c)_{\text{YP}}] \tag{4-4}$$

Knowing σ_{YP} and using the curves in Figs. 4-10 and 4-11, with a safety factor, for determining $(\sigma_c)_{\text{YP}}$, we can represent $Q_{\max}/A$ as a function of

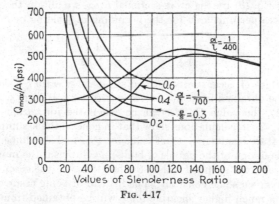

FIG. 4-17

the slenderness ratio l/r for any value of the initial deflection a. In Fig. 4-17 are shown two curves of this kind calculated for $a/l = \frac{1}{400}$ and $a/l = \frac{1}{700}$. It is assumed that all the material of the column is concentrated in the flanges; hence we have $s = r$.

In case (d), the angle of rotation θ of the middle cross section of the column is found by considering each half of the column as a compressed beam of span $l/2$ simply supported at the ends and bent by a couple Pe. Then, from Eq. (1-25),

$$\theta = \frac{e}{l} \left(\frac{kl}{\sin kl/2} - 2 \right)$$

where
$$k = \sqrt{\frac{P}{EI}}$$

The maximum shearing force is

$$Q_{max} = \frac{2Pe}{l} + P\theta = \frac{Pe}{l}\frac{kl}{\sin kl/2} \qquad (b)$$

In the case of short columns, failure occurs at a load that is small in comparison with Euler's load; then $kl/2$ is small and we can assume that $\sin kl/2 \approx kl/2$. Equation (b) gives in this case

$$Q_{max} = \frac{2Pe}{l}$$

That is, the maximum shearing force is equal to the lateral reaction at either end (Fig. 4-16d). In the case of slender columns, in which the load P may reach the Euler load before the maximum fiber stress reaches the yield-point stress, $kl/2$ approaches the value $\pi/2$ and we see from Eq. (b) that the maximum shearing force may become 57 per cent higher than the value of the lateral reactions $2Pe/l$.

Dividing Eq. (b) by the cross-sectional area, we obtain the maximum shearing force per unit area for the beginning of yielding in the extreme fiber:

$$\frac{Q_{max}}{A} = (\sigma_c)_{YP}\frac{e}{l}\frac{kl}{\sin kl/2} \qquad (4\text{-}5)$$

For any values of eccentricity e/l and slenderness ratio l/r the value of $(\sigma_c)_{YP}$ can be obtained from the curves in Fig. 1-31. Then, with the use of Eq. (4-5), the value Q_{max}/A can be calculated for various values of e/l and l/r. Such calculations show that in general the assumption of an initial curvature results in a larger value for Q_{max} and should be taken as a basis for the design of the details of built-up columns if the imperfections given by a or e are proportional to the length l. If the eccentricities in case (d) are given some constant value instead of being assumed proportional to l, a much higher shearing force will be obtained from Eq. (4-5) for small values of the slenderness ratio l/r. Curves are shown in Fig. 4-17, in which values of $e/s = 0.2, 0.3, 0.4, 0.6$ have been used.

The curves shown in Fig. 4-17 take account of shearing force due to initial imperfections only. When a compressed member in a truss is subjected to secondary end moments as discussed on p. 206, the shearing force may become very large, and it seems logical to design the details of such members to resist the shearing forces that actually arise because of end moments.[1]

With curves such as shown in Fig. 4-17 available, the procedure for the design of a built-up column will be as follows: Assume certain cross-sectional dimensions of the column and also dimensions for the details.

---

[1] This problem is discussed in the paper by D. H. Young, *loc. cit.*

Then the reduced length of the column will be calculated from Eq. (4-1) or (4-2) and the allowable average compressive stress will be obtained from the curves in Figs. 4-10 and 4-11, with a proper factor of safety. The use of this trial-and-error method will establish the necessary cross-sectional dimensions. The necessary strength of the lacing bars or batten plates and the necessary number of rivets at the joints should now be checked by using the curve in Fig. 4-17. The same curve can be used also for checking the distance between rivets in riveted columns.

It is assumed in this discussion that buckling of the column occurs in the plane parallel to the batten plates or lacing bars. Sometimes the

FIG. 4-18

possibility of distortion of the cross section of a built-up column should be considered. For instance, in the case of a column consisting of four longitudinal bars connected by lacing bars (Fig. 4-18), a distortion may occur such as shown in the figure by dotted lines. To eliminate the possibility of such distortion, certain bracing in the cross-sectional planes of the column or use of diaphragms is necessary. Between the two planes with cross-sectional bracings or two diaphragms, each longitudinal bar can be considered as a strut with hinged ends elastically supported along the length by lacing bars. With the use of the energy method, the required distance between the braced cross sections can be checked.

In the case of built-up columns consisting of comparatively thin plates, local failure may occur owing to buckling of the compressed plates if the unsupported width of these plates exceeds certain limits. The requirements regarding unsupported width of plates and the methods of reinforcing plates by stiffeners will be discussed in Chap. 9.

CHAPTER 5

TORSIONAL BUCKLING

5.1. Introduction. In the previous discussions of buckling, it was assumed that a column would buckle by bending in a plane of symmetry of the cross section. However, there are cases in which a column will buckle either by twisting or by a combination of bending and twisting. Such torsional buckling failures occur if the torsional rigidity of the section is very low, as for a bar of thin-walled open cross section. In the next two articles of this chapter the subject of torsion of bars of thin-walled open section will be discussed. Then, in the remaining articles, the theory of torsional buckling will be presented.

5.2. Pure Torsion of Thin-walled Bars of Open Cross Section. If a bar is twisted by couples applied at the ends and acting in planes normal to the axis of the bar, and if the ends of the bar are free to warp, we have the case of *pure torsion.* The only stresses produced are the shearing stresses at each section of the bar. The distribution of these stresses depends on the shape of the cross section and is the same for all sections. For a beam of thin-walled open section it can be assumed with reasonable accuracy that the shearing stress at any point is parallel to the corresponding tangent to the middle line of the cross section and is proportional to the distance from that line.

The angle of twist per unit length θ is given by the formula

$$\theta = \frac{M_t}{C} \tag{5-1}$$

where M_t denotes the torque and C is the *torsional rigidity* of the bar. The torsional rigidity can be represented in the form

$$C = GJ \tag{5-2}$$

where G is the shearing modulus of elasticity and J is the *torsion constant.* For a bar of thin-walled open section of constant thickness t, we can take the torsion constant[1] as

$$J = \tfrac{1}{3}mt^3 \tag{5-3}$$

[1] See Timoshenko, "Strength of Materials," 3d ed., part II, pp. 240–246, D. Van Nostrand Company, Inc., Princeton, N.J., 1956.

where m is the length of the middle line of the cross section. If the cross section consists of several portions of different thicknesses, we can assume that

$$J = \tfrac{1}{3} \sum m_i t_i{}^3 \tag{5-4}$$

where the summation is extended over all portions of the cross section. Formulas for J are given in Table A-3, Appendix, for several shapes of cross section.

The initially straight longitudinal fibers of the bar are deformed during twist into helices which, for small angles of twist, can be considered as straight lines inclined to the axis of rotation. If ρ denotes the distance of

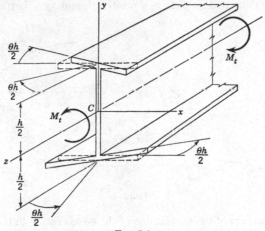

FIG. 5-1

the fiber from the axis of rotation, the angle of inclination of the fiber to the axis is $\rho\theta$.

Warping of the cross section in the case of a thin-walled open section can be visualized readily if we observe that there is no shearing stress along the middle line of the cross section. This indicates that the elements of the middle line remain normal to the longitudinal fibers after torsion. For example, the warping of the cross section of a twisted I beam is shown in Fig. 5-1.[1] During twisting with respect to the z axis, the central fibers of the flanges, distance $h/2$ from the z axis, become inclined to the z axis by the angle $\theta h/2$. The middle lines of the flange cross sections will therefore make the same angle with the x axis, as shown in the figure.

[1] The rotation of the cross sections is not shown.

Let us consider next a more general case in which the middle line of the cross section is of arbitrary shape (Fig. 5-2). Assuming that during torsion the cross sections of the bar rotate with respect to an axis through point A parallel to the longitudinal axis, we find that any longitudinal fiber N in the middle surface of the wall becomes inclined to the axis of rotation by the angle $\rho\theta$. The fiber N is defined by the distance s measured along the middle line of the cross section. The tangent to the middle line at N remains perpendicular to the longitudinal fiber, and the small angle between this tangent and the xy plane, after torsion, is $\rho\theta\cos\alpha = r\theta$. The distance r from the tangent at N to the axis of rotation is taken positive if a vector along the tangent and pointing in the direction of increasing s acts counterclockwise about the axis of rotation. Thus the distance r shown in Fig. 5-2a is a positive quantity. Letting w denote

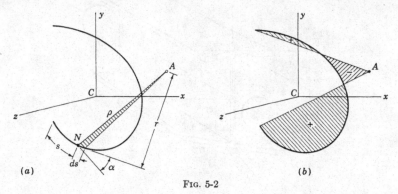

FIG. 5-2

the displacements of the middle line of the cross section in the z direction and considering the torque to be positive as shown in Fig. 5-1, we have the relation

$$\frac{\partial w}{\partial s} = -r\theta \tag{a}$$

By integration we then obtain

$$w = w_0 - \theta \int_0^s r\,ds \tag{5-5}$$

where w_0 denotes the displacement in the z direction of the point from which s is measured. Since the area of the shaded triangle in Fig. 5-2a is $r\,ds/2$, it is seen that the integral on the right-hand side of Eq. (5-5) represents the doubled sectorial area swept by the radius ρ as we move along the middle line of the cross section from the point $s = 0$ up to the point N under consideration. The swept area is taken positive when the radius ρ is rotating in the positive direction, that is, counterclockwise about A. The value of the integral for $s = m$ will then be represented by twice the algebraic sum of the three shaded areas in Fig. 5-2b.

In the preceding discussion it was assumed that the cross section rotated with respect to an arbitrary point A. Let us now investigate the effect on warping of a displacement of the center of rotation. Assume, for example, that the center of rotation is moved from A to B (Fig. 5-3). Considering an element ds of the middle line of the cross section and denoting by x, y the coordinates of point N and by x_a, y_a the coordinates of the center of rotation A, we see from the figure that

$$r\,ds = (y_a - y)\,dx - (x_a - x)\,dy$$

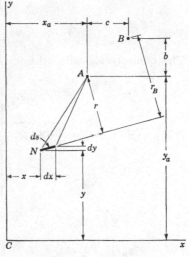

If the center of rotation is moved from A to B, the coordinates of the center of rotation become $x_a + c$ and $y_a + b$; hence

$$r_B\,ds = r\,ds + b\,dx - c\,dy$$

To calculate the warping produced by rotation about B, we have to substitute $r_B\,ds$ in place of $r\,ds$ in Eq. (5-5). This gives

$$\int_0^s r_B\,ds = \int_0^s r\,ds + \int_0^s (b\,dx - c\,dy)$$
$$= \int_0^s r\,ds + bx - cy + a$$

where a is a constant. It is seen that a change in the position of the cen-

Fig. 5-3

ter of rotation results in a change of the previously calculated displacements [Eq. (5-5)] by an amount

$$\theta(bx - cy + a)$$

Since this displacement is a linear function of x and y, it does not require any additional deformation of the bar and is accomplished by moving the bar as a rigid body. Therefore, we conclude that in the case of pure torsion of a bar with free ends, the selection of the axis of rotation is immaterial and any line parallel to the centroidal axis can be taken as the axis of rotation.

The average value $\bar{w}$ of the warping displacement can be calculated from Eq. (5-5) as follows:

$$\bar{w} = \frac{1}{m}\int_0^m w\,ds = w_0 - \frac{\theta}{m}\int_0^m \left[\int_0^s r\,ds\right]ds \qquad (b)$$

Subtracting this value from the displacement given by Eq. (5-5) gives the warping of the cross section with respect to the plane of average warping. Continuing to use the symbol w for displacements with respect to the new reference plane, we obtain

$$w = \frac{\theta}{m} \int_0^m \left[\int_0^s r\, ds \right] ds - \theta \int_0^s r\, ds \tag{c}$$

To simplify the writing of this expression, we introduce the notation

$$\begin{aligned} \omega_s &= \int_0^s r\, ds \\ \bar{\omega}_s &= \frac{1}{m} \int_0^m \omega_s\, ds \end{aligned} \tag{5-6}$$

The quantity ω_s, called the *warping function*, represents the doubled

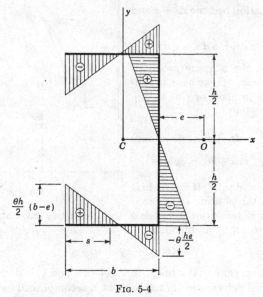

Fig. 5-4

sectorial area corresponding to the arc s of the middle line of the cross section, while $\bar{\omega}_s$ is the average value of ω_s. Using notation (5-6), we find that the expression for warping becomes

$$w = \theta(\bar{\omega}_s - \omega_s) \tag{5-7}$$

From this equation the warping displacements for pure torsion can be calculated for any bar of thin-walled open section.

As an example of the use of Eq. (5-7), let us consider a cross section in the form of a channel (Fig. 5-4). Assuming that rotation occurs about a longitudinal axis through O, the shear center,[1] we find the following expressions for the warping function:

$$\omega_s = \int_0^s r\,ds = \int_0^s \frac{h}{2}\,ds = \frac{sh}{2} \qquad\qquad 0 \le s \le b$$

$$\omega_s = \frac{bh}{2} - \int_b^s e\,ds = \frac{bh}{2} + be - se \qquad b \le s \le b+h$$

$$\omega_s = \frac{bh}{2} - he + \int_{b+h}^s \frac{h}{2}\,ds$$

$$\qquad = -he - \frac{h^2}{2} + \frac{sh}{2} \qquad\qquad b+h \le s \le 2b+h$$

Using the above expressions for ω_s, we obtain the average value of the warping function as

$$\bar{\omega}_s = \frac{1}{m}\int_0^m \omega_s\,ds$$

$$= \frac{1}{m}\left[\int_0^b \frac{sh}{2}\,ds + \int_b^{b+h}\left(\frac{bh}{2} + be - se\right)ds\right.$$

$$\left. + \int_{b+h}^{2b+h}\left(-he - \frac{h^2}{2} + \frac{sh}{2}\right)ds\right]$$

$$= \frac{1}{m}\left[\frac{h}{2}(b-e)(2b+h)\right]$$

and since $m = 2b + h$, this becomes

$$\bar{\omega}_s = \frac{h(b-e)}{2}$$

Substituting into Eq. (5-7), we obtain the following expressions for the warping displacements:

$$w = \theta\frac{h}{2}(b - e - s) \qquad\qquad 0 \le s \le b \qquad\qquad (d)$$

$$w = \theta e\left(-b - \frac{h}{2} + s\right) \qquad b \le s \le b+h \qquad (e)$$

$$w = \theta\frac{h}{2}(b + e + h - s) \qquad b+h \le s \le 2b+h \qquad (f)$$

The variation of w along the middle line of the cross section is shown by the shaded areas in Fig. 5-4.

If the cross section consists of thin rectangular elements which intersect at a common point (Fig. 5-5), and if the axis of rotation is taken through the shear center O, the distance r vanishes for all points of the middle line and hence there is no warping of this line during torsion.

---

[1] For a discussion of shear center, see Timoshenko, *op. cit.*, part I, p. 235.

5.3. Nonuniform Torsion of Thin-walled Bars of Open Cross Section.
In the preceding article we discussed the case of pure torsion, in which it is
assumed that the torque is applied at the ends of the bar only and that
cross sections of the bar are free to warp.   Under such conditions warping
is the same for all cross sections and takes place without any axial strain
of the longitudinal fibers.   The case of *nonuniform torsion* occurs if any
cross sections are not free to warp or if the torque varies along the length
of the bar.   In these cases warping will vary along the bar during torsion,
and hence there will be tension or compression of the longitudinal fibers.
In addition, the rate of change θ of the angle of twist will no longer be
constant but will vary along the axis of the bar.

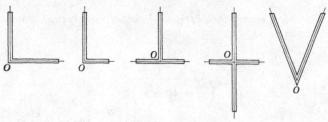

FIG. 5-5

Let us begin the discussion by considering the simple case of nonuniform torsion of a symmetrical I beam (Fig. 5-6).   One end of the beam is
assumed to be rigidly built in; thus, there is no warping of the cross
section at the support.   The torque M_t is applied at the free end.   It is
evident that the resistance of the beam to torsion is greater when the end
of the beam is built in than when the ends are free to warp, since in the
former case torsion is accompanied by bending of the flanges.   The
torque M_t is balanced partially by shearing stresses due to pure torsion,
as discussed in the preceding article, and partially by the resistance of
the flanges to bending.   These two parts of the torque will be denoted by
M_{t1} and M_{t2}, respectively.   The torque M_{t1} is proportional to the rate of
change of the angle of twist along the axis of the beam.   Denoting this
angle by ϕ and using Eq. (5-1), we obtain

$$M_{t1} = C \frac{d\phi}{dz} \tag{a}$$

In this equation the angle ϕ is assumed positive according to the right-
hand rule of signs; that is, the direction of positive rotation is the same
as the direction of positive M_t (shown in Fig. 5-6).

The second part of the torque is found by considering the bending of
the flanges.   Since the beam cross section is symmetrical, we conclude
that each cross section will rotate with respect to the centroidal z axis,

and hence the lateral deflection of the lower flange of the beam is

$$u = \phi \frac{h}{2}$$

The bending moment M_f in the lower flange is

$$M_f = EI_f \frac{d^2u}{dz^2} = \frac{EI_f h}{2} \frac{d^2\phi}{dz^2} \qquad (b)$$

where I_f is the moment of inertia of one flange about the y axis. The shearing force in the lower flange is, therefore,

$$V_f = \frac{dM_f}{dz} = \frac{EI_f h}{2} \frac{d^3\phi}{dz^3}$$

In the top flange there will be a shearing force of the same magnitude but opposite in direction. The couple formed by these two shearing forces represents the second part of the torque; that is,

$$M_{t2} = -V_f h = -\frac{EI_f h^2}{2} \frac{d^3\phi}{dz^3} \qquad (c)$$

The equation for nonuniform torsion of an I beam becomes, then,

$$M_t = M_{t1} + M_{t2} = C \frac{d\phi}{dz} - \frac{EI_f h^2}{2} \frac{d^3\phi}{dz^3} \qquad (5\text{-}8)$$

The angle of twist ϕ can be found by integration of this equation, provided M_t is a known function of z. Then, when ϕ is known, the two portions M_{t1} and M_{t2} of the torque can be found and, finally, the stresses produced in the beam by each of these portions can be calculated.

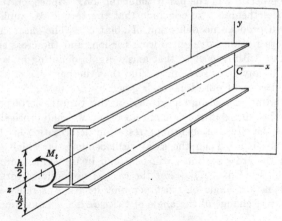

FIG. 5-6

It should be noted that Eq. (b) was used in analyzing bending of the flanges of the I beam. This means that the effect of shearing stresses in the flanges on the curvature was neglected and only the effect of normal stresses σ_z was considered, which is the usual practice in analyzing bending of beams.

The method used above for the analysis of nonuniform torsion of an I beam can be applied to the analysis of a thin-walled bar of any open cross section. Assume that a bar of arbitrary shape (Fig. 5-7) is built in at one end and subjected to a torque M_t at the free end. If a transverse force is applied at the shear center O' of the cross section at the free end,

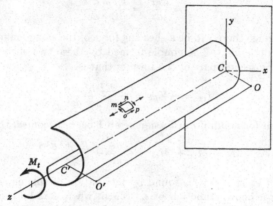

FIG. 5-7

there will be bending of the bar without torsion. Hence, on the basis of the reciprocal theorem, we conclude that the torque M_t applied at the free end will produce no deflection of point O'. The shear center axis OO' therefore remains straight during torsion, and the cross sections of the bar rotate with respect to that axis. Again denoting by ϕ the angle of rotation of any cross section, we find that the part M_{t1} of the torque, producing stresses of pure torsion, is given by Eq. (a).

In calculating the second part M_{t2} of the torque, corresponding to bending of the flanges in the case of an I beam, we shall proceed as before and neglect the effect of shearing stresses on deformation of the middle surface of the bar. Then the axial displacements w, which define the warping of the cross sections, will be found in the same manner as for pure torsion [see Eq. (5-7)]. For the case of nonuniform torsion, however, the constant angle of twist per unit length θ is replaced by the variable rate of change of the angle of twist $d\phi/dz$. Thus we obtain

$$w = (\bar{\omega}_s - \omega_s)\frac{d\phi}{dz} \qquad (5\text{-}9)$$

Since $d\phi/dz$ varies along the length of the bar, adjacent cross sections will not be warped equally and there will be axial strain ϵ_z of the longitudinal fibers of the bar. Observing that $\bar{\omega}_s$ and ω_s in Eq. (5-9) are independent of z, we obtain for the axial strain the expression

$$\epsilon_z = \frac{\partial w}{\partial z} = (\bar{\omega}_s - \omega_s) \frac{d^2\phi}{dz^2} \tag{d}$$

Assuming that there is no lateral pressure between the longitudinal fibers, we obtain the normal stresses produced during nonuniform torsion from Hooke's law:

$$\sigma_z = E\epsilon_z = E(\bar{\omega}_s - \omega_s) \frac{d^2\phi}{dz^2} \tag{5-10}$$

This expression shows that the normal stresses on any cross section are proportional to the warping displacements w, and hence a diagram showing the variation in warping along the middle line will also represent, to a suitable scale, the distribution of the stresses σ_z. For example, the diagram of Fig. 5-4 represents the variation in axial stresses during nonuniform torsion of a channel section.

In order to show that the stresses σ_z give no resultant force in the axial direction and give no moments about the x and y axes, we can use the reciprocal theorem. Let us assume that normal stresses of intensity p are distributed uniformly over the end cross section of the bar in Fig. 5-7. These stresses produce no rotation of the bar, and as a result, there will be no work done by the torque M_t. From the reciprocal theorem we conclude that the work done by the stresses p on the displacements w produced by the torque must vanish also. Thus we have

$$\int_0^m wpt\, ds = p\frac{d\phi}{dz} \int_0^m (\bar{\omega}_s - \omega_s)t\, ds = 0$$

and hence
$$\int_0^m (\bar{\omega}_s - \omega_s)t\, ds = 0 \tag{e}$$

which shows that the axial resultant of the normal stresses (5-10) vanishes. Let us now apply to the end of the bar bending stresses of intensity $p_{max}y/c$, where y is the distance from the x axis to any point in the cross section and c is the distance to the extreme fiber. Thus the stresses are proportional to the distance from the x axis and have a maximum intensity p_{max}. Such stresses produce pure bending of the bar with no rotation about the z axis. Since the torque M_t produces no work during this bending, we conclude that the work of the bending stresses during torsion must vanish also; hence

$$\int_0^m w\frac{p_{max}y}{c} t\, ds = \frac{p_{max}}{c}\frac{d\phi}{dz} \int_0^m (\bar{\omega}_s - \omega_s)yt\, ds = 0$$

This result shows that the moment with respect to the x axis of the stresses (5-10) vanishes. In a similar way it can be proved that the axial stresses give no moment about the y axis.

The normal stresses σ_z produce shearing stresses of the same type as those discussed in considering bending of the flanges of an I beam. These shearing stresses constitute the second part M_{t2} of the torque. To calculate the shearing stresses, let us consider an element $mnop$ (see Fig. 5-8) cut out from the wall of the bar in Fig. 5-7. If the walls of the bar are thin, it can be assumed that the shearing stresses τ are uniformly distributed over the thickness t and are parallel to the tangent to the middle line of the cross section. Along the middle line the stresses vary with the distance s from the edge of the section and can be calculated from an

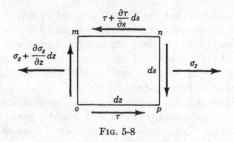

FIG. 5-8

equation of static equilibrium of the element $mnop$ (Fig. 5-8). Projecting all forces onto the z axis and observing that the thickness t may vary with s but is independent of z, we obtain the equation

$$\frac{\partial(\tau t)}{\partial s}\, ds\, dz + t\, \frac{\partial \sigma_z}{\partial z}\, ds\, dz = 0$$

or
$$\frac{\partial(\tau t)}{\partial s} = -t\, \frac{\partial \sigma_z}{\partial z} \qquad (f)$$

Substitution of expression (5-10) for σ_z into Eq. (f) gives

$$\frac{\partial(\tau t)}{\partial s} = -Et(\bar{\omega}_s - \omega_s)\, \frac{d^3\phi}{dz^3} \qquad (g)$$

Integrating (g) with respect to s and observing that ϕ is independent of s and that τ vanishes for $s = 0$, we obtain

$$\tau t = -E\, \frac{d^3\phi}{dz^3} \int_0^s (\bar{\omega}_s - \omega_s) t\, ds \qquad (5\text{-}11)$$

The portion M_{t2} of the torque is obtained by summation along the middle line of the section of the moments of the elemental shear forces $\tau t\, ds$ about the shear center. Thus we obtain

$$M_{t2} = \int_0^m \tau t r \, ds$$

$$= -E\frac{d^3\phi}{dz^3} \int_0^m \left[\int_0^s (\bar{\omega}_s - \omega_s)t \, ds \right] r \, ds \tag{h}$$

This equation can be simplified if we observe from the first of Eqs. (5-6) that

$$r \, ds = d(\omega_s)$$

or, since $\bar{\omega}_s$ is independent of s,

$$r \, ds = -\frac{d(\bar{\omega}_s - \omega_s)}{ds} \, ds \tag{i}$$

From this last expression it is seen that

$$\int_0^m \left[\int_0^s (\bar{\omega}_s - \omega_s)t \, ds \right] r \, ds = -\int_0^m \left[\int_0^s (\bar{\omega}_s - \omega_s)t \, ds \right] \frac{d(\bar{\omega}_s - \omega_s)}{ds} \, ds$$

Integrating the right-hand side of the equation by parts and also using Eq. (e), we obtain

$$\int_0^m \left[\int_0^s (\bar{\omega}_s - \omega_s)t \, ds \right] r \, ds = \int_0^m (\bar{\omega}_s - \omega_s)^2 t \, ds$$

Substitution into Eq. (h) then yields the following expression for M_{t2}:

$$M_{t2} = -E\frac{d^3\phi}{dz^3} \int_0^m (\bar{\omega}_s - \omega_s)^2 t \, ds \tag{j}$$

Introducing the notation

$$C_1 = E \int_0^m (\bar{\omega}_s - \omega_s)^2 t \, ds \tag{5-12}$$

we can write Eq. (j) in the form

$$M_{t2} = -C_1 \frac{d^3\phi}{dz^3} \tag{k}$$

This is the portion of the torque due to nonuniform torsion and nonuniform warping of the cross sections and will be referred to as the *warping torque*. The constant C_1 is called the *warping rigidity* and for convenience can be expressed in the form

$$C_1 = EC_w \tag{5-13}$$

where the quantity C_w, called the *warping constant*, is given by the expression[1]

$$C_w = \int_0^m (\bar{\omega}_s - \omega_s)^2 t \, ds \tag{5-14}$$

It is seen that C_w has units of length to the sixth power.

[1] Several symbols, including C_{BT}, C_{BD}, C_S, and Γ, have been used in the literature to denote the warping constant.

The differential equation for nonuniform torsion, obtained by combining Eqs. (a) and (k), is

$$M_t = C \frac{d\phi}{dz} - C_1 \frac{d^3\phi}{dz^3} \tag{5-15}$$

This equation applies to any bar of thin-walled open cross section. Equation (5-8) for an I beam is a particular case[1] of Eq. (5-15) in which the warping rigidity is $C_1 = EI_f h^2/2$. When Eq. (5-15) is solved and the expression for the angle of twist ϕ is known, the torques M_{t1} and M_{t2} can be obtained from Eqs. (a) and (k), respectively. The stresses produced by M_{t1} are calculated in the same way as for pure torsion. The normal and shearing stresses produced by M_{t2} can be found from Eqs. (5-10) and (5-11), respectively.

As an example of the calculation of the warping constant C_w, let us consider again the channel section shown in Fig. 5-4. The values of the warping displacements w, given by Eq. (5-7), are shown in the figure. The values of the quantity $\bar{\omega}_s - \omega_s$ are found from Eqs. (d), (e), and (f), Art. 5.2, by dividing the expressions for w by θ. In this way we obtain for C_w the expression[2]

$$\begin{aligned}
C_w &= \int_0^b \frac{h^2}{4} (b - e - s)^2 t \, ds \\
&+ \int_b^{b+h} e^2 \left(-b - \frac{h}{2} + s \right)^2 t \, ds \\
&+ \int_{b+h}^{2b+h} \frac{h^2}{4} (b + e + h - s)^2 t \, ds \\
&= \frac{th^2}{12} [he^2 + 2b^3 - 6eb(b - e)]
\end{aligned}$$

Substituting for e its expression from Table A-3, Appendix, we obtain

$$C_w = \frac{th^2 b^3}{12} \frac{3b + 2h}{6b + h}$$

For cross-sectional shapes consisting of thin rectangular elements which intersect at a common point (see Fig. 5-5), the warping constant C_w can be taken equal to zero. Formulas for C_w for other shapes of cross section are given in the Appendix.

---

[1] The equation for this particular case was derived by Timoshenko, *Bull. Polytech. Inst., St. Petersburg*, 1905. Extension of the equation to I beams with unequal flanges was made by C. Weber, *Z. angew. Math. u. Mech.*, vol. 6, p. 85, 1926. Further extension of the equation to all thin-walled open sections is due to H. Wagner, *Tech. Hochschule, Danzig, 25th Anniv. Publ.*, 1929; translated in *NACA Tech. Mem.* 807, 1936.

[2] Note that the first and last integrals in this particular expression for C_w have the same value.

5.4. Torsional Buckling. There are cases in which a thin-walled bar subjected to uniform axial compression will buckle torsionally while its longitudinal axis remains straight. In order to show how an axial compressive load may cause purely torsional buckling, let us consider as an example the doubly symmetric bar in Fig. 5-9. The bar is of cruciform section with four identical flanges of width b and thickness t. The x and y axes are the symmetry axes of the cross section. Under compression, a torsional buckling, as shown in Fig. 5-9, may occur. The axis of the bar remains straight, while each flange buckles by rotating about the z axis. In order to determine the compressive force which produces torsional buckling, it is necessary to consider the deflection of the flanges during buckling.

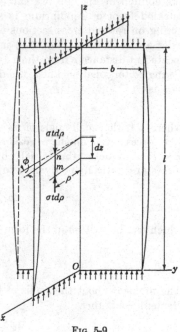

FIG. 5-9

To explain the method to be used in the analysis of the flanges, let us return to the simple case of the buckling of a pin-end strut, Fig. 5-10. Initially, the strut is straight and subjected to the centrally applied force P. Now let us assume that the force P reaches its critical value so that the strut can have a slightly deflected form of equilibrium. Because of this deflection there will be bending stresses superposed on the initial uniformly distributed compressive stresses. At the same time, the initial compressive stresses will act on slightly rotated cross sections, such as m and n in Fig. 5-10a and b. The differential equation for the deflection curve in this case is found from Eq. (1-5) by substituting $q = 0$. Denoting the deflection of the strut in the y direction by v, we can write Eq. (1-5) in the form

$$EI_z \frac{d^4v}{dz^4} = -P \frac{d^2v}{dz^2} \tag{5-16}$$

This equation was used in Art. 2.2 for calculating the critical value of the compressive force P. We see from Eq. (5-16) that the deflection curve of the strut and the corresponding bending stresses can be found by assuming that the strut is loaded by a fictitious lateral load of intensity $-P d^2v/dz^2$.

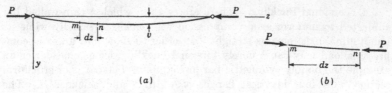

$$(a) \qquad (b)$$

FIG. 5-10

In an approximate discussion of the problem of torsional buckling of the column in Fig. 5-9, we can state that at the critical condition the buckled form of equilibrium is sustained by the compressive stresses acting on rotated cross sections of the longitudinal fibers. Let us consider an element mn (Fig. 5-9) in the form of a thin strip of length dz, located at distance ρ from the z axis and having cross-sectional area $t\,d\rho$. Owing to torsional buckling, the deflection of this element in the y direction is

$$v = \rho\phi \qquad (a)$$

where ϕ is the small angle of twist of the cross section.[1] The compressive forces acting on the rotated ends of the element mn are $\sigma t\,d\rho$, where $\sigma = P/A$ denotes the initial compressive stress. These compressive forces are statically equivalent to a lateral load of intensity

$$-(\sigma t\,d\rho)\frac{d^2v}{dz^2}$$

which can be written in the form [see Eq. (a)]

$$-\sigma t\rho\,d\rho\,\frac{d^2\phi}{dz^2}$$

The moment about the z axis of the fictitious lateral load acting on the element mn is then

$$-\sigma\frac{d^2\phi}{dz^2}\,dz\,t\rho^2\,d\rho$$

Summing up the moments for the entire cross section, we obtain the torque acting on an element of the buckled bar included between two consecutive cross sections. This torque is

$$-\sigma\frac{d^2\phi}{dz^2}\,dz\int_A t\rho^2\,d\rho = -\sigma\frac{d^2\phi}{dz^2}\,dz\,I_0$$

where I_0 is the polar moment of inertia of the cross section about the shear center O, coinciding in this case with the centroid. Finally, using

[1] It is assumed that the shape of the cross section does not change during twisting.

the notation m_z for the torque per unit length of the bar, we obtain

$$m_z = -\sigma \frac{d^2\phi}{dz^2} I_0 \qquad (b)$$

Expression (b) holds for any shape of cross section provided the shear center and centroid coincide.

To establish the differential equation for torsional buckling, we can use Eq. (5-15) for nonuniform torsion of a bar of thin-walled open section. Differentiation of this equation with respect to z gives

$$\frac{dM_t}{dz} = C \frac{d^2\phi}{dz^2} - C_1 \frac{d^4\phi}{dz^4} \qquad (c)$$

The positive directions of M_t and m_z are given by the right-hand rule, and hence these torques act on an element of a twisted bar in the directions shown in Fig. 5-11. Consideration of the equilibrium of this element gives

$$m_z = -\frac{dM_t}{dz} \qquad (d)$$

and therefore Eq. (c) takes the form

$$C_1 \frac{d^4\phi}{dz^4} - C \frac{d^2\phi}{dz^2} = m_z \qquad (5\text{-}17)$$

Substituting for m_z the value given by expression (b) we obtain

$$C_1 \frac{d^4\phi}{dz^4} - (C - \sigma I_0) \frac{d^2\phi}{dz^2} = 0 \qquad (5\text{-}18)$$

The critical value of the compressive stress σ, and hence also the critical load, can be calculated from Eq. (5-18). This equation holds for any shape of cross section as long as the shear center and centroid coincide.

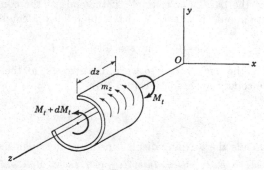

FIG. 5-11

For the column shown in Fig. 5-9 the warping rigidity C_1 vanishes; therefore, it is seen that, in the case of torsional buckling, Eq. (5-18) is satisfied if

$$C - \sigma I_0 = 0$$

which gives

$$\sigma_{cr} = \frac{C}{I_0} = \frac{\frac{1}{3}bt^3 G}{\frac{1}{3}tb^3} = \frac{Gt^2}{b^2} \tag{e}$$

The above result shows that in this case the critical value of the compressive stress is independent of the length of the bar. This conclusion is reached because in the preceding derivation the resistance of the flanges to bending in the directions perpendicular to the flanges was neglected. To obtain a more accurate result, it is necessary to consider each flange as a uniformly compressed plate simply supported along three sides and completely free along the fourth side. This more accurate investigation[1] shows that the critical stress is

$$\sigma_{cr} = \left(0.456 + \frac{b^2}{l^2}\right)\frac{\pi^2}{6(1 - \nu)}\frac{Gt^2}{b^2} \tag{f}$$

The second term in parentheses gives the influence of the length of the bar on the critical stress. For bars of considerable length this term can be neglected, and we obtain

$$\sigma_{cr} = \frac{0.75}{1 - \nu}\frac{Gt^2}{b^2} \tag{g}$$

For $\nu = 0.3$ this value is about 7 per cent greater than the value from Eq. (e).

For cases in which C_1 does not vanish, the critical compressive stress is obtained from the solution of Eq. (5-18). Introducing the notation

$$p^2 = \frac{\sigma I_0 - C}{C_1} \tag{5-19}$$

we find that this solution is

$$\phi = A_1 \sin pz + A_2 \cos pz + A_3 z + A_4 \tag{5-20}$$

The constants of integration A_1, A_2, A_3, and A_4 are found from the end conditions of the bar. For example, if the ends of the compressed bar cannot rotate about the z axis, we have the following conditions at the ends:

$$\phi = 0 \quad \text{at } z = 0 \text{ and } z = l \tag{5-21}$$

If the ends of the bar are free to warp, the stress σ_z at the ends will be zero and the conditions are [see Eq. (5-10)]

$$\frac{d^2\phi}{dz^2} = 0 \quad \text{at } z = 0 \text{ and } z = l \tag{5-22}$$

For built-in ends the warping displacements w must vanish, and hence,

[1] See Timoshenko, *Bull. Polytech. Inst., Kiev*, 1907, and *Z. Math. u. Physik*, vol. 58, p. 337, 1910.

from Eq. (5-9), we have

$$\frac{d\phi}{dz} = 0 \qquad \text{at } z = 0 \text{ and } z = l \qquad (5\text{-}23)$$

As a first example, let us consider the case of a bar with simple supports for which the ends cannot rotate about the z axis but are free to warp. Applying the conditions (5-21) and (5-22) to the general solution (5-20), we find that

$$A_2 = A_3 = A_4 = 0$$

and also $\qquad \sin pl = 0$

from which $\qquad pl = n\pi$

Substituting for p its value from Eq. (5-19), we obtain[1]

$$\sigma_{cr} = \frac{1}{I_O}\left(C + \frac{n^2\pi^2}{l^2}C_1\right) \qquad (5\text{-}24)$$

The smallest critical stress is found for $n = 1$, which corresponds to a buckled shape of the form

$$\phi = A_1 \sin\frac{\pi z}{l}$$

Equation (5-24) gives the critical stress for torsional buckling of a column in which the ends do not rotate but are free to warp.

As a second example, consider the case in which both ends of the bar are rigidly built in and cannot warp. Then the conditions at the ends are given by (5-21) and (5-23), and we find that

$$A_4 = -A_2 \qquad A_1 = A_3 = 0$$
$$pl = 2n\pi$$

The critical compressive stress for this case is

$$\sigma_{cr} = \frac{1}{I_O}\left(C + \frac{4n^2\pi^2}{l^2}C_1\right) \qquad (5\text{-}25)$$

It should be kept in mind that the column also may buckle because of lateral bending about the x or y axes at a stress given by Euler's column formula. Thus there are three critical values of the axial load, and only the lowest value is of practical interest. In general, torsional buckling is important for columns having wide flanges and short lengths.

5.5. Buckling by Torsion and Flexure. In the general case of a column of thin-walled open cross section, buckling failure usually occurs by a combination of torsion and bending. In order to investigate this type of buckling, let us consider the unsymmetrical cross section shown in Fig. 5-12. The x and y axes are the principal centroidal axes of the cross

[1] This solution was obtained by Wagner, *loc. cit.*

section and x_O, y_O are the coordinates of the shear center O. During buckling, the cross section will undergo translation and rotation. The translation is defined by the deflections u and v in the x and y directions, respectively, of the shear center O. Thus, during translation of the cross section, point O moves to O' and point C to C'. The rotation of the cross section about the shear center is denoted by the angle ϕ, as before, and the final position of the centroid is C''. Therefore the final deflections of the centroid C during buckling are[1]

$$u + y_O\phi \qquad v - x_O\phi$$

If the only load acting on the column is a central thrust P, as in the case of a pin-end column, the bending moments with respect to the principal axes at any cross section are

$$M_z = -P(v - x_O\phi)$$
$$M_y = -P(u + y_O\phi)$$

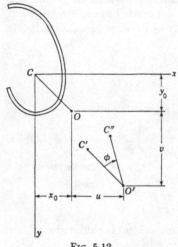

Fig. 5-12

The sign convention for moments M_z and M_y is shown in Fig. 5-13, where positive moments are shown acting on an element dz of the bar. The differential equations for the deflection curve of the shear-center axis become

$$EI_y \frac{d^2u}{dz^2} = +M_y = -P(u + y_O\phi) \qquad (5\text{-}26)$$

$$EI_x \frac{d^2v}{dz^2} = +M_z = -P(v - x_O\phi) \qquad (5\text{-}27)$$

These two equations for bending of the bar contain u, v, and ϕ as unknown quantities. A third equation is found by considering the twisting of the bar.

To obtain the equation for the angle of twist ϕ, we can follow the same method as in the preceding article and take a longitudinal strip of cross section $t\,ds$ defined by coordinates x, y in the plane of the cross section. The components of its deflection in the x and y directions during buckling are, respectively,

$$u + (y_O - y)\phi \qquad v - (x_O - x)\phi$$

Taking the second derivatives of these expressions with respect to z and again considering an element of length dz, we find that the compres-

[1] The angle ϕ is considered to be a small quantity.

sive forces $\sigma t\, ds$ acting on the slightly rotated ends of the element give forces in the x and y directions of intensity

$$-(\sigma t\, ds)\, \frac{d^2}{dz^2}\, [u + (y_0 - y)\phi] \tag{a}$$

$$-(\sigma t\, ds)\, \frac{d^2}{dz^2}\, [v - (x_0 - x)\phi] \tag{b}$$

Taking the moment about the shear-center axis of the above forces, we

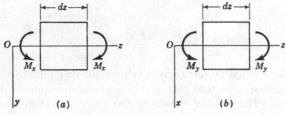

FIG. 5-13

obtain for one longitudinal strip the following torque per unit length of the bar:

$$dm_z = -(\sigma t\, ds)(y_0 - y)\left[\frac{d^2u}{dz^2} + (y_0 - y)\frac{d^2\phi}{dz^2}\right]$$
$$+ (\sigma t\, ds)(x_0 - x)\left[\frac{d^2v}{dz^2} - (x_0 - x)\frac{d^2\phi}{dz^2}\right]$$

Integrating over the entire cross-sectional area A and observing that

$$\sigma \int_A t\, ds = P \qquad \int_A xt\, ds = \int_A yt\, ds = 0$$

$$\int_A y^2 t\, ds = I_x \qquad \int_A x^2 t\, ds = I_y$$

$$I_0 = I_x + I_y + A(x_0{}^2 + y_0{}^2)$$

we obtain

$$m_z = \int_A dm_z = P\left[x_0\, \frac{d^2v}{dz^2} - y_0\, \frac{d^2u}{dz^2}\right] - \frac{I_0}{A}\, P\, \frac{d^2\phi}{dz^2} \tag{c}$$

In these expressions, I_x and I_y are the principal centroidal moments of inertia of the cross section and I_0 is the polar moment of inertia about the shear center O. Substituting expression (c) into Eq. (5-17) for non-uniform torsion, we find that

$$C_1\, \frac{d^4\phi}{dz^4} - \left(C - \frac{I_0}{A}\, P\right)\frac{d^2\phi}{dz^2} - Px_0\, \frac{d^2v}{dz^2} + Py_0\, \frac{d^2u}{dz^2} = 0 \tag{5-28}$$

Equations (5-26), (5-27), and (5-28) are the three simultaneous differential equations[1] for buckling by bending and torsion and can be used to determine the critical loads.   It is seen that the angle of rotation appears in all three equations, so that, in the general case, torsional buckling and bending of the axis occur simultaneously.

In the particular case when the shear center coincides with the centroid, we have $x_O = y_O = 0$ and Eqs. (5-26) to (5-28) become

$$EI_v \frac{d^2u}{dz^2} = -Pu$$

$$EI_z \frac{d^2v}{dz^2} = -Pv$$

$$C_1 \frac{d^4\phi}{dz^4} - \left(C - \frac{I_O}{A} P \right) \frac{d^2\phi}{dz^2} = 0$$

Each of these equations contains only one unknown quantity and can be treated separately, so that torsional buckling is independent of buckling by bending.   The first two equations give the values of the Euler critical loads for buckling in the two principal planes.   The third equation is the same as Eq. (5-18) and gives the critical load for purely torsional buckling, as discussed in the preceding article.   Only the lowest of the three values of the critical load is of interest in practical applications.

Returning now to the general case [Eqs. (5-26) to (5-28)], let us assume that the bar has simple supports, so that the ends of the bar are free to warp and to rotate about the x and y axes but cannot rotate about the z axis or deflect in the x and y directions.   In such a case the end conditions are

$$u = v = \phi = 0 \qquad \text{at } z = 0 \text{ and } z = l$$

$$\frac{d^2u}{dz^2} = \frac{d^2v}{dz^2} = \frac{d^2\phi}{dz^2} = 0 \qquad \text{at } z = 0 \text{ and } z = l$$

These end conditions will be satisfied if the solutions of Eqs. (5-26) to (5-28) are taken in the form

$$u = A_1 \sin \frac{\pi z}{l} \qquad v = A_2 \sin \frac{\pi z}{l} \qquad \phi = A_3 \sin \frac{\pi z}{l} \qquad (d)$$

Substituting expressions (d) into Eqs. (5-26) to (5-28) gives the following

---

[1] A system of equations equivalent to Eqs. (5-26) to (5-28) was obtained first by R. Kappus; see "Jahrbuch der Deutschen Luftfahrt-Forschung," 1937, and *Luftfahrt-Forsch.*, vol. 14, p. 444, 1937 (translated in *NACA Tech. Mem.* 851, 1938).   Torsional buckling has also been discussed by J. N. Goodier, *Cornell Univ. Eng. Expt. Sta. Bull.* 27, December, 1941, and 28, January, 1942; see also the book by V. Z. Vlasov, "Thin-walled Elastic Bars," Moscow, 1940, and Timoshenko, *J. Franklin Inst.*, vol. 239, nos. 3, 4, and 5, April and May, 1945.   Regarding experiments with torsional buckling, see A. S. Niles, *NACA Tech. Note* 733, 1939, and H. Wagner and W. Pretschner, *Luftfahrt-Forsch.*, vol. 11, p. 174, 1934 (translated in *NACA Tech. Mem.* 784, 1936).

three equations for determining the constants A_1, A_2, and A_3:

$$\left(P - EI_y \frac{\pi^2}{l^2}\right) A_1 + Py_0 A_3 = 0$$

$$\left(P - EI_x \frac{\pi^2}{l^2}\right) A_2 - Px_0 A_3 = 0 \qquad (5\text{-}29)$$

$$Py_0 A_1 - Px_0 A_2 - \left(C_1 \frac{\pi^2}{l^2} + C - \frac{I_0}{A} P\right) A_3 = 0$$

One solution of these equations is $A_1 = A_2 = A_3 = 0$, which corresponds to the straight form of equilibrium. For a buckled form of equilibrium, the constants A_1, A_2, A_3 must not vanish simultaneously, which is possible only if the determinant of Eqs. (5-29) vanishes. To simplify the expressions, let us introduce the notation

$$P_x = \frac{\pi^2 EI_x}{l^2} \qquad P_y = \frac{\pi^2 EI_y}{l^2} \qquad P_\phi = \frac{A}{I_0}\left(C + C_1 \frac{\pi^2}{l^2}\right) \qquad (5\text{-}30)$$

where P_x and P_y are the Euler critical loads for buckling about the x and y axes, respectively, and P_ϕ is the critical load for purely torsional buckling [see Eq. (5-24)]. Then, equating to zero the determinant of Eqs. (5-29), we obtain

$$\begin{vmatrix} P - P_y & 0 & Py_0 \\ 0 & P - P_x & -Px_0 \\ Py_0 & -Px_0 & \dfrac{I_0}{A}(P - P_\phi) \end{vmatrix} = 0$$

Expanding the determinant gives the following cubic equation for calculating the critical values of P:

$$\frac{I_0}{A}(P - P_y)(P - P_x)(P - P_\phi) - P^2 y_0^2 (P - P_x) - P^2 x_0^2 (P - P_y) = 0 \qquad (5\text{-}31)$$

or

$$\frac{I_c}{I_0} P^3 + \left[\frac{A}{I_0}(P_x y_0^2 + P_y x_0^2) - (P_x + P_y + P_\phi)\right] P^2$$
$$+ (P_x P_y + P_x P_\phi + P_y P_\phi)P - P_x P_y P_\phi = 0 \qquad (5\text{-}32)$$

where $I_c = I_x + I_y$ denotes the polar moment of inertia about the centroid C of the cross section.

To find the critical load in any particular case, we begin by calculating the numerical values of the coefficients in Eq. (5-32). Solution of the cubic equation then gives three values of the critical load P, of which the smallest will be used in practical application. Substitution of the three values of the critical loads into Eqs. (5-29) yields the ratios A_1/A_3 and A_2/A_3 for each of the corresponding three forms of buckling. These ratios establish the relations between rotation and translation of the cross sections and define the deflected shape of the shear-center axis.

An important conclusion regarding the relative magnitudes of the critical loads can be obtained from Eq. (5-31). Considering the left-hand side of this equation as a function $f(P)$, we wish to determine the sign of this function for various values of P in order to obtain information concerning the values of P which make $f(P)$ vanish. For very large values of P the polynomial $f(P)$ takes the sign of the term with highest power. This term is $P^3 I_0/A$ and is positive. If $P = 0$, the value of $f(P)$ is $-P_x P_y P_\phi I_0/A$, which is negative. Now let us assume that $P_x < P_y$, that is, that the smaller Euler load corresponds to bending in the yz plane. If $P = P_x$, we have

$$f(P) = -P_x^2 x_0^2 (P_x - P_y)$$

which is positive, and if $P = P_y$, we find that

$$f(P) = -P_y^2 y_0^2 (P_y - P_x)$$

which is negative. Thus we see that Eqs. (5-31) and (5-32) have three positive roots, one of which is less than P_x, one greater than P_y, and one between P_x and P_y. A similar result is obtained if we assume that $P_x > P_y$. It can be shown also that the smallest value of P is less than P_ϕ, for if P_ϕ is less than both P_x and P_y, we find that $f(P)$ is positive for $P = P_\phi$. Likewise, the largest root must always be greater than P_ϕ. Thus we finally conclude that, in all cases, one critical load is less than P_x, P_y, or P_ϕ and one is greater. The third critical load is always intermediate between P_x and P_y. This means that when we take into consideration the possibility of torsion during buckling, we always obtain a critical load smaller than either the Euler load or the purely torsional buckling load.

If the bar has wide flanges and short length l, P_ϕ may become small compared with P_x and P_y. In such a case the smallest root of Eq. (5-32) approaches the value P_ϕ. Substituting this root in Eqs. (5-29), we find that A_1 and A_2 are small in comparison with the rotational displacements, which indicates that the buckling approaches purely torsional buckling. In the case of narrow flanges and large length, P_ϕ will be large in comparison with P_x and P_y and the smallest root of Eq. (5-32) approaches either P_x or P_y. The effect of torsion on the critical load in such a case is small, and the Euler column formula gives satisfactory results.

The preceding discussion was based on the solution (d). Without any complication we can take the solution in a more general form and assume that

$$u = A_1 \sin \frac{n\pi z}{l} \qquad v = A_2 \sin \frac{n\pi z}{l} \qquad \phi = A_3 \sin \frac{n\pi z}{l} \qquad (e)$$

which corresponds to the assumption that during buckling, the bar subdivides into n half sine waves. Our previous conclusions will hold in this case if the values $n^2\pi^2/l^2$ are substituted for π^2/l^2 in expressions (5-30). The corresponding critical loads are larger than those obtained for $n = 1$ and are of practical interest only if the bar has intermediate equidistant lateral supports.

Bar with Built-in Ends. If the ends of the bar are built in, the end conditions become

$$u = v = \phi = 0 \qquad \text{at } z = 0 \text{ and } z = l$$

$$\frac{du}{dz} = \frac{dv}{dz} = \frac{d\phi}{dz} = 0 \qquad \text{at } z = 0 \text{ and } z = l$$

Since there are moments at the ends of the bar during buckling, we shall have, in place of Eqs. (5-26) and (5-27), the following equations:

$$EI_y \frac{d^2u}{dz^2} = -P(u + y_0\phi) + EI_y \left(\frac{d^2u}{dz^2}\right)_{z=0} \qquad (5\text{-}33)$$

$$EI_x \frac{d^2v}{dz^2} = -P(v - x_0\phi) + EI_x \left(\frac{d^2v}{dz^2}\right)_{z=0} \qquad (5\text{-}34)$$

These equations, together with Eq. (5-28),[1] define the buckled shapes of the bar and the corresponding critical loads. The three equations and the end conditions are satisfied by taking a solution in the form

$$u = A_1\left(1 - \cos\frac{2\pi z}{l}\right) \qquad v = A_2\left(1 - \cos\frac{2\pi z}{l}\right) \qquad \phi = A_3\left(1 - \cos\frac{2\pi z}{l}\right)$$

Substituting these expressions into Eqs. (5-28), (5-33), and (5-34), we again obtain the cubic equation (5-32) for calculating the critical loads; it is only necessary to use $4\pi^2/l^2$ instead of π^2/l^2 in the notation (5-30).

Cross Section with One Axis of Symmetry. Let us assume that the x axis is an axis of symmetry, as illustrated by the channel section in Fig. 5-14. In this case we have $y_0 = 0$ and Eqs. (5-26) to (5-28) become

$$EI_y \frac{d^2u}{dz^2} = -Pu \qquad (5\text{-}35)$$

$$EI_x \frac{d^2v}{dz^2} = -P(v - x_0\phi) \qquad (5\text{-}36)$$

$$C_1 \frac{d^4\phi}{dz^4} - \left(C - \frac{I_0}{A}P\right)\frac{d^2\phi}{dz^2} - Px_0 \frac{d^2v}{dz^2} = 0 \qquad (5\text{-}37)$$

[1] Equation (5-28) was developed from a consideration of an element of the bar between two adjacent cross sections and is not affected by changes in the end conditions.

The first equation does not contain ϕ and shows that buckling in the plane of symmetry is independent of torsion and that the corresponding critical load is given by Euler's formula. Buckling perpendicular to the plane of symmetry is combined with torsion and is given by Eqs. (5-36) and (5-37).

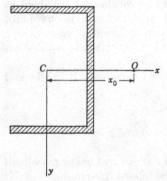

FIG. 5-14

If we assume that the ends of the bar are simply supported, that is, free to warp and to rotate about the x axis but restrained against rotation about the z axis, the end conditions become

$$v = \phi = 0 \qquad \text{at } z = 0 \text{ and } z = l$$

$$\frac{d^2v}{dz^2} = \frac{d^2\phi}{dz^2} = 0 \qquad \text{at } z = 0 \text{ and } z = l$$

Proceeding as before and taking a solution in the form

$$v = A_1 \sin \frac{\pi z}{l} \qquad \phi = A_2 \sin \frac{\pi z}{l}$$

we obtain for calculating the critical loads the equation

$$\begin{vmatrix} P - P_x & -Px_0 \\ -Px_0 & \dfrac{I_0}{A}(P - P_\phi) \end{vmatrix} = 0$$

which gives

$$\frac{I_0}{A}(P - P_x)(P - P_\phi) - P^2 x_0{}^2 = 0 \tag{5-38}$$

or

$$\frac{I_c}{I_0} P^2 - (P_x + P_\phi)P + P_x P_\phi = 0 \tag{5-39}$$

This quadratic equation gives two solutions for the critical load P, one of which is smaller than either P_x or P_ϕ, while the other is larger than either. The smaller of these roots or the Euler load for buckling in the plane of symmetry represents the critical load for the column. A graph of the two critical loads obtained from Eq. (5-39) is shown in Fig. 5-15. Note that when P_ϕ/P_x is small, the lower critical load is very close to P_ϕ and the mode of buckling is essentially torsional, whereas the upper critical load represents buckling which is primarily bending. For large P_ϕ/P_x the lower critical load corresponds to a form of buckling which is primarily bending. For an equal-legged angle section the value of I_0/I_c is 1.6, and the corresponding curve in Fig. 5-15 can be used for this case. For other sections with one axis of symmetry, such as a channel section, I_0/I_c must be computed for each individual case.

5.6. Combined Torsional and Flexural Buckling of a Bar with Continuous Elastic Supports. Let us consider the stability of a centrally compressed bar which is supported elastically throughout its length in such a way that lateral reactions proportional to the deflection will develop during buckling. Let us assume that these reactions are distributed along an axis N parallel to the axis of the bar (Fig. 5-16) and defined by coordinates h_x and h_y. Again denoting the components of the deflection of the shear-center axis by u and v and the angle of rotation with respect to that axis

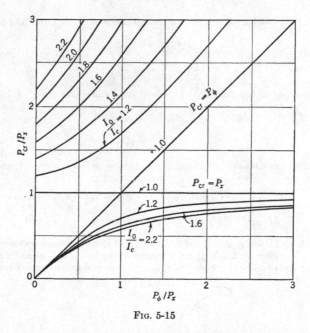

FIG. 5-15

by ϕ (see Fig. 5-12), we find that the components of deflection of the N axis, along which the reactions are distributed, are

$$u + (y_0 - h_y)\phi \qquad v - (x_0 - h_x)\phi$$

The corresponding reactions per unit length, assumed positive in the positive directions of the x and y axes, will be

$$-k_x[u + (y_0 - h_y)\phi] \qquad -k_y[v - (x_0 - h_x)\phi] \tag{a}$$

where k_x and k_y are constants defining the rigidity of the elastic support in the x and y directions. These constants, or moduli, represent the reactions per unit length when the deflections are equal to unity and have dimensions of force divided by length squared. To the above reactions we must add the lateral forces obtained from the action of the initial compressive forces acting on slightly rotated cross sections of the longitudinal fibers. These forces give reactions per unit length [see expressions (a) and (b) of Art. 5.5] equal to ·

$$-\int_A \sigma t \, ds \left[\frac{d^2 u}{dz^2} + (y_O - y) \frac{d^2 \phi}{dz^2} \right]$$

and

$$-\int_A \sigma t \, ds \left[\frac{d^2 v}{dz^2} - (x_O - x) \frac{d^2 \phi}{dz^2} \right]$$

Integrating the above two expressions and again observing that

$$\sigma \int_A t \, ds = P \qquad \int_A x t \, ds = \int_A y t \, ds = 0$$

we obtain the following expressions for the intensities of lateral force distribution:

$$-P \left(\frac{d^2 u}{dz^2} + y_O \frac{d^2 \phi}{dz^2} \right) \tag{b}$$

$$-P \left(\frac{d^2 v}{dz^2} - x_O \frac{d^2 \phi}{dz^2} \right) \tag{c}$$

The equations for bending of the bar about the y and x axes are

$$EI_y \frac{d^4 u}{dz^4} = q_x \tag{5-40}$$

$$EI_x \frac{d^4 v}{dz^4} = q_y \tag{5-41}$$

and using for the intensities of the distributed loads expressions (a), (b), and (c), we obtain

$$EI_y \frac{d^4 u}{dz^4} + P \left(\frac{d^2 u}{dz^2} + y_O \frac{d^2 \phi}{dz^2} \right) + k_x [u + (y_O - h_y) \phi] = 0 \tag{5-42}$$

$$EI_x \frac{d^4 v}{dz^4} + P \left(\frac{d^2 v}{dz^2} - x_O \frac{d^2 \phi}{dz^2} \right) + k_y [v - (x_O - h_x) \phi] = 0 \tag{5-43}$$

Since the lateral loads q_x and q_y are not distributed along the shear-center axis, there will be, in addition to bending, some torsion of the bar. The intensity m_z of the torque distributed along the shear-center axis will be equal to the couple developed by

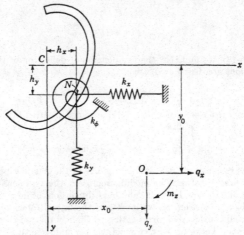

Fig. 5-16

the loads given by expressions (a), (b), and (c) plus the torsional reaction developed by the elastic support. Denoting by k_ϕ the torsional modulus of the elastic support, we find this latter torque to be

$$-k_\phi\phi \qquad (d)$$

The torque due to the lateral reactions (a), which act at point N, will be

$$-k_x[u + (y_0 - h_y)\phi](y_0 - h_y) + k_y[v - (x_0 - h_x)\phi](x_0 - h_x) \qquad (e)$$

The torque due to the forces given by expressions (b) and (c) was evaluated in the preceding article and is given by Eq. (c), p. 231. Adding this value to (d) and (e) above gives the total torque

$$m_z = P\left(x_0\frac{d^2v}{dz^2} - y_0\frac{d^2u}{dz^2}\right) - \frac{I_0}{A}P\frac{d^2\phi}{dz^2} - k_x[u + (y_0 - h_y)\phi](y_0 - h_y)$$
$$+ k_y[v - (x_0 - h_x)\phi](x_0 - h_x) - k_\phi\phi \qquad (f)$$

Substituting expression (f) for m_z into Eq. (5-17) for nonuniform torsion, we obtain the following equation for the angle of twist:

$$C_1\frac{d^4\phi}{dz^4} - \left(C - \frac{I_0}{A}P\right)\frac{d^2\phi}{dz^2} - P\left(x_0\frac{d^2v}{dz^2} - y_0\frac{d^2u}{dz^2}\right) + k_x[u + (y_0 - h_y)\phi](y_0 - h_y)$$
$$-k_y[v - (x_0 - h_x)\phi](x_0 - h_x) + k_\phi\phi = 0 \qquad (5\text{-}44)$$

Equations (5-42), (5-43), and (5-44) are three simultaneous differential equations for the buckling of a bar supported elastically along its length.[1]

If the ends of the bar are simply supported, that is, free to warp and to rotate about the x and y axes but with no rotation about the z axis, we can take the solution of Eqs. (5-42) to (5-44) in the form

$$u = A_1\sin\frac{n\pi z}{l} \qquad v = A_2\sin\frac{n\pi z}{l} \qquad \phi = A_3\sin\frac{n\pi z}{l} \qquad (g)$$

Substitution of these expressions into the differential equations leads to a cubic equation for the critical loads, in the same manner as described in the preceding article. Solving for the lowest root of the cubic equation will give the smallest critical load. Let us consider now some particular cases.

Cross Section with Two Axes of Symmetry. In the special case where the cross section has two axes of symmetry, the centroid and shear center coincide and $x_0 = y_0 = 0$. Let us assume also that the elastic reactions are distributed along the centroidal axis. Then we have $h_x = h_y = 0$ and Eqs. (5-42) to (5-44) take the simple form

$$EI_y\frac{d^4u}{dz^4} + P\frac{d^2u}{dz^2} + k_xu = 0 \qquad (5\text{-}45)$$

$$EI_x\frac{d^4v}{dz^4} + P\frac{d^2v}{dz^2} + k_yv = 0 \qquad (5\text{-}46)$$

$$C_1\frac{d^4\phi}{dz^4} - \left(C - \frac{I_0}{A}P\right)\frac{d^2\phi}{dz^2} + k_\phi\phi = 0 \qquad (5\text{-}47)$$

These equations show that buckling of the bar in the planes of symmetry is independent of torsion and the three forms of buckling can be treated separately.

Taking the solution in the form (g), we find from the first equation that

$$EI_y\frac{n^4\pi^4}{l^4} - P\frac{n^2\pi^2}{l^2} + k_x = 0$$

[1] These equations were obtained first by Vlasov, *loc. cit.*

or
$$P_{cr} = \frac{\pi^2 EI_y}{l^2} \left(n^2 + \frac{l^4 k_x}{n^2 \pi^4 EI_y} \right) \tag{5-48}$$

This result agrees with the value previously obtained for a bar on elastic foundation [see Eq. (2-37)]. A similar result is obtained also from Eq. (5-46). The last of the three equations gives the critical load for torsional buckling as

$$P_{cr} = \frac{(n^2 \pi^2 / l^2) C_1 + C + (l^2 / n^2 \pi^2) k_\phi}{I_0 / A} \tag{5-49}$$

In each particular case, knowing C_1 and k_ϕ, it is necessary to select the integer value of n which makes expression (5-49) a minimum. When $k_\phi = 0$, the lowest critical load is obtained by taking $n = 1$ and Eq. (5-49) gives the value obtained previously for purely torsional buckling [see Eq. (5-24)].

Cross Section with One Axis of Symmetry. If the x axis is taken as the axis of symmetry, we shall have $y_0 = 0$. Let us assume further that the elastic reactions are distributed along the shear-center axis so that $h_x = x_0$ and $h_y = 0$. Then Eqs. (5-42) to (5-44) become

$$EI_y \frac{d^4 u}{dz^4} + P \frac{d^2 u}{dz^2} + k_x u = 0 \tag{5-50}$$

$$EI_x \frac{d^4 v}{dz^4} + P \frac{d^2 v}{dz^2} + k_y v - P x_0 \frac{d^2 \phi}{dz^2} = 0 \tag{5-51}$$

$$C_1 \frac{d^4 \phi}{dz^4} - \left(C - \frac{I_0}{A} P \right) \frac{d^2 \phi}{dz^2} + k_\phi \phi - P x_0 \frac{d^2 v}{dz^2} = 0 \tag{5-52}$$

From the first equation we see that buckling in the plane of symmetry is independent of torsion and can be treated separately. The last two equations are simultaneous, and hence buckling in the y direction is combined with torsion.

If the end conditions are such that the solution of the differential equations can be taken in the form (g), we obtain, from Eqs. (5-51) and (5-52), the following determinant for calculating the critical loads:

$$\begin{vmatrix} EI_x \dfrac{n^4 \pi^4}{l^4} - P \dfrac{n^2 \pi^2}{l^2} + k_y & P x_0 \dfrac{n^2 \pi^2}{l^2} \\ P x_0 \dfrac{n^2 \pi^2}{l^2} & C_1 \dfrac{n^4 \pi^4}{l^4} + \left(C - \dfrac{I_0}{A} P \right) \dfrac{n^2 \pi^2}{l^2} + k_\phi \end{vmatrix} = 0 \tag{5-53}$$

This equation is quadratic in P and can be solved in each particular case for the two values of the critical load, of which only the lower value is normally of importance. If both k_ϕ and k_y vanish, the smaller load will occur with $n = 1$ and Eq. (5-53) gives the same result as derived previously [see Eq. (5-39)].

Bar with Prescribed Axis of Rotation. Using the differential equations (5-42) to (5-44), we can investigate buckling of a bar for which the axis about which the cross sections rotate during buckling is prescribed. To obtain a rigid axis of rotation, we have only to assume that $k_x = k_y = \infty$. Then the N axis (Fig. 5-16) will remain straight during buckling and the cross sections will rotate with respect to this axis. For this case, Eqs. (5-42) and (5-43) give

$$u + (y_0 - h_y)\phi = 0 \qquad v - (x_0 - h_x)\phi = 0$$
and hence
$$u = -(y_0 - h_y)\phi \qquad v = (x_0 - h_x)\phi$$

Differentiating these expressions gives

$$\frac{d^2 u}{dz^2} = -(y_0 - h_y) \frac{d^2 \phi}{dz^2} \qquad \frac{d^4 u}{dz^4} = -(y_0 - h_y) \frac{d^4 \phi}{dz^4} \tag{h}$$

$$\frac{d^2 v}{dz^2} = (x_0 - h_x) \frac{d^2 \phi}{dz^2} \qquad \frac{d^4 v}{dz^4} = (x_0 - h_x) \frac{d^4 \phi}{dz^4} \tag{i}$$

We also have, from Eqs. (5-42) and (5-43), the relations

$$k_z[u + (y_0 - h_y)\phi] = -EI_y \frac{d^4u}{dz^4} - P\left(\frac{d^2u}{dz^2} + y_0 \frac{d^2\phi}{dz^2}\right) \qquad (j)$$

$$k_y[v - (x_0 - h_z)\phi] = -EI_x \frac{d^4v}{dz^4} - P\left(\frac{d^2v}{dz^2} - x_0 \frac{d^2\phi}{dz^2}\right) \qquad (k)$$

An equation for the angle of rotation ϕ can be obtained now by substituting (h) and (i) into (j) and (k) and then substituting (j) and (k) into Eq. (5-44). The resulting equation is

$$[C_1 + EI_y(y_0 - h_y)^2 + EI_x(x_0 - h_z)^2]\frac{d^4\phi}{dz^4}$$

$$- \left[C - \frac{I_0}{A}P + P(x_0^2 + y_0^2) - P(h_z^2 + h_y^2)\right]\frac{d^2\phi}{dz^2} + k_\phi\phi = 0 \qquad (5-54)$$

Taking the solution of this equation in the form (g), we can calculate in each particular case the critical buckling load.

If the bar has two planes of symmetry, we have $x_0 = y_0 = 0$ and Eq. (5-54) becomes

$$(C_1 + EI_yh_y^2 + EI_xh_z^2)\frac{d^4\phi}{dz^4} - \left(C - \frac{I_0}{A}P - Ph_z^2 - Ph_y^2\right)\frac{d^2\phi}{dz^2} + k_\phi\phi = 0 \qquad (5-55)$$

Taking the solution in the form $\phi = A_3 \sin(n\pi z/l)$ and substituting into Eq. (5-55), we find that

$$P_{cr} = \frac{(C_1 + EI_yh_y^2 + EI_xh_z^2)(n^2\pi^2/l^2) + C + k_\phi(l^2/n^2\pi^2)}{h_z^2 + h_y^2 + (I_0/A)} \qquad (5-56)$$

In each particular case we must take the value of n which makes expression (5-56) a minimum.

If the fixed axis of rotation is the shear-center axis, we shall have $h_z = x_0$, $h_y = y_0$, and Eq. (5-54) becomes

$$C_1 \frac{d^4\phi}{dz^4} - \left(C - \frac{I_0}{A}P\right)\frac{d^2\phi}{dz^2} + k_\phi\phi = 0 \qquad (5-57)$$

Again taking a solution in the form $\phi = A_3 \sin(n\pi z/l)$, we obtain

$$P_{cr} = \frac{C_1(n^2\pi^2/l^2) + C + k_\phi(l^2/n^2\pi^2)}{I_0/A} \qquad (5-58)$$

This expression is valid for cross-sectional shapes which may be either symmetrical or unsymmetrical, provided the shear-center axis is restrained against deflection. When k_ϕ vanishes, the minimum value of P_{cr} occurs for $n = 1$, and Eq. (5-58) gives the same result as Eq. (5-24), as expected.

As another special case, let us assume that the fixed axis of rotation is infinitely distant from the bar. For example, if h_y becomes infinite, Eq. (5-54) reduces to

$$EI_y \frac{d^4\phi}{dz^4} + P \frac{d^2\phi}{dz^2} = 0$$

and this equation gives for P_{cr} the known Euler load for buckling in the xz plane.

Bar with Prescribed Plane of Deflection. In practical design of columns, the situation arises in which certain fibers of the bar must deflect in a known direction during buckling. For example, if a bar is welded to a thin sheet, as in Fig. 5-17, the fibers of the bar in contact with the sheet cannot deflect in the plane of the sheet. Instead, the fibers along the contact plane nn must deflect only in the direction perpendicular to the sheet.

In discussing problems of this type it is advantageous to take the centroidal axes x and y parallel and perpendicular to the sheet. Usually this means that the axes are no longer principal axes of the cross section, and therefore the differential equations of the deflection curve must be obtained in a different form. For this more general case, when the x and y axes are not principal axes, the bending moments in the bar are given by the expressions[1]

$$M_x = EI_x \frac{d^2v}{dz^2} + EI_{xy} \frac{d^2u}{dz^2} \tag{l}$$

$$M_y = EI_y \frac{d^2u}{dz^2} + EI_{xy} \frac{d^2v}{dz^2} \tag{m}$$

In these expressions I_{xy} represents the product of inertia of the cross section, and the

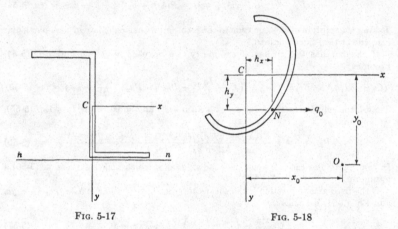

FIG. 5-17   FIG. 5-18

assumed positive directions for the bending moments M_x and M_y are given in Fig. 5-13. Observing that

$$q_x = \frac{d^2M_y}{dz^2} \qquad q_y = \frac{d^2M_x}{dz^2}$$

where q_x and q_y are taken positive in the positive directions of the x and y axes, we obtain from Eqs. (l) and (m) the differential equations for bending:

$$q_y = EI_x \frac{d^4v}{dz^4} + EI_{xy} \frac{d^4u}{dz^4} \tag{5-59}$$

$$q_x = EI_y \frac{d^4u}{dz^4} + EI_{xy} \frac{d^4v}{dz^4} \tag{5-60}$$

Considering a bar of arbitrary cross section, Fig. 5-18, let us assume that a longitudinal fiber N with coordinates h_x and h_y is prevented from deflecting in the x direction. Again denoting by u and v the deflections of the shear-center axis O, we find for the deflections of N the following expressions:

[1] For a discussion of bending of an unsymmetrical bar when the axes are not principal axes, see Timoshenko, "Strength of Materials," 3d ed., part I, pp. 230–231, D. Van Nostrand Company, Inc., Princeton, N.J., 1955.

$$u_N = u + (y_O - h_y)\phi = 0 \tag{n}$$

$$v_N = v - (x_O - h_x)\phi \tag{o}$$

Owing to the restraint at fiber N, there will be reactions of intensity q_O distributed continuously along N and acting in the direction parallel to the x axis (Fig. 5-18). The quantity q_x in Eq. (5-60) will be found by adding to q_O the lateral force obtained from the action of the compressive force in the bar acting on the rotated cross sections of longitudinal fibers. This latter value is found from expression (b) and therefore, we have

$$q_x = -P \left(\frac{d^2 u}{dz^2} + y_O \frac{d^2 \phi}{dz^2} \right) + q_O$$

The intensity of force in the y direction, from expression (c), is

$$q_y = -P \left(\frac{d^2 v}{dz^2} - x_O \frac{d^2 \phi}{dz^2} \right)$$

Substituting into Eqs. (5-59) and (5-60), we obtain

$$q_O = P \left(\frac{d^2 u}{dz^2} + y_O \frac{d^2 \phi}{dz^2} \right) + EI_y \frac{d^4 u}{dz^4} + EI_{xy} \frac{d^4 v}{dz^4} \tag{p}$$

$$EI_x \frac{d^4 v}{dz^4} + EI_{xy} \frac{d^4 u}{dz^4} + P \frac{d^2 v}{dz^2} - P x_O \frac{d^2 \phi}{dz^2} = 0 \tag{q}$$

We can eliminate u from the latter equation by solving for u from Eq. (n) and substituting into Eq. (q), which gives

$$EI_x \frac{d^4 v}{dz^4} + P \frac{d^2 v}{dz^2} - EI_{xy}(y_O - h_y) \frac{d^4 \phi}{dz^4} - P x_O \frac{d^2 \phi}{dz^2} = 0 \tag{5-61}$$

A second equation for v and ϕ is obtained by considering the torsion of the bar. For this purpose we can use Eq. (5-44). Assuming that there are no torsional reactions and no reaction in the y direction and substituting into Eq. (5-44) the following expressions:

$$k_y = k_\phi = 0$$

$$-k_x[u + (y_O - h_y)\phi] = q_O$$

we obtain

$$C_1 \frac{d^4 \phi}{dz^4} - \left(C - \frac{I_O}{A} P \right) \frac{d^2 \phi}{dz^2} - P \left(x_O \frac{d^2 v}{dz^2} - y_O \frac{d^2 u}{dz^2} \right) - q_O(y_O - h_y) = 0$$

Substituting in this equation the value of u from Eq. (n) and the value of q_O from Eq. (p), we find that

$$[C_1 + EI_y(y_O - h_y)^2] \frac{d^4 \phi}{dz^4} - \left(C - \frac{I_O}{A} P + P y_O^2 - P h_y^2 \right) \frac{d^2 \phi}{dz^2}$$
$$- EI_{xy}(y_O - h_y) \frac{d^4 v}{dz^4} - P x_O \frac{d^2 v}{dz^2} = 0 \tag{5-62}$$

Equations (5-61) and (5-62) can be used to find the critical buckling loads.[1]

As an example, let us consider again the case of simple supports and take the solution in the form

$$v = A_2 \sin \frac{\pi z}{l} \qquad \phi = A_3 \sin \frac{\pi z}{l}$$

----

[1] These equations were obtained by Goodier, *Cornell Univ. Eng. Expt. Sta. Bull.* 27, December, 1941.

After substitution into Eqs. (5-61) and (5-62), we obtain

$$\left(EI_z \frac{\pi^2}{l^2} - P\right) A_2 - \left[EI_{zy}(y_0 - h_y)\frac{\pi^2}{l^2} - Px_0\right] A_3 = 0$$

$$\left[-EI_{zy}(y_0 - h_y)\frac{\pi^2}{l^2} + Px_0\right] A_2$$

$$+ \left[C_1\frac{\pi^2}{l^2} + EI_y(y_0 - h_y)^2\frac{\pi^2}{l^2} + C - \frac{I_0}{A}P + Py_0^2 - Ph_y^2\right] A_3 = 0$$

Equating to zero the determinant of these equations, we obtain a quadratic equation for P from which the critical load can be calculated in each particular case.

If the bar is symmetrical with respect to the y axis, as in the case of the channel, Fig. 5-19, the x and y axes become principal axes. Then, with the substitution of $I_{zy} = 0$ and $x_0 = 0$, the above two equations become

$$\left(EI_z \frac{\pi^2}{l^2} - P\right) A_2 = 0$$

$$\left[C_1\frac{\pi^2}{l^2} + EI_y(y_0 - h_y)^2\frac{\pi^2}{l^2} + C - \frac{I_0}{A}P + Py_0^2 - Ph_y^2\right] A_3 = 0$$

The first of these equations gives the Euler load for buckling in the plane of symmetry. The second equation gives

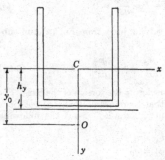

FIG. 5-19

$$P_{cr} = \frac{C_1(\pi^2/l^2) + EI_y(y_0 - h_y)^2(\pi^2/l^2) + C}{(I_0/A) - y_0^2 + h_y^2} \tag{5-63}$$

which represents the torsional buckling load for this case. The axis of rotation lies in the plane of the thin sheet. Equation (5-63) could be obtained also by substituting

$$x_0 = h_z = k_\phi = 0 \qquad \phi = A_3 \sin \frac{\pi z}{l}$$

into Eq. (5-54).

5.7. Torsional Buckling under Thrust and End Moments. In the previous articles we have considered the buckling of columns subjected to centrally applied compressive loads only. Let us now consider the case when the bar is subjected to the action of bending couples M_1 and M_2 at the ends (see Fig. 5-20) in addition to the central thrust P. The bending couples M_1 and M_2 are taken positive in the directions shown in the figure, that is, in the directions which cause positive bending moments in the bar (see Fig. 5-13).

It is assumed in our analysis that the effect of P on the bending stresses can be neglected. Then the normal stress at any point in the bar is independent of z and is given by the equation

$$\sigma = -\frac{P}{A} - \frac{M_1 y}{I_z} - \frac{M_2 x}{I_y} \tag{5-64}$$

in which x and y are centroidal principal axes of the cross section. The initial deflection of the bar due to the couples M_1 and M_2 will be considered as very small. In investigating the stability of this deflected form of equilibrium, we proceed as in the preceding articles and assume that additional deflections u and v of the shear-center axis and rotation ϕ with respect to that axis are produced. Thus the deflections u and v and the rotation ϕ result in a new shape of the axis of the bar, differing slightly from the initial curved shape produced by the couples M_1 and M_2. In writing equations of static equilibrium for this new form of the bar, we neglect the small initial deflections due to M_1 and M_2 and proceed as in the preceding cases in which the axis of the bar was initially straight. Thus the components of deflection of any longitudinal fiber of the bar, defined by coordinates x and y, are

$$u + (y_0 - y)\phi \qquad v - (x_0 - x)\phi$$

The intensities of the fictitious lateral loads and distributed torque obtained from the initial compressive forces in the fibers acting on their slightly rotated cross sections are obtained as in Art. 5.5 and are given by the equations

$$q_x = -\int_A (\sigma t \, ds) \frac{d^2}{dz^2} [u + (y_0 - y)\phi]$$

$$q_y = -\int_A (\sigma t \, ds) \frac{d^2}{dz^2} [v - (x_0 - x)\phi]$$

$$m_z = -\int_A (\sigma t \, ds)(y_0 - y) \frac{d^2}{dz^2} [u + (y_0 - y)\phi]$$

$$+ \int_A (\sigma t \, ds)(x_0 - x) \frac{d^2}{dz^2} [v - (x_0 - x)\phi]$$

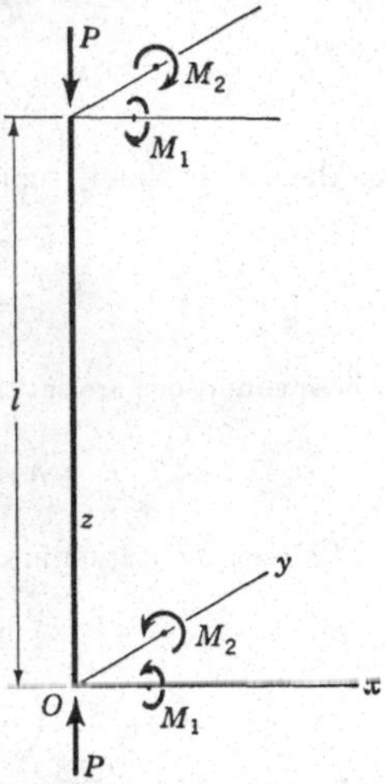

Substituting expression (5-64) for σ and integrating, we obtain

$$q_x = -P \frac{d^2 u}{dz^2} - (Py_0 - M_1) \frac{d^2 \phi}{dz^2}$$

$$q_y = -P \frac{d^2 v}{dz^2} + (Px_0 - M_2) \frac{d^2 \phi}{dz^2}$$

$$m_z = -(Py_0 - M_1) \frac{d^2 u}{dz^2} + (Px_0 - M_2) \frac{d^2 v}{dz^2} - \left(M_1\beta_1 + M_2\beta_2 + P \frac{I_0}{A} \right) \frac{d^2 \phi}{dz^2}$$

Fig. 5-20

where the following notation is introduced:

$$\beta_1 = \frac{1}{I_x} \left(\int_A y^3 \, dA + \int_A x^2 y \, dA \right) - 2y_0$$

$$\beta_2 = \frac{1}{I_y} \left(\int_A x^3 \, dA + \int_A xy^2 \, dA \right) - 2x_0 \tag{5-65}$$

The three equations for bending and torsion of the bar [Eqs. (5-40), (5-41), and (5-17)] then become

$$EI_y \frac{d^4 u}{dz^4} + P \frac{d^2 u}{dz^2} + (Py_0 - M_1) \frac{d^2 \phi}{dz^2} = 0 \tag{5-66}$$

$$EI_x \frac{d^4 v}{dz^4} + P \frac{d^2 v}{dz^2} - (Px_0 - M_2) \frac{d^2 \phi}{dz^2} = 0 \tag{5-67}$$

$$C_1 \frac{d^4 \phi}{dz^4} - \left(C - M_1\beta_1 - M_2\beta_2 - P \frac{I_0}{A} \right) \frac{d^2 \phi}{dz^2} + (Py_0 - M_1) \frac{d^2 u}{dz^2} - (Px_0 - M_2) \frac{d^2 v}{dz^2} = 0 \tag{5-68}$$

These are the three equations of equilibrium for the buckled form of the bar. From these equations the critical values of the external forces can be calculated for any given end conditions.

Eccentric Thrust. Let us begin by considering that the force P is applied eccentrically (Fig. 5-21). Denoting the coordinates of the point of application of P by e_x and e_y, we have

$$M_1 = Pe_y \qquad M_2 = Pe_x$$

and Eqs. (5-66) to (5-68) become

$$EI_y \frac{d^4u}{dz^4} + P \frac{d^2u}{dz^2} + P(y_O - e_y) \frac{d^2\phi}{dz^2} = 0 \tag{5-69}$$

$$EI_z \frac{d^4v}{dz^4} + P \frac{d^2v}{dz^2} - P(x_O - e_z) \frac{d^2\phi}{dz^2} = 0 \tag{5-70}$$

$$C_1 \frac{d^4\phi}{dz^4} - \left(C - Pe_y\beta_1 - Pe_z\beta_2 - P\frac{I_O}{A} \right) \frac{d^2\phi}{dz^2} + P(y_O - e_y) \frac{d^2u}{dz^2} - P(x_O - e_z) \frac{d^2v}{dz^2} = 0 \tag{5-71}$$

In the case of simply supported ends, the end conditions are

$$u = v = \phi = 0 \qquad \text{at } z = 0 \text{ and } z = l$$

$$\frac{d^2u}{dz^2} = \frac{d^2v}{dz^2} = \frac{d^2\phi}{dz^2} = 0 \qquad \text{at } z = 0 \text{ and } z = l$$

These conditions are satisfied by taking u, v, and ϕ in the form

$$u = A_1 \sin \frac{\pi z}{l} \qquad v = A_2 \sin \frac{\pi z}{l} \qquad \phi = A_3 \sin \frac{\pi z}{l}$$

and after substitution into Eqs. (5-69), (5-70) and (5-71), we obtain

$$\left(EI_y \frac{\pi^2}{l^2} - P \right) A_1 - P(y_O - e_y)A_3 = 0 \tag{5-72}$$

$$\left(EI_z \frac{\pi^2}{l^2} - P \right) A_2 + P(x_O - e_z)A_3 = 0 \tag{5-73}$$

$$-P(y_O - e_y)A_1 + P(x_O - e_z)A_2 + \left(C_1 \frac{\pi^2}{l^2} + C - Pe_y\beta_1 - Pe_z\beta_2 - P\frac{I_O}{A} \right) A_3 = 0 \tag{5-74}$$

Equating to zero the determinant of these equations gives a cubic equation for calculating P_{cr}. It is seen from these equations that, in general, buckling of the bar occurs

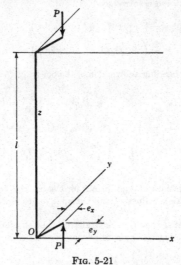

FIG. 5-21

by combined bending and torsion. In each particular case the coefficients in Eqs. (5-72) to (5-74) can be evaluated numerically and the cubic equation solved for the lowest value of the critical load.

The equations become very simple if the thrust P acts along the shear-center axis. We then have

$$e_x = x_O \qquad e_y = y_O$$

and Éqs. (5-69) to (5-71) become independent of one another. In this case lateral buckling in the two principal planes and torsional buckling may occur independently. The first two equations give the usual Euler formulas for critical buckling loads, and the third equation gives the critical load corresponding to purely torsional buckling of the column.

Another special case occurs when the bar has one plane of symmetry. Let us assume that the yz plane is the plane of symmetry and that the thrust P acts in that plane. Then $e_x = x_O = 0$ and Eqs. (5-72) to (5-74) become

$$\left(EI_y \frac{\pi^2}{l^2} - P\right) A_1 - P(y_O - e_y)A_3 = 0$$

$$\left(EI_x \frac{\pi^2}{l^2} - P\right) A_2 = 0 \qquad (5\text{-}75)$$

$$-P(y_O - e_y)A_1 + \left(C_1 \frac{\pi^2}{l^2} + C - Pe_y\beta_1 - P\frac{I_O}{A}\right) A_3 = 0$$

From the second of these equations we see that buckling in the plane of symmetry occurs independently and the corresponding critical load is the same as the Euler load. Lateral buckling in the xz plane and torsional buckling are coupled, and the corresponding critical loads are obtained by equating to zero the determinant of the first and third of Eqs. (5-75). This gives

$$\begin{vmatrix} EI_y \frac{\pi^2}{l^2} - P & -P(y_O - e_y) \\ -P(y_O - e_y) & C_1 \frac{\pi^2}{l^2} + C - Pe_y\beta_1 - P\frac{I_O}{A} \end{vmatrix} = 0$$

Using the notation

$$P_y = \frac{\pi^2 EI_y}{l^2} \qquad P_\phi = \frac{A}{I_O}\left(C + C_1 \frac{\pi^2}{l^2}\right)$$

and expanding the determinant, we obtain the following quadratic equation for calculating the critical loads:

$$(P_y - P)\left[\frac{I_O}{A}P_\phi - P\left(e_y\beta_1 + \frac{I_O}{A}\right)\right] - P^2(y_O - e_y)^2 = 0 \qquad (5\text{-}76)$$

It should be noted that the left-hand side of this equation is positive for very small values of P and negative when $P = P_y$. Hence there is a root of Eq. (5-76) smaller than P_y, i.e., smaller than the Euler load for buckling in the xz plane.

If the compressive force P is applied at the shear center, we have $e_y = y_O$ and Eq. (5-76) becomes

$$(P_y - P)\left[\frac{I_O}{A}P_\phi - P\left(e_y\beta_1 + \frac{I_O}{A}\right)\right] = 0$$

The two solutions of this equation are

$$P = P_y \qquad P = \frac{P_\phi}{1 + e_y\beta_1(A/I_O)}$$

The first solution corresponds to flexural buckling in the xz plane, and the second to purely torsional buckling.

Since the left-hand side of Eq. (5-76) is positive when P is small and gradually diminishes as P increases, we conclude that the smallest root of the equation will be increased by making the last term vanish, i.e., by taking $e_y = y_0$. Hence the critical load reaches its maximum value when the thrust is applied at the shear center.

If the load is applied at the centroid so that $e_y = 0$, Eq. (5-76) becomes

$$\frac{I_0}{A} (P - P_y)(P - P_\phi) - P^2 y_0^2 = 0$$

which is of the same form as Eq. (5-38) previously obtained for centrally applied thrust.

If the cross section of the bar has two axes of symmetry so that the shear center and the centroid coincide, we must substitute $y_0 = \beta_1 = 0$ into Eq. (5-76), yielding

$$(P - P_y)(P - P_\phi) - P^2 \frac{A e_y^2}{I_0} = 0$$

or
$$P^2 \left(1 - \frac{A e_y^2}{I_0}\right) - P(P_y + P_\phi) + P_y P_\phi = 0 \qquad (5\text{-}77)$$

The left-hand side of this equation is positive when $P = 0$ and negative when $P = P_y$ or $P = P_\phi$. Thus there is one critical load less than either P_y or P_ϕ. If $A e_y^2/I_0 < 1$, the left-hand side is positive for large values of P, which shows that the second critical load is greater than either P_y or P_ϕ. If the eccentricity e_y approaches zero, these two critical loads approach the values P_y and P_ϕ. If $A e_y^2/I_0 > 1$, the left-hand side is negative for large positive and negative values of P, indicating that the equation has one positive and one negative root. The negative root indicates that if the eccentricity is large, the bar may buckle under the action of eccentric tension. When $A e_y^2/I_0 = 1$, one root of Eq. (5-77) is

$$P = \frac{P_y P_\phi}{P_y + P_\phi}$$

and the other root becomes infinitely large.

The preceding discussion of eccentric thrust applied to bars with simply supported ends. If the ends are rigidly built in, the conditions are

$$u = v = \phi = 0 \qquad \text{at } z = 0 \text{ and } z = l$$
$$\frac{du}{dz} = \frac{dv}{dz} = \frac{d\phi}{dz} = 0 \qquad \text{at } z = 0 \text{ and } z = l$$

We can satisfy these end conditions by taking

$$u = A_1 \left(1 - \cos\frac{2\pi z}{l}\right) \qquad v = A_2 \left(1 - \cos\frac{2\pi z}{l}\right)$$
$$\phi = A_3 \left(1 - \cos\frac{2\pi z}{l}\right)$$

Substituting into Eqs. (5-69) to (5-71) and equating to zero the determinant of the resulting equations, we obtain an equation for calculating the critical loads. This equation will be similar to the equation derived for simply supported ends, the difference being only that $4\pi^2/l^2$ is substituted for π^2/l^2.

Pure Bending. If the axial force P becomes zero (Fig. 5-20), we have the case of pure bending of a bar by couples M_1 and M_2 at the ends. Substituting $P = 0$ into

Eqs. (5-66) to (5-68) gives the following three equations:

$$EI_y \frac{d^4u}{dz^4} - M_1 \frac{d^2\phi}{dz^2} = 0 \tag{5-78}$$

$$EI_x \frac{d^4v}{dz^4} + M_2 \frac{d^2\phi}{dz^2} = 0 \tag{5-79}$$

$$C_1 \frac{d^4\phi}{dz^4} - (C - M_1\beta_1 - M_2\beta_2) \frac{d^2\phi}{dz^2} - M_1 \frac{d^2u}{dz^2} + M_2 \frac{d^2v}{dz^2} = 0 \tag{5-80}$$

By taking the appropriate trigonometric expressions for u, v, and ϕ, we can derive readily the equation for calculating critical values of the moments M_1 and M_2.

Of particular interest is the case where the bar has flexural rigidity in one principal plane many times larger than in the other and is bent in the plane of greater rigidity

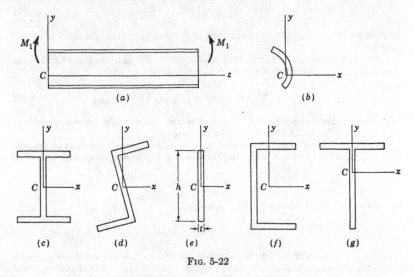

FIG. 5-22

Let us assume that the yz plane is the plane of greater rigidity and that the bar is bent in this plane by couples M_1 (Fig. 5-22a and b). The critical value of M_1 at which lateral buckling occurs is obtained from Eqs. (5-78) and (5-80). When $M_2 = 0$ is substituted these equations become

$$EI_y \frac{d^4u}{dz^4} - M_1 \frac{d^2\phi}{dz^2} = 0 \tag{5-81}$$

$$C_1 \frac{d^4\phi}{dz^4} - (C - M_1\beta_1) \frac{d^2\phi}{dz^2} - M_1 \frac{d^2u}{dz^2} = 0 \tag{5-82}$$

If the ends of the bar are simply supported, the expressions for u and ϕ can be taken again in the form

$$u = A_1 \sin \frac{\pi z}{l} \qquad \phi = A_2 \sin \frac{\pi z}{l}$$

By substituting these expressions into Eqs. (5-81) and (5-82) and equating to zero the determinant of the resulting equations, we obtain the following equation for calculating the critical values of the moment M_1:

$$\frac{\pi^2 E I_y}{l^2}\left(C + C_1\frac{\pi^2}{l^2} - M_1\beta\right) - M_1^2 = 0 \qquad (a)$$

Again using the notation

$$P_y = \frac{\pi^2 E I_y}{l^2} \qquad P_\phi = \frac{A}{I_0}\left(C + C_1\frac{\pi^2}{l^2}\right)$$

we can express Eq. (a) in the form

$$M_1^2 + P_y\beta_1 M_1 - \frac{I_0}{A}P_y P_\phi = 0 \qquad (5\text{-}83)$$

from which

$$(M_1)_{cr} = -\frac{P_y\beta_1}{2} \pm \sqrt{\left(\frac{P_y\beta_1}{2}\right)^2 + \frac{I_0}{A}P_y P_\phi} \qquad (5\text{-}84)$$

In each particular case, two critical values of the moment M_1 are obtained from this equation.

If the cross section of the bar has two axes of symmetry, as in the case of an I beam (Fig. 5-22c), β_1 vanishes and the critical moment is

$$(M_1)_{cr} = \pm\sqrt{\frac{I_0}{A}P_y P_\phi} = \pm\frac{\pi}{l}\sqrt{E I_y\left(C + C_1\frac{\pi^2}{l^2}\right)} \qquad (5\text{-}85)$$

This equation also holds for a bar with cross section having point symmetry, such as the z section in Fig. 5-22d, since $\beta_1 = 0$ for this case also.

For a bar of thin rectangular cross section (Fig. 5-22e), having thickness t and height h, the warping rigidity $C_1 = 0$ and the torsional rigidity is, approximately,

$$C = GJ = \frac{Ght^3}{3}$$

Hence the critical moment, from Eq. (5-85), is

$$(M_1)_{cr} = \pm\frac{\pi}{l}\frac{ht^3}{6}\sqrt{EG}$$

If the cross section of the bar has one axis of symmetry, and if the bending couple acts in the plane perpendicular to that axis (Fig. 5-22f), we again find that $\beta_1 = 0$ and Eq. (5-85) can be used. If the bending couple acts in the plane of symmetry, as in Fig. 5-22g, β_1 does not vanish and Eq. (5-84) must be used for calculating the critical value of M_1.

It was assumed in the above discussion of pure bending that $E I_y$ is small in comparison with $E I_x$. If C and C_1 are also small, buckling will occur at small values of M_1 and hence at small bending stresses in the bar. If $E I_z$ is of the same order of magnitude as $E I_y$, lateral buckling will occur at small stresses only if C and C_1 are very small. This condition can be fulfilled if the cross section is of cruciform shape (see Fig. 5-9), since C_1 vanishes in this case and C is very small if the flange thickness is small.

In this discussion we have considered the bending of a bar by couples applied at the ends. Only in this case are the normal stresses [see Eq. (5-64)] independent of z, so that we obtain a set of differential equations with constant coefficients [see Eqs. (5-66) to (5-68)]. If the bar is bent by lateral loads, the bending stresses vary with z and we obtain a set of equations with variable coefficients. The calculation of critical values of lateral loads on beams is discussed in the next chapter.

CHAPTER 6

LATERAL BUCKLING OF BEAMS

6.1. Differential Equations for Lateral Buckling. It was shown in the preceding article that a beam which is bent in the plane of greatest flexural rigidity may buckle laterally at a certain critical value of the load. This lateral buckling is of importance in the design of beams without lateral support, provided the flexural rigidity of the beam in the plane of bending is large in comparison with the lateral bending rigidity. As long as the load on such a beam is below the critical value, the beam will be stable. As the load is increased, however, a condition is reached at which a slightly deflected (and twisted) form of equilibrium becomes possible. The plane configuration of the beam is now unstable, and the lowest load at which this critical condition occurs represents the critical load for the beam.

Let us begin by considering the beam with two planes of symmetry shown in Fig. 6-1. This beam is assumed to be subjected to arbitrary loads acting in the yz plane, which is the plane of maximum rigidity. We assume that a small lateral deflection occurs under the action of these loads. Then from the differential equations of equilibrium for the deflected beam we can obtain the critical values of the loads. In deriving these equations, we shall use the fixed coordinate axes x, y, z, as shown in the figure. In addition, the coordinate axes ξ, η, ζ are taken at the centroid of the cross section at any section mn. The axes ξ and η are axes of symmetry and hence principal axes of the cross section, and ζ is in the direction of the tangent to the deflected axis of the beam after buckling. The deflection of the beam is defined by the components u and v of the displacement of the centroid of the cross section in the x and y directions, respectively, and by the angle of rotation ϕ of the cross section. The angle of rotation ϕ is taken positive about the z axis according to the right-hand rule of signs, and u and v are positive in the positive directions of the corresponding axes. Thus the displacements u and v of point C' in Fig. 6-1 are shown negative.

In later discussions, the expressions for the cosines of the angles between the coordinate axes x, y, z and ξ, η, ζ will be needed. When the quantities u, v, ϕ are considered as very small, the cosines of the angles between the positive directions of the axes have the values given in Table 6-1.

The curvatures of the deflected axis of the beam (Fig. 6-1) in the xz and yz planes can be taken as d^2u/dz^2 and d^2v/dz^2, respectively, for small

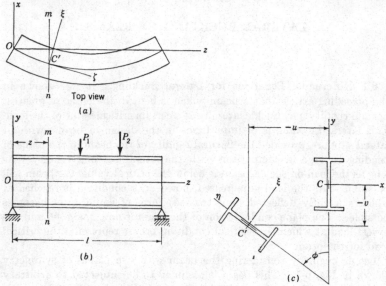

FIG. 6-1

deflections. For small angles of twist ϕ we can assume that the curvatures in the $\xi\zeta$ and $\eta\zeta$ planes have the same values. Thus the differential

TABLE 6-1. COSINES OF ANGLES BETWEEN AXES IN FIG. 6-1

	x	y	z
ξ	1	ϕ	$-\dfrac{du}{dz}$
η	$-\phi$	1	$-\dfrac{dv}{dz}$
ζ	$\dfrac{du}{dz}$	$\dfrac{dv}{dz}$	1

equations for bending of the beam become

$$EI_\xi \frac{d^2v}{dz^2} = M_\xi \qquad (6\text{-}1)$$

$$EI_\eta \frac{d^2u}{dz^2} = M_\eta \qquad (6\text{-}2)$$

In these equations I_ξ and I_η are the principal moments of inertia of the cross section about the ξ and η axes, respectively. The quantities M_ξ and M_η represent the bending moments about the same axes, with assumed positive directions as shown in Fig. 6-2.

The equation for twisting of the buckled bar [see Eq. (5-15)] is

$$C \frac{d\phi}{dz} - C_1 \frac{d^3\phi}{dz^3} = M_\zeta \qquad (6\text{-}3)$$

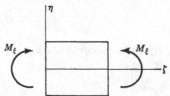

where $C = GJ$ is the torsional rigidity and $C_1 = EC_w$ is the warping rigidity.[1] The twisting moment M_ζ is taken positive in the directions shown in Fig. 6-3, which shows the twisting couples acting on an element of the beam.

Equation (6-3) is valid for a beam of thin-walled open cross section, such as the I beam in Fig. 6-1. The three differential equations (6-1), (6-2), and (6-3) represent the equations of equilibrium for the buckled beam and from them we can find the critical values of the load. In the following articles we shall consider various particular cases of loading.

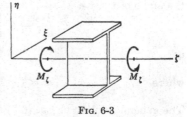

FIG. 6-2

6.2. Lateral Buckling of Beams in Pure Bending.[2] *I Beam.* If an I beam is subjected to couples M_0 at the ends (Fig. 6-4a and b), the bending and twist-

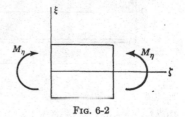

FIG. 6-3

ing moments at any cross section are found by taking the components of M_0 about the ξ, η, and ζ axes. Thus, using the values given in the first column of Table 6-1, and also considering the positive directions of the moments (Figs. 6-2 and 6-3), we obtain

$$M_\xi = M_0 \qquad M_\eta = \phi M_0 \qquad M_\zeta = -\frac{du}{dz} M_0$$

Substituting these expressions into Eqs. (6-1), (6-2), and (6-3) gives the following equations for u, v, and ϕ:

[1] Formulas for J and C_w are given in the Appendix.

[2] Various cases of lateral buckling of I beams were discussed by Timoshenko, *Bull. Polytech. Inst., St. Petersburg*, 1905. Further development is due to V. Z. Vlasov, "Thin-walled Elastic Bars," Moscow, 1940, and to J. N. Goodier, *Cornell Univ. Eng. Expt. Sta. Bull.* 27, December, 1941, and 28, January, 1942.

$$EI_\xi \frac{d^2v}{dz^2} - M_0 = 0 \tag{6-4}$$

$$EI_\eta \frac{d^2u}{dz^2} - \phi M_0 = 0 \tag{6-5}$$

$$C \frac{d\phi}{dz} - C_1 \frac{d^3\phi}{dz^3} + \frac{du}{dz} M_0 = 0 \tag{6-6}$$

By differentiating the last equation with respect to z, and eliminating

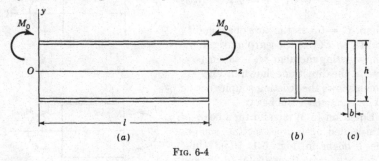

FIG. 6-4

d^2u/dz^2 by combining with Eq. (6-5), we obtain the following equation for the angle of twist ϕ:

$$C_1 \frac{d^4\phi}{dz^4} - C \frac{d^2\phi}{dz^2} - \frac{M_0^2}{EI_\eta} \phi = 0$$

or

$$\frac{d^4\phi}{dz^4} - 2\alpha \frac{d^2\phi}{dz^2} - \beta\phi = 0 \tag{6-7}$$

where

$$\alpha = \frac{C}{2C_1} \qquad \beta = \frac{M_0^2}{EI_\eta C_1} \tag{6-8}$$

The general solution of Eq. (6-7) is

$$\phi = A_1 \sin mz + A_2 \cos mz + A_3 e^{nz} + A_4 e^{-nz} \tag{a}$$

in which m and n are positive, real quantities defined by the relations

$$m = \sqrt{-\alpha + \sqrt{\alpha^2 + \beta}} \qquad n = \sqrt{\alpha + \sqrt{\alpha^2 + \beta}} \tag{b}$$

The constants of integration A_1, A_2, A_3, and A_4 must be determined from the conditions at the ends of the beam. Assuming that the ends of the beam cannot rotate about the z axis (Fig. 6-1) but are free to warp, we find that the conditions at the ends are (see p. 228)

$$\phi = \frac{d^2\phi}{dz^2} = 0 \qquad \text{at } z = 0 \text{ and } z = l \tag{c}$$

From the conditions at $z = 0$ we conclude that

$$A_2 = 0 \qquad A_3 = -A_4$$

and therefore the angle of twist ϕ can be represented in the form

$$\phi = A_1 \sin mz - 2A_4 \sinh nz$$

Now using the conditions at $z = l$ we obtain the equations

$$A_1 \sin ml - 2A_4 \sinh nl = 0$$
$$A_1 m^2 \sin ml + 2A_4 n^2 \sinh nl = 0 \qquad (d)$$

Setting the determinant of these equations equal to zero yields

$$(\sin ml)(n^2 \sinh nl + m^2 \sinh nl) = 0$$

Since m and n are positive, nonzero quantities, we conclude that

$$\sin ml = 0 \qquad (e)$$

and from Eqs. (d) we also obtain $A_4 = 0$. Therefore the form of buckling is given by the equation

$$\phi = A_1 \sin mz$$

and the beam buckles in the shape of a sine wave.

The smallest value of m satisfying Eq. (e) is

$$m = \frac{\pi}{l}$$

or, using expression (b),

$$-\alpha + \sqrt{\alpha^2 + \beta} = \frac{\pi^2}{l^2}$$

Substituting expressions (6-8) and solving for the critical value of the moment M_0 from the last equation, we find[1]

$$(M_0)_{cr} = \frac{\pi}{l} \sqrt{EI_\eta C \left(1 + \frac{C_1}{C} \frac{\pi^2}{l^2}\right)} \qquad (6\text{-}9)$$

This expression for the critical load can be represented in the form

$$(M_0)_{cr} = \gamma_1 \frac{\sqrt{EI_\eta C}}{l} \qquad (6\text{-}10)$$

where γ_1 is a dimensionless factor defined as

$$\gamma_1 = \pi \sqrt{1 + \frac{C_1}{C} \frac{\pi^2}{l^2}} \qquad (6\text{-}11)$$

Values of γ_1 are given in Table 6-2.

By taking higher roots of Eq. (e), we find larger values of the critical moment than given by Eq. (6-9). These critical moments correspond to

[1] This same result was obtained in Art. 5.7 [see Eq. (5-85)].

buckled shapes having one or more points of inflection, such as occur when the beam is supported laterally at intermediate points.

The magnitude of the critical moment given by Eq. (6-9) does not depend on the flexural rigidity EI_ξ of the beam in the vertical plane. This conclusion is obtained as a result of the assumption that the deflections in the vertical plane are small, which is justifiable when the flexural rigidity EI_ξ is very much greater than the rigidity EI_η. If the rigidities are of the same order of magnitude, the effect of bending in the vertical yz plane may be of importance and should be considered.[1] Some numerical results obtained from calculations in which this effect is considered are given in Art. 6.3 for a beam of narrow rectangular cross section.

TABLE 6-2. VALUES OF THE FACTOR γ_1 FOR I BEAMS IN PURE BENDING
[Eq. (6-11)]

$\dfrac{l^2 C}{C_1}$	0	0.1	1	2	4	6	8	10	12
γ_1	∞	31.4	10.36	7.66	5.85	5.11	4.70	4.43	4.24

$\dfrac{l^2 C}{C_1}$	16	20	24	28	32	36	40	100	∞
γ_1	4.00	3.83	3.73	3.66	3.59	3 55	3.51	3.29	π

After the critical value of the moment has been determined from Eq. (6-10), the critical stress is found from the flexure formula; that is,

$$\sigma_{cr} = \frac{(M_0)_{cr}}{Z_\xi} \tag{f}$$

where Z_ξ is the section modulus of the beam cross section taken with respect to the ξ axis. The stress calculated from Eq. (f) represents the true value of the critical stress only if it is below the proportional limit of the material.

Narrow Rectangular Beam. For a beam of narrow, rectangular section[2] (Fig. 6-4c) we can take the warping rigidity C_1 as zero (see p. 224) and instead of Eq. (6-7), we obtain the equation

$$\frac{d^2\phi}{dz^2} + \frac{M_0{}^2}{EI_\eta C}\phi = 0 \tag{6-12}$$

---

[1] This question was discussed by H. Reissner, *Sitzber. Berlin Math. Ges.*, 1904, p. 53. See also A. N. Dinnik, *Bull. Don. Polytech. Inst., Novotcherkassk*, vol. 2, 1913; and K. Federhofer, *Sitzber. Akad. Wiss. Wien.*, vol. 140, Abt. IIa, p. 237, 1931.

[2] This problem was discussed by L. Prandtl, "Kipperscheinungen," Dissertation, Munich, 1899, and A. G. M. Michell, *Phil. Mag.*, vol. 48, 1899.

This equation can be solved readily for ϕ, and since the angle ϕ must be zero at the ends of the beam, we obtain the following transcendental equation for determining the critical load:

$$\sin l \sqrt{\frac{M_0^2}{EI_\eta C}} = 0$$

The smallest root of this equation gives the lowest critical load:

$$(M_0)_{cr} = \frac{\pi}{l} \sqrt{EI_\eta C} \qquad (6\text{-}13)$$

In this equation, the torsional rigidity C for a thin rectangle usually can be taken with sufficient accuracy from the formula

$$C = GJ = \tfrac{1}{3}hb^3G \qquad (g)$$

6.3. Lateral Buckling of a Cantilever Beam. *I Beam.* Let us begin by considering a cantilever beam acted upon by a force P applied at the

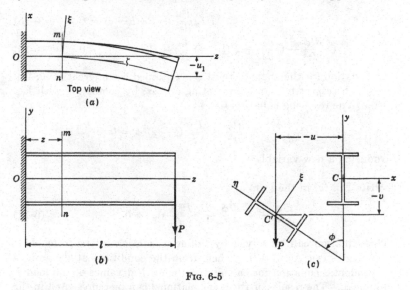

FIG. 6-5

centroid of the end cross section (Fig. 6-5). As the load P is increased gradually, we finally reach the critical condition where the deflected shape in the yz plane is unstable and lateral buckling occurs as shown in the figure. For determining the critical load, we again use the three equations of equilibrium, Eqs. (6-1) to (6-3). Considering the equilibrium of the part of the cantilever to the right of section mn (Fig. 6-5b), we find that the moments of the vertical load P with respect to axes through the centroid at section mn parallel to the x, y, and z axes are

$$M_x = -P(l - z) \qquad M_y = 0 \qquad M_z = P(-u_1 + u) \qquad (a)$$

The quantity u_1 represents the deflection of the free end of the beam, assumed positive when in the direction of the positive x axis. Taking components of the moments (a) about the ξ, η, and ζ axes by using Table 6-1 for the cosines of the angles between the axes and neglecting small quantities of higher order than the first, we obtain

$$M_\xi = -P(l - z) \qquad M_\eta = -P\phi(l - z)$$

$$M_\zeta = P(l - z) \frac{du}{dz} - P(u_1 - u)$$

Substituting these moments into Eqs. (6-1) to (6-3), we obtain the following three differential equations:

$$EI_\xi \frac{d^2v}{dz^2} + P(l - z) = 0 \qquad (6\text{-}14)$$

$$EI_\eta \frac{d^2u}{dz^2} + P\phi(l - z) = 0 \qquad (6\text{-}15)$$

$$C_1 \frac{d^3\phi}{dz^3} - C \frac{d\phi}{dz} + P(l - z) \frac{du}{dz} - P(u_1 - u) = 0 \qquad (6\text{-}16)$$

An equation for the angle of twist ϕ is obtained by differentiating Eq. (6-16) with respect to z and eliminating d^2u/dz^2 by combining with Eq. (6-15). The resulting equation for ϕ is

$$C_1 \frac{d^4\phi}{dz^4} - C \frac{d^2\phi}{dz^2} - \frac{P^2}{EI_\eta} (l - z)^2\phi = 0 \qquad (b)$$

Introducing a new variable

$$s = l - z$$

we write Eq. (b) in the form

$$\frac{d^4\phi}{ds^4} - \frac{C}{C_1} \frac{d^2\phi}{ds^2} - \frac{P^2}{EI_\eta C_1} s^2\phi = 0 \qquad (6\text{-}17)$$

This equation can be solved by taking a solution in the form of an infinite series (see Art. 2.12). Then, from the conditions at the ends, a transcendental equation for calculating the critical values of the load P is obtained. The results of these calculations[1] can be represented in the form

$$P_{cr} = \gamma_2 \frac{\sqrt{EI_\eta C}}{l^2} \qquad (6\text{-}18)$$

---

[1] The calculations were made by Timoshenko, Z. Math. u. Physik, vol. 58, pp. 337–385, 1910. This paper is reprinted in "The Collected Papers of Stephen P. Timoshenko," McGraw-Hill Book Company, Inc., New York, 1954.

in which γ_2 is a dimensionless factor depending on the ratio l^2C/C_1. Several values of γ_2 are given in Table 6-3. As the ratio l^2C/C_1 increases, the factor γ_2 approaches the limiting value 4.013, which corresponds to the critical load for a beam of narrow rectangular cross section ($C_1 = 0$), as discussed in the next section. For large values of l^2C/C_1, an approximate value of γ_2 is given by the equation

$$\gamma_2 = \frac{4.013}{(1 - \sqrt{C_1/l^2C})^2} \tag{6-19}$$

For example, if $l^2C/C_1 = 40$, the value of γ_2 from Eq. (6-19) is 5.66 as compared with the exact value of 5.64.

TABLE 6-3. VALUES OF THE FACTOR γ_2 FOR CANTILEVER BEAMS OF I SECTION
[Eq. (6-18)]

$\dfrac{l^2C}{C_1}$	0.1	1	2	3	4	6	8
γ_2	44.3	15.7	12.2	10.7	9.76	8.69	8.03
$\dfrac{l^2C}{C_1}$	10	12	14	16	24	32	40
γ_2	7.58	7.20	6.96	6.73	6.19	5.87	5.64

After the value of the critical load is determined from Eq. (6-18), the corresponding value of the critical stress is obtained from the equation

$$\sigma_{cr} = \frac{P_{cr}l}{Z_\xi} \tag{c}$$

This stress must be below the proportional limit of the material in order for Eq. (6-18) to be valid.

Narrow Rectangular Beam. If the cross section of the beam in Fig. 6-5 consists of a narrow rectangle with width b and height h, we obtain, instead of Eq. (b), the following equation for the angle of twist ϕ:

$$C\frac{d^2\phi}{dz^2} + \frac{P^2}{EI_\eta}(l - z)^2\phi = 0 \tag{d}$$

Again introducing the new variable $s = l - z$ and also using the notation

$$\beta_1 = \sqrt{\frac{P^2}{EI_\eta C}} \tag{6-20}$$

we find that Eq. (d) becomes

$$\frac{d^2\phi}{ds^2} + \beta_1^2 s^2\phi = 0 \tag{6-21}$$

The general solution of Eq. (6-21) is

$$\phi = \sqrt{s}\left[A_1 J_{\frac{1}{4}}\left(\frac{\beta_1}{2}s^2\right) + A_2 J_{-\frac{1}{4}}\left(\frac{\beta_1}{2}s^2\right)\right] \tag{6-22}$$

where $J_{\frac{1}{4}}$ and $J_{-\frac{1}{4}}$ represent Bessel functions of the first kind of order $\frac{1}{4}$ and $-\frac{1}{4}$, respectively.

The constants A_1 and A_2 in the general solution (6-22) are obtained from the end conditions. At the built-in end the angle of twist is zero, and hence the first condition is

$$\phi = 0 \qquad \text{at } s = l \tag{e}$$

At the free end the torque M_{ζ} is zero, and therefore the second condition [see Eq. (6-3)] is

$$\frac{d\phi}{ds} = 0 \qquad \text{at } s = 0 \tag{f}$$

Using condition. (f) with the general solution (6-22) gives $A_1 = 0$, and then, from condition (e), we find

$$J_{-\frac{1}{4}}\left(\frac{\beta_1}{2}l^2\right) = 0 \tag{g}$$

The lowest root of this equation[1] is

$$\frac{\beta_1}{2}l^2 = 2.0063$$

from which[2]
$$P_{cr} = \frac{4.013}{l^2}\sqrt{EI_\eta C} \tag{6-23}$$

The buckling shown in Fig. 6-5 corresponds to this value of the load. By taking larger roots of Eq. (g), we obtain deflection curves with one or more inflection points. These higher forms of unstable equilibrium are of no practical significance unless the beam is supported laterally at some intermediate points.

Formula (6-23) gives the correct value of the critical load only within the elastic region. Beyond the elastic limit buckling occurs at a load smaller than that given by the formula. To establish the proportions of cantilever beams for which formula (6-23) can be used, let us calculate the maximum bending stress in the cantilever. Observing that the maximum bending moment is $P_{cr}l$ and that the section modulus is $2I_\xi/h$, we obtain

$$\sigma_{cr} = \frac{P_{cr}lh}{2I_\xi} = 2.006\,\frac{h}{lI_\xi}\sqrt{EI_\eta C}$$

[1] See, for example, the tables of zeros of the Bessel functions, Jahnke and Emde, "Tables of Functions," 4th ed., p. 167, Dover Publications, New York, 1945.

[2] This result was obtained by Prandtl, loc. cit.

or, substituting $I_\eta = hb^3/12$, $I_\xi = bh^3/12$, and using expression (g), Art. 6.2, with Poisson's ratio taken as 0.3,

$$\sigma_{cr} = 2.487 \frac{b^2}{hl} E \tag{h}$$

It is seen from this result that, for a material such as steel, buckling in the elastic region may occur only if the quantity b^2/hl is very small.

Usually it is necessary to consider lateral buckling only in the case of a very narrow rectangular cross section where b/h is a small quantity. Theoretically [see Eq. (h)] buckling may occur also in the case where b/h is not very small but where the length l is very large. In this case a large deflection in the plane of the web will be produced before lateral buckling occurs, and this deflection should be considered in the derivation of the differential equations of equilibrium. More elaborate investigations[1] show that in this case we obtain in formula (6-23), instead of the constant numerical factor 4.013, a variable factor which depends on the ratio b/h. Taking b/h equal to $\frac{1}{10}$, $\frac{1}{5}$, and $\frac{1}{3}$, we obtain values of this factor equal to 4.085, 4.324, and 5.030, respectively.

It was assumed in the previous discussion that the load P was applied at the centroid of the cross section. The effect on the critical load of having the point of application of the load above or below the centroid of the end cross section has also been investigated.[2] If a denotes the distance of the point of application of the load vertically above the centroid (Fig. 6-5), the approximate formula for calculating the critical load can be put in the form

$$P_{cr} = \frac{4.013 \sqrt{EI_\eta C}}{l^2} \left(1 - \frac{a}{l} \sqrt{\frac{EI_\eta}{C}} \right) \tag{6-24}$$

It is seen that application of the load above the centroid of the cross section diminishes the critical value of the load. Formula (6-24) can be used also when the load is applied below the centroid. It is only necessary to change the sign of the displacement a.

If a distributed load acts on the cantilever beam, the same phenomenon of lateral buckling may occur when the load approaches a certain critical value. Assuming that a uniform load of intensity q is distributed along the center line of the cantilever, the critical value of this load, as obtained from the equations of equilibrium of the buckled cantilever,[3] is

$$(ql)_{cr} = \frac{12.85 \sqrt{EI_\eta C}}{l^2} \tag{6-25}$$

Comparing this result with formula (6-23), it can be concluded that the critical value of the total uniformly distributed load is approximately three times larger than the critical value of the concentrated load applied at the end.

If the intensity of the distributed load is given by the equation

$$q = q_0 \left(1 - \frac{z}{l} \right)^n \tag{i}$$

[1] See Dinnik and Federhofer, *loc. cit.*
[2] See Timoshenko, *Bull. Polytech. Inst.*, Kiev, 1910, and also A. Koroboff, *ibid.*, 1911.
[3] This result was obtained by Prandtl, *loc. cit.*

the critical value of the total load again can be represented by a formula analogous to formula (6-25).[1]   It is only necessary to replace the numerical factor 12.85 by another factor, the value of which depends on the exponent n in Eq. (i).   For $n = \frac{1}{4}, \frac{1}{2}, \frac{3}{4}$, and 1, these factors are 15.82, 19.08, 22.64, and 26.51, respectively.

The problem of buckling of a cantilever with a narrow rectangular cross section has been solved also for the case where the depth of the cross section varies according to the law[2]

$$h = h_0 \left(1 - \frac{z}{l} \right)^n \tag{j}$$

In all cases the critical value of the total load can be represented by the formula

$$Q_{cr} = \frac{\gamma_3 \sqrt{EI_\eta C}}{l^2} \tag{k}$$

in which the numerical value of the factor γ_3 depends on the type of loading and on the value of n in Eq. (j).   The quantities EI_η and C are the flexural and torsional rigidities for the fixed end of the cantilever.   Several values of γ_3 are given in Table 6-4.

TABLE 6-4. VALUES OF THE FACTOR γ_3 IN EQ. (k)

n	0	$\frac{1}{4}$	$\frac{1}{2}$	$\frac{3}{4}$	1
Uniform load............	12.85	12.05	11.24	10.43	9.62
Concentrated load at the free end	4.013	3.614	3.214	2.811	2.405

It is seen from this table that in the case of a cantilever with depth decreasing uniformly to zero at the free end [$n = 1$ in Eq. (j)], the critical value of the concentrated load at the end is approximately 60 per cent and the critical value of the uniformly distributed load is approximately 75 per cent of the load calculated for a cantilever of constant cross section [$n = 0$ in Eq. (j)].

6.4. Lateral Buckling of Simply Supported I Beams. *Concentrated Load at the Middle.*

Load at the Middle.   If a simply supported beam is bent in the yz plane by a load P applied at the centroid of the middle cross section (Fig. 6-6), lateral buckling may occur when the load reaches a certain critical value. It is assumed that during deformation the ends of the beam can rotate freely with respect to the principal axes of inertia parallel to the x and y axes, while rotation with respect to the z axis is prevented by some constraint (Fig. 6-6).   Thus the lateral buckling is accompanied by some twisting of the beam.   In calculating the critical value of the load, we assume that a small lateral buckling has occurred, and then we determine from the differential equations of equilibrium and the conditions at the ends the magnitude of the smallest load required to keep the beam in this slightly buckled form.

---

[1] Several cases of this kind were investigated by Dinnik and Federhofer, *loc. cit.*

[2] See K. Federhofer, *Repts. Intern. Congr. Appl. Mech.*, Stockholm, 1930.

Considering a portion of the beam to the right of any cross section mn, it is seen that the external forces acting on this portion reduce to a single vertical force $P/2$ acting at point B_1. The moments of this force with respect to axes through the centroid of cross section mn, parallel to the x, y, and z axes, are

$$M_x = \frac{P}{2}\left(\frac{l}{2} - z\right) \qquad M_y = 0 \qquad M_z = -\frac{P}{2}(-u_1 + u) \qquad (a)$$

In the expression for M_z, u_1 represents the lateral deflection of the centroid of the middle cross section and u the deflection at any cross section mn. Both of these quantities are taken positive in the positive direction of the

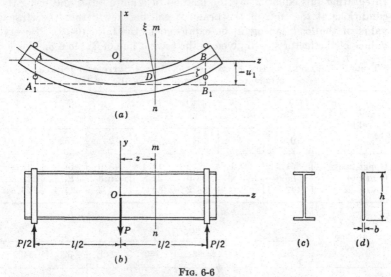

(a)

(b)

(c)   (d)

FIG. 6-6

x axis. Using the system of coordinate axes ξ, η, and ζ as in the preceding articles and projecting the moments (a) onto those axes by using the table of cosines (see p. 252), we obtain

$$M_\xi = \frac{P}{2}\left(\frac{l}{2} - z\right) \qquad M_\eta = \frac{P}{2}\left(\frac{l}{2} - z\right)\phi$$

$$M_\zeta = -\frac{P}{2}\left(\frac{l}{2} - z\right)\frac{du}{dz} + \frac{P}{2}(u_1 - u) \qquad (b)$$

Substituting expressions (b) into Eqs. (6-1) to (6-3), the following differential equations of equilibrium for the buckled beam (Fig. 6-6) are obtained:

$$EI_\xi \frac{d^2v}{dz^2} - \frac{P}{2}\left(\frac{l}{2} - z\right) = 0 \qquad (6\text{-}26)$$

$$EI_\eta \frac{d^2u}{dz^2} - \frac{P}{2}\left(\frac{l}{2} - z\right)\phi = 0 \qquad (6\text{-}27)$$

$$C \frac{d\phi}{dz} - C_1 \frac{d^3\phi}{dz^3} + \frac{P}{2}\left(\frac{l}{2} - z\right)\frac{du}{dz} - \frac{P}{2}(u_1 - u) = 0 \qquad (6\text{-}28)$$

Eliminating u from the second and third of these equations, we obtain

$$C_1 \frac{d^4\phi}{dz^4} - C \frac{d^2\phi}{dz^2} - \frac{P^2}{4EI_\eta}\left(\frac{l}{2} - z\right)^2 \phi = 0 \qquad (6\text{-}29)$$

Integrating this equation by the method of infinite series and using the conditions at the ends of the beam, it can be shown[1] that the critical value of the load can again be expressed in the form (6-18). Several values of the factor γ_2 are given in the second line of Table 6-5.

TABLE 6-5. VALUES OF THE FACTOR γ_2 FOR SIMPLY SUPPORTED I BEAMS WITH CONCENTRATED LOAD AT THE MIDDLE [EQ. (6-18)]

Load applied at	$\dfrac{l^2C}{C_1}$						
	0.4	4	8	16	24	32	48
Upper flange	51.5	20.1	16.9	15.4	15.0	14.9	14.8
Centroid	86.4	31.9	25.6	21.8	20.3	19.6	18.8
Lower flange	147	50.0	38.2	30.3	27.2	25.4	23.5

Load applied at	$\dfrac{l^2C}{C_1}$						
	64	80	96	160	240	320	400
Upper flange	15.0	15.0	15.1	15.3	15.5	15.6	15.8
Centroid	18.3	18.1	17.9	17.5	17.4	17.2	17.2
Lower flange	22.4	21.7	21.1	20.0	19.3	19.0	18.7

Instead of the critical load for lateral buckling of a beam being determined by integration of the differential equations, the strain-energy method can be used to advantage in many cases.[2] Let us apply this method to the beam of Fig. 6-6. When the beam buckles laterally the strain energy increases since bending in the lateral direction and twisting

---

[1] See Timoshenko, *Z. Math. u. Physik*, vol. 58, pp. 337–385, 1910.

[2] Several examples of the use of this method are given in Timoshenko, *Bull. Polytech. Inst., Kiev*, 1910 (in Russian), and later, Sur la stabilité des systèmes élastiques, *Ann. Ponts et chaussées*, Paris, 1913.

about the longitudinal axis are added to the bending in the plane of the web. At the same time the point of application of the load P is lowered and the load produces a certain amount of work. The critical value of the load is determined from the condition that this work is equal to the strain energy of lateral bending and twisting [see Eq. (2-31)]. The small change in the strain energy of bending of the beam in the plane of the web, which occurs during buckling, can be neglected in applying the energy method. This is equivalent to the previous assumption that the curvature in the plane of the web is infinitely small and can be neglected in deriving the differential equations of equilibrium. The results obtained on the basis of this assumption would be exact if the rigidity of the beam in the plane of the web were infinitely large. For beams in which the rigidity in the plane of the web is very large in comparison with the rigidity in the lateral direction, the assumption gives sufficiently accurate results for practical purposes.

In calculating the strain energy of bending and torsion, we use the general expression for strain energy

$$U = \frac{EI_\eta}{2} \int_0^l \left(\frac{d^2u}{dz^2}\right)^2 dz + \frac{C}{2} \int_0^l \left(\frac{d\phi}{dz}\right)^2 dz + \frac{C_1}{2} \int_0^l \left(\frac{d^2\phi}{dz^2}\right)^2 dz \quad (6\text{-}30)$$

in which the three terms represent, respectively, the strain energy due to lateral bending, twisting, and warping of the beam.[1] Taking into consideration the symmetry of the buckled form of the beam (Fig. 6-6), we find that the increase in strain energy due to lateral buckling is

$$\Delta U = EI_\eta \int_0^{l/2} \left(\frac{d^2u}{dz^2}\right)^2 dz + C \int_0^{l/2} \left(\frac{d\phi}{dz}\right)^2 dz + C_1 \int_0^{l/2} \left(\frac{d^2\phi}{dz^2}\right)^2 dz$$

$$(6\text{-}31)$$

For determining the lowering of the load P during lateral buckling, let us consider an element dz of the longitudinal axis of the beam at the point D (Fig. 6-6a). Owing to bending of this element in the $\xi\zeta$ plane and with the cross section mn considered as fixed, the end B of the beam describes an infinitely small arc

$$\frac{d^2u}{dz^2}\left(\frac{l}{2} - z\right) dz$$

in the $\xi\zeta$ plane, the vertical component of which is

$$\phi \frac{d^2u}{dz^2}\left(\frac{l}{2} - z\right) dz \qquad (c)$$

[1] The last term in Eq. (6-30) can be obtained by substituting Eq. (5-10) for σ_z into the expression for strain energy, $U = \int \frac{\sigma_z^2}{2E} dV$, and integrating over the volume V of the beam.

The lowering of the point of application of the load P due to lateral buckling of the beam is obtained by summation of the vertical components (c) for all elements of the beam between $z = 0$ and $z = l/2$. Thus we obtain the following expression for the work done by the load P during lateral buckling:

$$\Delta T = P \int_0^{l/2} \phi \frac{d^2u}{dz^2} \left(\frac{l}{2} - z \right) dz \tag{6-32}$$

The equation for determining the critical value of the load [see Eq. (2-31)] now becomes

$$P \int_0^{l/2} \phi \frac{d^2u}{dz^2} \left(\frac{l}{2} - z \right) dz = EI_\eta \int_0^{l/2} \left(\frac{d^2u}{dz^2} \right)^2 dz + C \int_0^{l/2} \left(\frac{d\phi}{dz} \right)^2 dz + C_1 \int_0^{l/2} \left(\frac{d^2\phi}{dz^2} \right)^2 dz$$

or, substituting for d^2u/dz^2 its expression from Eq. (6-27),

$$\frac{P^2}{4EI_\eta} \int_0^{l/2} \phi^2 \left(\frac{l}{2} - z \right)^2 dz = C \int_0^{l/2} \left(\frac{d\phi}{dz} \right)^2 dz + C_1 \int_0^{l/2} \left(\frac{d^2\phi}{dz^2} \right)^2 dz \tag{6-33}$$

To determine the critical value of the load, it is necessary to assume for ϕ a suitable expression, satisfying the conditions at the ends of the beam, and substitute it in Eq. (6-33). Taking this expression with one or more parameters and adjusting these parameters in such a way as to make the expression for P, as obtained from Eq. (6-33), a minimum, we can calculate the value of P_{cr} with great accuracy.

If the conditions of constraint are as shown in Fig. (6-6), we can take the angle of twist ϕ in the form of a trigonometric series

$$\phi = a_1 \cos \frac{\pi z}{l} + a_2 \cos \frac{3\pi z}{l} + \cdots \tag{d}$$

in which each term together with its second derivative vanishes at the ends of the beam as required by the conditions of constraint. By taking one, two, or more terms of the series (d), calculating the corresponding value of P_{cr} from Eq. (6-33), and comparing with the results obtained by the integration of Eqs. (6-26) to (6-28), we can investigate the accuracy of the energy method. In this manner it can be shown that, when only the first term of the series (d) is taken, the critical load obtained from Eq. (6-33) is only one-half of 1 per cent in error. When two terms of the series (d) are taken, the critical load is obtained with an error of less than one-tenth of 1 per cent. Thus the energy method, which simplifies considerably the calculation of the critical load, gives results which are accurate enough for practical applications. As explained in Art. 2.9, approximate values of the critical load found by the energy method are always larger than the exact value.[1]

It is assumed in the previous derivations that the load P is applied at the centroid of the middle cross section of the beam. It is apparent that the critical value of the load is decreased when the point of application is raised and increased when it is lowered. The extent of this effect on the critical load can be obtained by the energy method; it is only necessary to take into consideration the additional lowering or raising of the load P during lateral buckling due to rotation of the middle cross section. If ϕ_0 is this angle of rotation and a is the vertical distance of the point of application of the load from the centroid of the cross section, positive when above the centroid, the additional lowering of the load is

$$a(1 - \cos \phi_0) \approx \frac{a\phi_0{}^2}{2}$$

Then, instead of Eq. (6-33), we obtain

$$\frac{Pa\phi_0{}^2}{2} + \frac{P^2}{4EI_\eta} \int_0^{l/2} \phi^2 \left(\frac{l}{2} - z\right)^2 dz = C \int_0^{l/2} \left(\frac{d\phi}{dz}\right)^2 dz$$
$$+ C_1 \int_0^{l/2} \left(\frac{d^2\phi}{dz^2}\right)^2 dz \quad (6\text{-}34)$$

This equation can be solved for the critical value of P by taking ϕ in the form of the series (d). The results of such calculations are given in Table 6-5 for two cases: (1) The load is applied at the upper flange of the beam, and (2) the load is applied at the lower flange. It is seen that raising or lowering the point of application of the load has the greatest effect on the critical load when the quantity l^2C/C_1 is small.

Uniform Load. The method described above for the case of a concentrated load at the middle can be used also when the beam (Fig. 6-6) carries a uniformly distributed load. The critical value of this load can be expressed in the form

$$(ql)_{cr} = \gamma_4 \frac{\sqrt{EI_\eta C}}{l^2} \quad (6\text{-}35)$$

in which the numerical value of γ_4 depends on the ratio l^2C/C_1 and on the position of the load. Values of the factor γ_4 are tabulated in Table 6-6 for cases in which the load is applied along the upper flange, the centroidal axis, and the lower flange, respectively.

---

[1] A method of successive approximations, analogous to that discussed in the chapter on columns (see Art. 2.15), can also be used successfully in investigating lateral buckling of beams. See F. Stüssi, *Schweiz. Bauztg.*, vol. 105, p. 123, 1935, and *Publ. Intern. Assoc. Bridge Structural Eng.*, vol. 3, p. 401, 1935.

TABLE 6-6. VALUES OF THE FACTOR γ_4 FOR SIMPLY SUPPORTED I BEAMS
WITH UNIFORM LOAD [EQ. (6-35)]

Load applied at	$\dfrac{l^2 C}{C_1}$						
	0.4	4	8	16	24	32	48
Upper flange	92.9	36.3	30.4	27.5	26.6	26.1	25.9
Centroid	143	53.0	42.6	36 3	33.8	32.6	31.5
Lower flange	223	77.4	59.6	48.0	43.6	40.5	37.8

Load applied at	$\dfrac{l^2 C}{C_1}$						
	64	80	128	200	280	360	400
Upper flange	25.9	25.8	26.0	26.4	26.5	26.6	26.7
Centroid	30.5	30.1	29.4	29.0	28.8	28.6	28.6
Lower flange	36.4	35.1	33.3	32.1	31.3	31.0	30.7

6.5. Lateral Buckling of Simply Supported Beam of Narrow Rectangular Cross Section. If a beam of narrow rectangular cross section is bent by a load P applied at the centroid of the middle cross section (see Fig. 6-6d), we can use Eqs. (6-26) to (6-29) of the preceding article in investigating lateral buckling. It is only necessary to omit the terms in those equations containing the warping rigidity C_1. Thus, the equation for the angle of twist ϕ becomes

$$C \frac{d^2\phi}{dz^2} + \frac{P^2}{4EI_\eta} \left(\frac{l}{2} - z \right)^2 \phi = 0 \qquad (a)$$

Introducing the new variable $t = l/2 - z$ and using the notation

$$\beta_2 = \sqrt{\frac{P^2}{4EI_\eta C}} \qquad (b)$$

we find that Eq. (a) becomes

$$\frac{d^2\phi}{dt^2} + \beta_2{}^2 t^2 \phi = 0 \qquad (c)$$

for which the general solution is

$$\phi = \sqrt{t} \left[A_1 J_{\frac{1}{4}} \left(\frac{\beta_2}{2} t^2 \right) + A_2 J_{-\frac{1}{4}} \left(\frac{\beta_2}{2} t^2 \right) \right] \qquad (d)$$

where $J_{\frac{1}{4}}$ and $J_{-\frac{1}{4}}$ represent Bessel functions of the first kind of order $\frac{1}{4}$ and $-\frac{1}{4}$, respectively. For a beam with simply supported ends the conditions are

$$\phi = 0 \quad \text{at } t = 0 \qquad \frac{d\phi}{dt} = 0 \quad \text{at } t = \frac{l}{2}$$

From the first condition we find that $A_2 = 0$ and then, noting that

$$\frac{d\phi}{dt} = A_1 \beta_2 t^{\frac{3}{2}} J_{-\frac{3}{4}} \left(\frac{\beta_2}{2} t^2 \right)$$

we obtain from the second condition

$$J_{-\frac{1}{4}}\left(\frac{\beta_2 l^2}{8}\right) = 0$$

From a table[1] of zeros of the Bessel function of order $-\frac{3}{4}$, we find that

$$\frac{\beta_2 l^2}{8} = 1.0585$$

or [see notation (b)] $\qquad P_{cr} = \dfrac{16.94\ \sqrt{EI_\eta C}}{l^2}$ \hfill (6-36)

The critical value of the load can be found also by using the energy method, as described in the preceding article. It has been shown[2] that if one term of the series (d), Art. 6.4, is taken as a first approximation, the error is about 1.5 per cent. If two terms of the series are taken, the error is less than 0.1 per cent.

If the load P is applied at distance a above the centroid of the middle cross section of the beam, we can use Eq. (6-34) with C_1 set equal to zero. If the distance a is small, the first term on the left-hand side of this equation will be small in comparison with the other terms, and it is sufficiently accurate to substitute in that term the value (6-36) for P. Then, using one term of the series (d), Art. 6.4, we obtain the following approximate formula:

$$P_{cr} = \frac{16.94\ \sqrt{EI_\eta C}}{l^2}\left(1 - \frac{1.74a}{l}\sqrt{\frac{EI_\eta}{C}}\right) \tag{6-37}$$

When the load P is applied at distance c from one support, instead of at the middle of the span, we can represent the critical load by the formula

$$P_{cr} = \gamma_5 \frac{\sqrt{EI_\eta C}}{l^2} \tag{6-38}$$

in which the numerical factor γ_5 depends on the ratio c/l. Values[3] of γ_5 are given in Table 6-7. It is seen that the value of the critical load increases when its point of application is to one side of the middle of the span. However, this effect is not large as long as the load remains within the middle third of the span.

TABLE 6-7. VALUES OF THE FACTOR γ_5 IN EQ. (6-38)

$\dfrac{c}{l}$	0.05	0.10	0.15	0.20	0.25	0.30	0.35	0.40	0.45	0.50
γ_5	112	56.0	37.9	29.1	24.1	21.0	19.0	17.8	17.2	16.94

If a simply supported beam carries a load uniformly distributed along the centroidal axis, the critical value of the total load is given by the formula[4]

$$(ql)_{cr} = \frac{28.3\ \sqrt{EI_\eta C}}{l^2} \tag{6-39}$$

---

[1] Such a table is given on p. 384 of "Tables of Bessel Functions of Fractional Order," vol. 1, Columbia University Press, New York, 1948.

[2] See Timoshenko, *loc. cit.*

[3] See Koroboff, *loc. cit.*, and also A. N. Dinnik, *Bull. Don. Polytech. Inst., Novotcherkassk*, vol. 2, 1913.

[4] This solution is due to Prandtl, *loc. cit.*

6.6. Other Cases of Lateral Buckling. *Beams with Intermediate Lateral Support.* To increase the stability of a beam against buckling in the lateral direction, various kinds of lateral constraints are used. The effect of these constraints on the magnitude of the critical stress can be investigated by using the same methods discussed in the preceding articles. Let us consider the case where, owing to an additional constraint, the middle cross section of a beam is prevented from rotating with respect to the center line of the beam. Such a condition is obtained when two parallel beams are braced at the middle as shown in Fig. 6-7a. Owing to this bracing, the deflection curve of the beam buckled in the lateral direction must have an inflection point, as shown in the top view of the beam, Fig. 6-7b. Assuming that the beam undergoes pure bending in the *yz* plane and has a narrow rectangular cross section, the critical value of the bending moment can be obtained from Eq. (6-12). To take account of the lateral constraint, it will be necessary, in discussing the resulting transcendental equation, to take the root equal to 2π instead of the smallest root π, which was

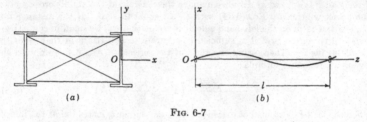

(a)   (b)

Fig. 6-7

taken before for the beam without lateral constraint at the middle. In this manner we obtain

$$(M_0)_{cr} = \frac{2\pi \sqrt{EI_\eta C}}{l} \qquad (a)$$

Hence, owing to lateral constraint, the critical value of the bending moment is doubled in this case. In an analogous way the problem of lateral buckling of a beam with a constraint at the middle can be treated for the case of I sections and for other kinds of loading.

If a load P is applied at the centroid of the middle cross section of the I beam shown in Fig. 6-7, Eq. (6-18) must be used for calculating the critical value of the load. The values of the numerical factor γ_2 in this equation are larger than in the case of a beam without lateral constraint and are given in Table 6-8.

TABLE 6-8. VALUES OF THE FACTOR γ_2 FOR SIMPLY SUPPORTED I BEAMS
WITH LATERAL SUPPORT AT THE MIDDLE AND WITH CONCENTRATED
LOAD AT THE MIDDLE [Eq. (6-18)]

$\dfrac{l^2 C}{C_1}$	0.4	4	8	16	32	96	128	200	400
γ_2	466	154	114	86.4	69.2	54.5	52.4	49.8	47.4

If the beam is uniformly loaded, the critical value of the total load is given by Eq. (6-35). Values of the factor γ_4 are given in Table 6-9 for cases in which the load is distributed along the upper flange, the centroidal axis, and the lower flange, respectively.

TABLE 6-9. VALUES OF THE FACTOR γ_4 FOR SIMPLY SUPPORTED I BEAMS WITH LATERAL SUPPORT AT THE MIDDLE AND WITH UNIFORM LOAD [EQ. (6-35)]

Load applied at	$\dfrac{l^2 C}{C_1}$							
	0.4	4	8	16	64	96	128	200
Upper flange	587	194	145	112	91.5	73.9	71.6	69.0
Centroid	673	221	164	126	101	79.5	76.4	72.8
Lower flange	774	251	185	142	112	85.7	81.7	76.9

Beams with Lateral Constraint at the Ends. Let us now consider the case in which the ends of the beam are prevented from rotating with respect to the vertical y axis during buckling, so that the deflection curve in the lateral direction has two inflection points (see Fig. 6-8). It is assumed that there are no constraints to prevent the ends from rotating about the horizontal x axis; hence the beam is simply supported when bending in the yz plane is considered.

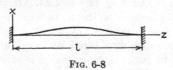

FIG. 6-8

In the case of an I beam with load P applied at the centroid of the middle cross section, the critical value of the load is given by Eq. (6-18). Values of the factor γ_2 are given in Table 6-10. A comparison of the figures in this table with those in Table 6-5 shows that the effect of lateral constraint at the ends is larger in the case of short beams than in the case of long beams.

TABLE 6-10. VALUES OF THE FACTOR γ_2 FOR I BEAMS WITH LATERAL CONSTRAINT AT THE ENDS (FIG. 6-8) AND WITH CONCENTRATED LOAD AT THE MIDDLE [EQ. (6-18)]

$\dfrac{l^2 C}{C_1}$	0.4	4	8	16	24	32	64	128	200	320
γ_2	268	88.8	65.5	50.2	43.6	40.2	34.1	30.7	29.4	28.4

In Table 6-11 are given data for an I beam uniformly loaded along the centroidal axis and having lateral constraint as in Fig. 6-8.

In the case of a beam of narrow rectangular cross section with load P applied at the centroid of the middle cross section, the critical value of the load is[1]

$$P_{cr} = \frac{26.6 \sqrt{EI_\eta C}}{l^2} \tag{6-40}$$

[1] See *ibid.*

$\dfrac{l^2 C}{C_1}$	0.4	4	8	16	32	96	128	200	400
γ_4	488	161	119	91.3	73.0	58.0	55.8	53.5	51.2

If the beam is loaded by couples M_0 applied at the ends in the yz plane, there will be inflection points at distance $l/4$ from the ends. Thus the middle portion of the beam, of length $l/2$, is in the same condition as the beam shown in Fig. 6-4. In this case the critical value of the couples M_0 can be found from Eq. (6-13) by substituting $l/2$ for l, which gives

$$(M_0)_{cr} = \frac{2\pi}{l} \sqrt{EI_\eta C} \qquad (6-41)$$

In each of the preceding discussions it is assumed that the maximum stress in the beam is below the proportional limit of the material. Otherwise it is necessary to consider inelastic lateral buckling, as discussed in the next article.

6.7. Inelastic Lateral Buckling of I Beams. If an I beam is stressed beyond the proportional limit of the material, the critical load can be calculated by using the tangent modulus E_t, varying with the stress, instead of the constant modulus of elasticity E. The method is analogous to that used before in investigating buckling of columns beyond the elastic limit (see Chap. 3). It was shown in the preceding articles that the critical load for lateral buckling within the elastic region depends on the magnitude of the lateral flexural rigidity EI_η, which is proportional to the modulus E in tension, and also on the magnitude of the torsional rigidity C, which is proportional to the shear modulus G. Beyond the proportional limit the lateral flexural rigidity diminishes in the ratio E_t/E. We assume in the following discussion that the torsional rigidity diminishes also in the same proportion,[1] and therefore the ratio $l^2 C/C_1$ remains unchanged.

Let us begin with pure bending. Since in this case the stress in the flanges is constant along the span, the tangent modulus will be the same for all cross sections of the beam bent beyond the proportional limit and the same differential equations of equilibrium as in the elastic region can be used. It is necessary only to replace the flexural and torsional rigidities of the beam by their corresponding values obtained by using the

[1] This assumption can be considered as being on the safe side. The lateral flexural rigidity is due primarily to the rigidity of the flanges; hence it diminishes beyond the proportional limit in the ratio E_t/E. The torsional rigidity depends on the rigidity of the web as well as the flanges, and since a portion of the web remains always elastic and retains its initial rigidity, we can expect that the torsional rigidity diminishes in a proportion smaller than E_t/E.

tangent modulus. The critical value of the bending moment is then given by the equation

$$(M_0)_{cr} = \frac{\pi}{l} \sqrt{E_t I_\eta C_t \left(1 + \frac{C_1}{C} \frac{\pi^2}{l^2}\right)}$$ (6-42)

which has the same form as Eq. (6-9) except that E is replaced by E_t and the notation

$$C_t = C \frac{E_t}{E}$$ (a)

is introduced.

Since the ratio C_1/Cl^2 remains unchanged beyond the proportional limit, it can be concluded from (6-42) that the critical value of the moment is less than the value of the same moment calculated on the assumption of perfect elasticity, in the ratio E_t/E. If the magnitude of this ratio for each value of the stress is known, the critical bending moment for each value of the stress can be calculated easily by the trial-and-error method. We assume a certain value for $(M_0)_{cr}$, calculate the value of the maximum bending stress, and take the corresponding value of the tangent modulus E_t. With this modulus the critical value of the bending moment is obtained from Eq. (6-42). If the value calculated in this way coincides with the assumed value, it represents the true value of $(M_0)_{cr}$. Otherwise a new assumption regarding $(M_0)_{cr}$ should be made and the calculations repeated. Such calculations should be repeated as many times as is necessary to obtain a satisfactory agreement between the assumed value of $(M_0)_{cr}$ and that calculated from Eq. (6-42).

In the case of bending of beams by concentrated or distributed loads, the bending moment and the stress in the flanges vary along the span of the beam. Hence, beyond the proportional limit, the tangent modulus E_t will vary also along the span and the differential equations of equilibrium for lateral buckling will be of the same kind as for beams of variable cross section. To simplify this problem and obtain an approximate value for the critical stress, we take a constant value for E_t, namely, the value corresponding to the maximum bending moment, and substitute it in the differential equations of equilibrium of the buckled beam. In this way the critical load will be obtained in the same form as before [see Eqs. (6-18) and (6-35)], it being only necessary to use the quantity $E_t I_\eta$ for the flexural rigidity and C_t for the torsional rigidity. It is evident that in calculating critical stresses in this way we shall always be on the safe side, since we reduce the lateral and torsional rigidities of the beam along the span in the constant proportion E_t/E, while this amount of reduction actually takes place only in the cross section with the maximum bending moment. In other cross sections which are stressed beyond the proportional limit, the reduction will be smaller, and there will be no reduction at all in portions of the beam where the stresses are below the proportional limit.   With the above assumption the critical value of the load can be obtained by the trial-and-error method in the same manner as was explained for the case of pure bending.

In practical applications it is usually convenient to work with critical stresses instead of critical loads. Let us assume that beyond the elastic limit the stresses in the flanges of an I beam can be obtained with satisfactory accuracy by dividing the maximum bending moment by the section modulus of the cross section. This assumption is on the safe side, since beyond the elastic limit the actual stress in the flanges and the actual reduction of the bending and torsional rigidities will be smaller than assumed.

The calculation of the critical stress σ_{cr} beyond the proportional limit will be illustrated now by an example. Let us take an I beam with a cross section assumed to be in the form of three narrow rectangles. The assumed proportions of the cross section are

$$\frac{t_f}{t_w} = 2 \qquad \frac{b}{t_f} = 10 \qquad \frac{h}{b} = 3 \qquad\qquad (b)$$

where t_f = flange thickness
  t_w = web thickness
  b = flange width
  h = depth of beam

The torsional and warping rigidities are, approximately (see Table A-3, Appendix),

$$C = \frac{G}{3}(2bt_f{}^3 + ht_w{}^3) \qquad C_1 = \frac{Et_f h^2 b^3}{24}$$

Substituting dimensions (b) in these formulas and also assuming that $G/E = 0.4$, we obtain

$$\frac{l^2 C}{C_1} = 0.076 \left(\frac{l}{h}\right)^2$$

Numerical values of this quantity are given in Table 6-12 for several values of the ratio l/h. If we assume that the beam carries a load uniformly distributed along the

TABLE 6-12. DATA FOR UNIFORMLY LOADED I BEAM WITH $t_f/t_w = 2$,
$b/t_f = 10$, $h/b = 3$, $E = 30 \times 10^6$ psi, $G = 0.4E$

$\dfrac{l}{h}$	4	6	8	10	12	14	16	18	20
$\dfrac{l^2 C}{C_1}$	1.22	2.74	4.86	7.60	10.9	14.9	19.5	24.6	30.4
σ_{cr} (psi)	181,000	85,600	52,200	36,400	27,700	22,100	18,400	15,900	13,900

centroidal axis of the beam, we can obtain the critical value of the load, provided buckling occurs within the elastic limit, from Eq. (6-35). Using this equation, we can write the expression for the critical stress in the form

$$\sigma_{cr} = (ql)_{cr}\frac{l}{8}\frac{h}{2I_\xi} = \gamma_4 \frac{h}{16l}\frac{\sqrt{EI_\eta C}}{I_\xi} \qquad\qquad (c)$$

in which values of γ_4 can be found from Table 6-6. Observing that the principal moments of inertia of the cross section are

$$I_\eta = \frac{1}{6} b^3 t_f \qquad I_\xi = \frac{1}{12} h^3 t_w + \frac{t_f}{2} (b - t_w)(h - t_f)^2$$

and also using dimensions (b), $E = 30 \times 10^6$ psi, and $G = 0.4E$, we note that expression (c) becomes

$$\sigma_{cr} = 8{,}416\gamma_4 \frac{h}{l} \qquad\qquad (d)$$

Values of σ_{cr} found from Eq. (d) are given in Table 6-12 and also are plotted as curve I in Fig. 6-9.

Curves II and III in Fig. 6-9 were calculated for $h/b = 4$ and 5, respectively, in a manner similar to that explained above for curve I. These curves show that the critical stress for an I beam decreases with an increase of the ratio l/h, provided h/b is held constant, and also decreases as h/b increases if l/h is constant. The points with

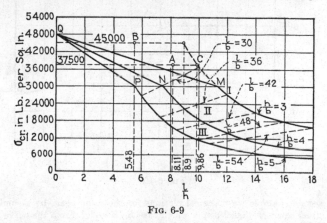

FIG. 6-9

constant values of the ratio l/b are connected in the figure by dotted lines and show that, for constant l/b, the critical stress decreases as l/h decreases.

If we use Table 6-12 in the case of structural steel having a proportional limit of 30,000 psi, we see that for $l/h \leq 10$ the calculated value of the critical stress is beyond the proportional limit. To take into account the plastic deformation of the material, information regarding the tangent modulus E_t must be given. Let us assume for this numerical example that the ratio E_t/E has the values $\frac{36}{48}$ and $\frac{4}{8}$ when the compressive stresses are 37,500 and 45,000 psi, respectively.[1] To determine what ratio l/h the beam should have if the critical stress is 37,500 psi, we note that on curve I (Fig. 6-9) the point C corresponds to this stress, and from the figure we obtain $l/h = 9.86$. It is known, however, that curve I gives exaggerated values of σ_{cr} beyond the elastic limit. The true value of σ_{cr}, for $l/h = 9.86$, will be less than 37,500 psi and the ratio l/h corresponding to $\sigma_{cr} = 37,500$ psi will be less than 9.86. Its value will be found now by trial and error. Assume, for instance, that this true value is $l/h = 8$.

[1] The true values of the ratio E_t/E can be obtained only by using the stress-strain diagram for the material, as explained in Art. 3.3.

Then, taking the critical stress for elastic buckling from Table 6-12 and multiplying by the ratio E_t/E, we obtain for the critical stress

$$\sigma_{cr} = 52,200(\tfrac{3\,6}{4\,9}) = 38,400 \text{ psi}$$

This stress is somewhat larger than the assumed stress of 37,500 psi.   Hence the true value of the ratio l/h is larger than the assumed value 8.   As a second trial, we assume that $l/h = 8.2$, and interpolating from Table 6-12, we find that

$$\sigma_{cr} = 50,000(\tfrac{3\,6}{4\,9}) = 36,800 \text{ psi}$$

This stress is now somewhat smaller than the assumed stress 37,500 psi.   Hence the true value of the ratio l/h is smaller than 8.2.   From these two trials we conclude that the true value of the ratio l/h lies between 8 and 8.2, and by linear interpolation we obtain $l/h = 8.11$.   This result is represented in Fig. 6-9 by the point A, which is much below the curve I, based on the assumption of perfect elasticity.

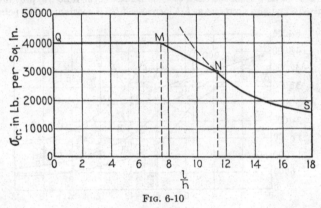

FIG. 6-10

As a second example we make calculations for the same beam by assuming the critical stress equal to 45,000 psi, in which case $E_t/E = \tfrac{4}{9}$.   From the curve I (Fig. 6-9) we have for this case $l/h = 8.9$.   The true value of this ratio, owing to plastic deformation, is much smaller.   Assuming $l/h = 5.6$ and proceeding as before, we obtain $\sigma_{cr} = 43,100$ psi, which is smaller than the assumed value; hence the true value of the ratio l/h is smaller than 5.6.   Taking, as the second trial, $l/h = 5.4$, we find for the critical stress the value 46,100 psi.   From these two trials we find, by interpolation, $l/h = 5.48$.   This result is shown in Fig. 6-9 by the point B.

In the same manner the value of l/h can be found for any other assumed value of the critical stress beyond the elastic limit.   Thus the curve representing the relation between σ_{cr} and the ratio l/h can be constructed provided the ratio E_t/E is known for any value of the stress σ_{cr}.

To simplify the calculations of critical stresses beyond the proportional limit in practical application, the curve representing the relation between σ_{cr} and the ratio l/h can be replaced by a straight line.   Thus a straight-line formula analogous to that applied for columns (see Art. 4.3) can be used also in the case of lateral buckling of beams.   Assuming that the elastic limit of structural steel is 30,000 psi, we find that the curves I, II, and III of Fig. 6-9 can be used only below the points M, N, and P.   For higher stresses the straight lines MQ, NQ, and PQ can be used.   These lines are obtained by taking the highest stress (for $l/h = 0$) equal to 48,000 psi, which is the value used in the straight-line formula for columns [see Eq. (d), Art. 4.3].

Curves analogous to those shown in Fig. 6-9 can be readily constructed by using Tables 6-5 and 6-6 for an I beam of any cross section; thus the problem of determining critical stresses can be solved for any value of the ratio l/h.

Instead of straight lines MQ, NQ, and PQ, as in Fig. 6-9, we can construct a diagram of the type shown in Fig. 6-10. To obtain such a diagram, we construct first the curve NS of critical stresses, assuming perfect elasticity. This curve can be used up to the point N, corresponding to the proportional limit of the material. Then we draw a horizontal line QM corresponding to the yield point of the material, which gives critical stresses for short beams. For beams of intermediate length we use, for calculating critical stresses, the inclined line MN, which joins the point N, mentioned before, with a point M arbitrarily chosen on the horizontal yield-point line. In Fig. 6-10 this point is taken so that the corresponding span of the beam is equal to 0.6 of the span at which the critical stress is equal to the proportional limit of the material. Having such a diagram and using a variable factor of safety, as in the case of columns, the safe stresses for any span of the beam of a given cross section can be readily obtained.[1]

[1] A calculation of the allowable load on the basis of an assumed initial lateral curvature of the I beam has been made by Stüssi, *loc. cit.*

CHAPTER 7

BUCKLING OF RINGS, CURVED BARS, AND ARCHES

7.1. Bending of a Thin Curved Bar with a Circular Axis. Let us consider a curved bar AB (Fig. 7-1) slightly bent in the plane of its initial curvature, and let us assume that this plane is the plane of symmetry of the bar. Denoting by R the initial radius of curvature of the center line of the bar and by ρ the radius of curvature after deformation at any point of the center line, defined by the angle θ, we can express the relation

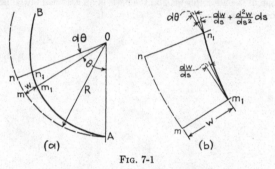

Fig. 7-1

between the change in curvature and the magnitude of the bending moment M in the case of a thin bar by the equation

$$EI\left(\frac{1}{\rho} - \frac{1}{R}\right) = -M \tag{a}$$

in which EI is the flexural rigidity of the bar in the plane of its initial curvature. The minus sign on the right-hand side of the equation follows from the sign of the bending moment, which is taken to be positive when it produces a decrease in the initial curvature of the bar.

The change in the curvature of the bar during bending will be found from a consideration of the deformation of a small element mn of the ring included between two radii with the angle $d\theta$ between them. The initial length of the element and its initial curvature are

$$ds = R\,d\theta \qquad \frac{d\theta}{ds} = \frac{1}{R} \tag{b}$$

278

The radial displacement w of a point m during bending, assumed to be a small quantity, is taken positive when it is directed toward the center. There will be also some displacement of the point m in a tangential direction, but this will be disregarded, and we shall assume that the curvature of the element mn after deformation is the same as the curvature of the element m_1n_1 included between the same radii mO and nO. This latter curvature is given by the equation

$$\frac{1}{\rho} = \frac{d\theta + \Delta \, d\theta}{ds + \Delta \, ds} \tag{c}$$

in which $d\theta + \Delta \, d\theta$ denotes the angle between the normal cross sections m_1 and n_1 of the deformed bar and $ds + \Delta \, ds$ denotes the length of the element m_1n_1. In calculating the small angle $\Delta \, d\theta$, we note that the angle between the tangent to the center line at m_1 and the perpendicular to the radius mO is dw/ds (Fig. 7-1b). The corresponding angle at the cross section n_1 is

$$\frac{dw}{ds} + \frac{d^2w}{ds^2} \, ds$$

and hence
$$\Delta \, d\theta = \frac{d^2w}{ds^2} \, ds \tag{d}$$

When the length of the element m_1n_1 is compared with that of the element mn, the small angle dw/ds is neglected and the length m_1n_1 is taken equal to $(R - w) \, d\theta$. Then

$$\Delta \, ds = -w \, d\theta = -\frac{w \, ds}{R} \tag{e}$$

Substitution of expressions (d) and (e) in Eq. (c) gives

$$\frac{1}{\rho} = \frac{d\theta + (d^2w/ds^2) \, ds}{ds \, (1 - w/R)}$$

or, with small quantities of higher order neglected,

$$\frac{1}{\rho} = \frac{1}{R}\left(1 + \frac{w}{R}\right) + \frac{d^2w}{ds^2}$$

Substituting this into Eq. (a), we obtain

$$\frac{d^2w}{ds^2} + \frac{w}{R^2} = -\frac{M}{EI} \tag{f}$$

or
$$\frac{d^2w}{d\theta^2} + w = -\frac{MR^2}{EI} \tag{7-1}$$

This is the differential equation for the deflection curve of a thin bar with a circular center line.[1] For an infinitely large radius R this equation coincides with that for a straight bar.

As an example of the application of Eq. (7-1), let us consider a ring of radius R compressed by two forces P acting along a diameter (Fig. 7-2a). Denoting the bending moment at A and B by M_0, we find that the

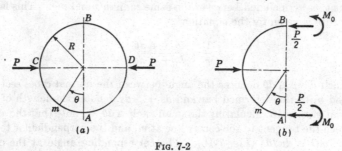

Fig. 7-2

moment at any cross section m is (see Fig. 7-2b)

$$M = M_0 + \frac{PR}{2}(1 - \cos\theta) \qquad (g)$$

and Eq. (7-1) becomes

$$\frac{d^2w}{d\theta^2} + w = -\frac{M_0 R^2}{EI} - \frac{PR^3}{2EI}(1 - \cos\theta) \qquad (h)$$

The general solution of Eq. (h) is

$$w = A_1 \sin\theta + A_2 \cos\theta - \frac{M_0 R^2}{EI} - \frac{PR^3}{2EI} + \frac{PR^3}{4EI}\theta\sin\theta \qquad (i)$$

The constants of integration are determined from the conditions of symmetry:

$$\frac{dw}{d\theta} = 0 \qquad \text{at } \theta = 0 \text{ and } \theta = \frac{\pi}{2}$$

From these conditions we find that

$$A_1 = 0 \qquad A_2 = \frac{PR^3}{4EI} \qquad (j)$$

The bending moment M_0 can be found from Castigliano's theorem.[2] The strain energy U for a thin ring is obtained by using the same formula

[1] This equation was established by J. Boussinesq, *Compt. rend.*, vol. 97, p. 843, 1883. See also H. Lamb, *Proc. London Math. Soc.*, vol. 19, p. 365, 1888, and R. Mayer, *Z. Math. u. Physik*, vol. 61, p. 246, 1913.

[2] See Timoshenko, "Strength of Materials," 3d ed., part I, pp. 328–330, D. Van Nostrand Company, Inc., Princeton, N.J., 1955.

as for a straight bar; hence, for the ring in Fig. 7-2, we have

$$U = \int_0^{2\pi} \frac{M^2 R \, d\theta}{2EI} = \frac{2R}{EI} \int_0^{\pi/2} M^2 \, d\theta$$

By Castigliano's theorem the partial derivative of the strain energy with respect to the bending moment M_0 is equal to the angle of rotation of the cross section A of the ring (Fig. 7-2b), which is zero in this case. Thus the equation becomes

$$\frac{\partial U}{\partial M_0} = \frac{2R}{EI} \int_0^{\pi/2} 2M \frac{\partial M}{\partial M_0} \, d\theta = 0$$

or

$$\int_0^{\pi/2} M \frac{\partial M}{\partial M_0} \, d\theta = 0$$

Substituting expression (g) for M and integrating, we obtain

$$M_0 = \frac{PR}{2} \left(\frac{2}{\pi} - 1 \right) \tag{k}$$

The final expression for the radial deflection is found by substituting relations (j) and (k) into (i), yielding

$$w = \frac{PR^3}{4EI} \left(\cos\theta + \theta \sin\theta - \frac{4}{\pi} \right)$$

From this expression the radial deflections at any point can be found. For example, at $\theta = 0$ and $\theta = \pi/2$ we obtain

$$(w)_{\theta=0} = -\frac{PR^3}{4EI} \left(\frac{4}{\pi} - 1 \right) \qquad (w)_{\theta=\pi/2} = \frac{PR^3}{4EI} \left(\frac{\pi}{2} - \frac{4}{\pi} \right)$$

and the lengthening of diameter AB is

$$\frac{PR^3}{2EI} \left(\frac{4}{\pi} - 1 \right) \tag{7-2}$$

while the shortening of diameter CD is

$$\frac{PR^3}{4EI} \left(\pi - \frac{8}{\pi} \right) \tag{7-3}$$

An equation analogous to Eq. (7-1) can be obtained also for the bending of a long circular tube if the load does not change along the axis of the tube. In such a case we consider an elemental ring cut out of the tube by two cross sections perpendicular to the axis of the tube and unit distance apart. Since the ring is a portion of the long tube, the rectangular cross sections of the ring do not distort during bending, as in the case of an isolated ring, and thus the quantity $E/(1 - \nu^2)$, where ν is Poisson's ratio, should be used in place of E in Eq. (7-1). If the thickness of the tube is denoted by h, the moment of inertia of the cross-sectional area of the elemental ring is $h^3/12$ and the differential equation for deflection of such a ring becomes

$$\frac{d^2w}{d\theta^2} + w = -\frac{12(1 - \nu^2)MR^2}{Eh^3} \qquad (7\text{-}4)$$

This equation, instead of Eq. (7-1), should be used in investigating bending of long circular tubes. In this case M denotes the bending moment per unit length of tube.

7.2. Application of Trigonometric Series in the Analysis of a Thin Circular Ring. In discussing small deflections of a ring in its plane, we resolve the displacement of any point m of the center line into two components, a radial displacement w, taken positive toward the center, and a tangential displacement v, taken positive in the direction of increasing θ (Fig. 7-3). In the most general case the radial displacement w can be represented in the form of the trigonometric series

$$w = a_1 \cos \theta + a_2 \cos 2\theta + \cdots + b_1 \sin \theta + b_2 \sin 2\theta + \cdots \qquad (a)$$

The tangential displacement will be taken in such a form as to make the extension of the center line of the ring zero. In this manner the actual ring is replaced by a certain hypothetical ideal ring with an inextensible center line. The discussion of *inextensional deformations* of rings and shells was originated by Lord Rayleigh.[1] He observed that, in the case of bending of thin rings, the displacements due to extension of the center line of a ring are very small in comparison with the displacements due to bending and usually can be neglected.

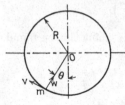

FIG. 7-3

The unit elongation of the center line of a ring during bending consists, in general, of two parts: (1) the part due to tangential displacement v, equal to dv/ds, and (2) the part due to radial displacements, equal to $-w/R$. Hence we have

$$\epsilon = \frac{1}{R}\left(\frac{dv}{d\theta} - w\right)$$

The condition of inextensional deformation of the ring is

$$\frac{dv}{d\theta} - w = 0 \qquad (b)$$

from which, by using Eq. (a), we obtain

$$v = a_1 \sin \theta + \tfrac{1}{2}a_2 \sin 2\theta + \cdots - b_1 \cos \theta - \tfrac{1}{2}b_2 \cos 2\theta - \cdots \qquad (c)$$

[1] Lord Rayleigh, "Theory of Sound," 2d ed., chap. X$_A$, 1894. See also Timoshenko, *Bull. Polytech. Inst., Kiev*, 1910 (in Russian).

Thus the problem of bending of a ring reduces to that of calculating the coefficients a_1, a_2, . . . and b_1, b_2, . . . in expressions (a) and (c). For this calculation the same method as in the case of straight beams (see Art. 1.11) will be used. We derive the expression for strain energy of the ring in bending and determine the coefficients of the above series by using the principle of virtual displacements.

The strain energy of bending of a thin ring is obtained by applying the same formula as in the case of a straight bar; then, by using Eq. (7-1),

$$U = \int_0^{2\pi} \frac{M^2 R\, d\theta}{2EI} = \frac{EI}{2R^3} \int_0^{2\pi} \left(\frac{d^2 w}{d\theta^2} + w \right)^2 d\theta$$

Substituting series (a) for w and integrating, we obtain

$$U = \frac{\pi EI}{2R^3} \sum_{n=2}^{n=\infty} (n^2 - 1)^2 (a_n^2 + b_n^2) \tag{7-5}$$

This series does not contain terms with coefficients a_1 and b_1, since the corresponding displacements

$$\begin{aligned} w &= a_1 \cos \theta + b_1 \sin \theta \\ v &= a_1 \sin \theta - b_1 \cos \theta \end{aligned} \tag{d}$$

represent the displacement of the ring as a rigid body in its plane. It can be seen readily that if a_1 and b_1 are vertical and horizontal components of the latter displacement, expressions (d) represent the radial and tangential displacements for any point m (Fig. 7-3).

Having expression (7-5) for the strain energy, we can, in each particular case, easily calculate the coefficients a_2, a_3, . . . ; b_2, b_3, ... Take, as an example, the case represented in Fig. 7-2. From the condition of symmetry it can be concluded that the coefficients b_1, b_2, . . . in expression (a) vanish. In calculating any coefficient a_n we assume that this coefficient is increased by δa_n. The corresponding deflection of the ring, from Eq. (a), is $\delta a_n \cos n\theta$, and during this displacement the forces P produce the work

$$P\, \delta a_n \left(\cos \frac{n\pi}{2} + \cos \frac{3n\pi}{2} \right)$$

For n equal to an odd number, this work is equal to zero; for n equal to an even number, we have

$$\cos \frac{n\pi}{2} = \cos \frac{3n\pi}{2} = (-1)^{n/2}$$

and the above expression for the work is equal to

$$(-1)^{n/2} 2P\, \delta a_n \tag{e}$$

When the principle of virtual work is applied, the equation for determining the coefficient a_n, for n equal to an even number, becomes

$$\frac{\partial U}{\partial a_n} \delta a_n = (-1)^{n/2} 2P \, \delta a_n$$

Substituting for U its expression (7-5), we obtain

$$a_n = \frac{(-1)^{n/2} 2PR^3}{\pi(n^2 - 1)^2 EI} \tag{f}$$

In the same way it can be shown that all coefficients with n equal to an odd number vanish.[1] Thus the series (a) and (c), representing deflections of the center line of the ring, contain only terms with coefficients a_n where $n = 2, 4, 6, \ldots$ For radial displacements we obtain, from (a),

$$w = \frac{2PR^3}{\pi EI} \sum_{n=2,4,6,\ldots}^{n=\infty} \frac{(-1)^{n/2} \cos n\theta}{(n^2 - 1)^2} \tag{7-6}$$

The points C and D where the forces P are applied will approach each other by the amount

$$\delta = (w)_{\theta=\pi/2} + (w)_{\theta=3\pi/2} = \frac{4PR^3}{\pi EI} \sum_{n=2,4,6,\ldots}^{n=\infty} \frac{1}{(n^2 - 1)^2} \tag{7-7}$$

It can be shown that this series gives the same result as formula (7-3) obtained before.

As a second example, let us consider the bending of a circular ring by hydrostatic pressure (Fig. 7-4). The center of the ring is held at a constant depth d by the force P. Then, assuming that the width of the ring in the direction perpendicular to the plane of the figure is unity and denoting by γ the weight of the liquid per unit volume, we find that the intensity of hydrostatic pressure at any point m is

$$\gamma(d + R \cos \theta)$$

From an equation of static equilibrium we find that the force P is equal to $\pi R^2 \gamma$. Using again the principle of virtual displacements and noting that the hydrostatic pressure produces work only on the radial displacements w, we find that the equation for calculating the coefficients a_n in the series (a) and (c) becomes

$$\frac{\partial U}{\partial a_n} \delta a_n = \delta a_n \left[P \cos n\pi + R\gamma \int_0^{2\pi} (d + R \cos \theta) \cos n\theta \, d\theta \right]$$
$$= \delta a_n P(-1)^n \quad \text{for } n > 1$$

---

[1] The same conclusion follows from the condition of symmetry with respect to the horizontal diameter of the ring.

For $n = 1$, which corresponds to the displacement of the ring as a rigid body, the right-hand side of the equation becomes equal to zero. Substituting for U its expression (7-5), we obtain

$$a_n = \frac{(-1)^n P R^3}{\pi(n^2 - 1)^2 EI} \quad \text{for } n > 1; a_1 = 0$$

All the coefficients b_n in series (a) and (c) vanish, as can be seen from the conditions of symmetry. Hence we obtain

$$w = \frac{PR^3}{\pi EI} \sum_{n=2,3,4,\ldots}^{n=\infty} \frac{(-1)^n \cos n\theta}{(n^2 - 1)^2}$$

$$v = \frac{PR^3}{\pi EI} \sum_{n=2,3,4,\ldots}^{n=\infty} \frac{(-1)^n \sin n\theta}{n(n^2 - 1)^2}$$

The displacements for any value of θ can be calculated from these equations. In this analysis the effect of the compressive force in the ring on

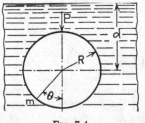

FIG. 7-4

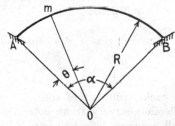

FIG. 7-5

bending is entirely neglected, which is legitimate only if this compressive force is small in comparison with the critical value of the same force at which the ring buckles. This question is discussed further in the following article.

If, instead of a complete circular ring, we have a curved bar with a circular center line and hinged ends (Fig. 7-5), the same method as in the previous discussion can be applied. When the central angle of the arch is designated by α, the radial displacement w can be taken in the form of a series

$$w = a_1 \sin \frac{\pi\theta}{\alpha} + a_2 \sin \frac{2\pi\theta}{\alpha} + a_3 \sin \frac{3\pi\theta}{\alpha} + \cdots \tag{g}$$

It is seen that this series and its second derivative become equal to zero at the ends, satisfying the conditions at the hinges with respect to radial displacements and bending moments. The tangential displacement v, from Eq. (b), is

$$v = -\frac{\alpha}{\pi}\left(a_1 \cos\frac{\pi\theta}{\alpha} + \frac{1}{2}a_2\cos\frac{2\pi\theta}{\alpha} + \frac{1}{3}a_3\cos\frac{3\pi\theta}{\alpha} + \cdots\right) \qquad (h)$$

This displacement, in general, does not vanish at the ends.

By taking only those terms in series (g) and (h) which have the even order coefficients, we find that

$$(v)_{\theta=0} = (v)_{\theta=\alpha} = -\frac{\alpha}{\pi}\left(\frac{1}{2}a_2 + \frac{1}{4}a_4 + \cdots\right)$$

We have in this case equal tangential displacements at both ends of the bar, and we can make these displacements vanish by superposing on the deformation represented by the series a rotation of the bar as a rigid body about the center O by an angle equal to

$$\frac{\alpha}{\pi R}\left(\frac{1}{2}a_2 + \frac{1}{4}a_4 + \cdots\right)$$

Thus, we obtain an inextensional deformation of a circular arch satisfying all end conditions:

$$w = a_2\sin\frac{2\pi\theta}{\alpha} + a_4\sin\frac{4\pi\theta}{\alpha} + \cdots$$
$$v = \frac{\alpha}{\pi}\left[\frac{1}{2}a_2\left(1 - \cos\frac{2\pi\theta}{\alpha}\right) + \frac{1}{4}a_4\left(1 - \cos\frac{4\pi\theta}{\alpha}\right) + \cdots\right] \qquad (i)$$

To each coefficient of these series there corresponds an inextensional deflection curve with an inflection point at the center of the bar. We shall use such deflection curves later in discussing the stability of compressed curved bars (see Art. 7:6).

If we take in series (g) only terms with coefficients of odd order a_1, a_3, . . . , we obtain deflection curves which are symmetrical with respect to the center of the bar. It is seen from series (h) that displacements v corresponding to each of the above terms do not vanish at the ends, and we can satisfy the conditions at the hinged ends only by taking a series of such terms with a definite relation between the coefficients a_1, a_3, a_5, ...

From the general requirement for inextensional deformation [Eq. (b)], it follows that the tangential displacements v vanish at the ends only if

$$\int_0^\alpha w\,d\theta = 0 \qquad (j)$$

from which it follows that, in the case of deflection curves symmetrical with respect to the center of the buckled bar, the half-waves cannot be equal. The simplest curve of this kind will be obtained by taking three

half-waves.   Then, from (j), the area of the middle half-wave must be
larger than that of either of the remaining two; if we represent w as a
function of θ, the curve shown in Fig. 7-6 will be obtained.   The area of
the middle wave in this curve is equal to the sum of the areas of the two
other waves.

7.3. Effect of Uniform Pressure on Bending of a Circular Ring.   In
the preceding articles we have neglected the effect of compression or
tension on bending of the ring.   For example, in the discussion of the
problem shown in Fig. 7-4, the
compression of the ring due to
hydrostatic pressure was entirely
disregarded, although it may become of great importance in the
case of thin rings, in which buck-

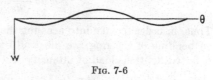

Fig. 7-6

ling may occur owing to the action of uniform pressure acting alone.   To
take into account the effect of the uniform pressure on the deformation of
the ring, we again consider only inextensional deformations and assume
that the internal forces over any cross section of the ring reduce to a constant axial force S and a bending moment.   Considering, for instance,
the ring of unit width under hydrostatic pressure (Fig. 7-4), we assume
first a uniform pressure of intensity γd, which produces in the ring a

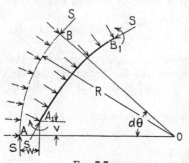

Fig. 7-7

uniform compressive force $S = \gamma Rd$.
On this uniform compression we superpose bending produced by a nonuniform pressure of a magnitude γR
$\cos \theta$, and we assume that this bending is inextensional, so that the axial
compressive force S remains unchanged and only bending moment
is produced.

The effect of the force S on the
bending of the ring can be replaced
by the action of an equivalent distributed load, acting radially, the inten-
sity of which can be found by considering an element AB (Fig. 7-7) of a
uniformly compressed ring.   The axial compressive forces S maintain in
equilibrium the uniformly distributed pressure that is acting on the element.   Assume now an inextensional deformation of the ring such that
the element AB takes the position A_1B_1.   There will be some bending
moments produced in the cross sections A_1 and B_1, the magnitude of
which can be found from Eq. (7-1) provided the displacements w are
known.   Owing to the change in curvature, represented by the left-hand
side of Eq. (f), Art. 7.1, cross section B of the ring rotates with respect to

cross section A by the angle

$$R \, d\theta \left(\frac{d^2 w}{ds^2} + \frac{w}{R^2} \right)$$

Because of this rotation the axial forces S are no longer in equilibrium with the uniform pressure and they give an additional force acting in a radial direction away from the center O, the magnitude of which is

$$SR \, d\theta \left(\frac{d^2 w}{ds^2} + \frac{w}{R^2} \right)$$

Thus, in order to take into account the effect of the compressive force S on bending of the ring, we must add to the given external loads a distributed fictitious load of intensity

$$S \left(\frac{d^2 w}{ds^2} + \frac{w}{R^2} \right) = \frac{S}{R^2} \left(\frac{d^2 w}{d\theta^2} + w \right) \qquad (a)$$

Let us apply this general reasoning to the case represented in Fig. 7-2 and consider the effect on the deflections of the ring of a uniform external pressure producing in the ring a compressive force S. Using for the displacement w the trigonometric series (a) of the preceding article and considering a virtual displacement $w = \delta a_n \cos n\theta$, we find that the equation for determining a coefficient a_n, for n equal to an even number, is

$$\frac{\partial U}{\partial a_n} \delta a_n = 2P \, \delta a_n \, (-1)^{n/2} - \frac{S}{R} \delta a_n \int_0^{2\pi} \left(\frac{d^2 w}{d\theta^2} + w \right) \cos n\theta \, d\theta \qquad (b)$$

The second term on the right-hand side of this equation represents the work done through the virtual displacement of the fictitious load (a). It is taken with negative sign, since the load is directed opposite to positive w.

Substituting series (a) of the previous article into Eq. (b) above and performing the integration, we obtain

$$a_n = \frac{2PR^3(-1)^{n/2}}{\pi EI(n^2 - 1)^2 \left[1 - \dfrac{SR^2}{(n^2 - 1)EI} \right]} \qquad (c)$$

where values of n are even numbers. The coefficients a_n of odd order and all the coefficients b_n in the series for w vanish for the symmetrical loading shown in Fig. 7-2. Thus we obtain

$$w = \frac{2PR^3}{\pi EI} \sum_{n=2,4,\ldots}^{n=\infty} \frac{(-1)^{n/2} \cos n\theta}{(n^2 - 1)^2 \left[1 - \dfrac{SR^2}{(n^2 - 1)EI} \right]} \qquad (7\text{-}8)$$

Comparing this result with Eq. (7-6), which was obtained before for an uncompressed ring, it can be seen that, owing to the compressive force S, each coefficient of the series for w increases in the ratio

$$\frac{1}{1 - \dfrac{SR^2}{(n^2 - 1)EI}}$$

We see that the deflection increases indefinitely when the compressive force S approaches the value

$$S = \frac{(n^2 - 1)EI}{R^2} \tag{d}$$

The smallest value of the force at which this occurs is obtained by taking $n = 2$, which gives the *critical value* of the compressive force as

$$S_{cr} = \frac{3EI}{R^2} \tag{7-9}$$

When S approaches S_{cr}, the first term in the series (7-8) becomes predominant and we obtain for the radial displacement, by taking this term alone, the expression

$$w = -\frac{2PR^3}{9\pi EI} \frac{\cos 2\theta}{1 - S/S_{cr}} \tag{7-10}$$

for which an equation, analogous to Eq. (a), Art. 1.11, for a compressed straight beam under lateral loading, can be obtained.

In the case of a long circular tube uniformly compressed by external pressure, we consider an elemental ring of unit width and we obtain the critical value of the compressive force S_{cr} in such a ring by using $E/(1 - \nu^2)$ instead of E and by taking $I = h^3/12$; then, from Eq. (7-9),

$$S_{cr} = \frac{Eh^3}{4(1 - \nu^2)R^2} \tag{7-11}$$

Observing that the compressive force in the elemental ring of unit width is equal to qR, where q is the uniform pressure, we obtain from Eq. (7-11) the critical value of this pressure:

$$q_{cr} = \frac{E}{4(1 - \nu^2)} \left(\frac{h}{R}\right)^3 \tag{7-12}$$

7.4. Buckling of Circular Rings and Tubes under Uniform External Pressure. In the discussion of combined uniform compression and bending of a circular ring by external forces, it was found in the previous article that the deflections increase indefinitely if the uniform pressure approaches a certain critical value, which can be calculated from Eq. (7-9). The same value of the critical pressure can be obtained in another way by considering an ideal uniformly compressed ring and assuming that some slight deflection from the circular form of equilibrium is produced. Then the critical value of the uniform pressure is that value which is necessary to keep the ring in equilibrium in the assumed slightly deformed shape.

In Fig. 7-8 half of the ring is shown. The dotted line indicates the initial circular shape of the ring, and the full line represents the slightly

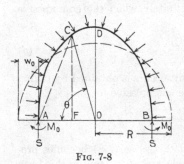

FIG. 7-8

deflected ring on which a uniformly distributed pressure is acting. It is assumed that AB and OD are axes of symmetry for the buckled ring; then the action of the removed lower portion of the ring on the upper portion can be represented by a longitudinal compressive force S and by a bending moment M_0 acting on each of the cross sections A and B. Let q be the uniform normal pressure per unit length of the center line of the ring and w_0 the radial displacement at A and B. Then the compressive force at A and B is[1]

$$S = q(R - w_0) = q\overline{AO}$$

and the bending moment at any cross section C of the buckled ring is

$$M = M_0 + q\overline{AO} \cdot \overline{AF} - \frac{q}{2}\,\overline{AC}^2 \qquad (a)$$

Now, considering the triangle ACO,

$$\overline{OC}^2 = \overline{AC}^2 + \overline{AO}^2 - 2\overline{AO} \cdot \overline{AF}$$

or $\qquad\qquad \tfrac{1}{2}\overline{AC}^2 - \overline{AO} \cdot \overline{AF} = \tfrac{1}{2}(\overline{OC}^2 - \overline{AO}^2)$

Substituting this in the expression (a) for the bending moment, we obtain

$$M = M_0 - \tfrac{1}{2}q(\overline{OC}^2 - \overline{AO}^2) \qquad (b)$$

Observing that $\overline{AO} = R - w_0$ and that $\overline{OC} = R - w$ and neglecting squares of the small quantities w and w_0, we find that the bending moment becomes

$$M = M_0 - qR(w_0 - w) \qquad (c)$$

With this expression for bending moment, the differential equation for

[1] We consider here the change in S due to buckling, whereas in the example of the preceding article S was assumed constant. A more elaborate investigation (see Timoshenko, *op. cit.*) showed that, for the problem of Fig. 7-8, the small changes in S must be considered in order to obtain a correct result for S_{cr}.

the deflection curve (7-1) is

$$\frac{d^2w}{d\theta^2} + w = -\frac{R^2}{EI}[M_0 - qR(w_0 - w)] \tag{d}$$

The critical value of the uniform pressure is now obtained by integrating this equation. Writing the equation in the form

$$\frac{d^2w}{d\theta^2} + w\left(1 + \frac{qR^3}{EI}\right) = \frac{-M_0R^2 + qR^3w_0}{EI}$$

and using the notation

$$k^2 = 1 + \frac{qR^3}{EI} \tag{e}$$

we obtain for the general solution

$$w = A_1 \sin k\theta + A_2 \cos k\theta + \frac{-M_0R^2 + qR^3w_0}{EI + qR^3} \tag{f}$$

Let us consider the conditions at the cross sections A and D of the buckled ring. From symmetry we conclude that

$$\left(\frac{dw}{d\theta}\right)_{\theta=0} = 0 \qquad \left(\frac{dw}{d\theta}\right)_{\theta=\pi/2} = 0$$

From the first of these conditions it follows that $A_1 = 0$, and from the second we obtain

$$\sin\frac{k\pi}{2} = 0 \tag{g}$$

The smallest root of this equation, different from zero, is $k\pi/2 = \pi$ and $k = 2$. Substituting this in expression (e), we obtain for the value of the critical pressure[1]

$$q_{cr} = \frac{3EI}{R^3} \tag{7-13}$$

The corresponding compressive force in the ring is seen to be that given by formula (7-9) obtained before.

The radial deflections of the buckled ring, from (f), are

$$w = \frac{1}{4}\left(\frac{M_0R^2}{EI} + w_0\right)\cos 2\theta - \frac{M_0R^2}{4EI} + \frac{3}{4}w_0 \tag{h}$$

Then, from the condition of inextensibility [see Eq. (b), Art. 7.2], we obtain

$$v = \frac{1}{8}\left(\frac{M_0R^2}{EI} + w_0\right)\sin 2\theta + \left(-\frac{M_0R^2}{4EI} + \frac{3}{4}w_0\right)\theta \tag{i}$$

---

[1] This result was obtained by Bresse; see his "Cours de mécanique appliquée," 2d ed., p. 334, 1866. See also M. Lévy, *J. math. pure et appl. (Liouville)*, series 3, vol. 10, p. 5, 1884, and A. G. Greenhill, *Math. Annal.*, vol. 52, p. 465, 1899.

For $\theta = 0$ and $\theta = \pi/2$ the conditions of symmetry require that the displacement v vanish. Hence we have

$$-\frac{M_0 R^2}{4EI} + \frac{3}{4} w_0 = 0$$

or
$$M_0 = \frac{3w_0 EI}{R^2} = q_{cr} w_0 R \qquad (j)$$

Substituting this result in (h) and (i), we obtain

$$w = w_0 \cos 2\theta \qquad v = \tfrac{1}{2} w_0 \sin 2\theta \qquad (7\text{-}14)$$

From Eq. (j) it is seen that the moment M_0 can be produced by applying at A and B (see Fig. 7-8) the compressive force S with an eccentricity w_0. In such a case the dotted circle in Fig. 7-8 can be considered as a funicular curve for the uniform pressure, and the area between this curve and the center line of the buckled ring represents the bending-moment diagram for the ring. The same result follows also by substituting expression (j) into Eq. (c). For $\theta = \pm\pi/4$ and $\theta = \pm 3\pi/4$ the radial displacement w is zero and the bending moment vanishes.

So far we have discussed only the solution of Eq. (d) corresponding to the smallest root of Eq. (g). By taking $k = 4, 6, \ldots$, we can obtain a series of possible shapes of a buckled ring with a larger and larger number of waves in them.

Another limitation introduced in the above discussion is the condition of symmetry of the buckled ring with respect to the horizontal and vertical axes. As a result of this we have obtained for k only even numbers. By assuming that only one axis, for instance, the horizontal axis AB (Fig. 7-8), is an axis of symmetry and that at the ends of the vertical axis the bending moments are zero, we obtain solutions with odd numbers for k, namely, $k = 3, 5, \ldots$. The case $k = 1$ has been discussed before [see Eqs. (d), Art. 7.2]; it was shown that this case represents a translation of the ring as a rigid body and should not be considered in discussing buckling of the ring. Thus $k = 2$ is the smallest root and the corresponding load [Eq. (7-13)] is the critical load. The buckling forms of higher order corresponding to larger roots can be obtained only by introducing certain additional constraints. Without such constraints, buckling will always be of the type shown in Fig. 7-8.[1]

The results obtained for a circular ring can be applied also in investigating the buckling of long[2] circular tubes submitted to the action of

[1] Buckling of a ring under a more general system of loads is discussed by C. B. Biezeno and J. J. Koch, *Konink. Ned. Akad. van Wetenschap. Proc.*, vol. 48, pp. 447–468, 1945.

[2] The case of shorter tubes is discussed in Art. 11.5.

uniform external pressure.[1]   To obtain q_{cr} for the tube it is only necessary to substitute $E/(1 - \nu^2)$ for E in formula (7-13) and $h^3/12$ for I; in this manner we obtain formula (7-12).

This formula can be used for calculating the critical value of the pressure q as long as the corresponding compressive stress does not exceed the proportional limit of the material.   The limiting value of the ratio $2R/h$, above which the formula can be used, is obtained by dividing Eq. (7-11) by the cross-sectional area of the elemental ring.   Then

$$\sigma_{cr} = \frac{E}{1 - \nu^2}\left(\frac{h}{2R}\right)^2 \tag{7-15}$$

Taking, for instance, steel with $E = 30 \times 10^6$ psi and $\nu = 0.3$ and plotting σ_{cr} against $2R/h$, we obtain a curve AB (Fig. 7-9).   This curve

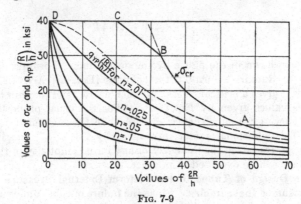

Fig. 7-9

gives the actual critical stress only if the magnitude of this stress does not exceed the proportional limit.   Beyond this limit the curve gives exaggerated values for the critical stress.   To find the true values of this stress we must proceed in the same manner as in the case of a column (Art. 3.3) and use in Eq. (7-15) the tangent modulus E_t instead of E.   If the compression-test diagram of the material is given, E_t can be calculated readily for any value of σ_{cr}, and then the corresponding value of $2R/h$ will be found from the equation

$$\sigma_{cr} = \frac{E_t}{1 - \nu^2}\left(\frac{h}{2R}\right)^2 \tag{7-16}$$

In this manner a curve of critical stresses beyond the proportional limit can be obtained.   For practical purposes this curve can be replaced by two straight lines.   In the case of materials with a pronounced yield

[1] See G. H. Bryan, *Proc. Cambridge Phil. Soc.*, vol. 6, p. 287, 1888.

point, the yield-point stress must be taken as the critical stress for thicker tubes. Taking, for instance, a steel with a yield point $\sigma_{YP} = 40,000$ psi and a proportional limit $\sigma_{PL} = 30,000$ psi, we find that the smallest value of $2R/h$ for which Eq. (7-15) can be used is about 33 (see Fig. 7-9). This corresponds to point B on the curve AB. For thicker tubes the horizontal line DC giving $\sigma_{cr} = \sigma_{YP}$ can be used. For intermediate thicknesses where $20 < 2R/h < 33$, we can use the inclined line BC for determining the critical stress. Thus the line $ABCD$ gives the critical values of compressive stress for all proportions of tubes, and if a factor of safety is decided upon, there is no difficulty in each particular case in finding the safe thickness of the tube.

Instead of the broken line $ABCD$, it is sometimes useful to have a continuous curve, such as the one given by the equation[1]

$$\sigma_{cr} = \frac{\sigma_{YP}}{1 + \dfrac{\sigma_{YP}(1 - \nu^2)}{E} \dfrac{4R^2}{h^2}} \tag{7-17}$$

This curve is shown in the figure by the dotted line. Formula (7-17) is analogous to Rankine's formula for columns [Eq. (a), Art. 4.3]. For thick tubes it gives a critical stress approaching σ_{YP}, and for thin tubes it approaches values given by Eq. (7-16). For the usual proportions of tubes the curve gives much lower stresses than those given by the line $ABCD$. This additional safety introduced by Eq. (7-17) can be considered as compensating for the effect of any initial ellipticity of the tube which always can be expected in practice.

7.5. The Design of Tubes under Uniform External Pressure on the Basis of Assumed Inaccuracies. Since the failure of tubes under uniform external pressure depends very much upon the various kinds of imperfections in them, it seems logical to derive a design formula for such tubes in which the quantities depending on the imperfections will appear explicitly. The most common imperfection in tubes is an initial ellipticity, the limiting value of which in each type of tube is usually known from numerous inspection measurements. The deviation of the shape of the tube from a perfect circular form can be defined by the initial radial deflections w_i. To simplify our investigation we assume that these deflections are given by the equation[2]

$$w_i = w_1 \cos 2\theta \tag{a}$$

in which w_1 is the maximum initial radial deviation from a circle and θ is the central angle measured as shown in Fig. 7-10.

[1] This curve was proposed by R. V. Southwell, *Phil. Mag.*, vol. 29, p. 67, 1915.
[2] See Timoshenko, *Trans. ASME*, Applied Mechanics Division, vol. 1, p. 173, 1933.

Under the action of external uniform pressure q, there will be an additional flattening of the tube. The corresponding additional radial displacements will be called w. To determine w, we shall use the differential equation (7-4). Since a positive bending moment produces a decrease in the initial curvature, we conclude that in the portions AB and CD there will be produced by a uniform external pressure a positive bending moment and in the portions AD and BC a negative bending moment. At points A, B, C, and D the bending moment is zero, and the actions between the parts of the tube are represented by forces S tangential to the dotted circle representing the ideal shape of the tube. This circle can be considered as a funicular curve for the external pressure (see p. 292). The compressive force along this curve remains constant

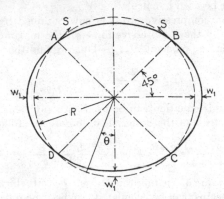

Fig. 7-10

and equal to S. Thus the bending moment at any cross section is obtained by multiplying S by the total radial displacement $w_i + w$ at this cross section. Then

$$M = qR(w + w_1 \cos 2\theta) \tag{b}$$

Substituting in Eq. (7-4), we obtain

$$\frac{d^2w}{d\theta^2} + w = -\frac{12(1 - \nu^2)}{Eh^3} qR^3(w + w_1 \cos 2\theta)$$

or $\quad \dfrac{d^2w}{d\theta^2} + w\left[1 + \dfrac{12(1 - \nu^2)}{Eh^3} qR^3\right] = -\dfrac{12(1 - \nu^2)}{Eh^3} qR^3 w_1 \cos 2\theta$

The solution of this equation satisfying the conditions of continuity at the points A, B, C, and D is

$$w = \frac{w_1 q}{q_{cr} - q} \cos 2\theta \tag{7-18}$$

in which q_{cr} is the critical value of the uniform pressure given by Eq. (7-12). It is seen that at points A, B, C, and D, w and $d^2w/d\theta^2$ are zero. Hence the bending moments at these points are zero, as was assumed above. The maximum moment occurs at $\theta = 0$ and at $\theta = \pi$, where

$$M_{\max} = qR\left(w_1 + \frac{w_1 q}{q_{cr} - q}\right) = qR\frac{w_1}{1 - q/q_{cr}} \qquad (7\text{-}19)$$

It is seen from Eq. (7-19) that for small values of the ratio q/q_{cr} the change in the ellipticity of the tube due to pressure q can be neglected and the maximum bending moment obtained by multiplying the compressive force qR by the initial deflection w_1. When the ratio q/q_{cr} is not small, the change in the initial ellipticity of the tube should be considered and Eq. (7-19) must be used in calculating $M_{\max}$.

The maximum compressive stress is now obtained by adding to the stress produced by the compressive force qR the maximum compressive stress due to bending moment $M_{\max}$. Thus we find that

$$\sigma_{\max} = \frac{qR}{h} + \frac{6qR}{h^2}\frac{w_1}{1 - q/q_{cr}} \qquad (c)$$

Assuming that this equation can be used with sufficient accuracy up to the yield-point stress of the material, we obtain the following equation:

$$\sigma_{YP} = \frac{q_{YP}R}{h} + 6q_{YP}\frac{R^2}{h^2}\frac{w_1}{R}\frac{1}{1 - q_{YP}/q_{cr}} \qquad (d)$$

from which the value of the uniform pressure q_{YP}, at which yielding in the extreme fibers begins, can be calculated. When the notations $R/h = m$ and $w_1/R = n$ are used, the equation for calculating q_{YP} becomes

$$q^2_{YP} - \left[\frac{\sigma_{YP}}{m} + (1 + 6mn)q_{cr}\right]q_{YP} + \frac{\sigma_{YP}}{m}q_{cr} = 0 \qquad (e)$$

It should be noted that the pressure q_{YP} determined in this manner is smaller than the pressure at which the collapsing of the tube occurs, and it becomes equal to the latter only in the case of a perfectly round tube. Hence, by using the value of q_{YP} calculated from Eq. (e) as the ultimate value of pressure, we are always on the safe side.[1]

In Fig. 7-9 several curves are shown giving the values of the average tangential compressive stress $q_{YP}R/h$ at which yielding begins, calculated from Eq. (e) by taking $n = 0.1$, 0.05, 0.025, 0.01, and $\sigma_{YP} = 40,000$ psi.

---

[1] Experiments with long tubes submitted to uniform external pressure were made by R. T. Stewart, *Trans. ASME*, vol. 27, 1906. See also H. A. Thomas, *Bull. Am. Petroleum Inst.*, vol. 5, p. 79, 1924, and B. V. Bulgakov, *Nauch.-Tehn. Upravl. V.S.N.H.*, Moscow, no. 343, 1930.

These curves can be used for calculating safe pressures on the tubes if the ellipticity of the tubes is known and if a suitable factor of safety is chosen.

7.6. Buckling of a Uniformly Compressed Circular Arch. If a curved bar with hinged ends and with its center line in the form of an arc of a circle is submitted to the action of a uniformly distributed pressure q, it will buckle as shown by the dotted line in Fig. 7-11. The critical value of the pressure q at which this buckling occurs can be found from the differential equation of the deflection curve of the buckled bar. Considering, as before, the initial circular arc as a funicular curve for the uniform pressure, Eq. (7-1) becomes

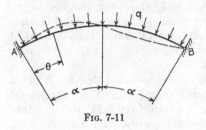

Fig. 7-11

$$\frac{d^2w}{d\theta^2} + w = -\frac{R^2Sw}{EI} \qquad (a)$$

where $S = qR$ is the axial compressive force and w is the radial displacement toward the center. Using the previous notation

$$k^2 = 1 + \frac{qR^3}{EI} \qquad (b)$$

we obtain
$$\frac{d^2w}{d\theta^2} + k^2w = 0$$

The general solution of this equation is

$$w = A \sin k\theta + B \cos k\theta$$

To satisfy the conditions at the left end ($\theta = 0$), we must take $B = 0$. The conditions at the right end ($\theta = 2\alpha$) will be satisfied if we take

$$\sin 2\alpha k = 0 \qquad (c)$$

The smallest root of this equation satisfying the condition of inextensibility of the center line of the bar [see Eq. (j), Art. 7.2] is

$$k = \frac{\pi}{\alpha}$$

and, using notation (b), we obtain[1]

$$q_{cr} = \frac{EI}{R^3}\left(\frac{\pi^2}{\alpha^2} - 1\right) \qquad (7\text{-}20)$$

---

[1] This solution was obtained by E. Hurlbrink, *Schiffbau*, vol. 9, p. 517, 1908; see also Timoshenko, Buckling of a Uniformly Compressed Circular Arch, *Bull. Polytech. Inst., Kiev*, 1910.

By taking $\alpha = \pi/2$, we find that Eq. (7-20) gives the same value of q_{cr} as for a complete ring [Eq. (7-13)]. This result should be expected, since at this value of α the bar represented in Fig. 7-11 is in exactly the same condition as each half of a buckled ring (Fig. 7-8) between the two opposite inflection points.

When α approaches the value π, i.e., when the arc approaches the complete ring, the value of q_{cr}, from Eq. (7-20), approaches zero. This can be explained if we observe that for $\alpha = \pi$ both hinges coincide and that the ring will be free to rotate as a rigid body about this common hinge.

When α is small in comparison with π, unity can be neglected in comparison with π^2/α^2 in the parentheses of formula (7-20). Then the critical compressive force $q_{cr}R$ becomes equal to the critical load for a prismatic bar with hinged ends and length $R\alpha$.

In the derivation of formula (7-20) it was assumed that the buckled arch had an inflection point at the middle (Fig. 7-11). From the general discussion of inextensional deflection of an arc (Art. 7.2), we know that it is possible to have also inextensional deflection curves symmetrical with respect to the middle of the bar. The simplest of these curves has two inflection points. By taking such a curve as a basis for calculating the critical load, we obtain a critical value larger than that given by Eq. (7-20).[1] Hence, this latter equation should be used for calculating q_{cr}.

If, instead of a circular arch, we have a flat parabolic arch and a vertical load uniformly distributed along the span AB (Fig. 7-11), the variation of the compressive force along the length of the arch can be neglected and its critical value can be calculated by taking half the length of the arch and applying Euler's formula as for a bar with hinged ends.[2]

In the derivation of Eq. (7-20) it was assumed that the curved bar, before buckling, had its center line in the form of an arc of a circle. This condition is fulfilled only if uniform unit compression $q_{cr}R/AE$ of the center line of the bar is produced before fastening the ends to the supports; otherwise some bending under the action of uniform pressure will start at the very beginning of loading. This bending is very small as long as the compression q is small in comparison with q_{cr}, and the con-

---

[1] Calculations of this kind were made by E. Chwalla, *Sitzber. Akad. Wiss. Wien.*, Abt. IIa, vol. 136, p. 645, 1927.

[2] The results of experiments are in satisfactory agreement with such calculations; see R. Mayer, *Der Eisenbau*, vol. 4, p. 361, 1913, and "Die Knickfestigkeit," Berlin, 1921. Here the case of an arch with three hinges is considered and the effect on the critical load of the compression of the center line of the arch and of the lowering of the middle hinge is discussed. Tests on uniformly loaded arches with three hinges were made by E. Gaber, *Bautechnik*, 1934, p. 646. The results of these tests are in agreement with Eq. (7-20). A discussion of buckling of arches with three hinges under the action of nonsymmetrical loading is given by E. Chwalla, *Der Stahlbau*, no. 16, 1935.

ditions are analogous to the bending that occurs in columns because of various kinds of imperfections.

Substituting $E/(1 - \nu^2)$ instead of E and $h^3/12$ instead of I in Eq. (7-20), we obtain the equation

$$q_{cr} = \frac{Eh^3}{12(1 - \nu^2)R^3} \left(\frac{\pi^2}{\alpha^2} - 1\right) \tag{7-21}$$

which can be used in calculating the critical load for a cylindrical shell hinged along the edges $\theta = 0$ and $\theta = 2\alpha$ (Fig. 7-11) and submitted to the action of uniform pressure.

If the ends of a uniformly compressed arch are built in[1] (Fig. 7-12), the shape of buckling will be as shown by the dotted line. At the middle point C there will act after buckling not only the horizontal compressive force S but also a vertical shearing force Q. Considering again the initial circular arc as the funicular curve

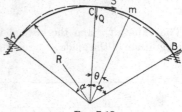

Fig. 7-12

for uniform pressure and designating with w the radial displacement toward the center, the bending moment at any cross section, defined by the angle θ, is

$$M = Sw - QR \sin \theta$$

and the differential equation (7-1) becomes

$$\frac{d^2w}{d\theta^2} + w = -\frac{R^2}{EI}(Sw - QR \sin \theta)$$

or, using notation (b),

$$\frac{d^2w}{d\theta^2} + k^2w = \frac{QR^3 \sin \theta}{EI}$$

The general solution of this equation is

$$w = A \sin k\theta + B \cos k\theta + \frac{QR^3 \sin \theta}{(k^2 - 1)EI} \tag{d}$$

The conditions for determining the constants A and B and the force Q are

$$w = \frac{d^2w}{d\theta^2} = 0 \qquad \text{at } \theta = 0 \tag{e}$$

$$w = \frac{dw}{d\theta} = 0 \qquad \text{at } \theta = \alpha \tag{f}$$

[1] It is assumed again that the arch is uniformly compressed before fixing the ends.

Conditions (e) are satisfied by taking $B = 0$ in solution (d). From conditions (f) we then obtain

$$A \sin k\alpha + Q \frac{R^3 \sin \alpha}{(k^2 - 1)EI} = 0$$

$$Ak \cos k\alpha + Q \frac{R^3 \cos \alpha}{(k^2 - 1)EI} = 0$$

$$(g)$$

The equation for calculating the critical value of uniform pressure q is obtained by equating to zero the determinant of Eqs. (g), which gives

$$\sin k\alpha \cos \alpha - k \sin \alpha \cos k\alpha = 0$$

or $$k \tan \alpha \cot k\alpha = 1 \qquad (h)$$

The value of k and the critical value of the pressure q depend on the magnitude of the angle α. Several solutions of Eq. (h) for various values

TABLE 7-1

α	30°	60°	90°	120°	150°	180°
k	8.621	4.375	3	2.364	2.066	2

of α are given in Table 7-1.[1] When k is substituted in Eq. (b), the critical value of the uniform pressure is found to be

$$q_{cr} = \frac{EI}{R^3} (k^2 - 1) \qquad (7\text{-}22)$$

This value of q_{cr} is always greater than that obtained from Eq. (7-20).

The problem of buckling of a uniformly compressed circular arch of constant cross section has been solved also for symmetrical three-hinged and one-hinged arches.[2] One of the forms of buckling in the case of three hinges is the same as that in the case of a two-hinged arch (Fig. 7-11). The presence of the hinge at the crown of the arch does not change the critical load for this case. The other possible form of buckling is symmetrical and is connected with a lowering of the central hinge as shown in Fig. 7-13. For smaller values of h/l this second form of buckling requires the smaller load and hence gives the limiting value of q_{cr}.

In all four cases under consideration the critical pressure q_{cr} can be represented in the form

$$q_{cr} = \gamma_1 \frac{EI}{R^3} \qquad (7\text{-}23)$$

[1] This solution is due to E. L. Nicolai, *Bull. Polytech. Inst.*, St. Petersburg, vol. 27, 1918; see also *Z. angew. Math. u. Mech.*, vol. 3, p. 227, 1923.

[2] See A. N. Dinnik, *Vestnik Inzhenerov*, no. 6, 1934; see also his book "Buckling and Torsion," pp. 160–163, Moscow, 1955.

Values of the factor γ_1 are given in Table 7-2 for various values of the central angle 2α. The values for arches with no hinges and two hinges are obtained from Eqs. (7-22) and (7-20), respectively. The remaining values were calculated by Dinnik.[1]

TABLE 7-2. VALUES OF THE FACTOR γ_1 FOR UNIFORMLY COMPRESSED
CIRCULAR ARCHES OF CONSTANT CROSS SECTION [EQ. (7-23)]

2α (deg)	No hinges	One hinge	Two hinges	Three hinges
30	294	162	143	108
60	73.3	40.2	35	27.6
90	32.4	17.4	15	12.0
120	18.1	10.2	8	6.75
150	11.5	6.56	4.76	4.32
180	8.0	4.61	3.00	3.00

For practical use it is convenient to represent the critical pressure as a function of the span l and the rise h of the arch (see Fig. 7-13). Then the formula for q_{cr} takes the form

$$q_{cr} = \gamma_2 \frac{EI}{l^3} \qquad (7\text{-}24)$$

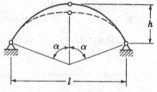

FIG. 7-13

in which γ_2 depends on the ratio h/l and the number of hinges. Numerical values of γ_2 are given in Table 7-3. It is seen from Tables 7-2 and 7-3 that the critical load decreases with an increase in the number of hinges. The only exception is when $2\alpha = 180°$ (or $h/l = 0.5$), in which case the critical load for two- and three-hinged arches is the same, since both arches have the same critical buckling form (see Fig. 7-11).

TABLE 7-3. VALUES OF THE FACTOR γ_2 FOR UNIFORMLY COMPRESSED
CIRCULAR ARCHES OF CONSTANT CROSS SECTION [EQ. (7-24)]

$\frac{h}{l}$	No hinges	One hinge	Two hinges	Three hinges
0.1	58.9	33	28.4	22.2
0.2	90.4	50	39.3	33.5
0.3	93.4	52	40.9	34.9
0.4	80.7	46	32.8	30.2
0.5	64.0	37	24.0	24.0

In the preceding discussion of buckling of circular arches it was assumed that during buckling, the external forces remained normal to the buckled
[1] *Ibid.*

axis of the arch, as in the case of hydrostatic pressure. Sometimes we encounter the case in which the forces retain their initial directions during buckling. An investigation of this problem has shown that the slight changes in the directions of the forces during buckling have only a small influence on the values of the critical pressure.[1]

The problem of elastic stability of the arch shown in Fig. 7-11 can be solved also in certain cases for a varying cross section. Assuming, for instance, that the cross-sectional moment of inertia for the left side of the symmetrical arch varies along the length of the arc following the law

$$I = I_0 \left[1 - \left(1 - \frac{I_1}{I_0} \right) \frac{\theta}{\alpha} \right] \qquad (7\text{-}25)$$

in which I_0 and I_1 are the moments of inertia for $\theta = 0$ and $\theta = \alpha$, respectively, we obtain for the critical pressure

$$q_{cr} = \gamma_3 \frac{EI_0}{\alpha^2 R^3} \qquad (7\text{-}26)$$

where γ_3 is a numerical factor depending on the angle α and on the ratio I_1/I_0. Several values of this factor[2] are given in Table 7-4. The first line in the table ($\alpha = 0$) gives the values of the coefficient γ_3 for the case of a straight bar of variable cross section when buckled with an inflection point at the middle. The last column in the table gives γ_3 for an arch of constant cross section [see Eq. (7-20)].

TABLE 7-4. VALUES OF THE FACTOR γ_3 FOR UNIFORMLY COMPRESSED CIRCULAR TWO-HINGED ARCHES OF VARYING CROSS SECTION
[EQ. (7-26)]

2α (deg) \\ I_1/I_0	0.1	0.2	0.4	0.6	0.8	1
0	4.67	5.41	6.68	7.80	8.85	π^2
60	4.54	5.20	6.48	7.58	8.62	9.60
120	4.16	4.82	5.94	6.94	7.89	8.77
180	3.53	4.08	5.02	5.86	6.66	7.40

7.7. Arches of Other Forms. In the preceding article we considered uniformly compressed arches with a circular axis. There are several other forms of arches for which the buckling problem is solved and some of the results are given in the following.[3]

Parabolic Arch. If a parabolic arch is submitted to the action of a load q uniformly

[1] *Ibid.*, pp. 163–165.

[2] See A. N. Dinnik, *Vestnik Inzhenerov*, nos. 8 and 12, 1933; see also I. J. Steurman, *Bull. Polytech. Inst.*, Kiev, 1929, and "Stability of Arches," Kiev, 1929.

[3] See K. Federhofer, *Sitzber. Akad. Wiss. Wien*, 1934, and *Bautechnik*, no. 41, 1936, A. N. Dinnik, *Vestnik Inzhenerov*, nos. 1 and 12, 1937, and "Buckling and Torsion," pp. 171–193, Moscow, 1955.

distributed along the span (Fig. 7-14), there will be axial compression but no bending of the arch, since a parabola is the funicular curve for a uniform load. By a gradual increase of the intensity of the load we can reach the condition in which the parabolic form of equilibrium becomes unstable and the arch buckles in a form similar to that for a circular arch. Considering symmetrical arches of uniform cross section with no hinges or with one, two, and three hinges, we can express the critical values of the intensity of the load by the formula

$$q_{cr} = \gamma_4 \frac{EI}{l^3} \qquad (7\text{-}27)$$

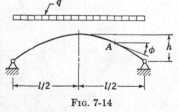

FIG. 7-14

The numerical factor γ_4 depends on the ratio h/l, where h is the rise of the arch and l is the span (Fig. 7-14). Values of the factor[1] γ_4 are given in Table 7-5. It is seen that for flat parabolic arches ($h/l < 0.2$) the values of γ_4 differ only slightly from those given in Table 7-3 for circular arches.

The numerical factor γ_4 is expressed graphically as a function of h/l in Fig. 7-15. The portions of the curves indicated by dotted lines correspond to symmetrical forms of buckling. In these cases unsymmetrical buckling will occur, and in obtaining values of γ_4, we have to use the curves for arches without a central hinge. For example, in the case of a three-hinged arch with $h/l > 0.3$, we take for γ_4 the value from the curve for a two-hinged arch. Experiments made with models of arches are in satisfactory agreement with the theoretical values given above.[2]

TABLE 7-5. VALUES OF THE FACTOR γ_4 FOR PARABOLIC ARCHES OF CONSTANT CROSS SECTION WITH UNIFORM LOAD [EQ. (7-27)]

$\dfrac{h}{l}$	No hinges	One hinge	Two hinges	Three hinges
0.1	60.7	33.8	28.5	22.5
0.2	101	59	45.4	39.6
0.3	115	....	46.5	46.5
0.4	111	96	43.9	43.9
0.5	97.4	....	38.4	38.4
0.6	83.8	80	30.5	30.5
0.8	59.1	59.1	20.0	20.0
1.0	43.7	43.7	14.1	14.1

In the case of a parabolic arch of rectangular cross section having constant width and depth proportional to sec ϕ, where ϕ is the angle between the horizontal and the tangent to the arch axis at any point A (Fig. 7-14), we can use again formula (7-27). In this case, I represents the moment of inertia of the cross section at the crown of the arch ($\phi = 0$). The values of the factor γ_4 are given in Table 7-6.[3]

[1] These values were calculated by Dinnik, op. cit.

[2] See E. Gaber, Bautechnik, 1934, pp. 646–656; C. F. Kollbrunner, Schweiz. Bauztg., vol. 120, 1942, p. 113; and Kollbrunner and M. Meister, "Knicken," pp. 191–200, Springer-Verlag, Berlin, 1955.

[3] See A. N. Dinnik, "Buckling and Torsion," Moscow, 1955.

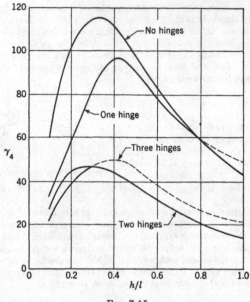

Fig. 7-15

Catenary Arch. Let us assume now that the load is uniformly distributed along the axis of the arch, as in the case of the dead load of an arch of uniform cross section. Then the catenary is the funicular curve for the load and no bending will be produced in an arch of that shape.[1] Buckling occurs when the intensity of the load reaches the critical value, which can be expressed again by formula (7-27). Table 7-7 gives values[2] of the factor γ_4. Comparing Tables 7-7 and 7-5, we see that for flat curves there is only a small difference in γ_4 for the two forms of arches.

TABLE 7-6. VALUES OF THE FACTOR γ_4 FOR PARABOLIC ARCHES OF VARYING CROSS SECTION WITH UNIFORM LOAD [EQ. (7-27)]

$\dfrac{h}{l}$	No hinges	One hinge	Two hinges	Three hinges
0.1	65.5	36.5	30.7	24
0.2	134	75.8	59.8	51.2
0.3	204	.....	81.1	81.1
0.4	277	187	101	101
0.6	444	332	142	142
0.8	587	497	170	170
1.0	700	697	193	193

[1] As in our previous discussions, it is assumed that the contraction of the arch axis takes place before the end constraints are applied.
[2] See *ibid.*

TABLE 7-7. VALUES OF THE FACTOR γ_4 FOR CATENARY ARCHES OF CONSTANT
CROSS SECTION WITH LOAD UNIFORMLY DISTRIBUTED ALONG THE
ARCH AXIS [EQ. (7-27)]

$\dfrac{h}{l}$	No hinges	Two hinges
0.1	59.4	28.4
0.2	96.4	43.2
0.3	112.0	41.9
0.4	92.3	35.4
0.5	80.7	27.4
1.0	27.8	7.06

In the design of arches we have to consider loads of several types, some of which produce only compression of the arch while others produce bending also. As long as the compressive force is small in comparison with its critical value, we can neglect its influence on bending and disregard the deformation of the arch in determining stresses. However, in the case of slender arches of long span, the axial compression may approach the critical value. Then the influence of axial force on bending becomes important and the deformation of the arch must be considered in carrying out a stress analysis.

7.8. Buckling of Very Flat Curved Bars.[1]

In the previous articles only extensionless forms of buckling of curved bars were considered. In the case of very flat curved bars, buckling in which axial strain is considered may occur at a smaller load than extensionless buckling and must be investigated. As an example of such buckling, let us consider a flat uniformly loaded arch with hinged ends (Fig. 7-16), the initial center line of which is given by the equation

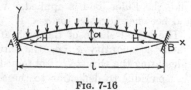

FIG. 7-16

$$y = a \sin \frac{\pi x}{l} \qquad (a)$$

If the rise a of the arch is large, the axial deformation of the arch under the action of the load can be neglected and the critical load can be obtained by assuming that during buckling there is an inflection point at the middle of the arch (see Art. 7.6). If a is very small, the axial deformation of the arch during loading cannot be neglected and the arch can buckle in a symmetrical form as shown in the figure by the dotted line.

In investigating the deformation of the arch, let us assume first that one of the hinges is on rollers; then the center line of the arch after loading can be represented with sufficient accuracy (see p. 27) by the equation

$$y_1 = \left(a - \frac{5}{384} \frac{q l^4}{EI} \right) \sin \frac{\pi x}{l} = a(1 - u) \sin \frac{\pi x}{l} \qquad (b)$$

[1] See Timoshenko, *J. Appl. Mech.*, vol. 2, p. 17, 1935.

in which q is the intensity of the uniform load, EI is the flexural rigidity of the bar in the plane of the center line, and

$$u = \frac{5}{384} \frac{ql^4}{EIa} \tag{c}$$

In the case of immovable hinges, a thrust H will be produced by the loading and the final equation of the center line will be (see Art. 1.12)

$$y_2 = \frac{a(1 - u)}{1 - \alpha} \sin \frac{\pi x}{l} \tag{d}$$

in which

$$\alpha = \frac{Hl^2}{\pi^2 EI} \tag{e}$$

This equation can be used not only for $u < 1$ but also for $u > 1$, i.e., in cases where the deflection of the curved bar, calculated as for a simple beam, is larger than the initial rise of the arch. The quantity α can also be larger than 1, but it must be smaller than 4, since for $\alpha = 4$ there occurs buckling of the arch with an inflection point at the middle, and our assumption [Eq. (d)] that the arch remains symmetrical with respect to the middle is no longer fulfilled.

Assuming first that $u < 1$, we find from Eq. (d) that y_2 is positive if $\alpha < 1$ and negative if $\alpha > 1$. This means that if the thrust is smaller than the Euler load for a bar with hinged ends, the arch has the form shown by the full line in Fig. 7-16. The same arch can be kept deflected downward, as shown in the figure by the dotted line, if a thrust larger than the Euler load is applied. When $u > 1$, y_2 is positive for $\alpha > 1$ and becomes negative for $\alpha < 1$.

The actual shape of the arch after loading can be determined only when the quantity α, i.e., the thrust H, is known. The equation for calculating the thrust H is obtained by equating the change in length of the span due to deflection to the compression of the bar due to thrust. Assuming for flat curves that the compressive force along the length of the bar is constant and equal to H, we obtain the equation

$$\frac{Hl}{AE} = \frac{1}{2} \int_0^l \left(\frac{dy}{dx}\right)^2 dx - \frac{1}{2} \int_0^l \left(\frac{dy_2}{dx}\right)^2 dx \tag{f}$$

in which A is the cross-sectional area of the bar. Substituting for y and y_2 their expressions (a) and (d) and integrating, we obtain

$$(1 - u)^2 = (1 - m\alpha)(1 - \alpha)^2 \tag{g}$$

in which

$$m = \frac{4I}{Aa^2} \tag{h}$$

For a given arch the quantity m is easily calculated, and if the load q is given, the quantity u can be determined from Eq. (c). Then the corresponding value of α and consequently the thrust H are obtained from Eq. (g). Since this equation is not linear, more than one real root for α can be obtained under certain conditions, which indicates that there are several possible forms of equilibrium and that the stability of these forms must be investigated.

Considering the right-hand side of Eq. (g) as a function of α, we find that for $m < 1$ this function has a minimum value equal to 0 for $\alpha = 1$ and a maximum value for $\alpha = (2 + m)/3m$. The magnitude of this maximum is

$$\frac{4}{27}\frac{(1-m)^3}{m^2} \tag{i}$$

In Fig. 7-17 the right-hand side of Eq. (g) is represented graphically for the case where $m = \frac{1}{2}$. For this value of m the maximum occurs at $\alpha = \frac{5}{3}$, and the magnitude of this maximum, from (i), is equal to $\frac{2}{27}$.

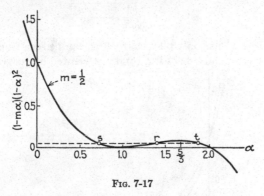

FIG. 7-17

If the load q is of such magnitude that the left-hand side of Eq. (g) is larger than the above maximum, we obtain only one real solution for α, which indicates that only one form of equilibrium is possible and the equilibrium is stable. If the left-hand side of Eq. (g) is smaller than the quantity (i), three solutions for α are obtained as shown by intersection points s, r, and t in Fig. 7-17 and the question of stability of the corresponding forms of equilibrium must be considered. Applying these conclusions in the above numerical example, we find that the equilibrium is always stable if

$$(1 - u)^2 > \tfrac{2}{27}$$

which is equivalent to the conditions

$$u < 1 - \sqrt{\tfrac{2}{27}}$$
$$u > 1 + \sqrt{\tfrac{2}{27}}$$

The first of these conditions corresponds to the form of equilibrium convex upward as shown in Fig. 7-16 by the full line, and the second to the form convex downward as indicated in the figure by the dotted line.

For any value of m smaller than 1 the conditions of stability, equivalent to conditions (j), are

$$u < 1 - \sqrt{\frac{4}{27}\frac{(1-m)^3}{m^2}}$$
$$u > 1 + \sqrt{\frac{4}{27}\frac{(1-m)^3}{m^2}} \tag{k}$$

Hence, conditions which make possible more than one form of equilibrium and for which an investigation of stability is necessary are

$$u > 1 - \sqrt{\frac{4}{27}\frac{(1-m)^3}{m^2}}$$
$$u < 1 + \sqrt{\frac{4}{27}\frac{(1-m)^3}{m^2}} \tag{l}$$

If $m \geq 1$, Eq. (g) has only one real root, as can be seen from Fig. 7-18, in which the right-hand side of Eq. (g) is represented by curves for $m = 2$

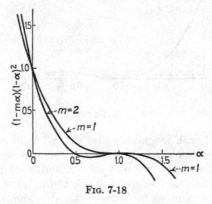

FIG. 7-18

and $m = 1$. It is seen that for any positive value of $(1 - u)^2$ we obtain only one value for α, and this value is less than unity. Hence for $m \geq 1$ there is only one possible form of equilibrium that will be stable. The question of instability will arise only if $m < 1$ and if the load is within the limits indicated by conditions (l).

In the investigation of the stability of the arch it is advantageous to represent graphically the rise of the arch as a function of the load or as a function of the quantity u. By using Eq. (d) we see that this rise is

$$a_1 = \frac{a(1-u)}{1-\alpha} \tag{m}$$

By taking in each particular case a series of values of α, we can calculate the corresponding values of u from Eq. (g) and the rise a_1 from Eq. (m). In Fig. 7-19 the values of a_1/a are plotted against u.   The full line represents the case where $m = \frac{1}{2}$, and the two dotted lines represent the cases where $m = \frac{1}{4}$ and $m = 1$.   Considering the case $m = \frac{1}{2}$, we see from the curve that the deflection gradually increases with an increase of the load up to the point A, which corresponds to the maximum ($\alpha = \frac{3}{5}$) on the curve shown in Fig. 7-17.   Beginning at this point, further deflection

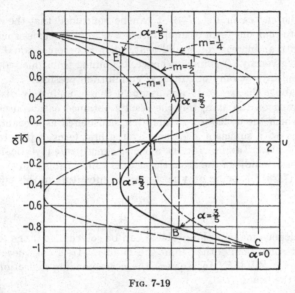

FIG. 7-19

continues with a decrease of u, i.e., with a decrease of the load q.   This fact indicates that at the point A, i.e., for

$$(1 - u)^2 = \tfrac{2}{27}$$

the form of equilibrium of the arch shown in Fig. 7-16 by the full line becomes unstable and the arch buckles downward as shown by the dotted line.   The sag of this new form of equilibrium is given in Fig. 7-19 by the position of the point B.   This new form is stable, and any further increase of the load produces a gradual increase in deflection as shown by the portion BC of the curve in Fig. 7-19.   The thrust H during this loading decreases, becoming zero at the point C and negative with further increase of the load.   If, starting from the point B, we begin to diminish the load, the deflection of the arch gradually diminishes up to the point D.   At this point the load becomes insufficient to keep the arch deflected convex down, and it buckles upward to the position given in Fig. 7-19 by the point E. From this discussion we see that there exists the possibility of having more than one form of equilibrium in the region limited by the verticals $\overline{ED}$ and $\overline{AB}$ which corresponds to the conditions (j). In the general case this region is defined by conditions (l), and we conclude that the critical load at which the arch will buckle convex down is determined from the equation[1]

$$u = 1 + \sqrt{\frac{4}{27} \frac{(1 - m)^3}{m^2}} \tag{7-28}$$

From the curves in Fig. 7-19 it can be concluded that the region in which more than one form of equilibrium is possible becomes smaller and smaller with an increase of m and that, when m becomes equal to 1, the two limits given by (l) coincide; hence, beginning from this value of m, there will always be only one form of equilibrium possible.

In the above discussion it was assumed that a uniformly distributed load is acting on the arch, but the results obtained can be used for all cases in which the deflection of the arch, considered as a beam, can be represented with sufficient accuracy by the first term of the sine series (see Art. 1.11). Taking, for instance, a concentrated vertical load P applied at the middle of the arch, the critical value of this load is obtained from Eq. (7-28),[2] it being only necessary to substitute in this equation

$$u = \frac{Pl^3}{48EI} \frac{1}{a} \tag{n}$$

In an analogous manner the problem can be solved also if the load P is applied at some point other than the middle. It is only necessary to substitute for $Pl^3/48EI$ in Eq. (n) the corresponding deflection at the middle of the beam.

7.9. Buckling of a Bimetallic Strip. The results of the preceding article can be used in investigating buckling of a bimetallic strip submitted to a variation of temperature. Such strips are used in thermostats for regulating temperature. Let us consider a simple case when a bimetallic strip of unit width consists of two metals of equal thickness and having the same modulus of elasticity but two different coefficients of thermal expansion ϵ_1 and ϵ_2 (Fig. 7-20). If $\epsilon_2 > \epsilon_1$, any increase of temperature from the initial temperature t_0 to a temperature t will produce bending of the strip convex downward. The corresponding curvature can be calculated if we consider the reac-

---

[1] The results obtained in this article are in satisfactory agreement with experiments made by Dinnik; see Dinnik, *op. cit.*, p. 231.

[2] This particular problem has been discussed by Navier, "Résumé des leçons sur l'application de la mécanique," 2d ed., p. 273, Paris, 1833. See also C. B. Biezeno, *Konink. Akad. Wetenschap. Amsterdam, Proc.*, vol. 32, p. 990, 1929, and *Z. angew. Math. u. Mech.*, vol. 18, p. 21, 1938. The case when the center line of the arch is not a flat curve has been considered by A. Nadai, *Tech. Blättern, Prague*, nos. 3 and 4, 1915.

tive forces P arising at the surface of contact of the two metals (Fig. 7-20c) because of the differences in expansion. Replacing these forces by forces centrally applied and by couples $M = Ph/4$, as shown in Fig. 7-20d, we find the curvature for the strip of each metal from the known equation

$$\frac{1}{\rho} = \frac{M}{EI} = \frac{24Ph}{Eh^3} \tag{a}$$

Another equation for determining ρ and P is obtained from the condition that the length of the longitudinal fibers of both metals at the surface of contact is the same hence

$$\epsilon_1(t - t_0) + \frac{2P}{Eh} + \frac{h}{4\rho} = \epsilon_2(t - t_0) - \frac{2P}{Eh} - \frac{h}{4\rho} \tag{b}$$

Eliminating P from Eqs. (a) and (b), we obtain

$$\frac{1}{\rho} = \frac{3}{2} \frac{(\epsilon_2 - \epsilon_1)(t - t_0)}{h} \tag{7-29}$$

Thus the curvature of the bimetallic strip produced by heating is proportional to the rise in temperature and to the difference between the two coefficients of expansion

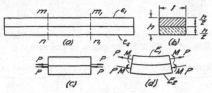

FIG. 7-20

and is inversely proportional to the thickness of the strip. Equation (7-29) can be used with sufficient accuracy also in the case where the materials have somewhat different moduli of elasticity, as, for instance, in the case of monel metal and nickel steel.[1]

With the curvature known, the deflection at the middle of the strip (supported at both ends), due to an increase in temperature $t - t_0$, is obtained as for a flat arc of a circle and is equal to

$$\delta = \frac{l^2}{8\rho} \tag{c}$$

in which l is the length of the strip and ρ the radius of curvature determined from Eq. (7-29).

Let us consider now a bimetallic strip which has a slight initial curvature, and let us assume that the ends of the strip are hinged as shown in Fig. 7-16 by the full line. If the metal on the concave side has a larger coefficient of thermal expansion than the metal on the convex side, the strip will deflect downward during heating and a certain thrust H will be produced owing to this deflection. At a certain temperature, depending on the proportions of the strip and on the difference of the coefficients of expansion of the two metals, the form of the strip, shown in Fig. 7-16 by the full line,

[1] A more general investigation of bending and buckling of bimetallic strips is given by Timoshenko, *J. Optical Soc. Am.*, vol. 11, p. 233, 1925; see also A. M. Wahl, *J. Appl. Mech.*, vol. 11, p. 183, 1944. The case of buckling of a circular bimetallic plate was discussed by W. H. Wittrick, *Quart. J. Mech. Appl. Math.*, vol. 6, 1953.

becomes unstable and the strip buckles suddenly downward as indicated by the dotted line. In a thermostat the source of heating is cut off at the instant of buckling, and cooling begins, which results finally in buckling of the strip in the upward direction, thus turning on the source of heat again. This phenomenon is analogous to that of buckling of a flat arch during loading and unloading as discussed in the preceding article. To determine the temperature at which buckling occurs we need use only the deflection (c) instead of the deflection produced by loading. Equation (7-28) becomes

$$\frac{l^2}{8\rho a} = 1 + \sqrt{\frac{4}{27}\frac{(1-m)^3}{m^2}}$$

Using expression (7-29) for the curvature, the temperature t_1, at which buckling of the strip in a downward direction occurs, is obtained from the equation

$$\frac{3}{16}\frac{l^2}{ah}(\epsilon_2 - \epsilon_1)(t_1 - t_0) = 1 + \sqrt{\frac{4}{27}\frac{(1-m)^3}{m^2}} \tag{7-30}$$

The temperature t_2, at which the strip buckles in an upward direction after cooling, is obtained from the first of the inequalities (k) of the previous article. Hence we obtain

$$\frac{3}{16}\frac{l^2}{ah}(\epsilon_2 - \epsilon_1)(t_2 - t_0) = 1 - \sqrt{\frac{4}{27}\frac{(1-m)^3}{m^2}} \tag{7-31}$$

Taking, for instance, $l/h = 100$, $\epsilon_2 - \epsilon_1 = 4 \times 10^{-6}$, and the values of $m = h^2/3a^2$ as given in the first line of Table 7-8, the ratios of the initial rise a of the strip to its

TABLE 7-8. NUMERICAL DATA ON BUCKLING OF BIMETALLIC STRIPS

m	$\frac{2}{3}$	$\frac{1}{2}$	$\frac{1}{3}$	$\frac{1}{4}$
a/h	0.707	0.814	1.000	1.154
$t_1 - t_0$	104°	137°	217°	307°
$t_2 - t_0$	83°	79°	50°	0°

thickness are as given in the next line of the table. The values of $t_1 - t_0$, calculated from Eq. (7-30), and $t_2 - t_0$, calculated from Eq. (7-31), expressed in degrees centigrade, are in the last two lines of the table. It is seen from this table that the temperature t_1 of buckling increases with an increase of the initial curvature and that at the same time the temperature t_2 decreases, so that the sensitivity of the thermostat, given by the difference $t_1 - t_2$, decreases rapidly with an increase of the temperature of action t_1. These conditions can be improved by introducing elastic supports for the bimetallic strip.

If, under the action of the thrust H, the distance between the supports increases by some quantity βH, proportional to H, the equation for calculating the thrust becomes

$$\frac{Hl}{AE} + \beta H = \frac{1}{2}\int_0^l \left(\frac{dy}{dx}\right)^2 dx - \frac{1}{2}\int_0^l \left(\frac{dy_2}{dx}\right)^2 dx$$

and, instead of Eq. (g) of the preceding article, we obtain

$$(1-u)^2 = (1 - m_1\alpha)(1-\alpha)^2 \tag{d}$$

in which
$$m_1 = m\left(1 + \frac{\beta AE}{l}\right) \tag{e}$$

It is seen that the operation of a bimetallic strip with elastic supports can be investigated in the same manner as before. It is only necessary to introduce, instead of the quantity m, a new quantity m_1 depending on the elasticity of supports. By a suitable adjustment of this elasticity the desired sensitivity of the thermostat can be obtained.[1]

7.10. Lateral Buckling of a Curved Bar with Circular Axis. Lateral buckling of a bar bent in the plane of its greatest rigidity, which has been discussed in Chap. 6 for the case of a straight bar, may occur also if the longitudinal axis of the bar is curved. To establish formulas for the critical loads in such cases, let us consider first bending of a curved strip out of its plane of initial curvature and let us develop the necessary differential equations for such bending.[2]

A simple problem of this kind is shown in Fig. 7-21. A curved bar AB of narrow rectangular cross section with its center line in the horizontal plane DAB is built in at A and is bent by a load distributed in an arbitrary way along the axis AB. If the deflections are small, the deformed shape of the bar is completely determined by the displacement of the centroid of each cross section and the rotation of each cross section about the tangent to the center line. For any cross section of the bar defined by the angle ψ, we take a system of rectangular coordinates with the origin at the centroid O and the axes so directed that x and y coincide with the principal axes of the cross

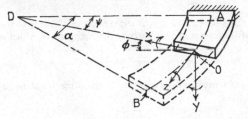

Fig. 7-21

section, while z coincides with the tangent to the center line. It is assumed that initially the plane xz coincides with the plane of curvature of the bar, that the positive direction of x is toward the center of curvature, that z is taken positive in the direction corresponding to an increase of the angle ψ, and that the arc s of the center line is measured from the built-in end A. The displacement of the centroid O is resolved into three components, u, v, and w in the directions of the x, y, and z axes, respectively. The angle of rotation of the cross section about the z axis is called ϕ and is taken positive when rotation is in the direction indicated in the figure. The deformation of an element of the curved bar cut out by two adjacent cross sections consists, generally, of bending in each of the two principal planes xz and yz and of twist about the z axis. Let $1/\rho_1$ and $1/\rho_2$ be the curvatures of the center line at O after deformation in the principal planes yz and xz, respectively, and let θ be the angle of twist per unit length at the same point. Then, if $1/R$ denotes the initial curvature of the center line of the bar, the equations for calculating the curvatures and the twist are

$$\frac{EI_x}{\rho_1} = M_x \qquad EI_y \left(\frac{1}{\rho_2} - \frac{1}{R} \right) = M_y \qquad C\theta = M_z \qquad (a)$$

[1] See Timoshenko, *loc. cit.*

[2] The theory of small deformation of naturally curved bars was developed by Saint-Venant; see his series of papers in *Compt. rend.*, Paris, vol. 17, 1843. See also A. E. H. Love, "Mathematical Theory of Elasticity," 4th ed., p. 444, 1927.

in which M_x, M_y, and M_z are the moments on the cross section at O about the x, y, and z axes, respectively, positive as shown in the figure; EI_x and EI_y are the two principal flexural rigidities; and C is the torsional rigidity of the bar.

To obtain the differential equations for calculating the displacements u, v, w and the angle ϕ, it is necessary to establish the expressions for the curvatures and the unit twist θ as functions of u, v, w, and ϕ and to substitute these expressions in Eqs. (a). In the case of small displacements, we can consider separately each component of the displacements and obtain the final change in curvature and unit twist by summing up the effects produced by the individual components.

The components u and w represent displacements in the plane of initial curvature of the bar. They produce only a change of curvature in the plane xz, which was discussed in Art. 7.1. From that discussion we have

$$\frac{1}{\rho_2} = \frac{1}{R} + \frac{u}{R^2} + \frac{d^2u}{ds^2} \qquad (b)$$

Let us consider now the angular displacement ϕ. It can be seen at once that this displacement produces a unit twist

$$\theta = \frac{d\phi}{ds} \qquad (c)$$

It also produces some bending in the principal plane yz. Owing to the rotation ϕ the surface of the bar becomes a conical surface, the curvature of which is

$$\frac{\sin \phi}{R} \approx \frac{\phi}{R} \qquad (d)$$

The displacement v produces a curvature in the principal plane yz of the amount

$$-\frac{d^2v}{ds^2} \qquad (e)$$

analogous to that for a straight bar. The minus sign follows from our assumption regarding the positive signs of moments and curvatures as given by Fig. 7-21 and by

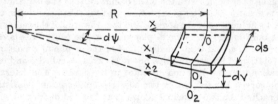

Fig. 7-22

Eqs. (a). It will be seen from these that for a positive M_z and a positive $1/\rho_1$ there is bending in the yz plane concave upward and positive values of d^2v/ds^2 correspond to bending with concavity downward; hence we use the minus sign in expression (e).

It can be seen also, from Fig. 7-22, that the displacement v produces a certain amount of twist. This figure represents an element of the curved bar between two adjacent cross sections O and O_1 a distance ds apart. Owing to the displacements v, this element rotates with respect to the axis Ox through the angle dv/ds. Because of this rotation, the axis O_1x_1 of the adjacent cross section O_1 comes into the position O_2x_2. The angle between O_1x_1 and O_2x_2 is equal to dv/R. Hence, the angle of twist per unit length, equal to

$$\frac{1}{R}\frac{dv}{ds} \qquad (f)$$

occurs owing to the displacements v if the angle ϕ is kept equal to zero along the length of the bar.

Summing up the results given by formulas (b) to (f), the curvatures and the twist of a curved bar after deformation are given, in the general case, by the following equations:

$$\frac{1}{\rho_1} = \frac{\phi}{R} - \frac{d^2v}{ds^2}$$
$$\frac{1}{\rho_2} = \frac{1}{R} + \frac{u}{R^2} + \frac{d^2u}{ds^2} \qquad (7\text{-}32)$$
$$\theta = \frac{d\phi}{ds} + \frac{1}{R}\frac{dv}{ds}$$

Substituting these into Eqs. (a), three equations for determining the displacements are obtained.

Having these equations, let us consider the problem of buckling of a bar of narrow rectangular cross section with a circular axis submitted to bending by two equal and opposite couples M_0 acting in the plane of the strip (Fig. 7-23).[1] We assume that the

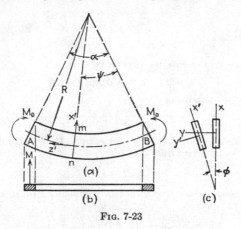

Fig. 7-23

ends of the bar are simply supported; that is, the ends can rotate freely with respect to their principal axes of inertia, but they cannot rotate with respect to tangents to the center line of the bar at A and B.   In calculating the critical value of the couples M_0, we assume that a small lateral buckling has occurred and we determine the value of M_0 necessary to keep the bar in such a buckled form.   Taking any cross section mn, which rotates during buckling through an angle ϕ, as shown in Fig. 7-23c, so that the axes x, y, and z take the directions x', y', and z', and considering the portion of the strip to the left of this cross section, the projections on the x', y', and z' axes of the vector M_0, applied at the left end of the strip and perpendicular to the initial plane of the strip, are, respectively,

$$M_{x'} = M_0\phi \qquad M_{y'} = M_0 \qquad M_{z'} = M_0\frac{dv}{ds}$$

[1] See Timoshenko, *Bull. Polytech. Inst., Kiev*, 1910.

Using these expressions with Eqs. (a) and using Eqs. (7-32), we obtain

$$\phi M_0 = EI_z\left(\frac{\phi}{R} - \frac{d^2v}{ds^2}\right)$$

$$M_0 = EI_y\left(\frac{u}{R^2} + \frac{d^2u}{ds^2}\right) \tag{g}$$

$$M_0\frac{dv}{ds} = C\left(\frac{d\phi}{ds} + \frac{1}{R}\frac{dv}{ds}\right)$$

Eliminating v from the first and third of these equations, we obtain the following equation for the angle ϕ:

$$EI_zC\frac{d^2\phi}{ds^2} - \left(M_0 - \frac{C}{R}\right)\left(\frac{EI_z}{R} - M_0\right)\phi = 0 \tag{h}$$

In the case where $R = \infty$, Eq. (h) coincides with Eq. (6-12) for a straight bar of narrow rectangular cross section. Using the notation

$$k^2 = -\frac{1}{EI_zC}\left(M_0 - \frac{C}{R}\right)\left(\frac{EI_z}{R} - M_0\right) \tag{i}$$

Eq. (h) becomes

$$\frac{d^2\phi}{ds^2} + k^2\phi = 0$$

and we obtain

$$\phi = A\sin ks + B\cos ks \tag{j}$$

From the conditions at the ends, we conclude that

$$\phi = 0 \qquad \text{at } s = 0 \text{ and } s = \alpha R$$

These conditions give $B = 0$ and

$$\sin k\alpha R = 0 \tag{k}$$

From this trigonometric equation the critical value of M_0 is obtained. The smallest root of this equation, different from zero, is

$$k\alpha R = \pi$$

from which, by using notation (i), we obtain the following equation for calculating M_{cr}:

$$M_{cr}^2 - \frac{EI_z + C}{R}M_{cr} + \frac{EI_zC}{R^2}\left(1 - \frac{\pi^2}{\alpha^2}\right) = 0$$

The two real roots of this equation are

$$M_{cr} = \frac{EI_z + C}{2R} \pm \sqrt{\left(\frac{EI_z - C}{2R}\right)^2 + \frac{EI_zC}{R^2}\frac{\pi^2}{\alpha^2}} \tag{7-33}$$

Substituting in this solution $R = \infty$ and $R\alpha = l$, we obtain

$$M_{cr} = \pm\frac{\pi}{l}\sqrt{EI_zC}$$

which coincides with formula (6-13) obtained for a straight bar.

When the angle α and the initial curvature of the center line of the bar (Fig. 7-23) are small, the first term under the radical in formula (7-33) can be neglected in comparison with the second term; then, by substituting $\alpha R = l$, we obtain

$$M_{cr} = \frac{EI_x + C}{2R} \pm \frac{\pi}{l} \sqrt{EI_x C} \tag{7-34}$$

The plus sign in this formula corresponds to the directions of the moments shown in Fig. 7-23, and the minus sign to the reversed directions of the moments. Thus a slight curvature in the direction indicated in the figure increases the value of the critical moment as compared with that for a straight bar of the same length. A curvature in the opposite direction decreases the critical value of the moment.

When $\alpha = \pi$, one of the two values of the moment given by formula (7-33) becomes zero. This result corresponds to the freedom of a semicircular strip to rotate about the diameter joining the ends. For $\alpha > \pi$ both values of M_{cr} from formula (7-33) become positive; to obtain negative values of M_{cr} it is necessary to consider higher roots of Eq. (k).

The problem of lateral buckling of a bar with a circular axis can be solved also[1] if the bar is submitted to the action of a continuous load of intensity q uniformly distributed along the center line and radially directed (Fig. 7-24). If the ends of the bar are simply supported, that is, free to rotate with respect to their principal axes but unable to rotate with respect to the tangents to the center line of the bar at A

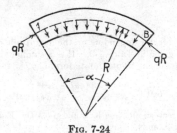

FIG. 7-24

and B, the critical value of the compressive force in the bar, at which lateral buckling occurs, is given by the following formula:

$$q_{cr}R = \frac{EI_x}{R^2} \frac{(\pi^2 - \alpha^2)^2}{\alpha^2[\pi^2 + \alpha^2(EI_x/C)]} \tag{7-35}$$

In the derivation of this formula it is assumed that the directions of the loads q do not change during lateral buckling and that they are displaced laterally only, remaining parallel to their initial direction.

When α is very small and we put $\alpha R = l$, formula (7-35) gives

$$q_{cr}R = \frac{\pi^2 EI_x}{l^2}$$

i.e., we obtain the known Euler formula. When $\alpha = \pi/2$, we obtain from (7-35)

$$q_{cr}R = \frac{EI_x}{R^2} \frac{9}{4 + EI_x/C} \tag{7-36}$$

[1] See Timoshenko, *Z. angew. Math. u. Mech.*, vol. 3, p. 358, 1923. Special problems in connection with the lateral stability of bridge arches are discussed by L. Östlund, *Trans. Roy. Inst. Technol. Stockholm*, no. 84, 1954.

This represents the critical compressive force for a complete ring which buckles under the action of radial pressure in four half-waves so that each half-wave corresponds to $\alpha = \pi/2$. When $\alpha = \pi$, the critical load becomes zero, since in this case the bar can rotate freely with respect to the diameter joining the ends of the bar.

If we assume that during buckling of the bar the loads q change their directions slightly so that they are directed always toward the center of the initial curvature of the bar, the critical value of the compressive force is

$$q_{cr}R = \frac{\pi^2 EI_x}{R^2} \frac{\pi^2 - \alpha^2}{\alpha^2[\pi^2 + \alpha^2(EI_x/C)]} \tag{7-37}$$

In the case of a complete circular ring buckling into four half-waves ($\alpha = \pi/2$), this formula gives[1]

$$q_{cr}R = \frac{EI_x}{R^2} \frac{12}{4 + EI_x/C} \tag{7-38}$$

It is seen that, owing to the assumed slight changes in the direction of the loads, the stability of the ring increases considerably.

If the ends of the bar (Fig. 7-24) are built in and if the loads retain their directions during buckling, as assumed in the first case, the critical value of the compressive force is given by the formula

$$q_{cr} = \gamma_5 \frac{EI_x}{R^3} \tag{7-39}$$

in which γ_5 is a numerical factor depending on the magnitude of the angle α. Several values of this factor are given in Table 7-9. If α is smaller than $\pi/2$, formula (7-39) can be replaced by the following approximate formula:

$$q_{cr}R = \frac{EI_x}{R^2} \frac{(4\pi^2 - \alpha^2)^2}{\alpha^2[4\pi^2 + \alpha^2(EI_x/C)]}$$

In the case where α is very small, this formula gives the Euler value of the critical load for a bar with built-in ends.

TABLE 7-9. VALUES OF THE FACTOR γ_5 IN EQ. (7-39)

$\dfrac{\alpha}{\pi}$	0.25	0.5	1	1.063	1.10	1.24	1.50	2
γ_5	60.1	12.6	1.85	1.54	1.40	1.00	0.69	0.60

In the treatment of problems of buckling of a bar with a circular axis, the energy method can be used also in the calculation of critical loads. It can be applied, for instance, in an approximate investigation of stability of a bar having I cross section, in which case the integration of the differential equations of equilibrium, analogous to Eqs. (g), becomes very complicated.

[1] This formula was obtained by H. Hencky, *Z. angew. Math. u. Mech.*, vol. 1, p. 451, 1921.

CHAPTER 8

BENDING OF THIN PLATES

8.1. Pure Bending of Plates. In the case of pure bending of a prismatic bar, a rigorous solution for the stress distribution is obtained by assuming that cross sections of the bar remain plane during bending and rotate only with respect to their neutral axes so as to be always normal to the deflection curve. A combination of such bending in two perpendicular directions occurs in pure bending of plates. Let us begin with pure bending of a rectangular plate by moments that are uniformly distributed along the sides of the plate as shown in Fig. 8-1. The plane midway between the faces of the plate, the so-called *middle plane* of the plate, will be taken as the xy plane, and the x and y axes are directed along the

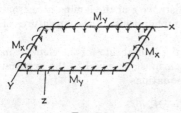

FIG. 8-1

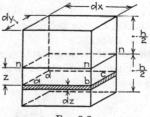

FIG. 8-2

edges as shown. The z axis is taken perpendicular to the middle plane and positive in a downward direction. The bending moment per unit length of the edges parallel to the y axis is denoted by M_x, and the moment per unit length of the edges parallel to the x axis is M_y. These moments are considered positive when they produce compression at the upper surface of the plate and tension at the lower. The thickness of the plate is denoted by h and is considered small in comparison with the other dimensions.

Let us consider an element cut out of the plate by two pairs of planes parallel to the xz and yz planes as shown in Fig. 8-2. Assuming that during bending of the plate the lateral sides of this element remain plane and rotate about the neutral axes n–n so as to remain normal to the deflection surface, it can be concluded that the middle plane of the plate does

319

not undergo any deformation during this bending and is therefore a *neutral surface*. Let $1/\rho_z$ and $1/\rho_y$ denote the curvatures of this neutral surface in sections parallel to the zx and yz planes, respectively, with positive curvature corresponding to bending which is convex down. Then the unit elongations in the x and y directions of an elemental lamina *abcd* (Fig. 8-2) at distance z from the neutral surface can be found as in the case of a beam. Thus we obtain

$$\epsilon_x = \frac{z}{\rho_x} \qquad \epsilon_y = \frac{z}{\rho_y} \qquad (a)$$

From Hooke's law we have the relations

$$\epsilon_x = \frac{1}{E}(\sigma_x - \nu\sigma_y) \qquad \epsilon_y = \frac{1}{E}(\sigma_y - \nu\sigma_x) \qquad (b)$$

where ν is Poisson's ratio, and therefore the corresponding stresses in the lamina *abcd* are

$$\sigma_x = \frac{Ez}{1 - \nu^2}\left(\frac{1}{\rho_x} + \nu\frac{1}{\rho_y}\right)$$

$$\sigma_y = \frac{Ez}{1 - \nu^2}\left(\frac{1}{\rho_y} + \nu\frac{1}{\rho_x}\right) \qquad (c)$$

These stresses are proportional to the distance z of the lamina *abcd* from the neutral surface and depend upon the magnitudes of the curvatures of the bent plate.

The normal stresses distributed over the lateral sides of the element in Fig. 8-2 can be reduced to couples which must be equal to the external moments. In this way we obtain the equations

$$\int_{-h/2}^{h/2} \sigma_x z \, dy \, dz = M_x \, dy$$

$$\int_{-h/2}^{h/2} \sigma_y z \, dx \, dz = M_y \, dx \qquad (d)$$

Substituting expressions (c) for σ_x and σ_y then gives

$$M_x = D\left(\frac{1}{\rho_x} + \nu\frac{1}{\rho_y}\right) \qquad (8\text{-}1)$$

$$M_y = D\left(\frac{1}{\rho_y} + \nu\frac{1}{\rho_x}\right) \qquad (8\text{-}2)$$

where $\qquad D = \dfrac{E}{1 - \nu^2}\displaystyle\int_{-h/2}^{h/2} z^2 \, dz = \dfrac{Eh^3}{12(1 - \nu^2)} \qquad (8\text{-}3)$

This quantity is called the *flexural rigidity of the plate*.

The assumption that there is no strain in the middle plane of a plate during bending is usually sufficiently accurate as long as the deflections of the plate are small in comparison with its thickness h. If this condition is not satisfied, some deformation in the middle surface of the plate usually will be produced[1] and should be considered in investigating the stress distribution in the plate.   This more complicated problem of bending of plates will be discussed later (see Art. 8.7).

The deflection of the plate being denoted by w, the approximate formulas for curvatures of the plate, analogous to the well-known formula for curvature of a beam, are

$$\frac{1}{\rho_x} = -\frac{\partial^2 w}{\partial x^2} \quad \frac{1}{\rho_y} = -\frac{\partial^2 w}{\partial y^2}$$

Substituting in Eqs. (8-1) and (8-2), we obtain

$$M_x = -D\left(\frac{\partial^2 w}{\partial x^2} + \nu \frac{\partial^2 w}{\partial y^2}\right) \tag{8-4}$$

$$M_y = -D\left(\frac{\partial^2 w}{\partial y^2} + \nu \frac{\partial^2 w}{\partial x^2}\right) \tag{8-5}$$

These equations define the deflection surface of the plate provided the moments M_x and M_y are given.   In the particular case where $M_y = 0$, the rectangular plate (Fig. 8-1) is bent as a beam.   From Eq. (8-5) we have for this case

$$\frac{\partial^2 w}{\partial y^2} = -\nu \frac{\partial^2 w}{\partial x^2}$$

The plate has two curvatures of opposite signs, so that it is bent into an *anticlastic surface*.

When $M_x = M_y = M$, the curvatures of the deflection surface in two perpendicular directions are equal and the surface is spherical.   The curvature of the sphere, from Eq. (8-1), is

$$\frac{1}{\rho} = \frac{M}{D(1 + \nu)} \tag{8-6}$$

Let us consider now, in the case of a bent plate, the stresses acting on a section parallel to the z axis and inclined to the x and y axes.   If acd (Fig. 8-3) represents a portion of the thin lamina $abcd$ (Fig. 8-2), cut by such a section, the stress acting on the side ac can be found from equations of static equilibrium.   When this stress is resolved into a normal component σ_n and a shearing component τ_{nt}, the magnitudes of these components are given by the known equations

$$\begin{aligned}\sigma_n &= \sigma_x \cos^2 \alpha + \sigma_y \sin^2 \alpha \\ \tau_{nt} &= \tfrac{1}{2}(\sigma_y - \sigma_x) \sin 2\alpha\end{aligned} \tag{e}$$

---

[1] Considerable deflections without strain in the middle surface of the plate are possible only if the deflection surface is a developable surface, such as a cylindrical or conical surface.

in which α is the angle between the normal n and the x axis or between the direction t and the y axis (Fig. 8-3a). This angle is considered positive if measured in a clockwise direction.

Considering all laminae, such as acd, over the thickness of the plate (Fig. 8-3b), the normal stresses σ_n give the bending moment acting on the section ac of the plate, the magnitude of which per unit length along ac is

$$M_n = \int_{-h/2}^{h/2} \sigma_n z \, dz = M_x \cos^2 \alpha + M_y \sin^2 \alpha \qquad (8\text{-}7)$$

The shearing stresses τ_{nt} give the twisting moment acting on the section ac of the plate, the magnitude of which per unit length of ac is

$$M_{nt} = -\int_{-h/2}^{h/2} \tau_{nt} z \, dz = \tfrac{1}{2} \sin 2\alpha (M_x - M_y) \qquad (8\text{-}8)$$

The signs of M_n and M_{nt} are chosen in such a manner that the positive values of these moments are represented by vectors in the positive directions of n and t if the right-hand rule is used, as shown in Fig. 8-3a.

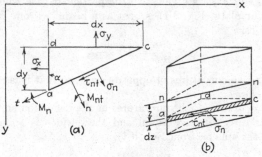

FIG. 8-3

When α is zero or π, Eq. (8-7) gives $M_n = M_x$. For $\alpha = \pi/2$ or $3\pi/2$, we obtain $M_n = M_y$. The moments M_{nt} become zero for these values of α. Thus we obtain the conditions shown in Fig. 8-1.

Equations (8-7) and (8-8) are similar to Eqs. (e) and can be used to calculate the bending and twisting moments for any value of α. We can solve also without any difficulty the problem in which M_n and M_{nt} are given for two sections perpendicular to each other and it is required to find the two values of α defining the *principal planes*, i.e., the planes on which only bending moments M_x and M_y act and on which the twisting moment is zero.

Let us now represent M_n and M_{nt} as functions of the deflections w of the plate. From the assumption that the sides of an element, as shown in Fig. 8-2, remain plane during bending of the plate and rotate only with respect to the neutral axes n–n, remaining normal to the deflection surface, it follows that every linear element perpendicular to the middle plane of the plate remains straight during bending and becomes normal to the deflection surface of the plate. The unit elongations of the fibers parallel to n and t (Fig. 8-3a) and at a distance z from the middle plane are expressed by the equations

$$\epsilon_n = \frac{z}{\rho_n} \qquad \epsilon_t = \frac{z}{\rho_t} \tag{f}$$

in which ρ_n and ρ_t denote the radii of curvature of the deflection surface in the nz and tz planes. Equations (f) are analogous to Eqs. (a), and together with Hooke's law [see Eqs. (b)], they yield

$$\sigma_n = \frac{Ez}{1 - \nu^2}\left(\frac{1}{\rho_n} + \nu\frac{1}{\rho_t}\right)$$

Substituting this in Eq. (8-7), we find that

$$M_n = D\left(\frac{1}{\rho_n} + \nu\frac{1}{\rho_t}\right)$$

or, by using approximate expressions for the curvatures,

$$M_n = -D\left(\frac{\partial^2 w}{\partial n^2} + \nu\frac{\partial^2 w}{\partial t^2}\right) \tag{8-9}$$

This is an equation analogous to Eqs. (8-4) and (8-5) obtained before. We arrive at the same result also by substituting expressions (8-4) and (8-5) for M_x and M_y in Eq. (8-7) and by using the known relations between the curvatures of a surface in two perpendicular directions:

$$\frac{1}{\rho_n} = \frac{1}{\rho_x}\cos^2\alpha + \frac{1}{\rho_y}\sin^2\alpha$$
$$\frac{1}{\rho_t} = \frac{1}{\rho_x}\sin^2\alpha + \frac{1}{\rho_y}\cos^2\alpha \tag{g}$$

To obtain an expression for the twisting moment M_{nt}, let us consider the distortion of a thin lamina $abcd$ with the sides ab and ad parallel to the n and t directions and at a distance z from the middle plane (Fig. 8-4a). During bending of the plate the points a, b, c, and d undergo small displacements. The components of the displacement of the point a in the n and t directions we denote by u_1 and v_1. Then the displacement of the adjacent point d in the n direction is $u_1 + (\partial u_1/\partial t)\,dt$ and the displacement of the point b in the t direction is $v_1 + (\partial v_1/\partial n)\,dn$. Owing to these displacements, we obtain for the shearing strain

$$\gamma_{nt} = \frac{\partial u_1}{\partial t} + \frac{\partial v_1}{\partial n} \tag{h}$$

The corresponding shearing stress is

$$\tau_{nt} = G \left(\frac{\partial u_1}{\partial t} + \frac{\partial v_1}{\partial n} \right) \tag{i}$$

From Fig. 8-4b, representing the section of the deflection surface made by a vertical plane through the n axis, it can be seen that the angle of rotation in this plane of an element pq, which initially was perpendicular to the xy plane, is equal to $\partial w/\partial n$. Owing to this rotation a point of the element at a distance z from the neutral surface has a displacement in the n direction equal to

$$u_1 = -z \frac{\partial w}{\partial n}$$

Considering the section of the plate made by a vertical plane through the

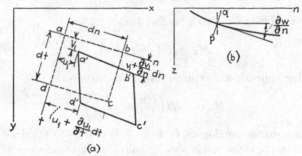

FIG. 8-4

t axis, it can be shown that the same point has a displacement in the t direction equal to

$$v_1 = -z \frac{\partial w}{\partial t}$$

Substituting these values of the displacements u_1 and v_1 in expression (i), we find that

$$\tau_{nt} = -2Gz \frac{\partial^2 w}{\partial n\,\partial t} \tag{8-10}$$

and the expression (8-8) for the twisting moment becomes

$$M_{nt} = -\int_{-h/2}^{h/2} \tau_{nt} z\, dz = \frac{Gh^3}{6} \frac{\partial^2 w}{\partial n\,\partial t} = D(1 - \nu) \frac{\partial^2 w}{\partial n\,\partial t} \tag{8-11}$$

The expressions (8-9) and (8-11) for bending and twisting moments will be used later in discussing more general cases of bending of plates.

In the above discussion of pure bending of plates, we started with the case of a rectangular plate with uniformly distributed bending moments acting along the edges. To obtain a general case of pure bending of plates, let us imagine that a portion of any shape is cut out from the plate discussed above by a cylindrical surface perpendicular to the plate. The conditions of bending of this portion will remain unchanged provided the bending and twisting moments in the amounts given by Eqs. (8-7) and (8-8) are distributed along the boundary of the isolated portion of the plate. Thus we arrive at the case of pure bending of a plate of any shape, and we conclude that pure bending of a plate is always produced if along the edges of the plate bending moments M_n and twisting moments M_{nt} are distributed in the manner given by formulas (8-7) and (8-8). Taking, for instance, the particular case where $M_x = M_y = M$, it can be concluded from Eqs. (8-7) and (8-8) that a plate of any shape will be bent to a spherical surface if uniformly distributed bending moments M act along the edges. Another particular case is obtained by taking $M_x = -M_y = M$. Cutting out from the plate shown in Fig. 8-1 a rectangular plate with sides which make angles of 45° with the x and y axes and substituting $\alpha = \pi/4$ or $3\pi/4$ in Eqs. (8-7) and (8-8), we obtain $M_n = 0$ for both directions, $M_{nt} = M$ for $\alpha = \pi/4$, and $M_{nt} = -M$ for $\alpha = 3\pi/4$. Hence in this case we produce pure bending in a rectangular plate by applying twisting moments uniformly distributed along the edges.

Regarding the stresses in a plate undergoing pure bending, it can be concluded from the first of Eqs. (e) that the maximum normal stress acts on those sections parallel to the xz or yz planes. The magnitudes of these stresses are obtained from Eqs. (c) by substituting $z = h/2$ and by using Eqs. (8-1), (8-2), and (8-3). In this way we find that

$$(\sigma_x)_{\max} = \frac{6M_x}{h^2} \qquad (\sigma_y)_{\max} = \frac{6M_y}{h^2} \qquad (8\text{-}12)$$

If these stresses are of opposite signs, the maximum shearing stress acts in the plane bisecting the angle between the xz and yz planes and is equal to

$$\tau_{\max} = \frac{1}{2}(\sigma_x - \sigma_y) = \frac{3(M_x - M_y)}{h^2}$$

If the stresses (8-12) are of the same sign, the maximum shear acts in the plane bisecting the angle between the xy and xz planes or in that bisecting the angle between the xy and yz planes and is equal to $\frac{1}{2}(\sigma_y)_{\max}$ or $\frac{1}{2}(\sigma_x)_{\max}$ depending on which of the two normal stresses $(\sigma_y)_{\max}$ or $(\sigma_x)_{\max}$ is greater.

8.2. Bending of Plates by Distributed Lateral Load. In considering the case of bending of a plate by a distributed load acting perpendicular to the middle plane of the plate, let us assume that this middle plane is horizontal and contains the x and y axes while the z axis is directed vertically downward. We denote by q the intensity of the load, which, in general, may vary along the surface of the plate and therefore is considered as a function of x and y. Cutting an element from the plate by two pairs of planes parallel to the xz and yz planes (Fig. 8-5), it can be concluded from considerations of statics that, owing to the action of the load q, there will be produced on the lateral sides of this element not only bending and twisting moments, as discussed in the previous article, but also vertical shearing forces, the magnitudes of which per unit of length will be defined by the following formulas:

$$Q_x = \int_{-h/2}^{h/2} \tau_{xz} \, dz \qquad Q_y = \int_{-h/2}^{h/2} \tau_{yz} \, dz \qquad (a)$$

The variation of τ_{xz} and τ_{yz} along the small distances dy and dx can be neglected, and it will be assumed that the resultant shearing forces $Q_x \, dy$ and $Q_y \, dx$ pass through the centroids of the sides of the element. For

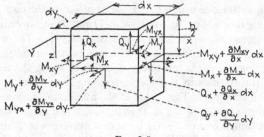

FIG. 8-5

the bending and twisting moments per unit length, we take the same definitions as in the previous article and assume that

$$M_x = \int_{-h/2}^{h/2} \sigma_x z \, dz \qquad M_y = \int_{-h/2}^{h/2} \sigma_y z \, dz \qquad (b)$$

$$M_{xy} = - \int_{-h/2}^{h/2} \tau_{xy} z \, dz \qquad M_{yx} = \int_{-h/2}^{h/2} \tau_{yz} z \, dz \qquad (c)$$

The shearing forces (a), bending moments (b), and twisting moments (c) are all functions of the coordinates x and y. Hence, in using the notations Q_x, M_x, and M_{xy} for the left side of the element in Fig. 8-5, we shall have for the right side of the element, distance dx from the left side, the corresponding quantities equal to

$$Q_x + \frac{\partial Q_x}{\partial x} \, dx \qquad M_x + \frac{\partial M_x}{\partial x} \, dx \qquad M_{xy} + \frac{\partial M_{xy}}{\partial x} \, dx$$

as shown in the figure. An analogous conclusion will be obtained also for the sides of the element parallel to the xz plane.

In considering the conditions of equilibrium of the element, we observe that all forces acting on it are parallel to the z axis and that the couples are represented by vectors perpendicular to the z axis. Hence we have only three equations of static equilibrium to consider: the projections of all forces on the z axis and the moments of all forces with respect to the x and y axes. Noting the directions of forces indicated in Fig. 8-5, we find that their projections on the z axis give the equation

$$\frac{\partial Q_x}{\partial x} dx\, dy + \frac{\partial Q_y}{\partial y} dy\, dx + q\, dx\, dy = 0$$

or, after simplification,

$$\frac{\partial Q_x}{\partial x} + \frac{\partial Q_y}{\partial y} + q = 0 \tag{d}$$

The weight of the plate itself can be considered as included in the value of q. By taking the moments of all forces acting on the element with respect to the x axis and observing the directions indicated in the figure, we obtain

$$\frac{\partial M_{xy}}{\partial x} dx\, dy - \frac{\partial M_y}{\partial y} dy\, dx + Q_y\, dx\, dy = 0$$

The moment of the load q and the moment due to change in the force Q_y are neglected in deriving this equation, since they are small quantities of a higher order than those which are retained. After simplification, the equation becomes

$$\frac{\partial M_{xy}}{\partial x} - \frac{\partial M_y}{\partial y} + Q_y = 0 \tag{e}$$

In the same way, by taking moments with respect to the y axis, we obtain

$$\frac{\partial M_{yx}}{\partial y} + \frac{\partial M_x}{\partial x} - Q_x = 0 \tag{f}$$

Determining Q_x and Q_y from Eqs. (f) and (e) and substituting in Eq. (d), we obtain

$$\frac{\partial^2 M_x}{\partial x^2} + \frac{\partial^2 M_{yx}}{\partial x\, \partial y} + \frac{\partial^2 M_y}{\partial y^2} - \frac{\partial^2 M_{xy}}{\partial x\, \partial y} = -q$$

Since $\tau_{xy} = \tau_{yx}$, we observe that $M_{yx} = -M_{xy}$, and hence we finally obtain the following equation of equilibrium:

$$\frac{\partial^2 M_x}{\partial x^2} - \frac{2\partial^2 M_{xy}}{\partial x\, \partial y} + \frac{\partial^2 M_y}{\partial y^2} = -q \tag{g}$$

Neglecting the effect of shearing forces Q_x and Q_y on the curvatures of the plate[1] and using Eqs. (8-9) and (8-11), developed for the case of pure bending, we obtain the following expressions for bending and twisting moments:

$$M_x = -D\left(\frac{\partial^2 w}{\partial x^2} + \nu\frac{\partial^2 w}{\partial y^2}\right) \qquad M_y = -D\left(\frac{\partial^2 w}{\partial y^2} + \nu\frac{\partial^2 w}{\partial x^2}\right) \qquad (8\text{-}13)$$

$$M_{xy} = -M_{yx} = D(1 - \nu)\frac{\partial^2 w}{\partial x\,\partial y} \qquad (8\text{-}14)$$

Substituting in Eq. (g), we obtain

$$\frac{\partial^4 w}{\partial x^4} + 2\frac{\partial^4 w}{\partial x^2\,\partial y^2} + \frac{\partial^4 w}{\partial y^4} = \frac{q}{D} \qquad (8\text{-}15)$$

Thus the determination of the deflection surface of a plate is reduced to the integration of Eq. (8-15).

If the solution of this equation is found for any particular case, the bending and twisting moments can be calculated from Eqs. (8-13) and (8-14). Then the shearing forces can be obtained from Eqs. (e) and (f). Substituting in the latter equations the expressions (8-13) and (8-14) for bending and twisting moments, we obtain

$$Q_x = \frac{\partial M_x}{\partial x} + \frac{\partial M_{yx}}{\partial y} = -D\frac{\partial}{\partial x}\left(\frac{\partial^2 w}{\partial x^2} + \frac{\partial^2 w}{\partial y^2}\right) \qquad (8\text{-}16)$$

$$Q_y = \frac{\partial M_y}{\partial y} - \frac{\partial M_{xy}}{\partial x} = -D\frac{\partial}{\partial y}\left(\frac{\partial^2 w}{\partial x^2} + \frac{\partial^2 w}{\partial y^2}\right) \qquad (8\text{-}17)$$

Having the bending and twisting moments, the normal stresses $(\sigma_x)_{max}$ and $(\sigma_y)_{max}$ are obtained from Eqs. (8-12). Shearing stresses parallel to the x and y axes are obtained from Eq. (8-10) by taking n and t in the x and y directions, respectively. Shearing stresses parallel to the z axis are obtained by assuming that the shearing forces Q_x and Q_y are distributed along the thickness of the plate following a parabolic law, as in the case of beams of rectangular cross section. Then

$$(\tau_{xz})_{max} = \frac{3}{2h}(Q_x)_{max} \qquad (\tau_{yz})_{max} = \frac{3}{2h}(Q_y)_{max}$$

Thus all stresses can be calculated provided the deflection surface of the plate is known.

---

[1] We know that in the case of a beam this effect is small if the depth of the beam is small in comparison with the span. An analogous conclusion can be made also in the case of a plate if the thickness of the plate is small in comparison with the other dimensions. A more exact theory of bending of plates, in which the effect of shearing stresses on deflections is considered, has been developed by J. H. Michell, *Proc. London Math. Soc.*, vol. 31, 1900, and by A. E. H. Love, "Theory of Elasticity," 4th ed., p. 465, 1927; see also the paper by E. Reissner, *J. Appl. Mech.*, vol. 12, pp. A–68, 1945.

The determination of the deflection surface of a plate requires in each particular case the integration of the partial differential equation (8-15) for a given distribution of the load q and for given conditions at the boundary of the plate. In the following discussion we shall be concerned primarily with rectangular plates, and the various boundary conditions for such plates will now be considered.

Built-in Edge. If the edge of a plate is built in, the deflection along this edge is zero and the tangent plane to the deflection surface along this

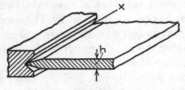

FIG. 8-6

edge coincides with the initial position of the middle plane of the plate. Taking in the middle plane of the plate x and y axes in the directions of the two edges of the plate, and assuming that the edge coinciding with the x axis is built in, we find the boundary conditions along this edge to be

$$(w)_{y=0} = 0 \qquad \left(\frac{\partial w}{\partial y}\right)_{y=0} = 0 \qquad (8\text{-}18)$$

Simply Supported Edge. If the edge $y = 0$ of the plate is simply supported, the deflection w along this edge must be zero. At the same time this edge can rotate freely with respect to the x axis, i.e., there

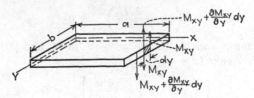

FIG. 8-7

are no bending moments M_y along this edge. This kind of support is represented in Fig. 8-6. The analytical expressions of the boundary conditions in this case are

$$(w)_{y=0} = 0 \qquad \left(\frac{\partial^2 w}{\partial y^2} + \nu \frac{\partial^2 w}{\partial x^2}\right)_{y=0} = 0 \qquad (8\text{-}19)$$

Free Edge. If an edge of a plate, say the edge $x = a$ (Fig. 8-7), is entirely free, it seems natural to assume that along this edge there will be no bending and twisting moments and also no vertical shearing forces; i.e.,

$$(M_x)_{x=a} = 0 \qquad (M_{xy})_{x=a} = 0 \qquad (Q_z)_{x=a} = 0$$

In this form the boundary conditions were discussed by Poisson.[1] But later, Kirchhoff[2] proved that three boundary conditions are too many and that two conditions are sufficient for the complete determination of the deflections w. He showed also that the two requirements regarding the twisting moment and shearing force can be replaced by one boundary condition. The physical meaning of this reduction in the number of boundary conditions has been explained by Thomson and Tait.[3] These authors point out that bending of the plate will not be changed if the horizontal forces giving the twisting couple $M_{xy}\,dy$ acting on an element of length dy of the edge $x = a$ are replaced by two vertical forces M_{xy}, distance dy apart, as shown in Fig. (8-7). Such replacement does not change the magnitude of the twisting moments and produces only local changes in stress distribution at the edge of the plate, leaving the stress condition in the rest of the plate unchanged. Proceeding with the above replacement of twisting couples along the edge of the plate, we shall find, as can be seen from a consideration of the two adjacent elements indicated in the figure, that the distribution of twisting moments M_{xy} is statically equivalent to a distribution of shearing forces of intensity

$$(Q'_x)_{x=a} = -\left(\frac{\partial M_{xy}}{\partial y}\right)_{x=a}$$

Hence the joint requirement regarding twisting moments and shearing forces along the free edge $x = a$ becomes

$$\left(Q_x - \frac{\partial M_{xy}}{\partial y}\right)_{x=a} = 0 \tag{h}$$

Substituting for Q_x and M_{xy} their expressions (8-16) and (8-14), we finally find for the free edge $(x = a)$

$$\left[\frac{\partial^3 w}{\partial x^3} + (2 - \nu)\frac{\partial^3 w}{\partial x\,\partial y^2}\right]_{x=a} = 0 \tag{8-20}$$

The condition that the bending moment along the free edge is zero requires that

$$\left(\frac{\partial^2 w}{\partial x^2} + \nu\frac{\partial^2 w}{\partial y^2}\right)_{x=a} = 0 \tag{8-21}$$

[1] See discussion of this subject in I. Todhunter and K. Pearson, "History of the Theory of Elasticity," vol. 1, p. 250. See also Saint Venant's discussion in "Théorie de l'élasticité des corps solides" by Clebsch, final note to §73, p. 689, 1883.

[2] See *J. de Crelle*, vol. 40, 1850.

[3] See "Natural Philosophy," vol. 1, part 2, p. 188, 1883.

Equations (8-20) and (8-21) represent the two necessary boundary conditions along the free edge $x = a$ of the plate.

Transforming the twisting couples as explained in the above discussion and shown in Fig. (8-7), we obtain not only shearing forces Q'_x distributed along the edge $x = a$ but also two concentrated forces at the ends of that edge, as indicated in Fig. (8-8). The magnitudes of these forces are equal to the magnitudes of the twisting couple M_{xy} at the corresponding

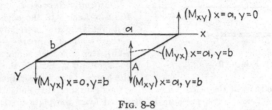

FIG. 8-8

corners of the plate. Making the analogous transformation of twisting couples M_{yx} along the edge $y = b$, we shall find that in this case again, in addition to the distributed shearing forces, there will be concentrated forces M_{yx} at the corners. This indicates that a rectangular plate supported in some way along the edges and laterally loaded will usually produce not only pressures distributed along the boundary but also concentrated pressures at the corners. Regarding the directions of concentrated pressures, a conclusion can be made if the general shape of the deflection surface is known. Take, for instance, a uniformly loaded square plate simply supported along the edges. The general shape of the deflection surface is indicated in Fig. 8-9a by dotted lines representing the sections of the deflection surface in planes parallel to the xz and yz coordinate planes. Considering these lines it is seen that near the

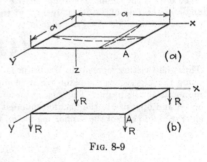

FIG. 8-9

corner A the derivative $\partial w/\partial x$, representing the slope of the deflection surface in the x direction, is negative and numerically decreases with increasing y. Hence $\partial^2 w/\partial x\, \partial y$ is positive at the corner A; from Eq. (8-14) we conclude that M_{xy} is positive and that M_{yx} is negative. From this and from the directions of M_{xy} and M_{yx} (Fig. 8-5), it follows that both concentrated forces, indicated at the corner A in Fig. 8-8, have a downward direction. From symmetry we conclude also that at all four corners of the plate the forces have the same direction and magnitude. Hence the conditions will be as indicated in Fig. 8-9b in which

$$R = 2D(1 - \nu)\left(\frac{\partial^2 w}{\partial x\,\partial y}\right)$$

When a square plate is uniformly loaded, the corners have a tendency to rise, and this is prevented by concentrated reactions at the corners as indicated in the figure.

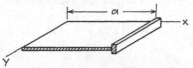

FIG. 8-10

Elastically Supported and Elastically Built-in Edge. If the edge $x = a$ of a rectangular plate is rigidly joined together with a supporting beam (Fig. 8-10), the deflection along this edge is not zero and is equal to the deflection of the beam. Also the rotation of the edge is equal to the twist of the beam. Let EI be the flexural rigidity and C the torsional rigidity of the beam. The pressure transmitted from the plate to the supporting beam, from Eq. (h), is

$$-\left(Q_x - \frac{\partial M_{xy}}{\partial y}\right)_{x=a} = D\,\frac{\partial}{\partial x}\left[\frac{\partial^2 w}{\partial x^2} + (2 - \nu)\,\frac{\partial^2 w}{\partial y^2}\right]_{x=a}$$

and the differential equation of the deflection curve of the beam is

$$EI\left(\frac{\partial^4 w}{\partial y^4}\right)_{x=a} = D\,\frac{\partial}{\partial x}\left[\frac{\partial^2 w}{\partial x^2} + (2 - \nu)\,\frac{\partial^2 w}{\partial y^2}\right]_{x=a} \tag{8-22}$$

This equation represents one of the boundary conditions of the plate along the edge $x = a$.

To obtain the second condition, twisting of the beam should be considered. The angle of rotation of any cross section of the beam is $-(\partial w/\partial x)_{x=a}$, and the rate of change of this angle along the edge is

$$-\left(\frac{\partial^2 w}{\partial x\,\partial y}\right)_{x=a}$$

Thus the twisting moment in the beam is $-C(\partial^2 w/\partial x\,\partial y)_{x=a}$. This moment varies along the edge, since the plate, rigidly connected with the beam, transmits twisting moments to the beam. The magnitude of these moments per unit length must be equal and opposite to the bending moment M_x in the plate. Hence, from the consideration of twist of the supporting beam, we obtain

$$-C\,\frac{\partial}{\partial y}\left(\frac{\partial^2 w}{\partial x\,\partial y}\right)_{x=a} = -(M_x)_{x=a}$$

or, substituting for M_x its expression (8-13),

$$-C\,\frac{\partial}{\partial y}\left(\frac{\partial^2 w}{\partial x\,\partial y}\right)_{x=a} = D\left(\frac{\partial^2 w}{\partial x^2} + \nu\,\frac{\partial^2 w}{\partial y^2}\right)_{x=a} \tag{8-23}$$

This is the second boundary condition at the edge $x = a$ of the plate.

8.3. Combined Bending and Tension or Compression of Plates. In the preceding discussion it was assumed that the plate was bent by lateral loads and that deflections were so small that any stretching of the middle plane of the plate could be neglected. Thus this plane was considered as the neutral plane of the plate. If in addition to lateral load there are forces acting in the middle plane of the plate, stretching of this plane is produced and the corresponding stresses should be considered. We shall distinguish two possible cases: (1) These stresses are small in comparison with the critical buckling stresses, and we can neglect their effect on bending of the plate and assume that the total stresses are obtained with sufficient accuracy by superposing the stresses due to stretching of the middle plane on the stresses produced by lateral load. (2) The stresses in the middle plane of the plate are not small, and their effect on bending of the plate must be considered. In deriving the corresponding differential equation of the deflection surface for this latter case, we consider

again the equilibrium of a small element cut out from the plate by two pairs of planes parallel to the xz and yz coordinate planes. In addition to the forces considered in the previous article (see Fig. 8-5), we shall now have forces acting in the middle plane of the plate, the notations for which per unit length are shown in Fig. 8-11. Projecting these forces on the x and y axes and assuming that there are no body forces acting in those directions, we obtain the following equations of equilibrium:

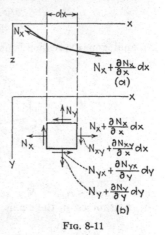

Fig. 8-11

$$\frac{\partial N_x}{\partial x} + \frac{\partial N_{yx}}{\partial y} = 0$$

$$\frac{\partial N_y}{\partial y} + \frac{\partial N_{xy}}{\partial x} = 0$$

(8-24)

It can be seen that these equations are entirely independent of the three equations of equilibrium considered in the previous article and can be treated separately as shown later in Art. 8.7.

In considering the projection of the forces shown in Fig. 8-11 on the z axis, we must take into account the deflection of the plate. Owing to curvature of the plate in the xz plane (Fig. 8-11a), the projection of the normal forces N_x on the z axis gives

$$-N_x \, dy \, \frac{\partial w}{\partial x} + \left(N_x + \frac{\partial N_x}{\partial x} \, dx \right)\left(\frac{\partial w}{\partial x} + \frac{\partial^2 w}{\partial x^2} \, dx \right) dy$$

After simplification and neglecting the small quantities of higher order, this projection will be

$$N_x \, \frac{\partial^2 w}{\partial x^2} \, dx \, dy + \frac{\partial N_x}{\partial x} \, \frac{\partial w}{\partial x} \, dx \, dy$$

(a)

In the same way the projection on the z axis of the normal forces N_y gives

$$N_y \frac{\partial^2 w}{\partial y^2}\, dx\, dy + \frac{\partial N_y}{\partial y} \frac{\partial w}{\partial y}\, dx\, dy \qquad (b)$$

In discussing the projection on the z axis of the shearing forces N_{xy}, let us consider the deflection of an element $dx\, dy$ of the middle plane shown in Fig. 8-12. It is seen that, owing to the angles

$$\frac{\partial w}{\partial y} \quad \text{and} \quad \frac{\partial w}{\partial y} + \frac{\partial^2 w}{\partial x\, \partial y}\, dx$$

the shearing forces N_{xy} have a projection on the z axis equal to

$$N_{xy} \frac{\partial^2 w}{\partial x\, \partial y}\, dx\, dy + \frac{\partial N_{xy}}{\partial x} \frac{\partial w}{\partial y}\, dx\, dy$$

An analogous expression is obtained for the projection on the z axis of the

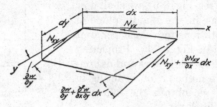

FIG. 8-12

shearing forces $N_{yx} = N_{xy}$, and the final expression for the projection of all shearing forces on the z axis is

$$2N_{xy} \frac{\partial^2 w}{\partial x\, \partial y}\, dx\, dy + \frac{\partial N_{xy}}{\partial x} \frac{\partial w}{\partial y}\, dx\, dy + \frac{\partial N_{xy}}{\partial y} \frac{\partial w}{\partial x}\, dx\, dy \qquad (c)$$

Adding expressions (a), (b), and (c) to the load $q\, dx\, dy$ acting on the element and using Eqs. (8-24), we obtain, instead of Eq. (g) of the previous article, the following equation of equilibrium:

$$\frac{\partial^2 M_x}{\partial x^2} - 2 \frac{\partial^2 M_{xy}}{\partial x\, \partial y} + \frac{\partial^2 M_y}{\partial y^2} = -\left(q + N_x \frac{\partial^2 w}{\partial x^2} + N_y \frac{\partial^2 w}{\partial y^2} + 2N_{xy} \frac{\partial^2 w}{\partial x\, \partial y} \right)$$

Substituting for M_x, M_y, and M_{xy} their expressions (8-13) and (8-14), we obtain[1]

$$\frac{\partial^4 w}{\partial x^4} + 2 \frac{\partial^4 w}{\partial x^2\, \partial y^2} + \frac{\partial^4 w}{\partial y^4}$$
$$= \frac{1}{D}\left(q + N_x \frac{\partial^2 w}{\partial x^2} + N_y \frac{\partial^2 w}{\partial y^2} + 2N_{xy} \frac{\partial^2 w}{\partial x\, \partial y} \right) \qquad (8\text{-}25)$$

[1] This differential equation has been derived by Saint Venant; see *op. cit.*, p. 704.

This differential equation, instead of Eq. (8-15), should be used in determining the deflection surface of a plate if the forces N_x, N_y, and N_{xy} are not small in comparison with the critical values of these forces.

If there are body forces acting in the middle plane of the plate, the differential equations of equilibrium of the element represented in Fig. 8-11 become

$$\frac{\partial N_x}{\partial x} + \frac{\partial N_{xy}}{\partial y} + X = 0$$
$$\frac{\partial N_{xy}}{\partial x} + \frac{\partial N_y}{\partial y} + Y = 0$$

(8-26)

Here X and Y represent the two components of the body force per unit area of the middle plane of the plate.

Adding as before expressions (a), (b), and (c) to the load $q\, dx\, dy$ and using Eqs. (8-26) instead of Eqs. (8-24), we obtain, instead of Eq. (8-25), the following equation:

$$\frac{\partial^4 w}{\partial x^4} + 2\frac{\partial^4 w}{\partial x^2\, \partial y^2} + \frac{\partial^4 w}{\partial y^4}$$
$$= \frac{1}{D}\left(q + N_x \frac{\partial^2 w}{\partial x^2} + N_y \frac{\partial^2 w}{\partial y^2} + 2N_{xy}\frac{\partial^2 w}{\partial x\, \partial y} - X\frac{\partial w}{\partial x} - Y\frac{\partial w}{\partial y} \right)$$

(8-27)

This differential equation should be used instead of Eq. (8-25) when there are body forces to be considered.

8.4. Strain Energy in Bending of Plates. In an investigation of the stability of thin plates, the energy method (see Art. 2.8) frequently can be used to advantage. We shall derive here expressions for the strain energy of a bent plate under various conditions of loading.

Pure Bending. If a plate is bent by uniformly distributed bending moments M_x and M_y (Fig. 8-1), the strain energy accumulated in an element, such as shown in Fig. 8-2, is obtained by calculating the work done by the moments $M_x\, dy$ and $M_y\, dx$ on the element during bending of the plate. Since the sides of the element remain plane, the work done by the moments $M_x\, dy$ is obtained by taking one-half of the product of the moment and the angle between the sides of the element after bending. Since $-\partial^2 w/\partial x^2$ represents approximately the curvature of the plate in the xz plane, the angle corresponding to the moments $M_x\, dy$ is $-(\partial^2 w/\partial x^2)\, dx$ and the work done by these moments is

$$-\frac{1}{2} M_x \frac{\partial^2 w}{\partial x^2}\, dx\, dy$$

An analogous expression is obtained also for the work produced by the moments $M_y\, dx$. Then the total work, equal to the potential energy of the element, is

$$dU = -\frac{1}{2}\left(M_x \frac{\partial^2 w}{\partial x^2} + M_y \frac{\partial^2 w}{\partial y^2} \right) dx\, dy$$

Substituting for the moments their expressions (8-4) and (8-5), the strain energy of the element will be represented in the following form:

$$dU = \frac{1}{2} D \left[\left(\frac{\partial^2 w}{\partial x^2} \right)^2 + \left(\frac{\partial^2 w}{\partial y^2} \right)^2 + 2\nu \frac{\partial^2 w}{\partial x^2} \frac{\partial^2 w}{\partial y^2} \right] dx\, dy \qquad (a)$$

The total strain energy of the plate will then be obtained by integration of expression (a):

$$U = \frac{1}{2} D \iint \left[\left(\frac{\partial^2 w}{\partial x^2} \right)^2 + \left(\frac{\partial^2 w}{\partial y^2} \right)^2 + 2\nu \frac{\partial^2 w}{\partial x^2} \frac{\partial^2 w}{\partial y^2} \right] dx\, dy \qquad (8\text{-}28)$$

where the integration must be extended over the entire surface of the plate.

Bending of a Plate by Lateral Load. Considering again an element of the plate, such as shown in Fig. 8-5, and neglecting the strain energy due to shearing forces Q_x and Q_y, we find that the strain energy of the element is equal to the work done on the element by the bending moments $M_x\, dy$ and $M_y\, dx$ and by the twisting moments $M_{xy}\, dy$ and $M_{yx}\, dx$. Since we neglect the effect of vertical shearing forces on the curvature of the deflection surface, the strain energy due to bending moments will be represented by expression (8-28) derived above for the case of pure bending.

In deriving the expression for the strain energy due to twisting moments $M_{xy}\, dy$, we note that the corresponding angle of twist (Fig. 8-12) is $(\partial^2 w / \partial x\, \partial y)\, dx$; hence the strain energy due to $M_{xy}\, dy$ is

$$\frac{1}{2} M_{xy} \frac{\partial^2 w}{\partial x\, \partial y} dx\, dy = \frac{1}{2} D\, (1 - \nu) \left(\frac{\partial^2 w}{\partial x\, \partial y} \right)^2 dx\, dy$$

The same amount of energy will be produced also by the couples $M_{yx}\, dx$, so that the strain energy due to both twisting couples is

$$D(1 - \nu) \left(\frac{\partial^2 w}{\partial x\, \partial y} \right)^2 dx\, dy$$

Since the twist does not affect the work produced by the bending moments, the total strain energy of an element of a plate is obtained by adding together the energy of bending and the energy of twist. Thus we obtain

$$dU = \frac{1}{2} D \left[\left(\frac{\partial^2 w}{\partial x^2} \right)^2 + \left(\frac{\partial^2 w}{\partial y^2} \right)^2 + 2\nu \frac{\partial^2 w}{\partial x^2} \frac{\partial^2 w}{\partial y^2} \right] dx\, dy$$
$$+ D(1 - \nu) \left(\frac{\partial^2 w}{\partial x\, \partial y} \right)^2 dx\, dy \qquad (b)$$

The strain energy of the entire plate is now obtained by integration

$$U = \frac{1}{2} D \iint \left[\left(\frac{\partial^2 w}{\partial x^2} \right)^2 + \left(\frac{\partial^2 w}{\partial y^2} \right)^2 + 2\nu \frac{\partial^2 w}{\partial x^2} \frac{\partial^2 w}{\partial y^2} \right.$$
$$\left. + 2(1 - \nu) \left(\frac{\partial^2 w}{\partial x \, \partial y} \right)^2 \right] dx \, dy$$

or

$$U = \frac{1}{2} D \iint \left\{ \left(\frac{\partial^2 w}{\partial x^2} + \frac{\partial^2 w}{\partial y^2} \right)^2 \right.$$
$$\left. -2(1 - \nu) \left[\frac{\partial^2 w}{\partial x^2} \frac{\partial^2 w}{\partial y^2} - \left(\frac{\partial^2 w}{\partial x \, \partial y} \right)^2 \right] \right\} dx \, dy \quad (8\text{-}29)$$

Combined Bending and Tension or Compression of Plates. In the case of bending of a plate submitted to the simultaneous action of transverse loads and of forces applied in the middle plane of the plate, we assume that the forces in the middle plane are applied first. In this way we obtain a two-dimensional problem of theory of elasticity. Solving this problem, we determine the forces N_x, N_y, and N_{xy} (Fig. 8-11) and the components of strain

$$\epsilon_x = \frac{1}{hE} (N_x - \nu N_y) \qquad \epsilon_y = \frac{1}{hE} (N_y - \nu N_x)$$
$$\gamma_{xy} = \frac{N_{xy}}{hG}$$

Then the energy U_0 due to deformation of the middle plane of the plate by forces applied in this plane is

$$U_0 = \frac{1}{2} \iint (N_x \epsilon_x + N_y \epsilon_y + N_{xy} \gamma_{xy}) \, dx \, dy$$
$$= \frac{1}{2hE} \iint [N_x^2 + N_y^2 - 2\nu N_x N_y + 2(1 + \nu) N_{xy}^2] \, dx \, dy \quad (8\text{-}30)$$

In the further discussion of small deflections of plates we consider N_x, N_y, and N_{xy} as remaining constant, so that the portion U_0 of the strain energy remains constant during bending and we do not need to consider it in the following discussion.

Let us apply now some lateral load, producing bending of the plate. The three components in the x, y, and z directions of the displacement of any point in the middle surface of the plate during bending will be denoted by u, v, and w, respectively. Considering a linear element AB of that plane (Fig. 8-13) in the x direction, it can be seen that the elongation of the element due to the displacement u is equal to $(\partial u / \partial x) \, dx$. The elongation of the same element due to the displacement w is $\frac{1}{2}(\partial w / \partial x)^2 \, dx$ as can be seen by comparison of the length of the element $A_1 B_1$ in Fig.

8-13 with the length of its projection on the x axis. Thus the total unit elongation in the x direction of an element taken in the middle plane of the plate is

$$\epsilon'_x = \frac{\partial u}{\partial x} + \frac{1}{2}\left(\frac{\partial w}{\partial x}\right)^2 \tag{c}$$

Similarly the strain in the y direction is

$$\epsilon'_y = \frac{\partial v}{\partial y} + \frac{1}{2}\left(\frac{\partial w}{\partial y}\right)^2 \tag{d}$$

The shearing strain due to displacements u and v will be found from Eq. (h), p. 323. For the determination of the shearing strain due to displacements w, we take two infinitely small linear elements OA and OB in the x and y directions, as shown in Fig. 8-14. Because of displacements

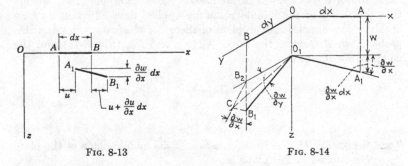

Fig. 8-13   Fig. 8-14

w these elements take the positions O_1A_1 and O_1B_1. The difference between the angle $A_1O_1B_1$ and $\pi/2$ is the shearing strain corresponding to displacements w. To determine this difference we consider the right angle $B_2O_1A_1$. Rotating this angle with respect to O_1A_1 by the small angle $\partial w/\partial y$, we bring the plane $B_2O_1A_1$ into coincidence with the plane $B_1O_1A_1$ and the point B_2 takes the position C. The displacement B_2C is equal to $(\partial w/\partial y)\,dy$ and is inclined to the vertical B_2B_1 by the small angle $\partial w/\partial x$. Then, from triangle B_2CB_1, we see that CB_1 is equal to

$$\frac{\partial w}{\partial x}\frac{\partial w}{\partial y}\,dy$$

and the angle CO_1B_1, representing the shearing strain corresponding to displacements w, is $(\partial w/\partial x)(\partial w/\partial y)$. Adding this to the shear due to displacements u and v, we obtain

$$\gamma'_{xy} = \frac{\partial u}{\partial y} + \frac{\partial v}{\partial x} + \frac{\partial w}{\partial x}\frac{\partial w}{\partial y} \tag{e}$$

We assume now that the strains ϵ'_x, ϵ'_y, and γ'_{xy} in the middle plane, due to bending, are small in comparison with ϵ_x, ϵ_y, and γ_{xy}. Assuming also that the forces N_x, N_y, and N_{xy} remain constant during bending, the strain energy due to the additional stretching of the middle surface is

$$\iint (N_x \epsilon'_x + N_y \epsilon'_y + N_{xy} \gamma'_{xy})\, dx\, dy$$

Adding to this the energy of bending represented by Eq. (8-29) and substituting for ϵ'_x, ϵ'_y, and γ'_{xy} their expressions (c), (d), and (e), we can represent the total change in strain energy of the plate during bending in the following form:

$$U = \iint \left[N_x \frac{\partial u}{\partial x} + N_y \frac{\partial v}{\partial y} + N_{xy}\left(\frac{\partial u}{\partial y} + \frac{\partial v}{\partial x}\right) \right] dx\, dy$$
$$+ \frac{1}{2}\iint \left[N_x \left(\frac{\partial w}{\partial x}\right)^2 + N_y \left(\frac{\partial w}{\partial y}\right)^2 + 2N_{xy}\frac{\partial w}{\partial x}\frac{\partial w}{\partial y} \right] dx\, dy$$
$$+ \frac{1}{2} D \iint \left\{ \left(\frac{\partial^2 w}{\partial x^2} + \frac{\partial^2 w}{\partial y^2}\right)^2 - 2(1-\nu)\left[\frac{\partial^2 w}{\partial x^2}\frac{\partial^2 w}{\partial y^2} - \left(\frac{\partial^2 w}{\partial x\, \partial y}\right)^2\right] \right\} dx\, dy$$
$$(8\text{-}31)$$

It can be shown, by integrating by parts, that the first integral on the right-hand side of this equation represents the work done during bending by the forces acting in the middle plane of the plate. Taking, for example, a rectangular plate, as shown in Fig. 8-9a, we obtain

$$\int_0^b \int_0^a \left[N_x \frac{\partial u}{\partial x} + N_y \frac{\partial v}{\partial y} + N_{xy}\left(\frac{\partial u}{\partial y} + \frac{\partial v}{\partial x}\right) \right] dx\, dy$$
$$= \int_0^b \left(\left| N_x u \right|_0^a + \left| N_{xy} v \right|_0^a \right) dy + \int_0^a \left(\left| N_y v \right|_0^b + \left| N_{xy} u \right|_0^b \right) dx$$
$$- \int_0^b \int_0^a u \left(\frac{\partial N_x}{\partial x} + \frac{\partial N_{xy}}{\partial y}\right) dx\, dy - \int_0^b \int_0^a v \left(\frac{\partial N_{xy}}{\partial x} + \frac{\partial N_y}{\partial y}\right) dx\, dy$$

The first two integrals on the right side of this equation represent the work done by the forces applied at the boundary of the plate and acting in its middle plane. The last two integrals vanish if Eqs. (8-24) hold, i.e., if there are no body forces acting in the middle plane of the plate. Otherwise these integrals represent the work of the body forces during bending of the plate [see Eqs. (8-26)].

If the first integral of expression (8-31) represents the work of forces acting in the middle plane of the plate, then the rest of the same expression must be equal to the work produced by the load normal to the plate. Using notations T_1 and T_2 for the work of the above two systems of forces and assuming that there are no body forces, we then obtain

$$T_1 = \iint \left[N_x \frac{\partial u}{\partial x} + N_y \frac{\partial v}{\partial y} + N_{xy} \left(\frac{\partial u}{\partial y} + \frac{\partial v}{\partial x} \right) \right] dx\, dy \qquad (f)$$

$$T_2 = \frac{1}{2} \iint \left[N_x \left(\frac{\partial w}{\partial x} \right)^2 + N_y \left(\frac{\partial w}{\partial y} \right)^2 + 2N_{xy} \frac{\partial w}{\partial x} \frac{\partial w}{\partial y} \right] dx\, dy$$

$$+ \frac{1}{2} D \iint \left\{ \left(\frac{\partial^2 w}{\partial x^2} + \frac{\partial^2 w}{\partial y^2} \right)^2 - 2(1 - \nu) \left[\frac{\partial^2 w}{\partial x^2} \frac{\partial^2 w}{\partial y^2} - \left(\frac{\partial^2 w}{\partial x\, \partial y} \right)^2 \right] \right\} dx\, dy$$
$$(g)$$

If, in discussing bending of plates, we neglect any stretching in the middle plane, we conclude from expressions (c), (d), and (e) that

$$\frac{1}{2} \left(\frac{\partial w}{\partial x} \right)^2 = - \frac{\partial u}{\partial x} \qquad \frac{1}{2} \left(\frac{\partial w}{\partial y} \right)^2 = - \frac{\partial v}{\partial y}$$

$$\frac{\partial w}{\partial x} \frac{\partial w}{\partial y} = - \frac{\partial u}{\partial y} - \frac{\partial v}{\partial x}$$

From this it follows that the work of the forces acting in the middle plane of the plate can be represented in the following form:

$$T_1 = - \frac{1}{2} \iint \left[N_x \left(\frac{\partial w}{\partial x} \right)^2 + N_y \left(\frac{\partial w}{\partial y} \right)^2 + 2N_{xy} \frac{\partial w}{\partial x} \frac{\partial w}{\partial y} \right] dx\, dy \quad (8\text{-}32)$$

We also obtain

$$T_1 + T_2 = \frac{1}{2} D \iint \left\{ \left(\frac{\partial^2 w}{\partial x^2} + \frac{\partial^2 w}{\partial y^2} \right)^2 - 2(1 - \nu) \left[\frac{\partial^2 w}{\partial x^2} \frac{\partial^2 w}{\partial y^2} \right. \right.$$
$$\left. \left. - \left(\frac{\partial^2 w}{\partial x\, \partial y} \right)^2 \right] \right\} dx\, dy \quad (8\text{-}33)$$

By using Eq. (8-33), together with the principle of virtual displacements, we can obtain the deflections of the plate in the same manner as in the case of beams (see Art. 1.11). Several examples of the application of this method will be discussed in the next two articles.[1]

8.5. Deflections of Rectangular Plates with Simply Supported Edges. In the case of a rectangular plate with simply supported edges (Fig. 8-15), the deflection surface can be represented by the double trigonometric series

$$w = \sum_{m=1}^{\infty} \sum_{n=1}^{\infty} a_{mn} \sin \frac{m\pi x}{a} \sin \frac{n\pi y}{b} \qquad (8\text{-}34)$$

Each term of this series vanishes for $x = 0$, $x = a$ and also for $y = 0$, $y = b$. Hence the deflection w is zero along the boundary as required.

---

[1] Equation (8-33) was established by G. H. Bryan, *Proc. London Math. Soc.*, vol. 22, p. 54, 1891. Further discussion of this equation was given by H. Reissner, *Z. angew. Math. u. Mech.*, vol. 5, p. 475, 1925.

Calculating the derivatives $\partial^2w/\partial x^2$ and $\partial^2w/\partial y^2$, we find again that each term of the calculated series becomes zero at the boundary. From this it can be concluded that the bending moments [see Eqs. (8-13)] are zero along the boundary as they should be in the case of simply supported edges. Thus expression (8-34) satisfies all boundary conditions. The expression for the potential energy of bending for this case is [see Eq. (8-29)]

$$U = \frac{1}{2}D\int_0^a\int_0^b\left\{\left(\frac{\partial^2w}{\partial x^2}+\frac{\partial^2w}{\partial y^2}\right)^2 - 2(1-\nu)\left[\frac{\partial^2w}{\partial x^2}\frac{\partial^2w}{\partial y^2}-\left(\frac{\partial^2w}{\partial x\,\partial y}\right)^2\right]\right\}dx\,dy$$

Substituting for w expression (8-34), it can be shown that the integral of the term in the brackets vanishes and we obtain

$$U = \frac{1}{2}D\int_0^a\int_0^b\left\{\sum_{m=1}^{\infty}\sum_{n=1}^{\infty}a_{mn}\left(\frac{m^2\pi^2}{a^2}+\frac{n^2\pi^2}{b^2}\right)\sin\frac{m\pi x}{a}\sin\frac{n\pi y}{b}\right\}^2 dx\,dy$$

Only the squares of the terms of the infinite series give integrals different

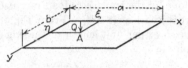

FIG. 8-15

from zero (see p. 28). Then observing that

$$\int_0^a\int_0^b \sin^2\frac{m\pi x}{a}\sin^2\frac{n\pi y}{b}\,dx\,dy = \frac{ab}{4}$$

we obtain

$$U = \frac{ab}{8}D\sum_{m=1}^{\infty}\sum_{n=1}^{\infty}a_{mn}^2\left(\frac{m^2\pi^2}{a^2}+\frac{n^2\pi^2}{b^2}\right)^2 \tag{8-35}$$

Having this expression for U, we can obtain the deflection of the plate for any kind of loading by using the principle of virtual displacements in the same manner as in the case of bending of beams (see Art. 1.11). Assume, for instance, that a concentrated force Q is acting at a point A with coordinates ξ and η (Fig. 8-15). To determine any coefficient a_{mn} of the series (8-34) in this case, we give to this coefficient a small increment δa_{mn}. The corresponding virtual deflection of the plate is

$$\delta a_{mn}\sin\frac{m\pi x}{a}\sin\frac{n\pi y}{b}$$

The work done by the load Q during this displacement is

$$Q\,\delta a_{mn}\sin\frac{m\pi\xi}{a}\sin\frac{n\pi\eta}{b}$$

and from the principle of virtual displacements we obtain the following equation:

$$Q\, \delta a_{mn} \sin \frac{m\pi\xi}{a} \sin \frac{n\pi\eta}{b} = \frac{\partial U}{\partial a_{mn}} \delta a_{mn} = \frac{ab}{4}\, D a_{mn} \left(\frac{m^2\pi^2}{a^2} + \frac{n^2\pi^2}{b^2}\right)^2 \delta a_{mn} \quad (a)$$

from which

$$a_{mn} = \frac{4Q \sin \dfrac{m\pi\xi}{a} \sin \dfrac{n\pi\eta}{b}}{ab D\pi^4 \left(\dfrac{m^2}{a^2} + \dfrac{n^2}{b^2}\right)^2} \quad (b)$$

When this is substituted in expression (8-34), the deflection of the plate produced by a concentrated load Q is obtained. Having this deflection and using the principle of superposition, we can determine the deflections for any kind of loading.

If the plate of Fig. 8-15 is uniformly compressed in the x direction, the same method as above can be used in calculating deflections. It is necessary in applying the principle of virtual displacements to consider not only the work done by the transverse load but also the work done by the compressive forces [see Eq. (8-33)]. From Eq. (8-32), by assuming that the compressive force per unit length of the edges $x = 0$ and $x = a$ is equal to N_x, we obtain

$$T_1 = \frac{1}{2} \int_0^a \int_0^b N_x \left(\frac{\partial w}{\partial x}\right)^2 dx\, dy$$

or, substituting for w its expression (8-34),

$$T_1 = \frac{ab}{8} N_x \sum_{m=1}^{\infty} \sum_{n=1}^{\infty} a_{mn}^2 \frac{m^2\pi^2}{a^2} \quad (c)$$

Giving to a coefficient a_{mn} an increase δa_{mn}, the corresponding work of the compressive forces is

$$\frac{\partial T_1}{\partial a_{mn}} \delta a_{mn} = \frac{\pi^2 b N_x}{4a} m^2 a_{mn}\, \delta a_{mn}$$

Adding this work to the left side of Eq. (a), we find that, in the case of bending of the compressed plate by a concentrated force Q, the coefficients a_{mn} in the general expression (8-34) for deflection are determined from the following formula:

$$a_{mn} = \frac{4Q \sin \dfrac{m\pi\xi}{a} \sin \dfrac{n\pi\eta}{b}}{ab D\pi^4 \left[\left(\dfrac{m^2}{a^2} + \dfrac{n^2}{b^2}\right)^2 - \dfrac{m^2 N_x}{\pi^2 a^2 D}\right]} \quad (d)$$

Comparing this with expression (b), it can be concluded that, owing to the compressive forces N_x, all the coefficients increase. Hence the deflections of a compressed plate are larger than those of the same plate without compression in the middle plane. It is seen also that by a gradual increase of the compressive forces we shall arrive at a value of N_x for which one of the coefficients (d) becomes infinity. The smallest of these values of N_x is called the *critical value*. It is determined from the equation

$$\left(\frac{m^2}{a^2} + \frac{n^2}{b^2}\right)^2 - \frac{m^2 N_x}{\pi^2 a^2 D} = 0 \tag{e}$$

from which

$$N_x = \frac{\pi^2 D}{a^2}\left(m + \frac{n^2}{m}\frac{a^2}{b^2}\right)^2 \tag{f}$$

To find the critical value of N_x, i.e., the smallest value satisfying Eq. (e), it is necessary to take $n = 1$ and to take for m the integer value such that the quantity in parentheses in expression (f) is a minimum. For instance, in the case of a square plate we must take $m = 1$ to make expression (f) a minimum. The result is

$$N_{cr} = \frac{4\pi^2 D}{a^2} \tag{8-36}$$

This formula is analogous to Euler's formula for buckling of a column. Since N_{cr} is the compressive load for a strip of the plate of unit width and D is the flexural rigidity of the strip, it can be concluded that, owing to continuity of the plate, each longitudinal strip can carry a load four times larger than Euler's load for an isolated strip. The question of lateral buckling of compressed plates will be discussed fully in the next chapter.

In the above discussion it was assumed that compressive forces acted on the plate. If, instead of compression, we have tensile forces N_x, expression (d) for the coefficients a_{mn} can be used again. It is only necessary to substitute $-N_x$ instead of N_x. It is seen that tensile forces decrease the deflection of the plate.

In the more general case in which normal forces N_x and N_y and shearing forces N_{xy} are acting on the boundary of the plate, the same general method can be used in calculating deflections. There will be no difficulty in such calculations provided the forces are uniformly distributed along the edges of the plate. If the distribution is not uniform, there may be difficulty in solving the corresponding two-dimensional problem[1] and determining N_x, N_y, and N_{xy}, as functions of x and y. But if the two-dimensional problem is solved and the forces N_x, N_y, and N_{xy} are determined, we can use the same expression (8-34) for the deflection surface of the plate and can determine the coefficients a_{mn} by applying the principle

[1] For a discussion of two-dimensional problems see S. P. Timoshenko and J. N. Goodier, "Theory of Elasticity," chap. 2, McGraw-Hill Book Company, Inc., New York, 1951.

of virtual displacements in the same manner as illustrated in the above example.

8.6. Bending of Plates with a Small Initial Curvature.[1]   Assume that a plate has some initial warp of the middle surface so that at any point there is initial deflection of magnitude w_0, which is small in comparison with the thickness of the plate.   If such a plate is submitted to the action of transverse loading, some additional deflection w_1 will be produced and the total deflection at any point of the middle surface of the plate will be $w_0 + w_1$.   In calculating the deflection w_1 we use Eq. (8-15) derived before for flat plates.   This procedure is justifiable if the initial deflection w_0 is small, since we can consider it as produced by some fictitious lateral load and then we can use the principle of superposition.[2]

If in addition to lateral load there are forces acting in the middle plane of the plate, the effect of these forces on bending depends not only on w_1 but also on w_0.   To take this into account in using Eq. (8-25), we take on the right side of this equation the total deflection $w = w_0 + w_1$.   The left side of the same equation is obtained from the expressions for bending moments, and since these moments depend not on the total curvature but only on the change in curvature of the plate, the deflection w_1, instead of w, should be used on that side.   Hence Eq. (8-25) for the case of an initially curved plate becomes

$$\frac{\partial^4 w_1}{\partial x^4} + 2\frac{\partial^4 w_1}{\partial x^2 \partial y^2} + \frac{\partial^4 w_1}{\partial y^4} = \frac{1}{D}\left[q + N_x \frac{\partial^2(w_0 + w_1)}{\partial x^2} \right.$$
$$\left. + N_y \frac{\partial^2(w_0 + w_1)}{\partial y^2} + 2N_{xy}\frac{\partial^2(w_0 + w_1)}{\partial x\,\partial y} \right] \quad (8\text{-}37)$$

It is seen that the effect of an initial curvature on the deflection is equivalent to the effect of a fictitious lateral load of intensity

$$N_x \frac{\partial^2 w_0}{\partial x^2} + N_y \frac{\partial^2 w_0}{\partial y^2} + 2N_{xy}\frac{\partial^2 w_0}{\partial x\,\partial y}$$

Thus a plate will undergo bending under the action of forces in the xy plane alone provided there is an initial curvature.

Take as an example the case of a simply supported rectangular plate (Fig. 8-15) and assume that the initial deflection of the plate is defined by the equation

$$w_0 = a_{11} \sin\frac{\pi x}{a} \sin\frac{\pi y}{b} \quad (a)$$

If on the edges $x = 0$ and $x = a$ of this plate uniformly distributed compressive forces N_x are acting, Eq. (8-37) becomes

[1] See Timoshenko's paper in *Mem. Inst. Engrs. Ways of Commun.*, St. Petersburg, vol. 89, 1915 (Russian).

[2] In the case of large deflections the magnitude of the deflections is no longer proportional to the load and the principle of superposition is not applicable.

$$\frac{\partial^4 w_1}{\partial x^4} + 2\frac{\partial^4 w_1}{\partial x^2 \partial y^2} + \frac{\partial^4 w_1}{\partial y^4} = \frac{1}{D}\left(N_x \frac{a_{11}\pi^2}{a^2} \sin\frac{\pi x}{a} \sin\frac{\pi y}{b} - N_x \frac{\partial^2 w_1}{\partial x^2}\right) \quad (b)$$

Let us take the solution of this equation in the form

$$w_1 = A \sin\frac{\pi x}{a} \sin\frac{\pi y}{b} \quad (c)$$

Substituting this value of w_1 in Eq. (b), we obtain

$$A = \frac{a_{11}N_x}{\dfrac{\pi^2 D}{a^2}\left(1 + \dfrac{a^2}{b^2}\right)^2 - N_x}$$

With this value of A expression (c) gives the deflection of the plate produced by compressive forces N_x. Adding this deflection to the initial deflection (a), we obtain for the total deflection of the plate the following expression:

$$w = w_0 + w_1 = \frac{a_{11}}{1 - \alpha} \sin\frac{\pi x}{a} \sin\frac{\pi y}{b} \quad (d)$$

in which

$$\alpha = \frac{N_x}{\dfrac{\pi^2 D}{a^2}\left(1 + \dfrac{a^2}{b^2}\right)^2} \quad (e)$$

The maximum deflection will be at the center, and we obtain

$$w_{\text{max}} = \frac{a_{11}}{1 - \alpha} \quad (f)$$

This formula is analogous to formula (a), Art. 1.11, derived for the deflection of an initially curved beam.

In a more general case we can take the initial deflection surface of the rectangular plate in the form of the following series:

$$w_0 = \sum_{m=1}^{\infty} \sum_{n=1}^{\infty} a_{mn} \sin\frac{m\pi x}{a} \sin\frac{n\pi y}{b} \quad (g)$$

Substituting this series in Eq. (8-37), we find that the additional deflection of any point of the plate is

$$w_1 = \sum_{m=1}^{\infty} \sum_{n=1}^{\infty} b_{mn} \sin\frac{m\pi x}{a} \sin\frac{n\pi y}{b} \quad (h)$$

in which

$$b_{mn} = \frac{a_{mn}N_x}{\dfrac{\pi^2 D}{a^2}\left(m + \dfrac{n^2}{m}\dfrac{a^2}{b^2}\right)^2 - N_x} \quad (i)$$

It is seen that all coefficients b_{mn} increase with an increase of N_x. When N_x approaches the critical value [see Eq. (f), p. 343] the term in the series (h) which corresponds to the shape of lateral buckling of the plate becomes the most important term.　We have here a complete analogy with the case of bending of initially curved compressed bars (see Art. 1.12). Hence, in experimental determinations of critical values of compressive forces, the method recommended by Southwell (see p. 191) can be used.

The problem can be handled in the same manner if, instead of compression, we have tension in the middle plane of the plate.　It is only necessary to change the sign of N_x in the previous equations.　Without any difficulty we can obtain also the deflection in the case when there are not only forces N_x but also forces N_y and N_{xy} uniformly distributed along the edges of the plate.

8.7. Large Deflections of Plates.　If the deflection of a plate is not small, the assumption regarding the inextensibility of the middle surface of the plate holds only if the deflection surface is a *developable* surface.　In the general case large deflections of a plate are accompanied by some stretching of the middle surface.　To give some idea of the magnitude of the stresses due to this stretching, it may be mentioned that in the case of pure bending of a circular plate having a maximum deflection equal to $0.6h$ the maximum stress due to stretching of the middle plane is about 18 per cent of the maximum bending stress.[1]　For a uniformly loaded circular plate with clamped edges and with maximum deflection equal to h, the maximum stress due to stretching is about 23 per cent of the maximum bending stress calculated by neglecting stretching.[2]　Thus, in these cases, it is only for small deflections, not exceeding, say, $0.4h$, that the stretching of the middle surface can be neglected without a substantial error in the magnitude of maximum stresses.

In considering large deflection[3] of plates, we can continue to use Eq. (8-25), derived from the condition of equilibrium of an element in the direction normal to the plate, but the forces N_x, N_y, and N_{xy} depend now, not only on the external forces applied in the xy plane, but also on the stretching of the middle surface of the plate due to bending.　Assuming that there are no body forces in the xy plane, the equations of equilibrium in this plane are [see Eqs. (8-24)]

$$\frac{\partial N_x}{\partial x} + \frac{\partial N_{xy}}{\partial y} = 0$$

$$\frac{\partial N_y}{\partial y} + \frac{\partial N_{xy}}{\partial x} = 0$$

$$(a)$$

The third equation for determining N_x, N_y, and N_{xy} will be obtained from the consideration of strain in the middle surface of the plate during bending.　The corresponding strain components are (see p. 338)

$$\epsilon_x = \frac{\partial u}{\partial x} + \frac{1}{2}\left(\frac{\partial w}{\partial x}\right)^2$$

$$\epsilon_y = \frac{\partial v}{\partial y} + \frac{1}{2}\left(\frac{\partial w}{\partial y}\right)^2$$

$$\gamma_{xy} = \frac{\partial u}{\partial y} + \frac{\partial v}{\partial x} + \frac{\partial w}{\partial x}\frac{\partial w}{\partial y}$$

$$(b)$$

[1] See *ibid.*

[2] See S. Way, *Trans. ASME*, 1934, p. 627.

[3] That is, deflections which are no longer small in comparison with the thickness, but which are at the same time small enough to justify the application of simplified formulas for curvatures of a plate (see p. 321).

By differentiating these expressions, it can be shown that

$$\frac{\partial^2 \epsilon_x}{\partial y^2} + \frac{\partial^2 \epsilon_y}{\partial x^2} - \frac{\partial^2 \gamma_{xy}}{\partial x \, \partial y} = \left(\frac{\partial^2 w}{\partial x \, \partial y}\right)^2 - \frac{\partial^2 w}{\partial x^2} \frac{\partial^2 w}{\partial y^2} \qquad (c)$$

By substituting for the strain components their expressions in terms of the stresses, we obtain the third equation in terms of N_x, N_y, and N_{xy}.

The solution of these three equations can be greatly simplified by introducing a *stress function*.[1] It is seen that Eqs. (a) are identically satisfied by taking

$$N_x = h \frac{\partial^2 F}{\partial y^2} \qquad N_y = h \frac{\partial^2 F}{\partial x^2} \qquad N_{xy} = -h \frac{\partial^2 F}{\partial x \, \partial y} \qquad (d)$$

where F is a function of x and y.

With these expressions for forces, the strain components become

$$\epsilon_x = \frac{1}{hE} (N_x - \nu N_y) = \frac{1}{E} \left(\frac{\partial^2 F}{\partial y^2} - \nu \frac{\partial^2 F}{\partial x^2}\right)$$

$$\epsilon_y = \frac{1}{hE} (N_y - \nu N_x) = \frac{1}{E} \left(\frac{\partial^2 F}{\partial x^2} - \nu \frac{\partial^2 F}{\partial y^2}\right) \qquad (e)$$

$$\gamma_{xy} = \frac{1}{hG} N_{xy} = -\frac{2(1 + \nu)}{E} \frac{\partial^2 F}{\partial x \, \partial y}$$

Substituting these expressions in Eq. (c), we obtain

$$\frac{\partial^4 F}{\partial x^4} + 2 \frac{\partial^4 F}{\partial x^2 \, \partial y^2} + \frac{\partial^4 F}{\partial y^4} = E \left[\left(\frac{\partial^2 w}{\partial x \, \partial y}\right)^2 - \frac{\partial^2 w}{\partial x^2} \frac{\partial^2 w}{\partial y^2}\right] \qquad (8\text{-}38)$$

By using expressions (d), Eq. (8-25) becomes

$$\frac{\partial^4 w}{\partial x^4} + 2 \frac{\partial^4 w}{\partial x^2 \, \partial y^2} + \frac{\partial^4 w}{\partial y^4} = \frac{h}{D} \left(\frac{q}{h} + \frac{\partial^2 F}{\partial y^2} \frac{\partial^2 w}{\partial x^2} + \frac{\partial^2 F}{\partial x^2} \frac{\partial^2 w}{\partial y^2} - 2 \frac{\partial^2 F}{\partial x \, \partial y} \frac{\partial^2 w}{\partial x \, \partial y}\right) \qquad (8\text{-}39)$$

Equations (8-38) and (8-39), together with the boundary conditions, determine[2] the two functions F and w. Having the stress function F, we determine the stresses in the middle surface of a plate by using Eqs. (d). From the function w, defining the deflection surface of the plate, the bending and shearing stresses are obtained by using the same formulas as in the case of plates with small deflections (see Art. 8.2). Thus the investigation of large deflections of plates reduces to the solution of the two Eqs. (8-38) and (8-39). Some approximate solutions of these equations have been obtained only for the simplest cases, namely, for uniformly loaded rectangular plates and for uniformly loaded circular plates.[3]

---

[1] See Timoshenko and Goodier, *op. cit.*

[2] The theory of large deflections of plates was developed by G. R. Kirchhoff (see his book, "Mechanik," 2d ed., p. 450, 1877). The use of the stress function F was introduced by A. Föppl (see his book "Technische Mechanik," vol. 5, p. 132, 1907). The final form of Eqs. (8-38) and (8-39) was given by T. V. Kármán, "Encyklopädie der Mathematischen Wissenschaften," vol. IV$_4$, p. 349, 1910.

[3] For additional information on the theory of large deflections of plates see Timoshenko, "Theory of Plates and Shells," 2d ed., in collaboration with S. Woinowsky-Krieger, pp. 415-428, McGraw-Hill Book Company, Inc., New York, 1959.

CHAPTER 9

BUCKLING OF THIN PLATES

9.1. Methods of Calculation of Critical Loads. In the calculation of *critical values* of forces applied in the middle plane of a plate at which the flat form of equilibrium becomes unstable and the plate begins to buckle, the same methods as in the case of compressed bars can be used.

From the discussion of the previous chapter (see Arts. 8.5 and 8.6) it can be seen that the critical values of the forces acting in the middle plane of a plate can be obtained by assuming that from the beginning the plate has some initial curvature or some lateral loading. Then those values of the forces in the middle plane at which deflections tend to grow indefinitely are usually the critical values.

Another way of investigating such stability problems is to assume that the plate buckles slightly under the action of forces applied in its middle plane and then to calculate the magnitudes that the forces must have in order to keep the plate in such a slightly buckled shape. The differential equation of the deflection surface in this case is obtained from Eq. (8-25) by putting $q = 0$, i.e., by assuming that there is no lateral load. If there are no body forces,[1] the equation for the buckled plate then becomes

$$\frac{\partial^4 w}{\partial x^4} + 2 \frac{\partial^4 w}{\partial x^2 \, \partial y^2} + \frac{\partial^4 w}{\partial y^4} = \frac{1}{D}\left(N_x \frac{\partial^2 w}{\partial x^2} + N_y \frac{\partial^2 w}{\partial y^2} + 2N_{xy} \frac{\partial^2 w}{\partial x \, \partial y}\right) \quad (9\text{-}1)$$

The simplest case is obtained when the forces N_x, N_y, and N_{xy} are constant throughout the plate. Assuming that there are given ratios between these forces so that $N_y = \alpha N_x$ and $N_{xy} = \beta N_x$, and solving Eq. (9-1) for the given boundary conditions, we shall find that the assumed buckling of the plate is possible only for certain definite values of N_x. The smallest of these values determines the desired critical value.

If the forces N_x, N_y, and N_{xy} are not constant, the problem becomes more involved, since Eq. (9-1) has in this case variable coefficients, but the general conclusion remains the same. In such cases we can assume that the expressions for the forces N_x, N_y, and N_{xy} have a common factor γ, so that a gradual increase of loading is obtained by an increase of this factor. From the investigation of Eq. (9-1), together with the given boundary

[1] In the case of body forces acting in the middle plane of the plate, Eq. (8-27) must be used.

conditions, it will be concluded that curved forms of equilibrium are possible only for certain values of the factor γ and that the smallest of these values will define the critical loading.

The energy method also can be used in investigating buckling of plates. This method is especially useful in those cases where a rigorous solution of Eq. (9-1) is unknown or where we have a plate reinforced by stiffeners and it is required to find only an approximate value of the critical load.[1] In applying this method we proceed as in the case of buckling of bars (see Art. 2.8) and assume that the plate, which is stressed by forces acting in its middle plane, undergoes some small lateral bending consistent with given boundary conditions. Such limited bending can be produced without stretching of the middle plane, and we need consider only the energy of bending and the corresponding work done by the forces acting in the middle plane of the plate. If the work done by these forces is smaller than the strain energy of bending for every possible shape of lateral buckling, the flat form of equilibrium of the plate is stable. If the same work becomes larger than the energy of bending for any shape of lateral deflection, the plate is unstable and buckling occurs. Denoting by ΔT_1 the above-mentioned work of external forces and by ΔU the strain energy of bending, we find the critical values of forces from the equation

$$\Delta T_1 = \Delta U \tag{a}$$

Substituting for the work ΔT_1 expression (8-32) and for ΔU expression (8-29), we obtain

$$-\frac{1}{2} \int\int \left[N_x \left(\frac{\partial w}{\partial x}\right)^2 + N_y \left(\frac{\partial w}{\partial y}\right)^2 + 2N_{xy} \frac{\partial w}{\partial x}\frac{\partial w}{\partial y} \right] dx\,dy$$
$$= \frac{D}{2} \int\int \left\{ \left(\frac{\partial^2 w}{\partial x^2} + \frac{\partial^2 w}{\partial y^2}\right)^2 - 2(1-\nu)\left[\frac{\partial^2 w}{\partial x^2}\frac{\partial^2 w}{\partial y^2} - \left(\frac{\partial^2 w}{\partial x\,\partial y}\right)^2\right] \right\} dx\,dy \tag{9-2}$$

Assuming that forces N_x, N_y, and N_{xy} are represented by certain expressions with a common factor γ, so that

$$N_x = \gamma N'_x \qquad N_y = \gamma N'_y \qquad N_{xy} = \gamma N'_{xy} \tag{b}$$

a simultaneous increase of these forces is obtained by increasing γ. The critical value of this factor is then obtained from Eq. (9-2), from which

$$\gamma = \frac{D \int\int \left\{ \left(\frac{\partial^2 w}{\partial x^2} + \frac{\partial^2 w}{\partial y^2}\right)^2 - 2(1-\nu)\left[\frac{\partial^2 w}{\partial x^2}\frac{\partial^2 w}{\partial y^2} - \left(\frac{\partial^2 w}{\partial x\,\partial y}\right)^2\right] \right\} dx\,dy}{-\int\int \left[N'_x \left(\frac{\partial w}{\partial x}\right)^2 + N'_y \left(\frac{\partial w}{\partial y}\right)^2 + 2N'_{xy} \frac{\partial w}{\partial x}\frac{\partial w}{\partial y} \right] dx\,dy} \tag{9-3}$$

For the calculation of γ it is necessary to find, in each particular case, an expression for w which satisfies the given boundary conditions and makes expression (9-3) a

[1] A variety of problems of this type were discussed by Timoshenko; see his papers Sur la stabilité des systèmes élastiques, *Ann. ponts et chaussées*, 1913; Stability of Rectangular Plates with Stiffeners, *Mem. Inst. Engrs. Ways of Commun.*, vol. 89, 1915; and *Der Eisenbau*, vol. 12, 1921.

minimum, i.e., the variation of the fraction (9-3) must be zero. Denoting the numerator by I_1 and the denominator by I_2, we find that the variation of expression (9-3) is

$$\delta\gamma = \frac{(I_2\,\delta I_1 - I_1\,\delta I_2)}{I_2{}^2}$$

which, when equated to zero and with $I_1/I_2 = \gamma$, gives

$$\frac{1}{I_2}(\delta I_1 - \gamma\,\delta I_2) = 0 \qquad\qquad (c)$$

By calculating the indicated variations and assuming that there are no body forces, we shall arrive at Eq. (9-1). Thus the energy method brings us in this way to the integration of the same equation, which we have discussed before.

For an approximate calculation of critical loads by the energy method, we shall proceed as in the case of struts (see p. 91) and assume w in the form of a series

$$w = a_1 f_1(x,y) + a_2 f_2(x,y) + \cdots \qquad\qquad (d)$$

in which the functions $f_1(x,y)$, $f_2(x,y)$, . . . satisfy the boundary conditions for w and are chosen so as to be suitable for the representation of the buckled surface of the plate. In each particular case we shall be guided in choosing these functions by experimental data regarding the shape of a buckled plate. The coefficients a_1, a_2, . . . of the series must now be chosen so as to make the expression (9-3) a minimum. Using this condition of minimum and proceeding as in the derivation of Eq. (c) above, we obtain the following equations:

$$\frac{\partial I_1}{\partial a_1} - \gamma\,\frac{\partial I_2}{\partial a_1} = 0$$
$$\frac{\partial I_1}{\partial a_2} - \gamma\,\frac{\partial I_2}{\partial a_2} = 0 \qquad\qquad (9\text{-}4)$$
$$\cdots\cdots\cdots\cdots$$

It can be seen from Eq. (9-3) that the expressions for I_1 and I_2, after integration, will be represented by homogeneous functions of the second degree in terms of $a_1, a_2, \ldots$ Hence Eqs. (9-4) will be a system of homogeneous linear equations in $a_1, a_2, \ldots$ Such equations may yield for $a_1, a_2, \ldots$ solutions different from zero only if the determinant of these equations is zero. When this determinant is put equal to zero, an equation for determining the critical value for γ will be obtained. This method of calculation of critical loads will be illustrated by several examples in the following articles.

We can arrive at Eq. (9-2) in another way by assuming that during buckling the edges of the plate are prevented from movement in the xy plane. Then the displacements u and v at the boundary vanish and the work ΔT_1 vanishes also. In such a case the lateral buckling is connected with some stretching in the middle plane of the plate, and we have to use Eq. (g), Art. 8.4. Since there are no lateral forces, T_2 vanishes and we obtain

$$\frac{1}{2}\iint\left[N_x\left(\frac{\partial w}{\partial x}\right)^2 + N_y\left(\frac{\partial w}{\partial y}\right)^2 + 2N_{xy}\frac{\partial w}{\partial x}\frac{\partial w}{\partial y}\right]dx\,dy$$
$$+ \frac{D}{2}\iint\left\{\left(\frac{\partial^2 w}{\partial x^2} + \frac{\partial^2 w}{\partial y^2}\right)^2 - 2(1-\nu)\left[\frac{\partial^2 w}{\partial x^2}\frac{\partial^2 w}{\partial y^2} - \left(\frac{\partial^2 w}{\partial x\,\partial y}\right)^2\right]\right\}dx\,dy = 0$$
$$(9\text{-}5)$$

The first integral in this equation represents the change in strain energy due to stretching of the middle plane of the plate during buckling, and the second represents the energy of bending of the plate. Equation (9-5) is identical with Eq. (9-2), but instead of discussing the work ΔT_1, which vanishes in this case, we have to state that the flat condition of equilibrium of the plate becomes critical when the strain energy of stretching of the plate, released during buckling, becomes equal to the strain energy of bending of the plate.

9.2. Buckling of Simply Supported Rectangular Plates Uniformly Compressed in One Direction.[1] Assume that a rectangular plate (Fig. 9-1) is compressed in its middle plane by forces uniformly distributed along the sides $x = 0$ and $x = a$. Let the magnitude of this compressive force per unit length of the edge be denoted by N_x. By gradually increasing N_x we arrive at the condition where the flat form of equilibrium of the compressed plate becomes unstable and buckling occurs. The corresponding critical value of the compressive force can be found in this case by integration of Eq. (9-1).[2] The same result is obtained also from a considera-

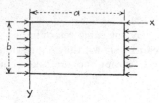

FIG. 9-1

tion of the energy of the system. The deflection surface of the buckled plate can be represented, in the case of simply supported edges, by the double series (8-34) (see Art. 8.5):

$$w = \sum_{m=1}^{\infty} \sum_{n=1}^{\infty} a_{mn} \sin \frac{m\pi x}{a} \sin \frac{n\pi y}{b} \tag{a}$$

The strain energy of bending in this case, from expression (8-35), is

$$\Delta U = \frac{\pi^4 ab}{8} D \sum_{m=1}^{\infty} \sum_{n=1}^{\infty} a_{mn}^2 \left(\frac{m^2}{a^2} + \frac{n^2}{b^2} \right)^2 \tag{b}$$

The work done by the compressive forces during buckling of the plate, from Eq. (8-32) and Eq. (c) (p. 342), will be

$$\frac{1}{2} N_x \int_0^a \int_0^b \left(\frac{\partial w}{\partial x} \right)^2 dx \, dy = \frac{\pi^2 b}{8a} N_x \sum_{m=1}^{\infty} \sum_{n=1}^{\infty} m^2 a_{mn}^2 \tag{c}$$

[1] Solution of this problem was given by G. H. Bryan, *Proc. London Math. Soc.*, vol. 22, p. 54, 1891.

[2] See Timoshenko, *Bull. Polytech. Inst., Kiev*, 1907.

Thus Eq. (9-2), for determining the critical value of compressive forces, becomes

$$\frac{\pi^2 b}{8a} N_x \sum_{m=1}^{\infty} \sum_{n=1}^{\infty} m^2 a_{mn}^2 = \frac{\pi^4 ab}{8} D \sum_{m=1}^{\infty} \sum_{n=1}^{\infty} a_{mn}^2 \left(\frac{m^2}{a^2} + \frac{n^2}{b^2}\right)^2$$

from which

$$N_x = \frac{\pi^2 a^2 D \displaystyle\sum_{m=1}^{\infty} \sum_{n=1}^{\infty} a_{mn}^2 \left(\frac{m^2}{a^2} + \frac{n_i^2}{b^2}\right)^2}{\displaystyle\sum_{m=1}^{\infty} \sum_{n=1}^{\infty} m^2 a_{mn}^2} \qquad (d)$$

By the same reasoning as in the case of compressed bars (see p. 92) it can be shown that expression (d) becomes a minimum if all coefficients a_{mn}, except one, are taken equal to zero.   Then

$$N_x = \frac{\pi^2 a^2 D}{m^2} \left(\frac{m^2}{a^2} + \frac{n^2}{b^2}\right)^2$$

It is obvious that the smallest value of N_x will be obtained by taking n equal to 1.   The physical meaning of this is that a plate buckles in such a way that there can be several half-waves in the direction of compression but only one half-wave in the perpendicular direction.   Thus the expression for the critical value of the compressive force becomes

$$(N_x)_{cr} = \frac{\pi^2 D}{a^2} \left(m + \frac{1}{m}\frac{a^2}{b^2}\right)^2 \qquad (e)$$

The first factor in this expression represents the Euler load for a strip of unit width and of length a.   The second factor indicates in what proportion the stability of the continuous plate is greater than the stability of an isolated strip.   The magnitude of this factor depends on the magnitude of the ratio a/b and also on the number m, which gives the number of half-waves into which the plate buckles.   If a is smaller than b, the second term in the parenthesis of expression (e) is always smaller than the first and the minimum value of the expression is obtained by taking $m = 1$, i.e., by assuming that the plate buckles in one half-wave and that the deflection surface has the form

$$w = a_{11} \sin \frac{\pi x}{a} \sin \frac{\pi y}{b} \qquad (f)$$

The maximum deflection a_{11} remains indefinite provided we consider it small and neglect stretching of the middle plane of the plate during buckling. The method of calculation of this deflection when it is not small will be discussed later (see Art. 9.13).

The critical load, with $m = 1$, in expression (e), can be finally represented in the following form:

$$(N_x)_{\text{cr}} = \frac{\pi^2 D}{b^2}\left(\frac{b}{a} + \frac{a}{b}\right)^2 \tag{g}$$

If we keep the width of the plate constant and gradually change the length a, the factor before the parentheses in expression (g) remains constant and the factor in parentheses changes with the change of the ratio a/b. It can be seen that this factor acquires its minimum value when $a = b$, i.e.,

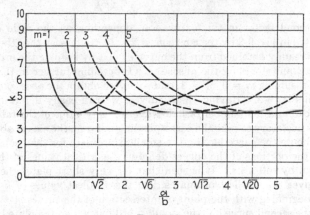

Fig. 9-2

for a plate of a given width the critical value of the load is the smallest if the plate is square. In this case

$$(N_x)_{\text{cr}} = \frac{4\pi^2 D}{b^2} \tag{h}$$

This is the same result obtained before [see Eq. (8-36)] in considering the simultaneous action on a plate of both bending and compression. For other proportions of the plate the expression (g) can be represented in the form

$$(N_x)_{\text{cr}} = k\,\frac{\pi^2 D}{b^2} \tag{9-6}$$

in which k is a numerical factor, the magnitude of which depends on the ratio a/b. This factor is represented in Fig. 9-2 by the curve marked $m = 1$. We see that it is large for small values of a/b and decreases as a/b increases, becoming a minimum for $a = b$ and then increasing again.

Let us assume now that the plate buckles into two half-waves and that the deflection surface is represented by the expression

$$w = a_{21} \sin \frac{2\pi x}{a} \sin \frac{\pi y}{b}$$

We have an inflection line dividing the plate in halves, and each half is in exactly the same condition as a simply supported plate of length $a/2$. For calculating the critical load we can again use Eq. (g) by substituting in it $a/2$ instead of a. Then

$$(N_x)_{cr} = \frac{\pi^2 D}{b^2} \left(\frac{2b}{a} + \frac{a}{2b} \right)^2 \qquad (i)$$

The second factor in this expression, depending on the ratio a/b, is represented in Fig. 9-2 by the curve $m = 2$. It can be seen that the curve $m = 2$ is readily obtained from the curve $m = 1$ by keeping the ordinates unchanged and doubling the abscissas. Proceeding further in the same way and assuming $m = 3, m = 4$, and so on, we obtain the series of curves shown in Fig. 9-2. Having these curves, we can easily determine the critical load and the number of half-waves for any value of the ratio a/b. It is only necessary to take the corresponding point on the axis of abscissas and to choose the curve having the smallest ordinate for that point. In Fig. 9-2 the portions of the curves defining the critical values of the load are shown by full lines. It is seen that for very short plates the curve $m = 1$ gives the smallest ordinates, i.e., the smallest values of k in Eq. (9-6). Beginning with the point of intersection of the curves $m = 1$ and $m = 2$, the second curve has the smallest ordinates; i.e., the plate buckles into two half-waves, and this holds up to the point of intersection of the curves $m = 2$ and $m = 3$. Beginning from this point, the plate buckles in three half-waves, and so on. The transition from m to $m + 1$ half-waves evidently occurs when the two corresponding curves in Fig. 9-2 have equal ordinates, i.e., when

$$\frac{mb}{a} + \frac{a}{mb} = \frac{(m + 1)b}{a} + \frac{a}{(m + 1)b}$$

From this equation we obtain

$$\frac{a}{b} = \sqrt{m(m + 1)} \qquad (j)$$

Substituting $m = 1$, we obtain

$$\frac{a}{b} = \sqrt{2} = 1.41$$

At this ratio we have transition from one to two half-waves. By taking

$m = 2$ we find that transition from two to three half-waves occurs when

$$\frac{a}{b} = \sqrt{6} = 2.45$$

It is seen that the number of half-waves increases with the ratio a/b, and for very long plates m is a large number. Then, from (j), we obtain

$$\frac{a}{b} \approx m$$

i.e., a very long plate buckles in half-waves, the lengths of which approach the width of the plate. Thus a buckled plate subdivides approximately into squares.

After the number of half-waves m in which a plate buckles has been determined from Fig. 9-2 or from Eq. (j), the critical load is calculated from Eq. (g). It is only necessary to substitute in Eq. (g) the length a/m of one half-wave, instead of a.

To simplify this calculation, Table 9-1 can be used; the values of the factor k in Eq. (9-6) are given for various values of the ratio a/b.

From Eq. (9-6) the critical value of the compressive stress is

$$\sigma_{cr} = \frac{(N_x)_{cr}}{h} = \frac{k\pi^2 E}{12(1 - \nu^2)} \frac{h^2}{b^2} \qquad (9-7)$$

For a given ratio a/b the coefficient k is constant, and σ_{cr} is proportional to the modulus of the material and to the square of the ratio h/b. In the

TABLE 9-1. VALUES OF FACTOR k IN EQ. (9-6) FOR UNIFORMLY COMPRESSED, SIMPLY SUPPORTED RECTANGULAR PLATES AND σ_{cr} IN PSI FOR $E = 30 \cdot 10^6$ PSI, $b/h = 100$, $\nu = 0.3$

a/b	0.2	0.3	0.4	0.5	0.6	0.7	0.8
k	27.0	13.2	8.41	6.25	5.14	4.53	4.20
σ_{cr}	73,200	35,800	22,800	16,900	13,900	12,300	11,400

a/b	0.9	1.0	1.1	1.2	1.3	1.4	1.41
k	4.04	4.00	4.04	4.13	4.28	4.47	4.49
σ_{cr}	11,000	10,800	11,000	11,200	11,600	12,100	12,200

third line of Table 9-1 the critical stresses are given for steel plates, assuming $E = 30 \cdot 10^6$ psi, $\nu = 0.3$, and $h/b = 0.01$. For any other material with a modulus E_1 and any other value of the ratio h/b, the critical stress is obtained by multiplying the values in the table by the factor[1]

$$\frac{E_1}{30 \cdot 10^2} \left(\frac{h}{b}\right)^2$$

[1] It is assumed that Poisson's ratio ν can be considered as a constant.

Comparing steel and duralumin plates of the same dimensions a and b, it is interesting to note that for the same weight the duralumin plate will be about three times thicker than the steel plate; since the modulus of elasticity of duralumin is about one-third that of steel, it can be concluded from Eq. (9-7) that the critical stress for the duralumin plate will be about three times larger and the critical load about nine times larger than for a steel plate of the same weight.   From this comparison it can be seen how important is the use of light aluminum alloy sheets in such structures as airplanes where the weight of the structure is of primary importance.

The critical values of σ_z, calculated by the use of Table 9-1, represent the true critical stresses provided they are below the proportional limit of the material.   Above this limit formula (9-7) gives an exaggerated value for σ_{cr}, and the true value of this stress can be obtained only by taking into consideration the plastic deformation of the material (see Art. 9.12).   In each particular case, assuming that formula (9-7) is accurate enough up to

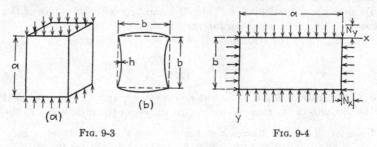

FIG. 9-3   FIG. 9-4

the yield point of the material, the limiting value of the ratio h/b, up to which formula (9-7) can be applied, is obtained by substituting in it $\sigma_{cr} = \sigma_{YP}$.   Taking, for instance, steel for which $\sigma_{YP} = 40,000$ psi, $E = 30 \cdot 10^6$ psi, and $\nu = 0.3$ and assuming that the plate is long enough so that $k \approx 4$, we find from Eq. (9-7) that $b/h \approx 52$.   Below this value of the ratio b/h the material begins to yield before the critical stress given by formula (9-7) is obtained.

The edge conditions assumed in the problem discussed above are realized in the case of uniform compression of a thin tube of square cross section (Fig. 9-3).   When compressive stresses become equal to their critical value (9-7), buckling begins and the cross sections of the tube become curved as shown in Fig. 9-3b.   There will be no bending moments acting between the sides of the buckled tube along the corners, and each side is in the condition of a compressed rectangular plate with simply supported edges.

9.3. Buckling of Simply Supported Rectangular Plates Compressed in Two Perpendicular Directions.   If a rectangular plate (Fig. 9-4) with simply supported edges is submitted to the action of uniformly distributed compressive forces N_x and N_y, the same expression for the deflection w can be used as in the previous article, and it can be proved again that only one term in the double series for w should be considered in calculating the critical values of N_x and N_y. Applying the energy method, we find that Eq. (9-2) then becomes

$$N_x \frac{m^2 \pi^2}{a^2} + N_y \frac{n^2 \pi^2}{b^2} = D\left(\frac{m^2 \pi^2}{a^2} + \frac{n^2 \pi^2}{b^2}\right)^2 \qquad (a)$$

in which m determines the number of half-waves in the x direction and n the number in the y direction. Dividing Eq. (a) by the thickness of the plate and introducing the notation

$$\frac{\pi^2 D}{a^2 h} = \sigma_e \qquad (b)$$

we obtain $\qquad \sigma_x m^2 + \sigma_y n^2 \dfrac{a^2}{b^2} = \sigma_e \left(m^2 + n^2 \dfrac{a^2}{b^2}\right)^2 \qquad (c)$

Taking any integer for m and n, the corresponding deflection surface of the buckled plate is given by the equation

$$w_{mn} = a_{mn} \sin \frac{m \pi x}{a} \sin \frac{n \pi y}{b}$$

and the corresponding values of σ_x and σ_y are such as to satisfy Eq. (c). Taking σ_x and σ_y as rectangular coordinates, Eq. (c) will be represented by a straight line. Several

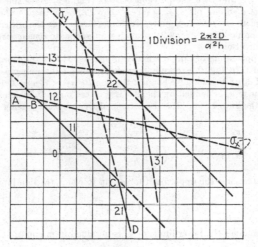

FIG. 9-5

lines of this kind for various values of m and n and for the case of a square plate ($a = b$) are shown in Fig. 9-5. The values of m and n are indicated on these lines and positive values of σ_x and σ_y indicate compressive stresses. Since we seek the smallest values of σ_x and σ_y at which buckling may occur, we need to consider only the portions of the straight lines shown in the figure by full lines and forming the polygon $ABCD$. By preparing a figure analogous to Fig. 9-5 for any given ratio a/b, we can obtain the corresponding critical values of σ_x and σ_y from that figure. It can be seen that the point of intersection of the line BC with the axis of abscissas gives the critical value of σ_x for the case where $\sigma_y = 0$, discussed in the previous article. The intersection of the same line with the vertical axis gives $(\sigma_y)_{cr}$, when $\sigma_x = 0$.

When $\sigma_x = \sigma_y = \sigma$, we draw through the origin O a line which makes an angle of $45°$ with the horizontal axis. Then the intersection of this line with the line BC determines the critical value of σ in this case.

Equation (c) in this case becomes

$$\sigma = \sigma_e \left(m^2 + n^2 \frac{a^2}{b^2} \right)$$

The smallest value of σ is obtained by taking $m = n = 1$. Hence

$$\sigma_{cr} = \sigma_e \left(1 + \frac{a^2}{b^2} \right) \tag{9-8}$$

In the particular case of a square plate

$$\sigma_{cr} = 2\sigma_e$$

i.e., the critical stress in this case is just half that for the case of compression of a square plate in one direction only.

For any value of σ_x the critical value of σ_y is obtained by drawing a vertical line through the corresponding point on the axis of abscissas. The ordinate of the point of intersection of this line with the polygon $ABCD$ (Fig. 9-5) gives the value of $(\sigma_y)_{cr}$. If, in the case of a square plate, σ_x is larger than $4\sigma_e$, $(\sigma_y)_{cr}$ becomes negative. This shows that the plate can stand a compressive stress larger than the critical value for the case of simple compression (Art. 9.2), provided an adequate tensile stress acts in the perpendicular direction.

In practical applications of Eq. (c) it is advantageous to know the coordinates of such points as B and C in Fig. 9-5 representing the apexes of the polygon $ABCD$. The largest value of σ_x up to which we can use the line $BC(m = 1, n = 1)$ in determining the critical value of σ_y is defined by the abscissa of the point C, which is the point of intersection of the lines 11 and 21. Equations of these lines, by using Eq. (c), are

$$\sigma_x + \sigma_y \frac{a^2}{b^2} = \sigma_e \left(1 + \frac{a^2}{b^2} \right)^2 \qquad m = 1 \qquad n = 1$$

$$4\sigma_x + \sigma_y \frac{a^2}{b^2} = \sigma_e \left(4 + \frac{a^2}{b^2} \right)^2 \qquad m = 2 \qquad n = 1$$

Solving these equations, the upper limit for σ_x, up to which the line 11 can be used, is

$$\sigma_x = \sigma_e \left(5 + 2 \frac{a^2}{b^2} \right) \tag{d}$$

In the same manner the lower limit for σ_x is obtained from equations

$$\sigma_x + \sigma_y \frac{a^2}{b^2} = \sigma_e \left(1 + \frac{a^2}{b^2} \right)^2 \qquad m = 1 \qquad n = 1$$

$$\sigma_x + 4\sigma_y \frac{a^2}{b^2} = \sigma_e \left(1 + \frac{4a^2}{b^2} \right)^2 \qquad m = 1 \qquad n = 2$$

from which $\qquad\qquad \sigma_x = \sigma_e \left(1 - 4 \frac{a^4}{b^4} \right) \tag{e}$

Hence the line $m = 1$, $n = 1$ should be used in calculating $(\sigma_y)_{cr}$ if the following inequality for σ_x holds:

$$\sigma_e \left(1 - 4\frac{a^4}{b^4} \right) < \sigma_x < \sigma_e \left(5 + 2\frac{a^2}{b^2} \right) \tag{f}$$

Let us take $a = 0.5b$ and $\sigma_x = \sigma_e$ and determine the corresponding critical value of σ_y. Substituting $a/b = 0.5$ in the inequality (f), we find

$$0.75\sigma_e < \sigma_x < 5.5\sigma_e$$

Since the given value of σ_x is within these limits, we substitute $m = n = 1$ in Eq. (c). Then

$$\sigma_x + \sigma_y\frac{a^2}{b^2} = \sigma_e \left(1 + \frac{a^2}{b^2} \right)^2$$

and by taking $a/b = 0.5$ and $\sigma_x = \sigma_e$ we obtain

$$(\sigma_y)_{\text{cr}} = 2.25\sigma_e = 2.25\frac{\pi^2 D}{a^2 h}$$

If the given value of σ_x is larger than the limiting value (d) it is necessary to consider straight lines obtained from Eq. (c) by taking $n = 1$ and putting $m = 2, 3, 4, \ldots$. Consider the line for which $n = 1$ and $m = i$. The lower limit for σ_x for which this line should be used is defined by the point of intersection of this line with the line for which $m = i - 1$, $n = 1$. The equations of these lines, from Eq. (c), are

$$\sigma_x i^2 + \sigma_y\frac{a^2}{b^2} = \sigma_e \left(i^2 + \frac{a^2}{b^2} \right)^2$$

$$\sigma_x(i - 1)^2 + \sigma_y\frac{a^2}{b^2} = \sigma_e \left[(i - 1)^2 + \frac{a^2}{b^2} \right]^2$$

from which the lower limit of σ_x is

$$\sigma_x = \sigma_e \left(2i^2 - 2i + 1 + 2\frac{a^2}{b^2} \right)$$

In the same manner the intersection point of the lines for which $m = i$, $n = 1$, and $m = i + 1$, $n = 1$ determines the upper limit for σ_x which is

$$\sigma_x = \sigma_e \left(2i^2 + 2i + 1 + 2\frac{a^2}{b^2} \right)$$

Thus the line $m = i$, $n = 1$ must be used for determining the critical value of σ_y if the given value of σ_x is within the following limits:

$$\sigma_e \left(2i^2 - 2i + 1 + 2\frac{a^2}{b^2} \right) < \sigma_x < \sigma_e \left(2i^2 + 2i + 1 + 2\frac{a^2}{b^2} \right) \tag{g}$$

In the same manner, if σ_x is smaller than the limiting value (e), the lines for which $m = 1$, $n = 2, 3, 4, \ldots$ must be considered. Proceeding as in the previous case, we will find that the line defined by $m = 1$, $n = i$ must be used if the following inequality[1] holds:

$$\sigma_e \left[1 - i^2(i - 1)^2\frac{a^4}{b^4} \right] > \sigma_x > \sigma_e \left[1 - i^2(i + 1)^2\frac{a^4}{b^4} \right] \tag{h}$$

Let us consider, as an example, the case where $a/b = 0.5$, and let us assume $\sigma_x = 7\sigma_e$. Since σ_x larger than the value (d), we use the general inequality (g) which will be

[1] Compressive stresses here must be considered as positive.

satisfied by taking $i = 2$. Hence, substituting $m = 2$, $n = 1$ in Eq. (c), the straight line which must be used in this case is

$$4\sigma_x + \sigma_y \frac{a^2}{b^2} = \sigma_e \left(4 + \frac{a^2}{b^2} \right)^2$$

Substituting in this equation $\sigma_x = 7\sigma_e$ and $a = 0.5b$, we find that

$$(\sigma_y)_{cr} = -39.75\sigma_e$$

This indicates that a tensile stress larger than $39.75\sigma_e$ must act in the y direction to prevent buckling of the plate under the action of the given value of the compressive stress σ_x.

If to the same plate there is applied a tensile stress in the x direction of magnitude $\sigma_x = -11\sigma_e$, we must use the inequality (h) which will be satisfied by taking $i = 4$. The corresponding critical value of σ_y will be determined from the following equation:

$$\sigma_x + 16\sigma_y \frac{a^2}{b^2} = \sigma_e \left(1 + 16 \frac{a^2}{b^2} \right)^2$$

Substituting in this equation $\sigma_x = -11\sigma_e$ and $a = 0.5b$, we find that

$$\sigma_{cr} = 9\sigma_e$$

9.4. Buckling of Uniformly Compressed Rectangular Plates Simply Supported along Two Opposite Sides Perpendicular to the Direction of Compression and Having Various Edge Conditions along the Other Two Sides.

FIG. 9-6

In the discussion of this problem both methods, the method of energy and the method of integration of the differential equation for the deflected plate, can be used.[1] In applying the method of integration we use Eq. (9-1), which for the case of uniform compression along the x axis (see Fig. 9-6), and with N_x considered positive for compression, becomes

$$\frac{\partial^4 w}{\partial x^4} + 2 \frac{\partial^4 w}{\partial x^2 \, \partial y^2} + \frac{\partial^4 w}{\partial y^4} = - \frac{N_x}{D} \frac{\partial^2 w}{\partial x^2} \tag{a}$$

Assuming that under the action of compressive forces the plate buckles in m sinusoidal half-waves, we take the solution of Eq. (a) in the form

$$w = f(y) \sin \frac{m\pi x}{a} \tag{b}$$

in which $f(y)$ is a function of y alone, which is to be determined later.

---

[1] The problem was discussed first by Timoshenko, *Bull. Polytech. Inst., Kiev*, 1907; see also *Z. Math. Physik*, vol. 58, p. 343, 1910. The use of the energy method was shown in Timoshenko, Sur la stabilité des systèmes élastiques, *Ann. ponts et chaussées*, 1913.

Expression (b) satisfies the boundary conditions along the simply supported sides $x = 0$ and $x = a$ of the plate, since

$$w = 0 \quad \text{and} \quad \frac{\partial^2 w}{\partial x^2} + \nu \frac{\partial^2 w}{\partial y^2} = 0 \quad \text{for } x = 0 \text{ and } x = a$$

Substituting (b) in Eq. (a), we obtain the following ordinary differential equation for determining the function $f(y)$:

$$\frac{d^4 f}{dy^4} - \frac{2m^2 \pi^2}{a^2} \frac{d^2 f}{dy^2} + \left(\frac{m^4 \pi^4}{a^4} - \frac{N_x}{D} \frac{m^2 \pi^2}{a^2} \right) f = 0 \qquad (c)$$

Noting that, owing to some constraints along the sides $y = 0$ and $y = b$, we always have

$$\frac{N_x}{D} > \frac{m^2 \pi^2}{a^2}$$

and using the notations

$$\alpha = \sqrt{\frac{m^2 \pi^2}{a^2} + \sqrt{\frac{N_x}{D} \frac{m^2 \pi^2}{a^2}}} \qquad \beta = \sqrt{-\frac{m^2 \pi^2}{a^2} + \sqrt{\frac{N_x}{D} \frac{m^2 \pi^2}{a^2}}} \qquad (d)$$

we can present the general solution of Eq. (c) in the following form:

$$f(y) = C_1 e^{-\alpha y} + C_2 e^{\alpha y} + C_3 \cos \beta y + C_4 \sin \beta y \qquad (e)$$

The constants of integration in this solution must be determined in each particular case from the conditions of constraint along the sides $y = 0$ and $y = b$. Several particular cases of constraint along these sides will now be discussed.

The side $y = 0$ is simply supported; the side $y = b$ is free (Fig. 9-6). From these conditions it follows [see Eqs. (8-19) to (8-21)] that

$$w = 0 \qquad \frac{\partial^2 w}{\partial y^2} + \nu \frac{\partial^2 w}{\partial x^2} = 0 \qquad \text{for } y = 0 \qquad (f)$$

$$\frac{\partial^2 w}{\partial y^2} + \nu \frac{\partial^2 w}{\partial x^2} = 0 \qquad \frac{\partial^3 w}{\partial y^3} + (2 - \nu) \frac{\partial^3 w}{\partial x^2 \partial y} = 0 \qquad \text{for } y = b \qquad (g)$$

The boundary conditions (f) will be satisfied if we take in the general solution (e)

$$C_1 = -C_2 \quad \text{and} \quad C_3 = 0$$

The function $f(y)$ can be written then in the form

$$f(y) = A \sinh \alpha y + B \sin \beta y$$

in which A and B are constants. From the boundary conditions (g) it follows that

$$A\left(\alpha^2 - \nu\frac{m^2\pi^2}{a^2}\right)\sinh\alpha b - B\left(\beta^2 + \nu\frac{m^2\pi^2}{a^2}\right)\sin\beta b = 0$$

$$A\alpha\left[\alpha^2 - (2-\nu)\frac{m^2\pi^2}{a^2}\right]\cosh\alpha b$$

$$-B\beta\left[\beta^2 + (2-\nu)\frac{m^2\pi^2}{a^2}\right]\cos\beta b = 0 \qquad (h)$$

These equations can be satisfied by taking $A = B = 0$. Then the deflection at each point of the plate is zero and we obtain the flat form of equilibrium of the plate. The buckled form of equilibrium of the plate becomes possible only if Eqs. (h) yield for A and B solutions different from zero, which requires that the determinant of these equations becomes zero; i.e.,

$$\beta\left(\alpha^2 - \nu\frac{m^2\pi^2}{a^2}\right)^2\tanh\alpha b = \alpha\left(\beta^2 + \nu\frac{m^2\pi^2}{a^2}\right)^2\tan\beta b \qquad (i)$$

Since α and β contain N_x [see notations (d)], Eq. (i) can be used for the calculation of the critical value of N_x if the dimensions of the plate and the elastic constants of the material are known. These calculations show that the smallest value of N_x is obtained by taking $m = 1$, i.e., by assuming that the buckled plate has only one half-wave. The magnitude of the corresponding critical compressive stress can be represented by the formula

$$(\sigma_x)_{cr} = \frac{(N_x)_{cr}}{h} = k\frac{\pi^2 D}{b^2 h} \qquad (j)$$

in which k is a numerical factor depending on the magnitude of the ratio a/b. Several values of this factor, calculated from Eq. (i) for $\nu = 0.25$, are given in the second line of Table 9-2. For long plates it can be assumed with sufficient accuracy that

$$k = \left(0.456 + \frac{b^2}{a^2}\right)$$

In the third line of Table 9-2 the critical stresses in pounds per square inch are given, calculated on the assumption that $E = 30 \cdot 10^6$ psi, $\nu = 0.25$, and $h/b = 0.01$. For any other material with a modulus E_1 and any other value of the ratio h/b, the critical stress is obtained by multiplying

TABLE 9-2. NUMERICAL VALUES OF THE FACTOR k IN EQ. (j) WHEN THE SIDE $y = 0$ IS SIMPLY SUPPORTED AND THE SIDE $y = b$ IS FREE (FIG. 9-6)

a/b	0.50	1.0	1.2	1.4	1.6	1.8	2.0	2.5	3.0	4.0	5.0
k	4.40	1.440	1.135	0.952	0.835	0.755	0.698	0.610	0.564	0.516	0.506
$(\sigma_x)_{cr}$	11,600	3,790	2,990	2,500	2,200	1,990	1,840	1,600	1,480	1,360	1,330

the numbers of the table by the factor

$$\frac{E_1}{30 \cdot 10^2} \left(\frac{h}{b}\right)^2$$

Edge conditions similar to those assumed above are realized in the case of compression of an angle as shown in Fig. 9-7. When the compressive stresses, uniformly distributed over the width of the sides of the angle, become equal to the critical stress given by formula (j), the free longitudinal edges of the angle buckle, as shown in the figure, while the line AB remains straight and the edge conditions along this line are the same as along a simply supported edge. Experiments made with compression of angles[1] are in good agreement with the theory. In the case of comparatively short angles buckling occurs as shown in Fig. 9-7. For a long strut with such an angular cross section, the critical compressive stress may become smaller than that given by formula (j), in which case the strut buckles like a compressed column.

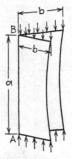

FIG. 9-7

The side $y = 0$ is built in; the side $y = b$ is free (Fig. 9-6). In this case the edge conditions for determining the constants in the general solution (e) are

$$w = 0 \qquad\qquad \frac{\partial w}{\partial y} = 0 \qquad \text{for } y = 0 \quad (k)$$

$$\frac{\partial^2 w}{\partial y^2} + \nu \frac{\partial^2 w}{\partial x^2} = 0 \qquad \frac{\partial^3 w}{\partial y^3} + (2 - \nu) \frac{\partial^3 w}{\partial x^2 \, \partial y} = 0 \qquad \text{for } y = b \quad (l)$$

From the conditions (k) it follows that

$$C_1 = -\frac{\alpha C_3 - \beta C_4}{2\alpha} \qquad C_2 = -\frac{\alpha C_3 + \beta C_4}{2\alpha}$$

and the function $f(y)$ can be represented in the form

$$f(y) = A(\cos \beta y - \cosh \alpha y) + B\left(\sin \beta y - \frac{\beta}{\alpha} \sinh \alpha y\right)$$

Substituting this expression in the conditions (l), two homogeneous equations linear in A and B are obtained. The critical value of the compressive stress is determined by equating to zero the determinant of these equations, which gives

$$2ts + (s^2 + t^2) \cos \beta b \cosh \alpha b = \frac{1}{\alpha\beta} (\alpha^2 t^2 - \beta^2 s^2) \sin \beta b \sinh \alpha b \quad (m)$$

---

[1] See F. J. Bridget, C. C. Jerome, and A. B. Vosseller, *Trans. ASME*, Applied Mechanics Division, vol. 56, p. 569, 1934. See also Dissertation by C. F. Kollbrunner, Zürich, 1935, and E. E. Lundquist, *NACA Tech. Note* 722, 1939.

where $\quad . \quad t = \beta^2 + \nu \dfrac{m^2\pi^2}{a^2} \qquad s = \alpha^2 - \nu \dfrac{m^2\pi^2}{a^2}$

For a given value of the ratio a/b and a given value of ν the critical value of compressive stress can be calculated from the transcendental equation (m) and can be represented by Eq. (j). Calculations show that for a comparatively short length a the plate buckles in one half-wave and we must take $m = 1$ in our calculations. Several values of the numerical factor k in Eq. (j) for various values of the ratio a/b are given in Table 9-3.

TABLE 9-3. NUMERICAL VALUES OF THE FACTOR k IN EQ. (j) WHEN THE SIDE
$y = 0$ IS BUILT IN AND THE SIDE $y = b$ IS FREE (FIG. 9-6); $\nu = 0.25$

a/b	1.0	1.1	1.2	1.3	1.4	1.5	1.6	1.7	1.8	1.9	2.0	2.2	2.4
k	1 70	1.56	1.47	1.41	1.36	1.34	1.33	1.33	1.34	1.36	1.38	1 45	1.47
$(\sigma_x)_{cr}$	4,470	4,110	3,870	3,710	3,580	3,520	3,500	3,500	3,520	3,580	3,630	3,820	3,870

The same values are also represented in Fig. 9-8 by the curve $m = 1$. It is seen that at the beginning the values of k decrease with an increase in the ratio a/b. The minimum value of k ($k = 1.328$) is obtained for $a/b = 1.635$, and beginning from this value, k increases with the ratio a/b.

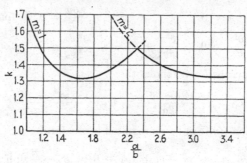

FIG. 9-8

Having the curve for $m = 1$, we can construct the curves for $m = 2$, $m = 3, \ldots$, as explained in Art. 9.2. With the use of such curves the number of half-waves in any particular case can be determined readily. In the case of a comparatively long plate we can take with sufficient accuracy $k = 1.328$ in Eq. (j).

In the third line of Table 9-3 are given the values of the critical stresses in pounds per square inch, calculated from Eq. (j) assuming $E = 30 \cdot 10^6$ psi, $\nu = 0.25$, and $h/b = 0.01$. By using these figures we can easily calculate the critical stresses for any other proportions of the plate and any value of the modulus.

The side $y = 0$ is elastically built in and the side $y = b$ is free (Fig. 9-6). In the previous discussions two extreme assumptions for the constraint along the side $y = 0$ have been considered, namely, a simply supported edge and a built-in edge. In practical cases we shall usually have some intermediate condition of constraint. Take, for instance, the case of a compression member of a T cross section (Fig. 9-9). While the upper edge of the vertical web cannot be assumed to rotate freely during buckling, neither can it be considered as rigidly built in since during buckling of the web some rotation of the horizontal flange will take place. We can consider in this case the upper edge of the plate as *elastically built in*, since the bending moments that appear during buckling along this edge are proportional at each point to the angle of rotation of the edge. To show this, let us consider torsion of the flange of the member shown in Fig. 9-9. The angle of rotation of this flange during buckling of the web is equal to $\partial w/\partial y$, and the rate of change of this angle is $\partial^2 w/\partial x\,\partial y$; hence the twisting moment at any cross section of the flange along the x axis is

$$C\,\frac{\partial^2 w}{\partial x\,\partial y}$$

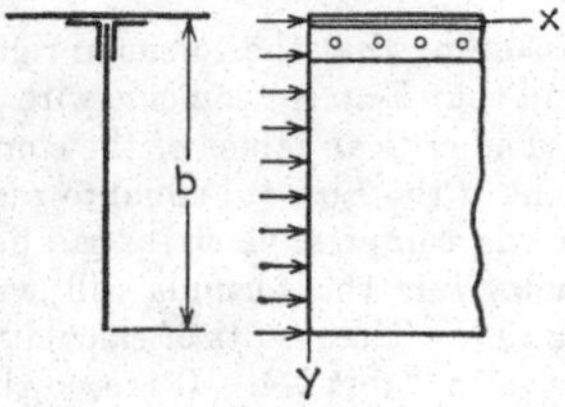

where C is the torsional rigidity of the flange.[1] The rate of change of this twisting moment is numerically equal to the bending moment M_y per unit length of the upper edge of the web. The signs of these two moments, with the assumed rule regarding the positive direction of M_y (see p. 319), are also the same. Hence the corresponding boundary condition along the upper edge of the web is

$$-D\left(\frac{\partial^2 w}{\partial y^2} + \nu\,\frac{\partial^2 w}{\partial x^2}\right) = C\,\frac{\partial^3 w}{\partial x^2\,\partial y}$$

FIG. 9-9

Using for w expression (b) and observing that along the upper edge of the web $w = 0$,[2] the above boundary condition can be put in the following form:[3]

$$D\,\frac{\partial^2 w}{\partial y^2} = C\,\frac{\pi^2}{a^2}\,\frac{\partial w}{\partial y} \qquad \text{for } y = 0 \tag{n}$$

That is, the bending moments M_y along the upper edge of the buckled web are proportional to the angle of rotation $\partial w/\partial y$ as stated before.

From the condition (n), together with the condition that $w = 0$ for $y = 0$, we find the following relations between the constants in expression (e):

$$C_1 = \frac{C_3(\alpha^2 + \beta^2 - r\alpha)}{2r\alpha} + \frac{C_4\beta}{2\alpha}$$

$$C_2 = -\frac{C_3(\alpha^2 + \beta^2 + r\alpha)}{2r\alpha} - \frac{C_4\beta}{2\alpha}$$

where

$$r = \frac{C}{D}\frac{\pi^2}{a^2} \tag{o}$$

---

[1] This rigidity can be calculated in each particular case with sufficient accuracy by considering the cross section of the flange as consisting of narrow rectangles (see Art. 5.2). The center of twist is assumed to coincide with the edge of the web. The warping rigidity of the flanges is neglected here.

[2] We assume here that the flexural rigidity of the flange in the xz plane is very large and we neglect deflection of the flange in that plane.

[3] It is assumed that $m = 1$, i.e., that there is only one half-wave formed by the buckled web.

With these relations between the constants, we obtain

$$f(y) = C_3 \left(\cos \beta y - \cosh \alpha y - \frac{\alpha^2 + \beta^2}{r\alpha} \sinh \alpha y \right) + C_4 \left(\sin \beta y - \frac{\beta}{\alpha} \sinh \alpha y \right)$$

Substituting this in the expression (b) for w and in the boundary conditions (l) along the free edge of the web, we obtain the following equations:

$$C_3(t \cos \beta b + s \cosh \alpha b + qs \sinh \alpha b) + C_4 \left(t \sin \beta b + \frac{\beta}{\alpha} s \sinh \alpha b \right) = 0$$

$$C_3(-\beta s \sin \beta b + \alpha t \sinh \alpha b + q\alpha t \cosh \alpha b) + C_4(\beta s \cos \beta b + \beta t \cosh \alpha b) = 0 \qquad (p)$$

where

$$s = \alpha^2 - \nu \frac{\pi^2}{a^2} \qquad t = \beta^2 + \nu \frac{\pi^2}{a^2} \qquad q = \frac{\alpha^2 + \beta^2}{\alpha r}$$

Assuming that the torsional rigidity of the flange is very large and taking $q = 0$, we find that Eqs. (p) coincide with those for a plate rigidly built in along the edge $y = 0$.

The critical values of the compressive forces will be found by setting the determinant of the Eqs. (p) equal to zero. In this way we find again that the critical values of the compressive stress can be represented by the formula (j). The values of the factor k in this formula will evidently depend on the magnitude of the *coefficient of fixity* r. The results of calculations made for $rb = 2$ and $rb = 8$ and for $\nu = 0.25$ are given in Table 9-4. It is seen that with an increase of r the factor k increases, and for

TABLE 9-4. NUMERICAL VALUES OF THE FACTOR k IN EQ. (j) WHEN THE SIDE $y = 0$ IS ELASTICALLY BUILT IN AND THE SIDE $y = b$ IS FREE; $\nu = 0.25$

a/b	1	1.3	1.5	1.8	2.0	2.3	2.5	2.7	3.0	4.0
$rb = 2$	1.49	1.13	1.01	0.92	0.90	0.89	0.90	0.93	0.98	0.90
$rb = 8$	1.58	1.25	1.16	1.11	1.12	1.18	1.23	1.30	1.16	1 12

$rb = 8$ the values of k approach the values given in Table 9-3, calculated for a rigidly built-in edge. It can be seen also that with an increase of r the ratio a/b, at which k becomes a minimum, decreases. This means that in the case of long plates the length of the waves into which the plate subdivides during buckling decreases as r increases. By using the values of k given in Table 9-4 we can construct curves analogous to those shown in Fig. 9-8, and from such curves the number of half-waves in which a plate subdivides at buckling can be determined in the same manner as explained before.

Both sides $y = 0$ and $y = b$ are built in. In this case the boundary conditions are

$$w = 0 \qquad \frac{\partial w}{\partial y} = 0 \qquad \text{for } y = 0 \text{ and } y = b$$

Proceeding as in the previous cases, we find for the calculation of the critical value of the compressive forces the following transcendental equation:

$$2(1 - \cos \beta b \cosh \alpha b) = \left(\frac{\beta}{\alpha} - \frac{\alpha}{\beta} \right) \sin \beta b \sinh \alpha b$$

The critical values of the compressive stress again are given by Eq. (j). Several values of the numerical factor k, calculated for various values of the ratio a/b, are given in Table 9-5.

TABLE 9-5. NUMERICAL VALUES OF THE FACTOR k IN EQ. (j) WHEN
BOTH SIDES $y = 0$ AND $y = b$ ARE BUILT IN; $\nu = 0.25$

a/b	0.4	0.5	0.6	0.7	0.8	0.9	1.0
k	9.44	7.69	7.05	7.00	7.29	7.83	7.69

It is seen that the smallest value of k is obtained when $0.6 < a/b < 0.7$; that is, a long compressed plate buckles in this case in comparatively short waves. The number of half-waves can be determined as before by plotting curves analogous to those shown in Fig. 9-8.

Both sides $y = 0$ and $y = b$ are supported by elastic beams. Taking the coordinate axes as shown in Fig. 9-10, we assume that the conditions at the edges $x = 0$ and $x = a$ of the plate are the same as before. Along the edges $y = b/2$ and $y = -b/2$ the plate is free to rotate during buckling, but deflections of the plate at these edges are resisted by the two equal elastic supporting beams. The condition of freedom of rotation requires that

$$\frac{\partial^2 w}{\partial y^2} + \nu \frac{\partial^2 w}{\partial x^2} = 0 \qquad \text{for } y = \pm \frac{b}{2} \quad (q)$$

To obtain an expression for the second condition, bending of the supporting beams must be considered. We assume that these beams are simply supported at the ends, that they have the same modulus of elasticity as the plate, and that they are compressed together with the plate so that the

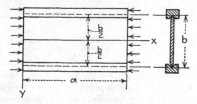

FIG. 9-10

compressive forces on each are equal to $A\sigma_x$, where A is the cross-sectional area of one beam. Denoting by EI the flexural rigidity of the beam, the differential equation of its deflection curve is

$$EI \frac{\partial^4 w}{\partial x^4} = q - A\sigma_x \frac{\partial^2 w}{\partial x^2}$$

where q is the intensity of the load transmitted from the plate to the beam. From expressions for shearing forces (see p. 330), this intensity is

$$q = D\left[\frac{\partial^3 w}{\partial y^3} + (2 - \nu)\frac{\partial^3 w}{\partial x^2\, \partial y}\right] \qquad \text{for } y = \frac{b}{2}$$

$$q = -D\left[\frac{\partial^3 w}{\partial y^3} + (2 - \nu)\frac{\partial^3 w}{\partial x^2\, \partial y}\right] \qquad \text{for } y = -\frac{b}{2}$$

Substituting these values of q in the above equation for the deflection curve, the following boundary conditions are obtained:

$$EI \frac{\partial^4 w}{\partial x^4} = D\left[\frac{\partial^3 w}{\partial y^3} + (2 - \nu)\frac{\partial^3 w}{\partial x^2\, \partial y}\right] - A\sigma_x \frac{\partial^2 w}{\partial x^2} \qquad \text{for } y = \frac{b}{2}$$

$$EI \frac{\partial^4 w}{\partial x^4} = -D\left[\frac{\partial^3 w}{\partial y^3} + (2 - \nu)\frac{\partial^3 w}{\partial x^2\, \partial y}\right] - A\sigma_x \frac{\partial^2 w}{\partial x^2} \qquad \text{for } y = -\frac{b}{2}$$

(r)

By using the four boundary conditions (q) and (r) for determining the constants in expression (e) and by equating to zero the determinant of these equations, we obtain the transcendental equation for calculating the critical values of the compressive stresses. Assuming that during buckling both supporting beams deflect in the same direction and that the deflection surface of the plate is symmetrical with respect to the x axis (Fig. 9-10), the equation for determining the critical stress σ_{cr} becomes[1]

$$\beta \left(1 - \nu + \frac{a}{m\pi} \sqrt{\frac{h\sigma_{cr}}{D}} \right)^2 \tan \frac{\beta b}{2} + \alpha \left(1 - \nu - \frac{a}{m\pi} \sqrt{\frac{h\sigma_{cr}}{D}} \right)^2 \tanh \frac{\alpha b}{2}$$
$$= \frac{2m\pi}{a} \sqrt{\frac{h\sigma_{cr}}{D}} \left(\frac{EI}{D} - \frac{a^2 A \sigma_{cr}}{m^2 \pi^2 D} \right) \quad (s)$$

where α and β are given, as before, by the formulas (d).

If we assume that the supporting beams will deflect during buckling in opposite directions, we obtain a transcendental equation which always gives for σ_{cr} larger values than those obtained from the assumption of symmetry. Thus, in this case, only the symmetrical case of buckling should be considered.

To simplify the solution of the equation we introduce the following notations:

$$\frac{m\pi b}{a} = \phi \qquad b \sqrt{\frac{h\sigma_{cr}}{D}} = \psi \qquad \frac{EI}{bD} - \frac{A}{bh} \frac{\psi^2}{\phi^2} = \theta$$

Then the equation becomes

$$\sqrt{\psi - \phi} \, [\psi + (1 - \nu)\phi]^2 \tan \tfrac{1}{2} \sqrt{\psi\phi - \phi^2}$$
$$+ \sqrt{\psi + \phi} \, [\psi - (1 - \nu)\phi]^2 \tanh \tfrac{1}{2} \sqrt{\psi\phi + \phi^2} = 2\phi^{\frac{3}{2}} \psi\theta \quad (t)$$

To simplify further the solution of this equation, curves representing ψ as a function of ϕ for various numerical values of θ are plotted in Fig. 9-11. The procedure of calculation in each particular case will then be as follows: we calculate EI/bD and A/bh by using the given dimensions of the plate and assume a certain value for the number of half-waves m. In this manner the quantity ϕ is determined, and the quantity ψ can be found by trial and error by using the curves in Fig. 9-11. In the case of very rigid supporting beams, we can take, as a first approximation, the same number of half-waves and the same value of σ_{cr} as for the case of a plate supported by absolutely rigid beams. To illustrate the application of the curves in Fig. 9-11, the critical stresses for the channels shown in Fig. 9-12 are calculated. Considering the web of the channel as a compressed rectangular plate and the flanges as the supporting beams and varying the width d of the flanges, we can obtain a variety of cases. By taking $m = 1$ and assuming $d = 1$ in., 2 in., 3 in., we obtain the curves, with always decreasing ordinates, as shown in Fig. 9-12. For comparison there is given also the curve $\theta = 0$ representing the case where the longitudinal edges of the plate are entirely free. For $d = 4$ in. the corresponding curve in Fig. 9-12 has a more complicated shape. It has a minimum for $a/b \approx 1$; then it increases and has a maximum for $a/b \approx 2.5$, after which the ordinates begin to decrease continuously. By making calculations with $m = 2$ and $m = 3$ and by plotting the corresponding curves in Fig. 9-12, it is shown that in the region $1.5 < a/b < 3.4$ these curves have smaller ordinates than the curve $m = 1$; hence, in calculating σ_{cr}, these curves should be used as shown by the full lines in Fig. 9-12. For

---

[1] The first calculations of critical stresses for a compressed plate supported by elastic beams were made by K. Cališev, *Mem. Inst. Engrs. Ways of Commun.*, St. Petersburg, 1914. The curves for calculating σ_{cr} given in this book are taken from the thesis of A. J. Miles, presented at the University of Michigan, January, 1935. Independently, the same problem was discussed by E. Melan, *Repts. Intern. Congr. Appl. Mech.*, vol. 3, p. 59, 1930, and by L. Rendulič, *Ingr.-Arch.*, vol. 3, p. 447, 1932.

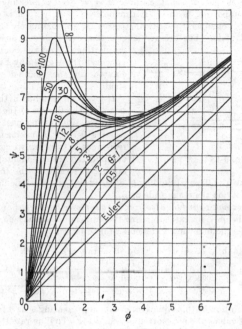

FIG. 9-11

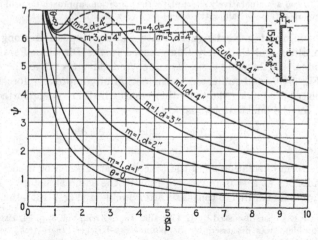

FIG. 9-12

comparison there is shown also the curve $\theta = \infty$ giving buckling conditions for absolutely rigid supporting beams. It can be seen that, with $d = 4$ in. and for a comparatively short channel, the critical stresses are approximately the same as for a rigidly supported plate. For greater lengths the curve $m = 1$ should be used in calculating σ_{cr} and the buckling conditions approach those given by Euler's curve calculated for the channel considered as a strut.

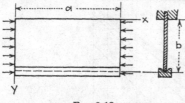

It has been assumed in the above discussion that the flanges of the member do not resist twisting during buckling of the web (see Fig. 9-10). By taking into account the resistance to twisting, we can obtain a more complicated transcendental equation, containing the ratio of the torsional rigidity C of the flanges to the flexural rigidity D of the web,[1] and can calculate the effect of the torsional rigidity on the magnitude of σ_{cr}.

FIG. 9-13

The edge $y = 0$ is built in and the edge $y = b$ is supported by an elastic beam (Fig. 9-13). Proceeding as in the previous case and using the same notations as before, we obtain for the critical stresses the following transcendental equation:

$$\theta \psi \phi^2 (\sqrt{\psi\phi - \phi^2} \tanh \sqrt{\psi\phi + \phi^2} - \sqrt{\psi\phi + \phi^2} \tan \sqrt{\psi\phi - \phi^2})$$
$$= \sqrt{\psi^2 - \phi^2} \left[\frac{\psi^2 - \phi^2(1 - \nu)^2}{\cos \sqrt{\psi\phi - \phi^2} \cosh \sqrt{\psi\phi + \phi^2}} + \psi^2 + \phi^2(1 - \nu)^2 \right]$$
$$+ \phi[\psi^2(1 - 2\nu) - \phi^2(1 - \nu)^2] \tan \sqrt{\psi\phi - \phi^2} \tanh \sqrt{\psi\phi + \phi^2}$$

To simplify the solution of this equation, curves representing ψ as a function of ϕ for various numerical values of θ are plotted in Fig. 9-14. To illustrate the application of these curves, the case represented in Fig. 9-15 has been calculated. With the use of the curves in Fig. 9-14 curves representing ψ as functions of the ratio a/b for various values of the width d of the flange have been plotted in Fig. 9-15. It can be seen that for $d = 3$ in. and a comparatively short plate, the conditions approach those of a plate rigidly supported ($\theta = \infty$) along the edge $y = b$ (Fig. 9-13).

9.5. Buckling of a Rectangular Plate Simply Supported along Two Opposite Sides and Uniformly Compressed in the Direction Parallel to Those Sides. If the plate is simply supported along the edges $x = 0$ and $x = a$ (Fig. 9-16) and subjected to the action of compressive forces uniformly distributed along the two other edges which are unsupported, the equation for the buckled plate is

$$\frac{\partial^4 w}{\partial x^4} + 2 \frac{\partial^4 w}{\partial x^2 \partial y^2} + \frac{\partial^4 w}{\partial y^4} = - \frac{N_y}{D} \frac{\partial^2 w}{\partial y^2} \qquad (a)$$

The same form of solution as in the preceding article can be used.[2] The edge conditions at $x = 0$ and $x = a$ remain the same as before. If the forces N_y remain parallel to the y axis after buckling (Fig. 9-16b), the bend-

---

[1] This case has been discussed by E. Chwalla, *Ingr.-Arch.*, vol. 5, p. 54, 1934.

[2] This problem was discussed by S. Woinowsky-Krieger, *Ingr.-Arch.*, vol. **19**, pp. 200–207, 1951.

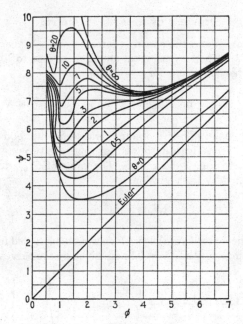

Fig. 9-14

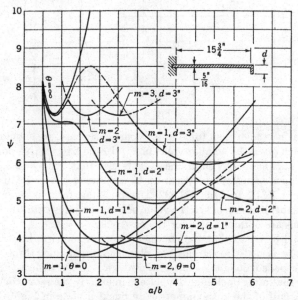

Fig. 9-15

ing moments vanish and the shearing force becomes $N_y\,\partial w/\partial y$ along the edge $y = -b/2$. Then by using an equation similar to Eq. (8-20), we find that the edge conditions for $y = \pm b/2$ are

$$\frac{\partial^2 w}{\partial y^2} + \nu\,\frac{\partial^2 w}{\partial x^2} = 0 \qquad \frac{\partial^3 w}{\partial y^3} + (2 - \nu)\,\frac{\partial^3 w}{\partial x^2\,\partial y} = -\,N_y\frac{\partial w}{\partial y} \qquad (b)$$

From these edge conditions we obtain a transcendental equation for calculating $(N_y)_{cr}$. Investigation shows that there are two possible forms of buckling, shown in Fig. 9-16b and c. The value of $(N_y)_{cr}$ for both cases can be represented by the equation

$$(N_y)_{cr} = k\,\frac{\pi^2 D}{a^2} \qquad (c)$$

The values of the factor k for the two forms of buckling and for two values of Poisson's ratio ($\nu = 0$ and $\nu = 0.3$) are given by the curves in Fig. 9-17.

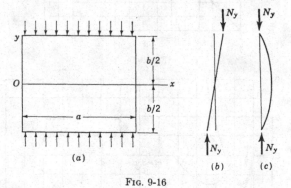

(a)

(b)   (c)

Fig. 9-16

In calculating the critical force in any particular case, we should use the curve which gives the smaller value of k. In the figure these portions of the curves are shown by solid lines. It is seen that, for $\nu = 0.3$, the first buckling form is antisymmetrical up to a value of $b/a = 1.316$. Then, from $b/a = 1.316$ to $b/a = 2.632$, the buckling form is symmetrical. With a further increase of b/a the value of k remains close to 2.31. Similar conclusions are obtained also for the case where $\nu = 0$.

It should be noted that the curves for antisymmetrical buckling (Fig. 9-17) can be used also in the case of a plate supported along three sides and uniformly loaded along the fourth side, which is free to deflect (Fig. 9-18). It is seen that after buckling, the plate is in exactly the same condition as the upper half of the antisymmetrically buckled plate in Fig. 9-16, since along the x axis there is no deflection and no bending moment. The values of k for the case shown in Fig. 9-18 are obtained from the curves for

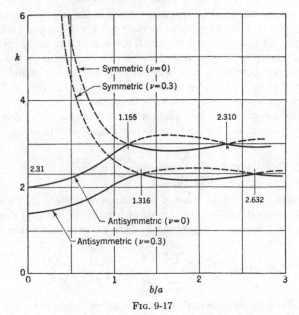

FIG. 9-17

antisymmetrical buckling in Fig. 9-17. It is necessary only to use $2b/a$ instead of b/a.[1]

9.6. Buckling of a Simply Supported Rectangular Plate under Combined Bending and Compression.[2] Let us consider a simply supported

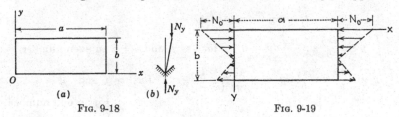

FIG. 9-18   FIG. 9-19

rectangular plate (Fig. 9-19) along whose sides $x = 0$ and $x = a$ distributed forces, acting in the middle plane of the plate, are applied, their intensity being given by the equation

$$N_x = N_0\left(1 - \alpha\frac{y}{b}\right) \qquad (a)$$

[1] The case when the plate in Fig. 9-18 is built in along the edge $y = 0$ is also discussed by Woinowsky-Krieger, *op. cit.*

[2] See Timoshenko, *loc. cit.* See also J. Boobnov, "Theory of Structure of Ships," vol. 2, p. 515, St. Petersburg, 1914, and J. H. Johnson and R. G. Noel, *J. Aeronaut. Sci.*, vol. 20, p. 535, 1953.

where N_0 is the intensity of compressive force at the edge $y = 0$ and α is a numerical factor. By changing α, we can obtain various particular cases. For example, by taking $\alpha = 0$ we obtain the case of uniformly distributed compressive force as discussed in Art. 9.2. Again, for $\alpha = 2$ we obtain the case of pure bending. If α is less than 2, we have a combination of bending and compression as indicated in Fig. 9-19. If $\alpha > 2$, there will be a similar combination of bending and tension.

The deflection of the buckled plate simply supported on all sides can be taken, as before, in the form of the double trigonometric series

$$w = \sum_{m=1}^{\infty} \sum_{n=1}^{\infty} a_{mn} \sin \frac{m\pi x}{a} \sin \frac{n\pi y}{b} \tag{b}$$

For calculating the critical value of the compressive force N_0 we shall use the energy method. Then for the strain energy of bending due to the deflections (b) we use the previously derived expression [see Eq. (b), Art. 9.2]

$$\Delta U = \frac{D}{2} \frac{ab\pi^4}{4} \sum_{m=1}^{m=\infty} \sum_{n=1}^{n=\infty} a_{mn}^2 \left(\frac{m^2}{a^2} + \frac{n^2}{b^2} \right)^2 \tag{c}$$

The work done by the external forces during buckling of the plate is [see Eq. (8-32)]

$$\Delta T = \frac{1}{2} \int_0^a \int_0^b N_0 \left(1 - \alpha \frac{y}{b} \right) \left(\frac{\partial w}{\partial x} \right)^2 dx\, dy \tag{d}$$

Substituting expression (b) for w and observing that

$$\int_0^b y \sin \frac{i\pi y}{b} \sin \frac{j\pi y}{b} dy = \frac{b^2}{4} \qquad \text{for } i = j$$

$$\int_0^b y \sin \frac{i\pi y}{b} \sin \frac{j\pi y}{b} dy = 0 \qquad \text{for } i \neq j, \text{ and } i \pm j \text{ an even number}$$

$$\int_0^b y \sin \frac{i\pi y}{b} \sin \frac{j\pi y}{b} dy = -\frac{4b^2}{\pi^2} \frac{ij}{(i^2 - j^2)^2} \qquad \text{for } i \neq j, \text{ and}$$

$$i \pm j \text{ an odd number}$$

we obtain, for the work done by external forces during buckling, the following expression:

$$\Delta T = \frac{N_0}{2} \frac{ab}{4} \sum_{m=1}^{m=\infty} \sum_{n=1}^{n=\infty} a_{mn}^2 \frac{m^2\pi^2}{a^2}$$

$$- \frac{N_0}{2} \frac{\alpha a}{2b} \sum_{m=1}^{m=\infty} \frac{m^2\pi^2}{a^2} \left[\frac{b^2}{4} \sum_{n=1}^{n=\infty} a_{mn}^2 - \frac{8b^2}{\pi^2} \sum_{n=1}^{n=\infty} \sum_i \frac{a_{mn} a_{mi} ni}{(n^2 - i^2)^2} \right]$$

where for i only such numbers are taken that $n \pm i$ is always odd.

Equating this work to the strain energy of bending (c), we obtain for the critical value of N_0 an equation which states that $(N_0)_{cr}$ is equal to

$$\frac{\pi^4 D \sum\limits_{m=1}^{m=\infty} \sum\limits_{n=1}^{n=\infty} a_{mn}{}^2 \left(\dfrac{m^2}{a^2} + \dfrac{n^2}{b^2}\right)^2}{\sum\limits_{m=1}^{m=\infty} \sum\limits_{n=1}^{n=\infty} a_{mn}{}^2 \dfrac{m^2\pi^2}{a^2} - \dfrac{\alpha}{2} \sum\limits_{m=1}^{m=\infty} \dfrac{m^2\pi^2}{a^2} \left[\sum\limits_{n=1}^{n=\infty} a_{mn}{}^2 - \dfrac{32}{\pi^2} \sum\limits_{n=1}^{n=\infty} \sum\limits_{i}^{\infty} \dfrac{a_{mn}a_{mi}ni}{(n^2 - i^2)^2}\right]}$$

$$(e)$$

The coefficients a_{mn} must be adjusted now so as to make the obtained expression for $(N_0)_{cr}$ a minimum. By taking the derivatives of this expression with respect to each coefficient a_{mn} and equating these derivatives to zero [see Eq. (9-4)], we finally obtain a system of linear equations of the following form:

$$Da_{mn}\pi^4 \left(\frac{m^2}{a^2} + \frac{n^2}{b^2}\right)^2$$

$$= (N_0)_{cr} \left\{ a_{mn} \frac{m^2\pi^2}{a^2} - \frac{\alpha}{2} \frac{m^2\pi^2}{a^2} \left[a_{mn} - \frac{16}{\pi^2} \sum_{i}^{\infty} \frac{a_{mi}ni}{(n^2 - i^2)^2} \right] \right\} \quad (f)$$

Let us collect all equations with a certain value of the number m. These equations will contain coefficients a_{m1}, a_{m2}, a_{m3}, ... All other coefficients we take equal to zero, i.e., instead of the general expression (b) we take, for the deflection of the plate, the expression

$$w = \sin \frac{m\pi x}{a} \sum_{n=1}^{n=\infty} a_{mn} \sin \frac{n\pi y}{b}$$

which is equivalent to the assumption that the buckled plate is subdivided along the x axis into m half-waves.[1]

We can consider one half-wave between two nodal lines as a simply supported plate which is buckled into one half-wave. Substituting $m = 1$ in Eqs. (f) and using the notation

$$\sigma_{cr} = \frac{(N_0)_{cr}}{h}$$

---

[1] This assumption is justified, since the system of Eqs. (f) can be subdivided into groups, each of which has a definite value of m, i.e., represents a buckling of the plate into a number of half-waves with nodal lines parallel to the y axis. The determinant of the entire system of Eqs. (f) becomes zero if the determinant of one of these groups is zero.

we obtain a system of equations of the following kind:

$$a_{1n}\left[\left(1 + n^2\frac{a^2}{b^2}\right)^2 - \sigma_{cr}\frac{a^2h}{\pi^2D}\left(1 - \frac{\alpha}{2}\right)\right] - 8\alpha\sigma_{cr}\frac{a^2h}{\pi^4D}\sum_i^\infty\frac{a_{1i}ni}{(n^2 - i^2)^2} = 0 \quad (g)$$

where the summation is taken over all numbers i such that $n \pm i$ is an odd number.

These are homogeneous linear equations in $a_{11}, a_{12}, \ldots$ which will be satisfied by putting $a_{11}, a_{12}, \ldots$ equal to zero, which corresponds to the flat form of equilibrium of the plate. To get for the coefficients a_{11}, $a_{12}, \ldots$ solutions different from zero, which indicates the possibility of buckling of the plate, the determinant of the Eqs. (g) must be zero. In this way an equation for calculating the critical values of compressive stresses is obtained. The calculation can be made by successive approximations. We begin by taking only the first of the Eqs. (g) and assuming that all coefficients except a_{11} are zero. In this way we obtain

$$\left(1 + \frac{a^2}{b^2}\right)^2 - \sigma_{cr}\frac{a^2h}{\pi^2D}\left(1 - \frac{\alpha}{2}\right) = 0$$

from which

$$\sigma_{cr} = \frac{\pi^2D}{a^2h}\left(1 + \frac{a^2}{b^2}\right)^2\frac{1}{1 - \alpha/2} = \frac{\pi^2D}{b^2h}\left(\frac{b}{a} + \frac{a}{b}\right)^2\frac{1}{1 - \alpha/2} \quad (h)$$

This first approximation gives a satisfactory result only for small values of α, i.e., in the cases when bending stresses are small in comparison with the uniform compressive stress [see Eq. (a)]. In the case of $\alpha = 0$, the expression (h) coincides with the expression for the critical stress of a uniformly compressed plate [see Eq. (g), Art. 9.2].

To obtain a second approximation, two equations of the system (g) with coefficients a_{11} and a_{12} should be taken, and we obtain

$$a_{11}\left[\left(1 + \frac{a^2}{b^2}\right)^2 - \sigma_{cr}\frac{a^2h}{\pi^2D}\left(1 - \frac{\alpha}{2}\right)\right] - 8\alpha\sigma_{cr}\frac{a^2h}{\pi^4D}\frac{2}{9}a_{12} = 0$$

$$-8\alpha\sigma_{cr}\frac{a^2h}{\pi^4D}\frac{2}{9}a_{11} + \left[\left(1 + 4\frac{a^2}{b^2}\right)^2 - \sigma_{cr}\frac{a^2h}{\pi^2D}\left(1 - \frac{\alpha}{2}\right)\right]a_{12} = 0$$

Equating to zero the determinant of these equations, we obtain

$$\left(\frac{\sigma_{cr}a^2h}{\pi^2D}\right)^2\left[\left(1 - \frac{\alpha}{2}\right)^2 - \left(\frac{8\alpha}{\pi^2}\frac{2}{9}\right)^2\right] - \frac{\sigma_{cr}a^2h}{\pi^2D}\left(1 - \frac{\alpha}{2}\right)\left[\left(1 + \frac{a^2}{b^2}\right)^2\right.$$
$$\left. + \left(1 + 4\frac{a^2}{b^2}\right)^2\right] + \left(1 + \frac{a^2}{b^2}\right)^2\left(1 + 4\frac{a^2}{b^2}\right)^2 = 0 \quad (i)$$

From this equation the second approximation for σ_{cr} can be calculated. The accuracy of this approximation decreases as α increases; for pure bending, when $\alpha = 2$, and for a square plate $(a = b)$ the error is about 8 per cent, so that the calculation of a further approximation is necessary to obtain a more satisfactory accuracy. By taking three equations of the system (g) and assuming $\alpha = 2$, we obtain

$$\left(1 + \frac{a^2}{b^2}\right)^2 a_{11} - 16\sigma_{cr}\frac{a^2h}{\pi^4D}\frac{2}{9}a_{12} = 0$$

$$-16\sigma_{cr}\frac{a^2h}{\pi^4D}\frac{2}{9}a_{11} + \left(1 + 4\frac{a^2}{b^2}\right)^2 a_{12} - 16\sigma_{cr}\frac{a^2h}{\pi^4D}\frac{6}{25}a_{13} = 0$$

$$-16\sigma_{cr}\frac{a^2h}{\pi^4D}\frac{6}{25}a_{12} + \left(1 + 9\frac{a^2}{b^2}\right)^2 a_{13} = 0$$

Equating to zero the determinant of these equations, we obtain an equation for calculating the third approximation which is sufficiently accurate for the case of pure bending.[1]

The final expression for σ_{cr} can be represented again by the equation

$$\sigma_{cr} = k\frac{\pi^2D}{b^2h} \qquad (j)$$

Several values of the numerical factor k for various values of the ratio a/b and for various values of α are given in Table 9-6. For the case of pure bending $(\alpha = 2)$ the third approximation was used; in other cases Eq. (i), representing the second approximation, was used in calculating σ_{cr}.

TABLE 9-6. NUMERICAL VALUES OF THE FACTOR k IN EQ. (j)

α \ a/b	0.4	0.5	0.6	0.667	0.75	0.8	0.9	1.0	1.5
2	29.1	25.6	24.1	23.9	24.1	24.4	25.6	25.6	24.1
$\frac{4}{3}$	18.7	....	12.9	....	11.5	11.2	....	11.0	11.5
1	15.1	....	9.7	....	8.4	8.1	....	7.8	8.4
$\frac{4}{5}$	13.3	....	8.3	....	7.1	6.9	....	6.6	7.1
$\frac{2}{3}$	10.8	....	7.1	....	6.1	6.0	....	5.8	6.1

It is seen from the table that in the case of pure bending, the value of σ_{cr} becomes a minimum when $a/b = \frac{2}{3}$. With a decrease of α the ratio a/b, at which σ_{cr} is a minimum, increases and approaches unity as obtained before for a uniformly compressed plate. The values of k for pure bending are shown in Fig. 9-20 by the curve $m = 1$.

It is assumed in the preceding discussion that the plate is buckled into one half-wave, but the results obtained can be applied also for the case of

---

[1] Calculation of a fourth approximation for the case of pure bending shows that the difference between the third and the fourth approximation is only about one-third of 1 per cent.

several half-waves, it being necessary only to construct the curves $m = 2$, $m = 3, \ldots$ as explained in Art. 9.2. The portions of these curves shown in Fig. 9-20 by a full line should be used in calculating σ_{cr} and in determining the number of half-waves into which a buckled plate is subdivided by nodal lines. It is seen that a long plate will buckle in such a way that the length of each half-wave will be approximately equal to $\frac{2}{3}b$. Curves analogous to those shown in Fig. 9-20 can be constructed also for other values of α, i.e., for various combinations of compressive and bending stresses.

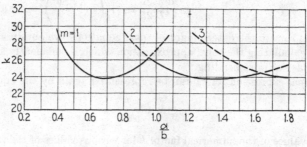

Fig. 9-20

The problem of buckling of the plate in Fig. 9-19 is solved also in the following cases: (1) when the edges of the plate $y = 0$ and $y = b$ are built in and (2) when the edge $y = 0$ is simply supported and the edge $y = b$ is built in.[1] The critical value of the maximum compressive stress again can be represented by formula (j). Values of the factor k for the above two cases, obtained by the use of the energy method, are given in Tables 9-7 to 9-9. Comparing values of k from Table 9-7 with the corresponding line of Table 9-6, it is seen that when the edges are fixed, the critical stress

TABLE 9-7. VALUES OF THE FACTOR k FOR PURE BENDING WITH
EDGES $y = 0$, $y = b$, BUILT IN

α \ a/b	0.30	0.35	0.40	0.45	0.47	0.48	0.50	0.60	0.70
2	47.3	43.0	40.7	39.7	39.6	39.6	39.7	41.8	45.8

TABLE 9-8. VALUES OF THE FACTOR k FOR TRIANGULAR LOADING WITH
EDGES $y = 0$, $y = b$, BUILT IN

α \ a/b	0.40	0.50	0.60	0.64	0.65	0.66	0.67	0.70	0.80	0.90
1	17.7	14.7	13.7	13.57	13.56	13.57	13.58	13.65	14.3	15.4

[1] See K. Nölke, *Ingr.-Arch.*, vol. 8, p. 403, 1937.

TABLE 9-9. VALUES OF THE FACTOR k FOR PURE BENDING WITH EDGE $y = 0$
SIMPLY SUPPORTED AND THE EDGE $y = b$ BUILT IN

$\dfrac{a/b}{\alpha}$	0.40	0.50	0.60	0.65	0.66	0.67	0.70	0.80	0.90	1.00
2	29.5	26.0	24.65	24.48	24.48	24.48	24.6	25.3	26.6	28.3

in the case of longer plates is increased by about 75 per cent and the wave length is diminished from $a/b = 0.667$ to $a/b = 0.47$. Comparing the values in Table 9-9 with the corresponding values in Table 9-6, we see that fixing the edge on the tension side of the plate (Fig. 9-19) has only a small effect on the critical stress and on the wave length.

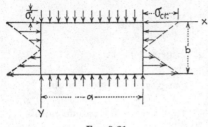

FIG. 9-21

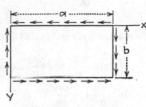

FIG. 9-22

If, in addition to bending stresses, uniformly distributed compressive stresses σ_y act on the plate in a perpendicular direction (Fig. 9-21), the critical value of the maximum bending stress σ_{cr} will be reduced and can be calculated on the basis of our preceding derivation by adding to the work done by bending stresses [see expression (d)] the work done by the compressive stresses σ_y. Calculations show that the effect of σ_y on the magnitude of σ_{cr} depends on the ratios $\sigma_y : 4\pi^2 D/b^2 h$ and $a:b$. Taking, for instance,

$$\frac{\sigma_y b^2 h}{4\pi^2 D} = \frac{1}{3} \quad \text{and} \quad a = b$$

the critical stress σ_{cr} becomes about 25 per cent smaller than in the case when bending stresses are acting alone.[1]

9.7. Buckling of Rectangular Plates under the Action of Shearing Stresses.

We begin with a simply supported rectangular plate submitted to the action of shearing forces N_{xy} uniformly distributed along the edges[2] (Fig. 9-22). In calculating the critical value of shearing stress τ_{cr} at which buckling of the plate occurs, we use again the energy method. The boundary conditions at the supported edges are satisfied by taking for the deflection surface of the buckled plate the previously used expression

[1] Several examples of such calculations are given by Timoshenko, *loc. cit.*

[2] See Timoshenko, *Ann. ponts et chaussées*, p. 372, 1913; *Mem. Inst. Engrs. Ways of Commun.*, vol. 89, p. 23, 1915; and *Der Eisenbau*, vol. 12, p. 147, 1921.

in the form of the double series

$$w = \sum_{m=1}^{m=\infty} \sum_{n=1}^{n=\infty} a_{mn} \sin \frac{m\pi x}{a} \sin \frac{n\pi y}{b} \qquad (a)$$

Then for the strain energy of bending of the buckled plate we will have the expression (see p. 341)

$$\Delta U = \frac{D}{2} \frac{\pi^4 ab}{4} \sum_{m=1}^{m=\infty} \sum_{n=1}^{n=\infty} a_{mn}^2 \left(\frac{m^2}{a^2} + \frac{n^2}{b^2} \right)^2 \qquad (b)$$

The work done by the external forces during buckling of the plate is [see Eq. (8-32)]

$$\Delta T = -N_{xy} \int_0^a \int_0^b \frac{\partial w}{\partial x} \frac{\partial w}{\partial y} \, dx \, dy \qquad (c)$$

Substituting expression (a) for w and observing that

$$\int_0^a \sin \frac{m\pi x}{a} \cos \frac{p\pi x}{a} \, dx = 0 \qquad \text{if } m \pm p \text{ is an even number}$$

$$\int_0^a \sin \frac{m\pi x}{a} \cos \frac{p\pi x}{a} \, dx = \frac{2a}{\pi} \frac{m}{m^2 - p^2} \qquad \text{if } m \pm p \text{ is an odd number}$$

we obtain

$$\Delta T = -4N_{xy} \sum_m \sum_n \sum_p \sum_q a_{mn} a_{pq} \frac{mnpq}{(m^2 - p^2)(q^2 - n^2)} \qquad (d)$$

in which m, n, p, q are such integers that $m \pm p$ and $n \pm q$ are odd numbers.

Equating the work (d) produced by external forces to the strain energy (b), we obtain, for determining the critical value of shearing forces, the following expression:

$$N_{xy} = -\frac{abD}{32} \frac{\displaystyle\sum_{m=1}^{m=\infty} \sum_{n=1}^{n=\infty} a_{mn}^2 \left(\frac{m^2 \pi^2}{a^2} + \frac{n^2 \pi^2}{b^2} \right)^2}{\displaystyle\sum_m \sum_n \sum_p \sum_q a_{mn} a_{pq} \frac{mnpq}{(m^2 - p^2)(q^2 - n^2)}} \qquad (e)$$

It is necessary now to select such a system of constants a_{mn} and a_{pq} as to make N_{xy} a minimum. Proceeding as before and equating to zero the derivatives of the expression (e) with respect to each of the coefficients $a_{mn}, \ldots,$ we obtain a system of homogeneous linear equations in $a_{mn}, \ldots.$ This system can be divided into two groups, one containing constants a_{mn} for which $m + n$ are odd numbers and the other for which $m + n$ are even numbers. Calculations show that for shorter plates ($a/b < 2$) this second group of equations gives for $(N_{xy})_{cr}$ the smallest value. For longer plates both groups of equations should be considered.[1]

Using the notations

$$\beta = \frac{a}{b} \qquad \lambda = -\frac{\pi^2}{32\beta}\frac{\pi^2 D}{b^2 h \tau_{cr}} \qquad (f)$$

we can write the group of equations for shorter plates in the following form:[2]

a_{11}	a_{22}	a_{13}	a_{31}	a_{33}	a_{42}	
$\dfrac{\lambda(1 + \beta^2)^2}{\beta^2}$	$\dfrac{4}{9}$	0	0	0	$\dfrac{8}{45}$	$= 0$
$\dfrac{4}{9}$	$\dfrac{16\lambda(1 + \beta^2)^2}{\beta^2}$	$-\dfrac{4}{5}$	$-\dfrac{4}{5}$	$\dfrac{36}{25}$	0	$= 0$
0	$-\dfrac{4}{5}$	$\dfrac{\lambda(1 + 9\beta^2)^2}{\beta^2}$	0	0	$-\dfrac{24}{75}$	$= 0$
0	$-\dfrac{4}{5}$	0	$\dfrac{\lambda(9 + \beta^2)^2}{\beta^2}$	0	$\dfrac{24}{21}$	$= 0$
0	$\dfrac{36}{25}$	0	0	$\dfrac{\lambda(9 + 9\beta^2)^2}{\beta^2}$	$-\dfrac{72}{35}$	$= 0$
$\dfrac{8}{45}$	0	$-\dfrac{24}{75}$	$\dfrac{24}{21}$	$-\dfrac{72}{35}$	$\dfrac{\lambda(16 + 4\beta^2)^2}{\beta^2}$	$= 0$

(g)

The equation for calculating τ_{cr} will now be obtained by equating to zero the determinant of the above system of equations. The calculation can be made by successive approximations. Limiting calculations to two equations with two constants a_{11} and a_{22} and equating to zero the determinant of these equations, we obtain

$$\lambda = \pm\frac{1}{9}\frac{\beta^2}{(1 + \beta^2)^2}$$

or, by using notations (f),

$$\tau_{cr} = \pm\frac{9\pi^2}{32}\frac{(1 + \beta^2)^2}{\beta^3}\frac{\pi^2 D}{b^2 h} \qquad (h)$$

The two signs indicate that in this case the critical value of the shearing stress does not depend on the direction of the stress.

The approximation (h) is not sufficiently accurate, since the error is about 15 per cent for square plates and increases as the ratio a/b increases. To get a more satisfactory approximation, a larger number of equations in the system (g) must be considered.

[1] See M. Stein and J. Neff, *NACA Tech. Note* 1222, 1947.

[2] For simplification we write only the factors with which the constants shown in the first row should be multiplied.

By taking five equations and equating to zero their determinant, we obtain

$$\lambda^2 = \frac{\beta^4}{81(1 + \beta^2)^4}\left[1 + \frac{81}{625} + \frac{81}{25}\left(\frac{1 + \beta^2}{1 + 9\beta^2}\right)^2 + \frac{81}{25}\left(\frac{1 + \beta^2}{9 + \beta^2}\right)^2\right] \quad (i)$$

Calculating λ and substituting it in Eq. (f), we obtain

$$\tau_{cr} = k\,\frac{\pi^2 D}{b^2 h} \quad (j)$$

where k is a constant depending on the ratio $a/b = \beta$. For a square plate we obtain in this manner $k = 9.4$. Calculations made with a larger number of equations of the system (g)[1] show that the exact value of k is about 9.34, so that in this case the error of the approximation represented by Eq. (i) is less than 1 per cent. Equation (i) gives a satisfactory approximation for k if the shape of the plate does not differ much from a square, say $a/b \lesssim 1.5$. For larger values of the ratio a/b a larger

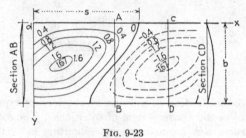

Fig. 9-23

number of equations must be considered. In Table 9-10 the results of such calculations are given.

TABLE 9-10. VALUES OF THE FACTOR k IN EQ. (j)

a/b	1.0	1.2	1.4	1.5	1.6	1.8	2.0	2.5	3	4
k	9.34	8 0	7.3	7.1	7.0	6.8	6.6	6.1	5.9	5.7

To get an approximate solution for long narrow plates, let us consider a limiting case of an infinitely long plate with simply supported edges. An approximate solution of the problem is obtained by taking for the deflection surface of the plate the following expression:

$$w = A \sin\frac{\pi y}{b} \sin\frac{\pi}{s}(x - \alpha y) \quad (k)$$

---

[1] Several authors have made such calculations; see, for instance, S. Bergmann and H. Reissner, *Z. Flugtech. Motorluftsch.*, vol. 23, p. 6, 1932; E. Seydel, *Ingr.-Arch.*, vol. 4, 1933, p. 169; and Stein and Neff, *loc. cit.*

This expression gives zero deflections for $y = 0, y = b$ and also along nodal lines, for which $(x - \alpha y)$ is a multiple of s. Here s represents the length of half-waves of the buckled plate and the factor α is the slope of nodal lines. The exact solution of the problem[1] shows that the nodal lines are not straight and that the deflection surface of the buckled plate has the form shown in Fig. 9-23. Expression (k) does not satisfy the boundary conditions regarding equality to zero of the bending moments along the longitudinal edges of the plate, since $\partial^2 w / \partial y^2$ is not zero along these edges; however, it can be used for an approximate solution of the problem. Substituting expression (k) in Eq. (8-29) for the strain energy of bending of the buckled plate and also in the equation for the work done by the external forces,[2] and equating these two quantities, we obtain

$$\tau_{cr} = \frac{\pi^2 D}{2\alpha b^2 h} \left[6\alpha^2 + 2 + \frac{s^2}{b^2} + \frac{b^2}{s^2}(1 + \alpha^2)^2 \right] \quad (l)$$

The smallest value for τ_{cr} is obtained by taking

$$s = b\sqrt{1 + \alpha^2} \quad \text{and} \quad \alpha = \frac{1}{\sqrt{2}}$$

Then we obtain

$$\tau_{cr} = 5.7 \frac{\pi^2 D}{b^2 h} \quad (m)$$

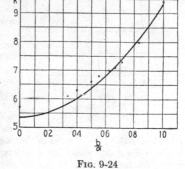

Fig. 9-24

The exact solution of the problem for an infinitely long strip with simply supported edges gives

$$\tau_{cr} = 5.35 \frac{\pi^2 D}{b^2 h} \quad (9\text{-}9)$$

so that the error of the approximate solution in this case is about $6\frac{1}{2}$ per cent.

Having the exact value for k in Eq. (j) for an infinitely long plate and a very accurate value of k for a square plate, a parabolic curve given by the equation $k = 5.35 + 4(b/a)^2$ and shown in Fig. 9-24 can be taken to approximate values of k for other proportions of plates.[3] For comparison there are shown also points corresponding to the figures given in Table 9-10. It is seen that for longer plates the values of k given in the table

---

[1] An exact solution of this problem was given by R. V. Southwell, *Phil. Mag.*, vol. 48, p. 540, 1924, and also *Proc. Roy. Soc., London*, series A, vol. 105, p. 582.

[2] The strain energy and the work done per wave should be considered.

[3] This is only a rough approximation since the exact values of k are defined by a system of intersecting curves similar to those in Fig. 9-2.

are always above the curve.   The values obtained from the curve can be used for practical applications.

The problem of buckling of an infinitely long plate under uniform shear has been solved also for the case of clamped edges,[1] and the following value found for the critical stress:

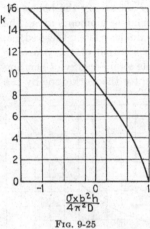

FIG. 9-25

$$\tau_{cr} = 8.98 \frac{\pi^2 D}{b^2 h} \qquad (9\text{-}10)$$

The combination of pure shear with uniform longitudinal compression or tension σ_x was also studied,[1] and the corresponding values of k, which must be substituted in Eq. (j) for determining τ_{cr}, are given by the curve in Fig. 9-25.   It is seen that any compressive stress[2] reduces the stability of a plate submitted to shear, while any tension increases this stability.   For $\sigma_x = 0$ the value $k = 8.98$, given before [see Eq. (9-10)], is obtained from the curve.

The case of combined shearing and bending stresses (Fig. 9-26) is also of practical interest and has been investigated in the case of a plate with simply supported edges by taking for the deflection surface the expression (a) and using the energy method: The results of this investigation[3] are represented by the curves I, II, III, and IV in Fig. 9-27.[4]   As abscissas in plotting these curves, the ratios of the actual shearing stress to the critical shearing stress determined by using Table 9-10 have been taken. The ordinates are the values of the factor k which must be substituted in Eq. (j) (see p. 377) for the critical value of the bending stresses.   It is seen that for small values of

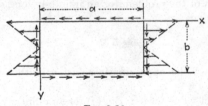

FIG. 9-26

τ/τ_{cr}, say (τ/τ_{cr}) < 0.4, the effect of shearing stress on the critical value of bending stress is small.   When τ/τ_{cr} approaches unity, the curves I, II, III, and IV become

[1] *Ibid.*

[2] Compressive stress σ_x is considered positive.

[3] See Timoshenko, *Engineering*, vol. 138, p. 207, 1934.   See also O. Stein, *Der Stahlbau*, vol. 7, p. 57, 1934.   In the latter paper a smaller number of terms in expression (a) was used than in the calculations represented in Fig. 9-27, and thus the results are less accurate.

[4] These curves were calculated by S. Way.   Eight equations of the system (g) were used in these calculations.

steep, which indicates that some bending stresses can be added to pure shear without producing a substantial reduction in the critical value of the shearing stress.

The case of shearing stress combined with tension or compression in two perpendicular directions has been discussed by H. Wagner.[1]

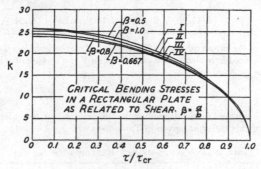

FIG. 9-27

The problem of buckling of rectangular plates under pure shear was solved also for other edge conditions,[2] namely, for the cases in which (1) the edges $y = 0$, $y = b$ are built in (Fig. 9-22) and the other two edges are simply supported and (2) all four edges are built in. The calculated values of k in Eq. (j) are given in Tables 9-11 and 9-12,

TABLE 9-11. VALUES OF THE FACTOR k FOR THE CASE IN WHICH THE EDGES $y = 0$, $y = b$ ARE BUILT IN, THE EDGES $x = 0$, $x = a$ ARE SIMPLY SUPPORTED

a/b	1.0	1.5	2.0	2.5	3.0	∞
k	12.28	11.12	10.21	9.81	9.61	8.99

TABLE 9-12. VALUES OF THE FACTOR k FOR THE CASE IN WHICH ALL FOUR EDGES ARE BUILT IN

a/b	1.0	1.5	2.0	2.5	∞
k	14.71	11.50	10.34	10.85	8.99

respectively.[3] If, in addition to τ, there are compressive stresses $\sigma_x = \sigma_y = \sigma$ applied to the edges of the plate in Fig. 9-22, the value of τ_{cr} depends on the ratio σ/τ. Several values of k in Eq. (j) for a square plate with all edges built in are given in Table 9-13.

9.8. Other Cases of Buckling of Rectangular Plates. *A rectangular plate clamped at two opposite sides, simply supported along the other two sides, and uniformly compressed*

[1] H. Wagner, *Jahrb. Wiss. Ges. Luftf.*, 1928, p. 113.

[2] See S. Iguchi, *Ingr.-Arch.*, vol. 9, pp. 1–12, 1938.

[3] In Table 9-12 some corrections based on calculations by B. Budiansky and R. Connor are introduced (see *NACA Tech. Note* 1559, 1948).

TABLE 9-13. VALUES OF THE FACTOR k FOR A SQUARE PLATE IF
σ AND τ ARE ACTING SIMULTANEOUSLY

σ/τ	0	0.5	1.0	1.5	2
k	14.71	7.09	4.50	3.24	2.51

in the direction of the supported sides[1] (Fig. 9-28). A method analogous to that used in Art. 9.4 can be used here also. The critical value of the compressive stress is given by the equation

$$\sigma_{cr} = k \frac{\pi^2 D}{b^2 h} \tag{a}$$

in which k is a numerical factor depending on the ratio a/b of the sides of the plate. Several values of this factor are given in Table 9-14.

TABLE 9-14. VALUES OF THE FACTOR k IN EQ. (a)

a/b	0.6	0.8	1.0	1.2	1.4	1.6	1.7	1.73	1.8	2.0	2.5	2.83	3.0
k	13.38	8.73	6.74	5.84	5.45	5.34	5.33	5.33	5.18	4.85	4.52	4.50	4.41

Up to the value $a/b = 1.73$ the plate buckles into one half-wave. From $a/b = 1.73$ to $a/b = 2.83$ there will be two half-waves and the shape of the buckled plate is as indicated in Fig. 9-28. Generally the transition from m to $m + 1$ half-waves occurs when $a/b = \sqrt{m(m + 2)}$. It can be seen that the effect of clamping the edges on the magnitude of σ_{cr} decreases as the ratio a/b increases; for $a/b = 3$ the value of k in Table 9-14 is only 10 per cent higher than the value 4 obtained for a plate with all four edges simply supported.

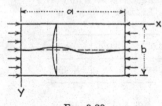

FIG. 9-28

A rectangular plate with clamped edges under pressure in two perpendicular directions. Taking the coordinate axes as shown in Fig. 9-28 and assuming that the shape of the plate does not differ much from a square and that the stresses σ_x and σ_y are about equal, we can expect that the deflection surface of the buckled plate is represented with sufficient accuracy by the equation

$$w = \frac{\delta}{4} \left(1 - \cos \frac{2\pi x}{a} \right) \left(1 - \cos \frac{2\pi y}{b} \right)$$

which satisfies the boundary conditions. With this expression for deflections, the strain energy of bending is

$$\Delta U = \frac{\pi^4 \delta^2 D}{8} \, ab \left(\frac{3}{a^4} + \frac{3}{b^4} + \frac{2}{a^2 b^2} \right)$$

The work done by the compressive forces during buckling of the plate is

$$\Delta T = \frac{\sigma_x h}{2} \int_0^a \int_0^b \left(\frac{\partial w}{\partial x} \right)^2 dx \, dy + \frac{\sigma_y h}{2} \int_0^a \int_0^b \left(\frac{\partial w}{\partial y} \right)^2 dx \, dy = \frac{3}{32} \pi^2 \delta^2 h \frac{b}{a} \left(\sigma_x + \frac{a^2}{b^2} \sigma_y \right)$$

[1] F. Schleicher, *Mitt. Forschungsanstalt. Gutehoffnungshütte Konzerns*, vol. 1, 1931.

TABLE 9-15. VALUES OF THE FACTOR k IN Eq. (a) FOR PLATES WITH
ALL SIDES CLAMPED

a/b	0.75	1.00	1.25	1.50	1.75	2.00	2.25	2.50	2.75	3.00	3.25	3.50	3.75	4.00
k	11.69	10.07	9.25	8.33	8.11	7.88	7.63	7.57	7.44	7.37	7.35	7.27	7.24	7.23

Equating this work to the strain energy of bending, we obtain the following equation for calculating critical values of the compressive stresses σ_x and σ_y:

$$\left(\sigma_x + \frac{a^2}{b^2}\sigma_y\right)_{cr} = \frac{4}{3}\frac{\pi^2 D a^2}{h}\left(\frac{3}{a^4} + \frac{3}{b^4} + \frac{2}{a^2 b^2}\right) \tag{9-11}$$

In the particular case of a square plate submitted to the action of uniform thrust, we obtain from the above equation

$$\sigma_{cr} = 5.33\frac{\pi^2 D}{a^2 h}$$

It is interesting to note that for this case we have another solution of the problem[1] which gives as a lower limit for σ_{cr} the value

$$\sigma_{cr} = 5.30\frac{\pi^2 D}{a^2 h} \tag{9-12}$$

Thus the approximate solution is very accurate in this case.

In the case of rectangular plates compressed only longitudinally, the buckling of the plate with one or several nodal lines perpendicular to the length of the plate should be considered. The solution of this problem was discussed by several authors. In Table 9-15 the values of the factor k in Eq. (a) are given[2] for plates clamped at all four edges and compressed longitudinally as shown in Fig. 9-28. For a plate of considerable length the critical value of compressive stresses evidently must approach that obtained from Table 9-5 for $a/b = 0.7$.

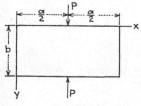

FIG. 9-29

A simply supported rectangular plate compressed by two equal and opposite forces (Fig. 9-29).[3] An approximate solution of this problem is obtained by taking for the deflection surface of the buckled plate the following series:

$$w = \sin\frac{\pi y}{b}\sum_{m=1,3,5,\ldots}^{\infty} a_m \sin\frac{m\pi x}{a} \tag{b}$$

[1] G. I. Taylor, Z. angew. Math. u. Mech., vol. 13, p. 147, 1933. See also A. Weinstein, Compt. rend., 1935.

[2] These values were calculated by S. Levy, J. Appl. Mech., vol. 9, p. 171, 1942. This paper contains also the results obtained by earlier authors and the references for their publications.

[3] See A. Sommerfeld, Z. Math. Physik, vol. 54, 1906, and Timoshenko, Z. Math. Physik, vol. 58, p. 357, 1910. For a further discussion of the problem see D. M. A. Leggett, Proc. Cambridge Phil. Soc., vol. 33, p. 325, 1937, and the paper by H. G. Hopkins presented at the Seventh International Congress of Applied Mechanics, London, 1948.

The expression for strain energy of bending becomes

$$\Delta U = \frac{abD}{8} \sum_{m=1,3,5,\ldots}^{\infty} a_m{}^2 \left(\frac{m^2\pi^2}{a^2} + \frac{\pi^2}{b^2} \right)^2 \tag{c}$$

The work done by the compressive forces during buckling is

$$\Delta T = \frac{P}{2} \int_0^b \left(\frac{\partial w}{\partial y} \right)_{z=a/2}^2 dy = \frac{\pi^2 P}{4b} (a_1 - a_3 + a_5 - \cdots)^2 \tag{d}$$

Equating this work to the strain energy of bending (c), we obtain for the critical value of compressive forces

$$P_{cr} = \frac{\pi^2 D a b^2}{2} \cdot \frac{\sum\limits_{m=1,3,5,\ldots} a_m{}^2 \left(\frac{m^2}{a^2} + \frac{1}{b^2} \right)^2}{(a_1 - a_3 + a_5 - \cdots)^2} \tag{e}$$

Equating to zero the derivatives of this expression with respect to each coefficient a_n, we obtain a system of linear equations of the following form:

$$a_n = \frac{2P_{cr}}{\pi^2 D a b^2} \frac{(-1)^{(n-1)/2} \sum\limits_{m=1,3,5,\ldots}^{\infty} a_m (-1)^{(m-1)/2}}{\left(\frac{n^2}{a^2} + \frac{1}{b^2} \right)^2} \tag{f}$$

Multiplying each of these equations by $(-1)^{(n-1)/2}$ and adding them together, we obtain

$$P_{cr} = \frac{\pi^2 D a b^2}{2} \frac{1}{\sum\limits_{n=1,3,5,\ldots}^{\infty} \frac{1}{\left(\frac{n^2}{a^2} + \frac{1}{b^2} \right)^2}} \tag{g}$$

or, by using notation $a/b = \beta$,

$$P_{cr} = \frac{\pi^2 D}{2b} \frac{1}{\beta^3 \sum\limits_{n=1,3,5,\ldots}^{\infty} \frac{1}{(\beta^2 + n^2)^2}} \tag{h}$$

For summation of the series in the denominator we note that

$$\frac{e^{\pi z/2} + e^{-\pi z/2}}{2} = (1 + z^2) \left(1 + \frac{z^2}{9} \right) \left(1 + \frac{z^2}{25} \right) \cdots$$

Taking the logarithm of each side and differentiating, we obtain

$$\frac{\pi}{4} \tanh \frac{\pi z}{2} = z \sum_{m=1,3,5,\ldots}^{\infty} \frac{1}{m^2 + z^2} \tag{i}$$

The second differentiation gives

$$\frac{\pi^2}{8} \frac{1}{\cosh^2 \pi z/2} = \sum_{m=1,3,5,\ldots}^{\infty} \frac{1}{m^2 + z^2} - \sum_{m=1,3,5,\ldots}^{\infty} \frac{2z^2}{(m^2 + z^2)^2}$$

Multiplying this by z and using Eq. (i), we obtain

$$z^3 \sum_{m=1,3,5,\ldots}^{\infty} \frac{1}{(m^2+z^2)^2} = \frac{\pi}{8}\left(\tanh\frac{\pi z}{2} - \frac{\pi z/2}{\cosh^2\frac{\pi z}{2}}\right)$$

The left side of this equation is the same series that we had in Eq. (h), and it is seen that the sum of this series can be readily calculated for each value of the ratio $\beta = a/b$ by using tables of hyperbolic functions. With increasing β the sum rapidly approaches the limiting value $\pi/8$ (for $a/b = 2$, this sum is $0.973\pi/8$), and for long plates we can assume that

$$P_{cr} = \frac{4\pi D}{b} \tag{9-13}$$

If the sides $y = 0$ and $y = b$ of the plate are clamped, we take the deflection surface of the buckled plate in the form of the following series:

$$w = \left(1 - \cos\frac{2\pi y}{b}\right) \sum_{m=1,3,5,\ldots}^{\infty} a_m \sin\frac{m\pi x}{a}$$

Then proceeding as in the previous case, we obtain for long plates

$$P_{cr} = \frac{8\pi D}{b} \tag{9-14}$$

9.9. Buckling of Circular Plates.[1] Let us begin with the simple case of a circular plate with clamped edges (Fig. 9-30). To determine the critical value of the compressive forces N_r uniformly distributed around the edge of the plate, we assume that a slight buckling has taken place and we use the differential equation of the deflection surface of the plate. Assuming that the deflection surface is a surface of revolution and denoting by ϕ the angle between the axis of revolution and any normal to the plate, the required equation is[2]

$$r^2\frac{d^2\phi}{dr^2} + r\frac{d\phi}{dr} - \phi = -\frac{Qr^2}{D} \tag{a}$$

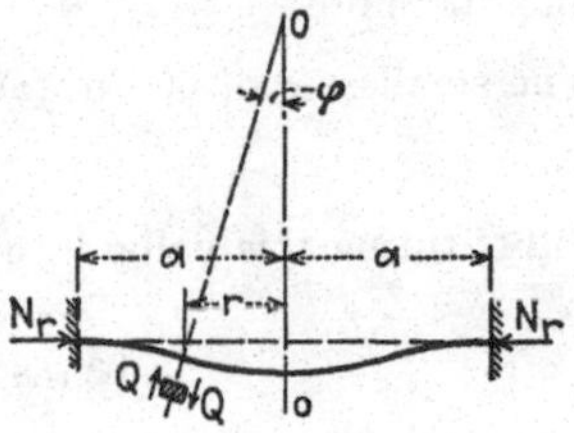

Fig. 9-30

In this equation r is the distance of any point measured from the center of the plate and Q is the shearing force per unit of length, the positive direction of which is shown in Fig. 9-30. Since there are no lateral loads acting on the plate, we have

$$Q = N_r\phi \tag{b}$$

[1] See G. H. Bryan, *Proc. London Math. Soc.*, vol. 22, p. 54, 1891. See also A. Nadai, *Z. Ver. deut. Ingr.*, vol. 59, p. 169, 1915.

[2] See Timoshenko, "Strength of Materials," 3d ed., part 2, p. 94, D. Van Nostrand Company, Inc., Princeton, N.J., 1956

and, using the notation

$$\frac{N_r}{D} = \alpha^2 \tag{c}$$

we obtain

$$r^2 \frac{d^2\phi}{dr^2} + r \frac{d\phi}{dr} + (\alpha^2 r^2 - 1)\phi = 0 \tag{d}$$

Let us introduce now a new variable

$$u = \alpha r \tag{e}$$

by the use of which Eq. (d) becomes

$$u^2 \frac{d^2\phi}{du^2} + u \frac{d\phi}{du} + (u^2 - 1)\phi = 0 \tag{f}$$

The general solution of this equation is

$$\phi = A_1 J_1(u) + A_2 Y_1(u) \tag{g}$$

where $J_1(u)$ and $Y_1(u)$ are Bessel functions of first order of the first and second kinds, respectively. At the center of the plate ($r = u = 0$) the angle ϕ must be zero in order to satisfy the condition of symmetry. Since the function $Y_1(u)$ becomes infinite as u approaches zero, the above condition requires that we take $A_2 = 0$. To satisfy the condition at the clamped edge of the plate, we must have

$$(\phi)_{r=a} = 0 \tag{h}$$

and therefore

$$J_1(\alpha a) = 0 \tag{i}$$

The smallest root of Eq. (i) is[1]

$$\alpha a = 3.832$$

Substituting this value into Eq. (c), we obtain

$$(N_r)_{cr} = \frac{(3.832)^2 D}{a^2} = \frac{14.68 D}{a^2} \tag{9-15}$$

For comparison we note that the critical compressive force for a strip of unit width with clamped ends and having the length equal to the diameter of the plate is

$$\frac{\pi^2 D}{a^2}$$

Thus for producing buckling of the plate compressive stresses about 50 per cent higher than for the strip should be applied.

Solution (g) can be used also for the case of buckling of a compressed

[1] See, for example, Jahnke and Emde, "Tables of Functions," 4th ed., p. 167, Dover Publications, New York, 1945.

circular plate with a simply supported edge. As in the previous example, the constant A_2 must be taken equal to zero in order to satisfy the condition at the center of the plate. A second condition is obtained by noting that the bending moment along the edge must be zero; hence[1]

$$\left(\frac{d\phi}{dr} + \nu \frac{\phi}{r}\right)_{r=a} = 0 \qquad (j)$$

Using the derivative formula

$$\frac{dJ_1}{du} = J_0 - \frac{J_1}{u} \qquad (k)$$

in which J_0 represents the Bessel function of zero order, we can express the boundary condition (j) in the form

$$\alpha a J_0(\alpha a) - (1 - \nu)J_1(\alpha a) = 0 \qquad (l)$$

Taking $\nu = 0.3$ and using tables of the functions J_0 and J_1, we find the smallest root of the transcendental equation (l) to be 2.05. Then, from Eq. (c),

$$(N_r)_{cr} = \frac{(2.05)^2 D}{a^2} = \frac{4.20 D}{a^2} \qquad (9\text{-}10)$$

That is, the critical compressive stress in this case is about three and a half times smaller than in the case of a plate with clamped edge.

In the case of a plate with a hole at the center,[2] the compressive stresses produced by forces N_r, uniformly distributed along the outer boundary of the plate, are no longer constant and are determined by the known Lamé's formula. Assuming a buckling of the plate symmetrical with respect to the center, the differential equation for the deflection surface of the plate can again be integrated by Bessel functions and the expression for $(N_r)_{cr}$ is

$$(N_r)_{cr} = k \frac{D}{a^2} \qquad (m)$$

in which k is a numerical factor, the magnitude of which depends on the ratio b/a, where b is the radius of the hole. The values of k for various values of b/a are given in Fig. 9-31a for a clamped plate and in Fig. 9-31b for a plate supported along the outer boundary. It is assumed in both cases that the boundary of the hole is free from forces.[3] It is seen that in the case of a plate with a clamped edge the factor k is a minimum for b/a approximately equal to 0.2, and for the ratio b/a larger than 0.2 it increases rapidly with this ratio and becomes larger than for a plate without a hole. It must be noted, however, that in this discussion buckling

[1] See Timoshenko, *op. cit.*, p. 93.
[2] This case was discussed by E. Meissner, *Schweiz. Bauztg.*, vol. 101, p. 87, 1933.
[3] Poisson's ratio ν is taken equal to $\frac{1}{3}$ in this problem.

symmetrical with respect to the center of the plate is assumed, while for b/a approaching unity the conditions for a compressed ring with outer boundary clamped are analogous to those of a long compressed rectangular plate clamped along one side and free along the other. Such a plate buckles in many waves (see p. 364); we should expect that in the case of a narrow ring also several waves along the circumference would be formed during buckling and that the values of k obtained on the assumption of symmetrical buckling would give exaggerated values for $(N_r)_{cr}$.

The problem of buckling of a circular plate, with and without a central hole, has been discussed for the case of a plate with varying thickness.[1] Also, the case of a circular plate with a central hole subjected to the action of shearing forces uniformly distributed along the inner and outer boundaries has been discussed.[2]

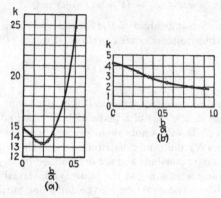

FIG. 9-31

Buckling of plates into several waves with radial and circular nodal lines has been discussed in the case of a plate without a hole, and it has been shown that the critical value of the compressive forces is always given by Eq. (m). With the use of these results, the problem of buckling of plates supported along two radii and two concentric circles can be solved.[3]

9.10. Buckling of Plates of Other Shapes. *Oblique Plates with Clamped Edges under Uniform Compression.*[4] In the case of an oblique plate compressed along the x axis

---

[1] See R. Gran Olsson, *Ingr.-Arch.*, vol. 9, p. 205, 1938; K. Federhofer, *ibid.*, vol. 11, p. 224, 1940; H. Egger, *ibid.*, vol. 12, 1941; and A. Schubert, *Z. angew. Math. u. Mech.*, vols. 25–27, 1947.

[2] The case of a plate with constant thickness was considered by W. R. Dean, *Proc. Roy. Soc.*, vol. 106, p. 268, 1924. The case of varying thickness was considered by K. Federhofer, *Ingr.-Arch.*, vol. 14, p. 155, 1943.

[3] B. Galerkin, *Compt. rend.*, vol. 179, p. 1392, 1924.

[4] See W. H. Wittrick, *Aeronaut. Quart.*, vol. 4, p. 151, 1953.

(Fig. 9-32) the critical value of the compressive stress is given by the formula

$$(\sigma_x)_{cr} = k \frac{\pi^2 D}{b^2 h} \qquad (9\text{-}17)$$

The values of the factor k for three different values of the angle α are given by curves in Fig. 9-33. The case $\alpha = 0$ corresponds to a rectangular plate (see p. 387).

Oblique Plate with Clamped Edges under Pure Shear.[1] We assume that shearing stresses τ_{xy} act along the sides of the plate parallel to the x axis (Fig. 9-34). Along the other sides there act shearing and normal stresses of such a magnitude as to produce in the plate pure shearing stress τ_{xy}. The calculations show that the critical value of τ_{xy} is smaller if it acts in the direction shown in the figure, i.e., producing an increase in the angle α. The value of $(\tau_{xy})_{cr}$ can be represented by the formula

$$(\tau_{xy})_{cr} = k \frac{\pi^2 D}{(b \cos \alpha)^2 h}$$

Values of k for $\alpha = 45°$, as a function of the ratio a/b, are given in Fig. 9-35. For comparison, values of k for rectangular plates ($\alpha = 0$) are given. Also a horizontal asymptote corresponding to an infinitely long plate is shown [see Eq. (9-10)].

Triangular Plate. The problem of buckling of a uniformly compressed equilateral triangular plate with simply supported edges also has been discussed,[2] and it was found that the critical value of the compressive forces is

$$N_{cr} = \frac{4\pi^2 D}{a^2} \qquad (9\text{-}18)$$

where a is the altitude of the triangle. It is seen that the critical stress in this case is about the same as for a simply supported and uniformly compressed circular plate whose boundary is an inscribed circle to the equilateral triangle of altitude a.

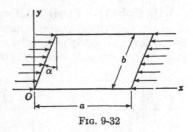

FIG. 9-32

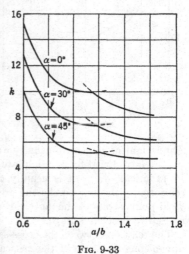

FIG. 9-33

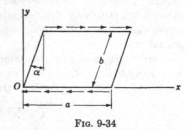

FIG. 9-34

[1] See *ibid.*, vol. 5, p. 39, 1954.

[2] See S. Woinowsky-Krieger, *Ingr.-Arch.*, vol. 4, p. 254, 1933. For some further data for triangular plates see Wittrick, *op. cit.*, vol. 5, p. 131, 1954; and J. M. Klitchieff, *Quart. J. Mech. Appl. Math.*, vol. 4, p. 257, 1951.

9.11. Stability of Plates Reinforced by Ribs. [1]   In all cases of buckling of plates, the critical values of normal or shearing forces are proportional to the flexural rigidity of the plates.   Hence, for a rectangular plate with given boundary conditions and a given ratio a/b, the magnitude of the critical stress is proportional to h^2/b^2.

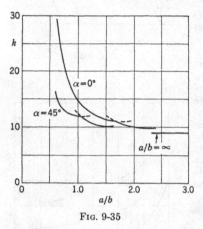

Fig. 9-35

The stability of the plate always can be increased by increasing its thickness, but such a design will not be economical in respect to the weight of material used.   A more economical solution is obtained by keeping the thickness of the plate as small as possible and increasing the stability by introducing reinforcing ribs.   In the case of a compressed plate, as shown in Fig. 9-36, its stability can be made about four times greater by adding a longitudinal rib of suitable cross section which bisects the width of the plate.

The weight of such a rib usually will be much smaller than the additional weight introduced by an adequate increase in the thickness of the plate. In practical design it is usually desirable to have such proportions of reinforcing ribs as to make the critical value of the stresses equal to the yield-point stress of the material.   Then the strength of all the material is used to best advantage. [2]

The relation between the cross-sectional dimensions of ribs and the critical value of stresses in a plate can be obtained as before by using the energy method.   Several cases of stiffening of rectangular plates under compression and shear will now be considered.

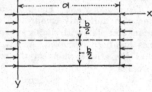

Fig. 9-36

Simply Supported Rectangular Plates with Longitudinal Ribs.   In this case we take the deflection surface of the buckled plate (Fig. 9-36) in the form of a double trigonometric series

$$w = \sum_{m=1}^{m=\infty} \sum_{n=1}^{n=\infty} a_{mn} \operatorname{s} n \frac{m\pi x}{a} \sin \frac{n\pi y}{b} \tag{a}$$

[1] See Timoshenko, *Mem. Inst. Engrs. Ways of Commun.*, St. Petersburg, vol. 89, 1915, and *Der Eisenbau*, vol. 12, p. 147, 1921.   See also P. Seide, *NACA Tech. Note* 2873, 1953, in which a list of publications on this subject is given.

[2] The case of light structures in which the strength of plates is taken into account after buckling has occurred will be discussed later (see Art. 9.13).

The corresponding strain energy of bending of the plate is

$$\Delta U = \frac{\pi^4 D}{2} \frac{ab}{4} \sum_{m=1}^{m=\infty} \sum_{n=1}^{n=\infty} a_{mn}{}^2 \left(\frac{m^2}{a^2} + \frac{n^2}{b^2}\right)^2 \qquad (b)$$

Assuming a general case of several longitudinal ribs and denoting by EI_i the flexural rigidity of a rib[1] at the distance c_i from the edge $y = 0$, we find that the strain energy of bending of this rib, when buckled together with the plate, is

$$\Delta U_i = \frac{EI_i}{2} \int_0^a \left(\frac{\partial^2 w}{\partial x^2}\right)^2_{y=c_i} dx$$

$$= \frac{\pi^4 EI_i}{4a^3} \sum_{m=1}^{m=\infty} m^4 \left(a_{m1} \sin \frac{\pi c_i}{b} + a_{m2} \sin \frac{2\pi c_i}{b} + \cdots\right)^2 \qquad (c)$$

The work done during buckling by the compressive forces N_x acting on the plate is

$$\Delta T = \frac{N_x}{2} \frac{ab}{4} \sum_{m=1}^{m=\infty} \sum_{n=1}^{n=\infty} \frac{m^2 \pi^2}{a^2} a_{mn}{}^2 \qquad (d)$$

The work done during buckling by the compressive force P_i acting on a rib is

$$\Delta T_i = \frac{P_i}{2} \int_0^a \left(\frac{\partial w}{\partial x}\right)^2_{y=c_i} dx = \frac{P_i}{2} \frac{\pi^2}{a^2} \frac{a}{2} \sum_{m=1}^{m=\infty} m^2 \left(a_{m1} \sin \frac{\pi c_i}{b} \right.$$
$$\left. + a_{m2} \sin \frac{2\pi c_i}{b} + \cdots\right)^2 \qquad (e)$$

The general equation for calculating the critical stress is[2]

$$\Delta U + \Delta \sum_i U_i = \Delta T + \Delta \sum_i T_i \qquad (f)$$

in which the summation must be extended over all the stiffening ribs. Using the notations

$$\frac{a}{b} = \beta \qquad \frac{EI_i}{bD} = \gamma_i \qquad \frac{P_i}{bN_x} = \frac{A_i}{bh} = \delta_i \qquad (g)$$

---

[1] Since the rib is rigidly connected with the plate, a portion of the plate must be taken in calculating I_i (see p. 399). See also Seide, *op. cit.*

[2] The energy of twist of the rib, which occurs during buckling, is neglected in this discussion. Its influence on the value of σ_{cr} was studied by F. W. Bornscheuer; see Dissertation, Darmstadt, 1946.

where bh is the cross-sectional area of the plate and A, that of one rib, we obtain from Eq. (f)

$$\sigma_{cr} = \frac{\pi^2 D}{b^2 h \beta^2} \cdot \frac{\sum\limits_{m=1}^{m=\infty} \sum\limits_{n=1}^{n=\infty} a_{mn}{}^2 (m^2 + n^2 \beta^2)^2 + 2 \sum\limits_i \gamma_i \sum\limits_{m=1}^{m=\infty} m^4 \left(\sum\limits_{n=1}^{n=\infty} a_{mn} \sin \frac{n \pi c_i}{b} \right)^2}{\sum\limits_{m=1}^{m=\infty} \sum\limits_{n=1}^{n=\infty} m^2 a_{mn}{}^2 + 2 \sum\limits_i \delta_i \sum\limits_{m=1}^{m=\infty} m^2 \left(\sum\limits_{n=1}^{n=\infty} a_{mn} \sin \frac{n \pi c_i}{b} \right)^2}$$

(h)

Proceeding as before and equating to zero the derivatives of this expression with respect to the coefficients a_{mn}, we obtain a system of homogeneous linear equations of the following type:

$$\frac{\pi^2 D}{b^2 h} \left[a_{mn}(m^2 + n^2 \beta^2)^2 + 2 \sum_i \gamma_i \sin \frac{n \pi c_i}{b} m^4 \sum_{p=1}^{p=\infty} a_{mp} \sin \frac{p \pi c_i}{b} \right]$$
$$- \beta^2 \sigma_{cr} \left(m^2 a_{mn} + 2 \sum_i \delta_i \sin \frac{n \pi c_i}{b} m^2 \sum_{p=1}^{p=\infty} a_{mp} \sin \frac{p \pi c_i}{b} \right) = 0 \quad (i)$$

By equating to zero the determinant of this system of equations, we obtain an equation for determining σ_{cr}.

Let us begin with the case of one longitudinal rib dividing the width of the plate in halves (Fig. 9-36); then $c_i = b/2$. Without limiting the generality of our conclusions, we can assume that the reinforced plate buckles into one half-wave and we can take $m = 1$. Then Eq. (i) can be represented in the following simplified form:[1]

$$\frac{\pi^2 D}{b^2 h \beta^2} [a_1(1 + \beta^2)^2 + 2\gamma(a_1 - a_3 + a_5 - \cdots)]$$
$$- \sigma_{cr}[a_1 + 2\delta(a_1 - a_3 + a_5 - \cdots)] = 0$$

$$\frac{\pi^2 D}{b^2 h \beta^2} (1 + 4\beta^2)^2 a_2 - \sigma_{cr} a_2 = 0$$

$$\frac{\pi^2 D}{b^2 h \beta^2} [a_3(1 + 9\beta^2)^2 - 2\gamma(a_1 - a_3 + a_5 - \cdots)] \qquad (j)$$
$$- \sigma_{cr}[a_3 - 2\delta(a_1 - a_3 + a_5 - \cdots)] = 0$$

$$\frac{\pi^2 D}{b^2 h \beta^2} (1 + 16\beta^2)^2 a_4 - \sigma_{cr} a_4 = 0$$

. . . .

We note that the equations of even order each contain only one coefficient. The corresponding values of σ_{cr} are the values for which the buckled plate

[1] The first subscript in the coefficients a_{mn} is omitted in the following derivations.

has a nodal line coinciding with the rib, and the rib remains straight during buckling of the plate. To establish the relation between the flexural rigidity of the rib and the critical value of the compressive stresses, the equations of odd order in the system (j) must be considered. The first approximation for σ_{cr} is obtained by taking the first equation of the system and assuming that only one coefficient a_1 is different from zero, i.e., by taking only the first term in the double series (a) representing the surface of the buckled plate. Then

$$\sigma_{cr} = \frac{\pi^2 D}{b^2 h} \frac{(1 + \beta^2)^2 + 2\gamma}{\beta^2(1 + 2\delta)} \qquad (9\text{-}19)$$

The following calculations, in which a larger number of Eqs. (j) is considered, show that this first approximation is a very accurate one for longer plates, say for $\beta > 2$. For shorter plates a larger number of Eqs. (j) must be considered. By taking the first and the third of those equations with coefficients a_1 and a_3 different from zero and by equating to zero the determinant of these two equations, we obtain for the second approximation of σ_{cr} the following quadratic equation:

$$(k\beta^2)^2(1 + 4\delta) - k\beta^2[(1 + 2\delta)(c + d) - 8\gamma\delta] + cd - 4\gamma^2 = 0 \qquad (k)$$

in which

$$k = \frac{\sigma_{cr} b^2 h}{\pi^2 D}$$

$$c = (1 + \beta^2)^2 + 2\gamma \qquad d = (1 + 9\beta^2)^2 + 2\gamma$$

By taking three equations of the system (j) with three coefficients a_1, a_3, and a_5 different from zero, we obtain for the third approximation of σ_{cr} a cubic equation. These calculations show that the difference between the second and the third approximations is small; hence Eq. (k) is accurate enough for calculating critical stresses in a compressed plate with one rib. These stresses always can be represented by the formula

$$\sigma_{cr} = k \frac{\pi^2 D}{b^2 h} \qquad (l)$$

in which the factor k depends on the proportions of the plate and of the rib which are defined by the ratios (g). Several values of this factor are given in Table 9-16. It is seen that for a definite value of γ and δ the factor k varies with the ratio a/b and becomes a minimum for a certain value of this ratio. This indicates that a long plate will usually buckle into several half-waves such that the ratio of the length of one half-wave to the width of the plate approaches the value at which k is a minimum. It can be seen from the table that the length of the half-waves increases as the flexural rigidity of the rib increases. By using for σ_{cr} the first approxima-

TABLE 9-16. NUMERICAL VALUES OF THE FACTOR k IN EQ. (l) FOR A PLATE STIFFENED BY ONE LONGITUDINAL RIB (FIG. 9-36)

β	$\gamma = 5$			$\gamma = 10$			$\gamma = 15$			$\gamma = 20$			$\gamma = 25$		
	$\delta = 0.05$	$\delta = 0.10$	$\delta = 0.20$	$\delta = 0.05$	$\delta = 0.10$	$\delta = 0.20$	$\delta = 0.05$	$\delta = 0.10$	$\delta = 0.20$	$\delta = 0.05$	$\delta = 0.10$	$\delta = 0.20$	$\delta = 0.05$	$\delta = 0.10$	$\delta = 0.20$
0.6	16.5	16.5	16.5	16.5	16.5	16.5	16.5	16.5	16.5	16.5	16.5	16.5	16.5	16.5	16.5
0.8	15.4	14.6	13.0	16.8	16.8	16.8	16.8	16.8	16.8	16.8	16.8	16.8	16.8	16.8	16.8
1.0	12.0	11.1	9.72	16.0	16.0	15.8	16.0	16.0	16.0	16.0	16.0	16.0	16.0	16.0	16.0
1.2	9.83	9.06	7.88	15.3	14.2	12.4	16.5	16.5	16.5	16.5	16.5	16.5	16.5	16.5	16.5
1.4	8.62	7.91	6.82	12.9	12.0	10.3	16.1	15.7	13.6	16.1	16.1	16.1	16.1	16.1	16.1
1.6	8.01	7.38	6.32	11.4	10.5	9.05	14.7	13.6	11.8	16.1	16.1	14.4	16.1	16.1	16.1
1.8	7.84	7.19	6.16	10.6	9.70	8.35	13.2	12.2	10.5	15.9	14.7	12.6	16.2	16.2	14.7
2.0	7.96	7.29	6.24	10.2	9.35	8.03	12.4	11.4	9.80	14.6	13.4	11.6	16.0	15.4	13.3
2.2	8.28	7.58	6.50	10.2	9.30	7.99	12.0	11.0	9.45	13.9	12.7	10.9	15.8	14.5	12.4
2.4	8.79	8.06	6.91	10.4	9.49	8.15	11.9	10.9	9.37	13.5	12.4	10.6	15.1	13.8	11.9
2.6	9.27	8.50	7.28	10.8	9.86	8.48	12.1	11.1	9.53	13.5	12.4	10.6	14.8	13.6	11.6
2.8	8.62	7.91	6.31	11.4	10.4	8.94	12.5	11.5	9.85	13.7	12.6	10.8	14.8	13.6	11.6
3.0	8.31	7.62	6.53	12.0	11.1	9.52	13.1	12.0	10.3	14.1	13.0	11.1	15.2	13.9	11.9
3.2	8.01	7.38	6.32	11.4	10.5	9.05	13.9	12.7	10.9	14.8	13.5	11.6	15.6	14.3	12.3
3.6	7.84	7.19	6.16	10.6	9.70	8.35	13.2	12.2	10.5	15.9	14.7	12.6	16.2	15.7	13.5
4.0	7.96	7.29	6.24	10.2	9.35	8.03	12.4	11.4	9.8	14.6	13.4	11.6	16.0	15.4	13.3

tion (9-19), it can be readily shown that k becomes a minimum when

$$\beta^2 = \sqrt{1 + 2\gamma}$$

The values of k above the horizontal lines in Table 9-16 are the same as for a simply supported plate of width equal to $b/2$. It indicates those proportions of the rib and of the plate for which the rib remains straight when the plate buckles.

From Table 9-16 the critical compressive stress for a plate stiffened by one rib (Fig. 9-36) can be calculated without difficulty in each particular case. Let a compressed steel plate with simply supported edges have the following dimensions: $a = 48$ in., $b = 80$ in., $h = \frac{9}{16}$ in., $E = 30 \cdot 10^6$ psi, and $\nu = 0.3$. Then $\beta = a/b = 0.6$, and for a plate without a stiffening rib, we obtain from Table 9-1 $\sigma_{cr} = 6,900$ psi. Let us assume now that an absolutely rigid rib subdivides the width of the plate in halves. In such case each half can be considered as a plate of the width $b/2 = 40$ in. For this plate, considered as simply supported, we find, by using Table 9-1, that

$$\sigma_{cr} = 22,100 \text{ psi}$$

To decide what cross-sectional dimensions the rib must have in order to remain straight during buckling of the plate, we use Table 9-16. It is seen from this table that, for $a/b = 0.6$, such a condition is obtained by using a stiffener for which

$$\gamma = \frac{EI}{bD} = 5$$

and $\delta = A/bh < 0.2$. In calculating the quantity EI, note that the stiffener is riveted or welded to a plate of great width; this results in a considerable increase of flexural rigidity of the rib. If the stiffener is taken in the form of a channel or a bar of Z section, riveted to the plate by one flange, the centroid of the cross section, consisting of the stiffener and the plate, will be very near to the surface of the plate; the moment of inertia of the cross section of the stiffener with respect to the axis coinciding with the outer surface of the flange should be taken in calculating EI.[1] Taking for the rib a standard channel of 4 in. depth and 1.56 in.[2] in cross-sectional area, we obtain $I = 3.8 + 1.56 \cdot 4 = 10.0$ in.[4], $\gamma = EI/bD \approx 7.7$, and $\delta = A/bh \approx 0.034$. For such proportions, as we see from Table 9-16 for $\beta = 0.6$, the rib can be considered as absolutely rigid.

If the length of the plate is twice as great, we have $a/b = 1.2$ and it is seen from the table that to eliminate buckling of the stiffener the ratio γ must be larger than 10. Taking for the stiffener a standard channel of 5 in. depth and 1.95 in.[2] cross-sectional area, we obtain $I = 7.4 + 1.95 \cdot 2.5^2 = 19.6$ in.[4], $\gamma = 15$, and $\delta = 0.043$. For such proportions, as seen from the table, the rib can be considered as absolutely rigid.

In the case of two equal longitudinal ribs dividing the width of the plate into three equal parts, the problem of stability can be analyzed in a similar way. Assuming that the buckling form is symmetrical about the middle axis ($y = b/2$), we obtain as the first approximation, from Eqs. (i):

$$\sigma_{cr} = \frac{\pi^2 D}{b^2 h} \frac{(1 + \beta^2)^2 + 3\gamma}{\beta^2(1 + 3\delta)} \tag{9-20}$$

---

[1] This question is discussed further by E. Chwalla and A. Novak, *Der Stahlbau*, nos. 10 and 12, 1937.

TABLE 9-17. VALUES OF THE FACTOR k IN EQ. (l) FOR THE CASE OF TWO LONGITUDINAL RIBS DIVIDING THE PLATE INTO THREE EQUAL PARTS

β	$\gamma = \frac{10}{3}$		$\gamma = 5$		$\gamma = \frac{20}{3}$		$\gamma = 10$	
	$\delta = 0.05$	$\delta = 0.10$	$\delta = 0.05$	$\delta = 0.10$	$\delta = 0.05$	$\delta = 0.10$	$\delta = 0.05$	$\delta = 0.10$
0.6	26.8	24.1	36.4	33.2	36.4	36.4	36.4	36.4
0.8	16.9	15.0	23.3	20.7	29.4	26.3	37.2	37.1
1.0	12.1	10.7	16.3	14.5	20.5	18.2	28.7	25.6
1.2	9.61	8.51	12.6	11.2	15.5	13.8	21.4	19.0
1.4	8.32	7.36	10.5	9.32	12.7	11.3	17.2	15.2
1.6	7.70	6.81	9.40	8.31	11.1	9.82	14.5	12.8
1.8	7.51	6.64	8.85	7.83	10.2	9.02	12.9	11.4
2.0	7.61	6.73	8.70	7.69	9.78	8.65	11.9	10.6

This formula has the same form as formula (l) given above for the case of one rib. Several values of the numerical factor k are given in Table 9-17. By using formula (9-20) we can establish in each particular case the value of β at which the factor k in Eq. (l) becomes a minimum and also the number of waves into which the plate will subdivide during buckling if the

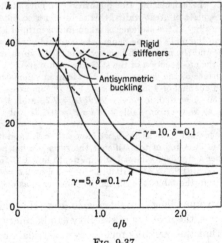

FIG. 9-37

proportions of the stiffeners are chosen. The values of k are represented in Fig. 9-37 by curves. For comparison there are given also the curves for the case when there are two absolutely rigid stiffeners. In discussing plates with smaller values of a/b and of γ, we should consider the possibility of an antisymmetrical form of buckling, in which the middle axis

$(y = b/2)$ is a nodal line.[1] From Fig. 9-37 it is seen that for $0.37 <$ $a/b < 0.66$ and $\gamma = 5$, the smallest σ_{cr} corresponds to this form of buckling.

Instead of the energy method, the method of integration of the differential equation, as shown in Art. 9.4, can be used in studying buckling of stiffened plates.[2]   The values of k for a plate with one stiffener at the middle and the longitudinal edges built in (Fig. 9-36), as obtained by integration, are given in Fig. 9-38.   For comparison there are given also curves for the cases (1) of no stiffener, and (2) of an absolutely rigid stiffener.   In the

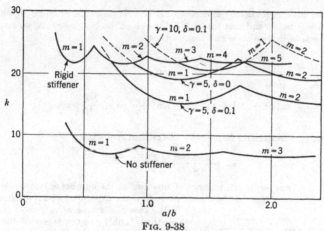

Fig. 9-38

latter case each half of the plate is considered as a plate of width $b/2$ clamped along one edge parallel to the x axis and simply supported along the other.

If the number of equidistant stiffening ribs is larger than two, and if the ribs are relatively flexible, the approximate expression for the critical stress, from the first of Eqs. (i), is[3]

$$\sigma_{cr} = \frac{\pi^2 D}{b^2 h} \frac{(1 + \beta^2)^2 + 2 \sum_i \gamma_i \sin^2 \frac{\pi c_i}{b}}{\beta^2 \left(1 + 2 \sum_i \delta_i \sin^2 \frac{\pi c_i}{b}\right)} \tag{9-21}$$

Simply Supported Compressed Plate with Transverse Stiffening Ribs (Fig. 9-39).   In this case we take the deflection surface of the plate in the form of the series[4]

[1] This form of buckling was considered by R. Barbré, *Ingr.-Arch.*, vol. 8, p. 117, 1937.

[2] This method was used by A. Lokshin, *J. Appl. Math. Mech.*, vol. 2, p. 225, 1935 (Russian), and by Barbré, *op. cit.* Figure 9-38 is taken from the latter paper.

[3] For another method of analysis, see p. 403.

[4] There will be formed during buckling only one half-wave in the y direction.

$$w = \sum_{m=1}^{m=\infty} a_m \sin \frac{m\pi x}{a} \sin \frac{\pi y}{b} \qquad (m)$$

Proceeding as before and using notations (g), we obtain a system of homogeneous linear equations of the following kind:

$$a_m(m^2 + \beta^2)^2 + \sum_i 2\gamma_i\beta^3 \sin \frac{m\pi c_i}{a} \left(a_1 \sin \frac{\pi c_i}{a} + a_2 \sin \frac{2\pi c_i}{a} + \cdots\right)$$
$$= \sigma_{cr} \frac{b^2 h}{\pi^2 D} \beta^2 m^2 a_m \qquad (n)$$

If there are many equidistant and comparatively flexible ribs of equal rigidity, so that there will be several ribs for each half-wave of the buckled plate, we can take only one term in expression m and assume that

$$w = a_m \sin \frac{m\pi x}{a} \sin \frac{\pi y}{b}$$

Equation (n) becomes

$$a_m(m^2 + \beta^2)^2 + a_m 2\gamma\beta^3 \sum_i \sin^2 \frac{m\pi c_i}{a} = \sigma_{cr} \frac{b^2 h}{\pi^2 D} \beta^2 m^2 a_m$$

In this case an approximate formula for the critical stress is

$$\sigma_{cr} = \frac{\pi^2 D}{b^2 h} \frac{(m^2 + \beta^2)^2 + r\gamma\beta^3}{\beta^2 m^2} \qquad (9\text{-}22)$$

in which $r - 1$ represents the number of ribs and m the number of half-waves. In each particular case m should be chosen so as to make expression (9-22) a minimum.

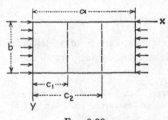

In the case of a comparatively short plate in which there is only one transverse rib bisecting the plate (Fig. 9-39), we assume that only the coefficient a_1 in expression (m) is different from zero, and from the first of Eqs. (n) we obtain as the first approximation

$$\sigma_{cr} = \frac{\pi^2 D}{b^2 h} \frac{(1 + \beta^2)^2 + 2\gamma\beta^3}{\beta^2} \qquad (9\text{-}23)$$

FIG. 9-39

This formula shows how the critical stress is affected by the presence of the rib if the plate buckles into one half-wave.

A better approximation is obtained by taking a_1 and a_3 different from zero and using the first and third of Eqs. (n). By gradually increasing γ, we finally arrive at the condition where the plate buckles into two half-waves and the rib becomes the nodal line for the buckled plate. Several limiting values of γ, at which the rib remains straight during buckling of the plate, have been calculated for various values of β by using two of Eqs. (n); these are given in Table 9-18.[1] It is seen that the effect of the rib on the magnitude of the critical compressive stress depends on the proportions of the plate. Taking a square plate and using such a rib that $\gamma = 1.19$, we obtain two half-waves instead of one and the critical stress increases, as compared with that for the unstiff-

[1] This table was computed by Miss N. Naerlovich of Belgrade and communicated by letter in October, 1952.

TABLE 9-18. LIMITING VALUES OF γ FOR ONE, TWO, AND
THREE TRANSVERSE RIBS

β	0.5	0.6	0.7	0.8	0.9	1.0	1.2	$\sqrt{2}$
One rib	12.8	7.25	4 42	2.82	1.84	1.19	0.435	0
Two ribs	65.5	37.8	23.7	15.8	11.0	7.94	4.43	2.53
Three ribs	177	102	64.4	43.1	30.2	21.9	12.6	7.44

ened plate (see Table 9-1) in the ratio 6.25:4. If the proportions of the plate are such that $\beta = 1.41$, the critical stress for the unstiffened plate is the same for buckling into one or two half-waves; hence in this case a transverse rib bisecting the plate has no effect whatsoever on the magnitude of the critical stress.

The case of stiffening the plate by using two or three equal and equidistant ribs can be discussed in a similar manner, and the limiting values of γ, at which the ribs remain straight when the plate buckles into three or four equal half-waves, respectively, can be calculated. Several limiting values of γ for this case are given in Table 9-18.

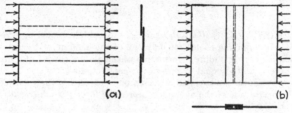

(a)   (b)

FIG. 9-40

The method used above in calculating the stiffening effect of ribs can be applied also if it is desired to investigate the effect of riveted joints (Fig. 9-40) on critical stresses. For example, this problem is encountered in the design of ship structures.[1]

In the case of a large number of equal and equidistant ribs parallel to one of the sides of a compressed rectangular plate, we can consider the stiffened plate as a plate having two different flexural rigidities in the two perpendicular directions.[2] The general differential equation for the deflection surface of such a plate, if submitted to the action of forces in its middle plane, is

$$D_1 \frac{\partial^4 w}{\partial x^4} + 2D_3 \frac{\partial^4 w}{\partial x^2 \partial y^2} + D_2 \frac{\partial^4 w}{\partial y^4} = N_s \frac{\partial^2 w}{\partial x^2} + N_y \frac{\partial^2 w}{\partial y^2} + 2N_{xy} \frac{\partial^2 w}{\partial x \partial y} \quad (9\text{-}24)$$

In this equation, $D_1 = (EI)_x/(1 - \nu_x\nu_y)$ is the average flexural rigidity of the stiffened plate corresponding to bending moments M_x, ν_x and ν_y are the values of Poisson's ratio corresponding to the directions x and y, $D_2 = (EI)_y/(1 - \nu_x\nu_y)$ is the average flexural

---

[1] Several examples of this kind have been discussed by G. Schnadel, *Werft, Reederei, Hafen*, vol. 11, 1930.

[2] Bending of such plates was considered by M. T. Huber, *Bauingenieur*, p. 354, 1923, and *Repts. Intern. Congr. Appl. Mech.*, Zürich, 1926. See also "Theory of Plates and Shells," by Timoshenko, 2d ed., in collaboration with S. Woinowsky-Krieger, McGraw-Hill Book Company, New York, 1959.

rigidity corresponding to bending moments M_y, and

$$D_3 = \tfrac{1}{2}(\nu_x D_2 + \nu_y D_1) + 2(GI)_{xy}$$

where $2(GI)_{xy}$ is the average torsional rigidity [see Eqs. (o) below]. Equation (9-24) can be derived by substituting in the equation of equilibrium (g) of Art. 8.2 the following expressions for moments:

$$
\begin{aligned}
M_x &= -\frac{(EI)_x}{1 - \nu_x \nu_y}\left(\frac{\partial^2 w}{\partial x^2} + \nu_y \frac{\partial^2 w}{\partial y^2}\right)\\
M_y &= -\frac{(EI)_y}{1 - \nu_x \nu_y}\left(\frac{\partial^2 w}{\partial y^2} + \nu_x \frac{\partial^2 w}{\partial x^2}\right)\\
M_{xy} &= 2(GI)_{xy}\frac{\partial^2 w}{\partial x\,\partial y}
\end{aligned}
\qquad (o)
$$

which are obtained from the equations

$$
\begin{aligned}
\frac{\partial^2 w}{\partial x^2} &= -\frac{M_x}{(EI)_x} + \frac{\nu_y}{(EI)_y}M_y\\
\frac{\partial^2 w}{\partial y^2} &= -\frac{M_y}{(EI)_y} + \frac{\nu_x}{(EI)_x}M_x\\
\frac{\partial^2 w}{\partial x\,\partial y} &= \frac{1}{2(GI)_{xy}}M_{xy}
\end{aligned}
\qquad (p)
$$

The quantities $(EI)_x$, $(EI)_y$, $(GI)_{xy}$, ν_x, and ν_y can be determined by direct tests of stiffened plates by applying each time only one of the bending or torsional moments and measuring the corresponding deformation of the plate. Such tests[1] show that we usually can take ν_x and ν_y equal to zero in our calculations.

In the case of uniform compression of such a plate parallel to the x axis (Fig. 9-1), if we denote the magnitude of the average compressive force per unit length by N_x, we obtain, from Eq. (9-24),

$$D_1 \frac{\partial^4 w}{\partial x^4} + 2D_3 \frac{\partial^4 w}{\partial x^2\,\partial y^2} + D_2 \frac{\partial^4 w}{\partial y^4} + N_x \frac{\partial^2 w}{\partial x^2} = 0 \qquad (9\text{-}25)$$

Assuming that the plate buckles into one half-wave and substituting in Eq. (9-25)

$$w = A \sin\frac{\pi x}{a}\sin\frac{\pi y}{b}$$

we obtain $\sigma_{cr} = \dfrac{\pi^2}{b^2 h}\left(D_1\dfrac{b^2}{a^2} + 2D_3 + D_2\dfrac{a^2}{b^2}\right)$

The smallest value for the critical stress is obtained when

$$\frac{a}{b} = \sqrt[4]{\frac{D_1}{D_2}} \qquad (9\text{-}26)$$

and the value is

$$\sigma_{cr} = \frac{2\pi^2}{b^2 h}\left(\sqrt{D_1 D_2} + D_3\right) \qquad (9\text{-}27)$$

From this we conclude that a long rectangular plate compressed longitudinally and stiffened by parallel longitudinal ribs buckles into many equal half-waves, the lengths of which satisfy Eq. (9-26). The critical stress is then determined from Eq. (9-27). For an isotropic plate $D_1 = D_2 = D_3$ and Eq. (9-27) coincides with Eq. (h) of Art. 9.2.

[1] See, for instance, E. Seydel's paper in "Jahrbuch 1930 der deutschen Versuchs-anstalt für Luftfahrt," p. 235, Berlin, 1930.

Stiffening of Simply Supported Rectangular Plates under Shearing Stresses.   Several simple cases of this kind have been investigated.   Let us begin with a simply supported rectangular plate submitted to the action of uniformly distributed shearing stresses and stiffened by one rib bisecting the plate (Fig. 9-41).   In studying the effect of ribs on the magnitude of critical shearing stresses we can use the energy method as before.   It can be shown in this way that, if the rigidity of the stiffener is not sufficient, the inclined waves of the buckled plate run across the stiffener and buckling of the plate is accompanied by bending of the rib. By subsequent increase of the rigidity of the rib, we may finally arrive at a condition in which each half of the plate will buckle as a rectangular plate with simply supported edges of dimensions $a/2$ and

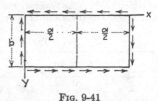

FIG. 9-41

b and the rib will remain straight.[1]   The corresponding limiting value of the flexural rigidity EI of the rib can be found from the consideration of strain energy of bending of the plate and of the rib.   Several values of the ratio γ of this flexural rigidity to the rigidity Da of the plate if bent into a cylindrical surface are given in Table 9-19.

TABLE 9-19. LIMITING VALUES OF THE RATIO γ IN THE CASE OF ONE RIB

a/b	1	1.25	1.5	2
$\gamma = EI/Da$	15	6.3	2.9	0.83

TABLE 9-20. LIMITING VALUES OF THE RATIO γ IN THE CASE OF TWO RIBS

a/b	1.2	1.5	2	2.5	3
$\gamma = EI/Da$	22.6	10.7	3.53	1.37	0.64

If there are two ribs dividing the plate into three equal portions, the limiting value of γ at which the ribs remain straight when the plate buckles can be determined in a similar manner.   Several such values of γ are given in Table 9-20.   Some applications of these results in determining the proper dimensions of stiffeners in the case of plate girders will be shown later (see Art. 9.16).[2]

If a long rectangular plate is stiffened by several longitudinal ribs, an approximate value of the critical shearing stress can be obtained by using for the deflection surface of the buckled plate the expression (k) of Art. 9.7.   Adding to the strain energy of bending of the buckled plate the strain energy of bending of the ribs and equating this sum to the work done by shearing forces, we obtain

$$\tau_{cr} = \frac{\pi^2 D}{b^2 h} \frac{1}{2\alpha} \left\{ 2 + 6\alpha^2 + \frac{s^2}{b^2} + \frac{b^2}{s^2} \left[\gamma + (1 + \alpha^2)^2 \right] \right\} \qquad (9\text{-}28)$$

[1] The line of the rib in this case is not a nodal line for the buckled plate and there will be some bending of the rib, but its effect on τ_{cr} is negligible; see paper by A. Kromm, *Der Stahlbau*, nos. 18 and 20, 1944.

[2] The cases of three and four stiffeners were discussed by Tsun Kuei Wang, *J. Appl. Mech.*, vol. 14, p. 269, 1947.   See also the paper by Vlatka Brčić, *Der Stahlbau*, vol. 25, p. 88, 1956.

$$\text{where} \qquad \gamma = \frac{2 \sum_i (EI)_i \sin^2 \frac{\pi c_i}{b}}{Db} \qquad (q)$$

in which $(EI)_i$ is the flexural rigidity of a rib at a distance c_i from the edge of the plate and where b is the width of the plate. For any assumed value of γ it is necessary to

TABLE 9-21. VALUES OF k IN EQ. (r)

γ	5	10	20	30	40	50	60	70	80	90	100
k	6.98	7.70	8.67	9.36	9.90	10.4	10.8	11.1	11.4	11.7	12.0

determine the quantities α and s so as to make expression (9-28) a minimum. In this way the critical value of shearing stress will be represented by the formula

$$\tau_{cr} = k \frac{\pi^2 D}{b^2 h} \qquad (r)$$

Several values of the factor k are given in Table 9-21. To show the effect of stiffeners on the critical stress, the values of k given in this table should be compared

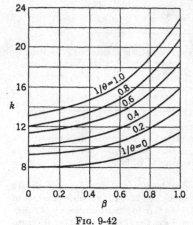

FIG. 9-42

with the approximate value 5.7 obtained for an unstiffened long plate [see Eq. (m), Art. 9.7].

Let us take, as an example, a long rectangular plate with $b = 84$ in., $h = \frac{3}{8}$ in., and stiffened by three equidistant ribs in the form of standard channels of depth 4 in. and cross-sectional area 1.56 in.[2] In such a case we have

$$I = 3.8 + 1.56 \cdot 4 = 10 \text{ in.}^4$$

and we obtain, from Eq. (q), $\gamma \approx 98$; hence k, in Eq. (r), is about 12.

The problem of shear buckling of long rectangular plates is of primary importance in aircraft design, and a considerable amount of experimental and theoretical work has been done in that field.[1]

In the case of a large number of parallel, equal, and equidistant ribs, the stiffened plate can again be considered as a plate having two different flexural rigidities in the two perpendicular directions and Eq. (9-24) can be used. In this way buckling of corrugated plates was discussed by Bergmann and Reissner.[2] Assuming that the corrugation waves are parallel to one of the sides of a simply supported rectangular plate (Fig. 9-22) and using the notations

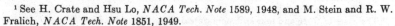

[1] See H. Crate and Hsu Lo, *NACA Tech. Note* 1589, 1948, and M. Stein and R. W. Fralich, *NACA Tech. Note* 1851, 1949.

[2] S. Bergmann and H. Reissner, *Z. Flugtech. u. Motorluftsch.*, vol. 20, p. 475, 1929. See also Seydel, *loc. cit.*

$$\theta = \frac{\sqrt{D_1 D_2}}{D_3} \quad \text{and} \quad \beta = \frac{b}{a} \sqrt[4]{\frac{D_1}{D_2}}$$

we obtain the critical value of the shearing force N_{xy} for $\theta > 1$ from the equation

$$(N_{xy})_{cr} = 4k \frac{\sqrt[4]{D_1 D_2{}^3}}{b^2} \tag{9-29}$$

in which k is a factor, depending on the values of θ and β, which can be taken from the curves[1] in Fig. 9-42. In the case of an infinitely long isotropic plate $\beta = 0$, $\theta = 1$ and

TABLE 9-22. VALUES OF THE FACTOR k IN EQ. (9-30) FOR AN INFINITELY LONG PLATE

θ	0	0.2	0.5	1.0
k	11.7	11.8	12.2	13.17

TABLE 9-23. VALUES OF THE FACTOR k IN EQS. (9-29) AND (9-30) FOR AN INFINITELY LONG PLATE WITH CLAMPED EDGES

θ	0	0.2	0.5	1	2	3	5	10	20	40	∞
k	18.6	18.9	19.9	22.15	18.8	17.6	16.6	15.9	15.5	15.3	15.1

we obtain from Fig. 9-42 a value for k which brings Eq. (9-29) in agreement with Eq. (9-9) obtained before. For $\theta < 1$ the critical value of the shearing force is obtained from the equation

$$(N_{xy})_{cr} = 4k \frac{\sqrt{D_2 D_3}}{b^2} \tag{9-30}$$

Several values of k for an infinitely long plate are given in Table 9-22.

Equations (9-29) and (9-30) can be used also in the case of a long plate with clamped edges, in which case the values of k must be taken from Table 9-23.

The problem of stiffening of a rectangular plate subjected to pure bending in its

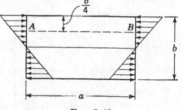

FIG. 9-43

plane (Fig. 9-43) was also investigated. The critical value of the maximum compressive stress again is given by Eq. (l). The values of k in this equation depend on the ratio a/b and on the flexural rigidity and the cross-sectional area of the stiffener as defined by expressions (g). The calculations were made[2] for the case of a stiffener AB (Fig. 9-43) located at a distance $b/4$ from the edge of maximum compression. The values of k obtained for various values of γ and δ are represented by curves in Fig. 9-44. For comparison there are shown also curves for an unstiffened plate ($\gamma = \delta = 0$), constructed by using values of k from Table 9-6, and for the case of an absolutely rigid stiffener. With the use of these curves the proper dimensions of the stiffener can be selected in each particular case. The curves in Fig. 9-45 give the values of γ at which

[1] See E. Seydel, Z. Flugtech. u. Motorluftsch., vol. 24, p. 78, 1933.
[2] See C. Massonnet, Intern. Assoc. Bridge Structural Eng., vol. 6, p. 233, 1940.

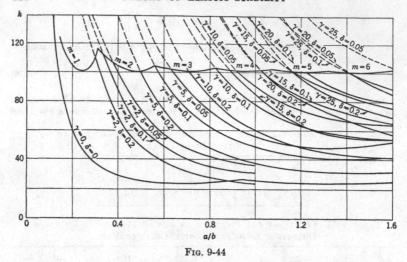

FIG. 9-44

values of k are the same as for an absolutely rigid stiffener. In Fig. 9-46 are given the values of k for the case when the stiffener is placed along the middle line of the plate.[1]

The problem of buckling of a plate subjected to nonuniform compression during bending of the ribs (Fig. 9-47) was also studied,[2] and the values of q_{cr} for uniformly loaded ribs were established for some assumed proportions of the structure.

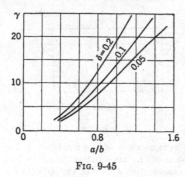

FIG. 9-45

9.12. Buckling of Plates beyond Proportional Limit.

So far in our discussion of buckling of plates we have assumed that the stresses remain within the elastic range. Beyond the proportional limit the formulas previously derived give exaggerated values for critical stresses, and in order to get more satisfactory results the behavior of the material beyond the proportional limit must be considered. Let us begin with the case of a rectangular plate simply supported along all edges and uniformly compressed parallel to one side (Fig. 9-1). The critical value of the compressive stress within the elastic limit and for buckling into one half-wave is given by the formula

$$\sigma_{cr} = \frac{\pi^2 D}{b^2 h} \left(\frac{a}{b} + \frac{b}{a} \right)^2 \tag{a}$$

[1] This case was discussed by M. Hampl, *Der Stahlbau*, vol. 10, p. 16, 1937.
[2] See V. Bogunović, *Der Stahlbau*, vol. 24, p. 8, 1955.

The smallest value of this stress, for a given value of the ratio h/b, is obtained in the case of a square plate for which

$$\sigma_{cr} = \frac{4\pi^2 D}{b^2 h} = \frac{\pi^2 E}{3(1 - \nu^2)} \frac{h^2}{b^2} \tag{b}$$

This value of σ_{cr} can be used also for long rectangular plates buckling into many waves. Taking the case of structural steel for which $E = 30 \cdot 10^6$

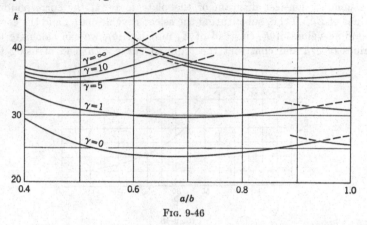

FIG. 9-46

psi and $\nu = 0.3$, we can represent σ_{cr} as a function of the ratio b/h by the curve AB shown in Fig. 9-48. This curve can be used for obtaining σ_{cr} within the elastic region.

Experiments show that, when the compressive stress reaches the yield point of the material, which in this case is assumed equal to 34,000 psi, the plate buckles for any value of the ratio b/h. This is shown in the figure by the horizontal line BC. If the material has a sharply defined yield point and follows Hooke's law up to that point, the horizontal line BC, together with the curve BA, determines the value of the critical compressive stress for any value of the ratio b/h. In such a material as

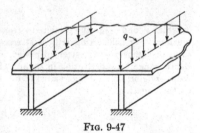

FIG. 9-47

structural steel some permanent set usually takes place at a stress lower than the yield point. Assuming that point D in the figure corresponds to the proportional limit of the material, i.e., to the stress at which the beginning of permanent set becomes noticeable, there must be in the figure, instead of a sharp corner at B, some intermediate curve joining curve AD, corresponding to the region of perfect elasticity, with the horizontal line representing plastic flow. To obtain such a curve we assume that compression of a plate beyond the proportional limit in one direction affects equally the mechanical properties of the material in all other directions. Thus the plate remains isotropic,[1] and we use the tangent modulus of the material (see Art. 3.3). We take several values of the initial stress between the yield point and the proportional limit and determine from the compression-test diagram of the plate material the corresponding values of E_t.   Then, substituting the selected values of σ_{cr} and the corresponding values of E_t (instead of E) into Eq. (b), we can calculate the values of b/h and construct the required curve.   Instead of using the

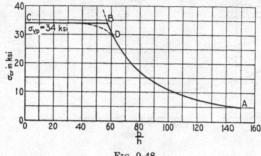

Fig. 9-48

compression-test diagram, we can determine E_t from Eq. (3-17) by using a proper value of the parameter c.

In the general case of a plate subjected to the action of a homogeneous plane stress, the yield-point stress is determined by using the equation[2]

$$\sigma_1{}^2 - \sigma_1\sigma_2 + \sigma_2{}^2 = \sigma_{YP}{}^2$$

in which σ_1 and σ_2 are the principal stresses and σ_{YP} is the yield-point stress for simple tension.   For calculating critical stresses beyond the proportional limit in various cases of homogeneously stressed plates, we use again the formulas developed for the elastic range by substituting in them E_t instead of E.   If there are no experimental data for determining proper values of E_t, the notion of equivalent tensile stress, defined by the equation

$$\sigma_e = \sigma_1 \sqrt{1 - \frac{\sigma_2}{\sigma_1} + \left(\frac{\sigma_2}{\sigma_1}\right)^2}$$

can be used.   Then E_t for the combined stress state will be taken equal to the tangent modulus for a simple tensile stress of magnitude σ_e.

[1] See E. Chwalla, *Repts. 2d Intern. Congr. Bridge Structural Eng.*, Vienna, p. 322, 1928, and also M. Rôs and A. Eichinger, *Repts. 3d Intern. Congr. Bridge Structural Eng.*, Paris, p. 144, 1932.

[2] See Timoshenko, "Strength of Materials," 3d ed., part 2, p. 454, D. Van Nostrand Company, Inc., Princeton, N.J., 1956.

Some experiments with aluminum alloy plates stressed beyond the proportional limit in shear showed[1] a satisfactory agreement with calculated values of τ_{cr} by using, instead of E_t/E, the values of the ratios G_s/G, where G is the modulus in shear and G_s is the secant modulus in shear. This latter quantity is determined for each value of τ from a shear-test diagram (Fig. 9-49) as the ratio τ/γ.

9.13. Large Deflections of Buckled Plates. So far in the calculation of critical stresses for various cases of buckling of plates we have assumed that the deflections are very small and that the strain in the middle plane of the plate due to buckling can be neglected. Experience has shown that after buckling, the behavior of plates is quite different from that of compressed struts. The critical load for a strut can be considered as the ultimate load, but a thin buckled plate can carry a much larger load than the critical load at which buckling begins. In airplane construction, where the weight of the structure is of primary importance, the extra strength of plates after buckling is usually utilized and the plates may undergo considerable buckling under service conditions. To investigate bending of plates when the stresses are above critical, the strain

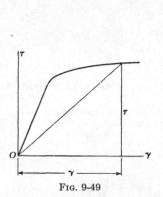

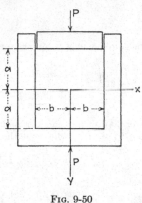

FIG. 9-49   FIG. 9-50

in the middle plane should be considered. The application of the general Eqs. (8-38) and (8-39) to this problem represents great difficulty, since these equations are very complicated. To get an approximate solution let us use the expression for the strain energy of the buckled plate and determine its deflection from the condition of minimum of this energy.[2]

Let us begin with the case in which a simply supported plate is compressed in the y direction and lateral expansion in the x direction is prevented by a rigid frame (Fig. 9-50). Taking the origin of coordinates as shown in the figure, an approximate expression for the deflection surface of the buckled plate, satisfying the boundary conditions, is

$$w = f \cos \frac{\pi x}{2b} \cos \frac{\pi y}{2a} \qquad (a)$$

[1] See G. Gerard, *J. Appl. Mech.*, vol. 15, p. 7, 1948. A comparison of various theories of plastic buckling of plates is given in the paper by R. A. Pride and G. J. Heimerl, *NACA Tech. Note* 1817, 1949. See also E. Z. Stowell, *NACA Tech. Note* 1681, 1948, and P. P. Bijlaard, *J. Aeronaut. Sci.*, vol. 16, p. 529, 1949, and vol. 24, p. 291, 1957.

[2] Another method of solution of the problem has been given by G. Schnadel, *Repts. 3d Intern. Congr. Appl. Mech.*, Stockholm, vol. 3, p. 73, 1930.

The components v and u of displacements in the middle plane of the plate, satisfying the boundary conditions, can be taken in the following form:

$$v = C_1 \sin \frac{\pi y}{a} \cos \frac{\pi x}{2b} - ey$$

$$u = C_2 \sin \frac{\pi x}{b} \cos \frac{\pi y}{2a}$$

(b)

where C_1, C_2, and e are constants. It is seen that the displacements u become zero at the edges $x = \pm b$ and $y = \pm a$. The displacements v at the boundaries $x = \pm b$ and $y = \pm a$ are equal to those due to the uniform compression strain e in the y direction.[1]

The components of strain in the middle plane of the plate are [see Eqs. (b), p. 346]

$$\epsilon_x = \frac{\partial u}{\partial x} + \frac{1}{2}\left(\frac{\partial w}{\partial x}\right)^2$$

$$\epsilon_y = \frac{\partial v}{\partial y} + \frac{1}{2}\left(\frac{\partial w}{\partial y}\right)^2$$

$$\gamma_{xy} = \frac{\partial v}{\partial x} + \frac{\partial u}{\partial y} + \frac{\partial w}{\partial x}\frac{\partial w}{\partial y}$$

(c)

and the corresponding strain energy of the plate is

$$U_1 = \frac{h}{2}\int_{-a}^{+a}\int_{-b}^{+b} (\sigma_x\epsilon_x + \sigma_y\epsilon_y + \tau_{xy}\gamma_{xy})\, dx\, dy$$

$$= \frac{Gh}{1-\nu}\int_{-a}^{+a}\int_{-b}^{+b}\left[\epsilon_x{}^2 + \epsilon_y{}^2 + 2\nu\epsilon_x\epsilon_y + \frac{1}{2}(1-\nu)\gamma_{xy}{}^2\right] dx\, dy$$

Using for the displacements u, v, and w expressions (a) and (b) and adding to the energy of strain U_1 of the middle plane the energy of bending

$$U_2 = \frac{\pi^4 abf^2 D}{32}\left(\frac{1}{a^2} + \frac{1}{b^2}\right)^2$$

we obtain

$$U = U_1 + U_2 = \frac{\pi^4 abf^2 D}{32}\left(\frac{1}{a^2} + \frac{1}{b^2}\right)^2 + \frac{Gh}{1-\nu}\left[4abe^2 - \frac{\pi^2 f^2 be}{4a}\right.$$

$$+ \frac{\pi^4 f^4}{1{,}024ab}\left(9\frac{a^2}{b^2} + 9\frac{b^2}{a^2} + 2\right) - C_1\frac{\pi^2 f^2}{6}\left(\frac{2b}{a^2} + \frac{1-3\nu}{2b}\right)$$

$$- C_2\frac{\pi^2 f^2}{6}\left(\frac{2a}{b^2} + \frac{1-3\nu}{2a}\right) + C_1{}^2\pi^2\left(\frac{b}{a} + \frac{1-\nu}{8}\frac{a}{b}\right)$$

$$\left. + C_2{}^2\pi^2\left(\frac{a}{b} + \frac{1-\nu}{8}\frac{b}{a}\right) + C_1 C_2\frac{16}{9}(1+\nu) - \nu\frac{\pi^2 f^2 ae}{4b}\right]$$

(d)

For a given unit compression e of the plate the constants C_1, C_2, and f in the expressions (a) and (b) will be found from the conditions that the strain energy U is a minimum; hence

$$\frac{\partial U}{\partial C_1} = 0 \qquad \frac{\partial U}{\partial C_2} = 0 \qquad \frac{\partial U}{\partial f} = 0$$

(e)

Let us consider now the case of a square plate. Then $a = b$, $C_1 = C_2 = C$ and the first two of Eqs. (e) become

$$-\frac{\pi^2 f^2}{6a}\frac{5-3\nu}{2} + 2\pi^2 C\frac{9-\nu}{8} + \frac{16}{9}C(1+\nu) = 0$$

[1] It is assumed that the center of the plate remains immovable.

from which, with $\nu = 0.3$, we obtain

$$C = 0.1418 \frac{f^2}{a}. \tag{f}$$

Substituting this in the third of Eqs. (e), we obtain

$$f(4.058h^2 - 6.42a^2e + 5.688f^2) = 0 \tag{g}$$

The solution $f = 0$ of this equation corresponds to the flat form of the compressed plate. The other solution giving the deflection of the buckled plate is obtained by equating to zero the term in parentheses in Eq. (g). Then

$$f = \sqrt{\frac{6.42a^2e - 4.058h^2}{5.688}}. \tag{h}$$

We obtain a real solution for f only if

$$6.42a^2e > 4.058h^2$$

The limiting condition

$$6.42a^2e = 4.058h^2$$

gives

$$e_{cr} = 0.632 \frac{h^2}{a^2} \tag{i}$$

and the corresponding compressive stress is

$$\sigma_{cr} = \frac{e_{cr}E}{1 - \nu^2} = \frac{0.632h^2E}{(1 - \nu^2)a^2} \tag{j}$$

This stress is equal to the critical compressive stress $(\sigma_y)_{cr}$ for a square plate compressed in two perpendicular directions as obtained from Eq. (c) (p. 357) by taking $m = n = 1$, $a = b$, and $\sigma_x = \nu\sigma_y = 0.3\sigma_y$.

If we take for e a value n times larger than that given by (i), we obtain from (h)

$$f = 0.845h \sqrt{n - 1} \tag{k}$$

Taking $n = 10$, i.e., making the compression of the plate 10 times larger than its critical value, given by (i), we find $f = 2.535h$, i.e., deflection at the center is about 2.5 times larger than the thickness of the plate.

Substituting the values of the constants f and C from (k) and (f) in expressions (a) and (b) for the displacements, we can find the corresponding strain in the middle plane of the plate from expressions (c) and calculate the corresponding stresses. For $y = a$ we obtain in this way

$$(\sigma_x)_{y=a} = \nu(\sigma_y)_{y=a}$$

$$(\sigma_y)_{y=a} = \frac{E}{1 - \nu^2} [(\epsilon_y)_{y=a} + \nu(\epsilon_x)_{y=a}]$$

$$= 0.714 \frac{E}{1 - \nu^2} \frac{h^2}{a^2} (n - 1) \cos \frac{\pi x}{2a} \left(\frac{\pi^2}{8} \cos \frac{\pi x}{2a} - 0.142\pi \right) - 0.632 \frac{E}{1 - \nu^2} n \frac{h^2}{a^2}$$

The stresses at the edges of the plate calculated for $n = 10$ are shown in Fig. 9-51. It is seen that for larger deflections of the plate the distribution of compressive stresses is no longer uniform and the larger portion of the load is taken by the portions of the plate near the edges.

The total compressive force taken by the plate at a unit compression ne_{cr} is

$$P = -h \int_{-a}^{a} \sigma_y \, dx = 2ah \left(0.623 + \frac{0.377}{n} \right) n\sigma_{cr}$$

The factor in parentheses gives the relative diminishing of the resistance to compression of the plate due to buckling.   We can state that this resistance is equivalent to that of a flat plate having a width equal to

$$c = 2a \left(0.623 + \frac{0.377}{n} \right) \tag{9-31}$$

The width c is called the *effective width* of the compressed plate and was calculated here on the assumption that lateral expansion of the plate is prevented.

For comparatively thin plates as used in airplane construction, for which the ultimate load may be many times larger than the critical load, we cannot expect this

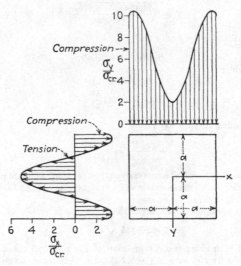

Fig. 9-51

approximate solution to be sufficiently accurate.   The solution can be improved by taking more terms in expressions (a) and (b) for the displacements, but in such case the amount of work required in making the calculation increases immensely.   Calculations were made by adding one term to the expression (a) for deflections and taking

$$w = f \cos \frac{\pi y}{2a} \cos \frac{\pi x}{2b} + f_1 \cos \frac{3\pi y}{2a} \cos \frac{3\pi x}{2b} \tag{l}$$

Then, retaining the previous expressions for u and v, we obtain a second approximation.   These calculations showed that the difference between the first and second approximations is small and that the difference in the magnitude of $(\sigma_y)_{max}$ becomes noticeable only for a very large value of n (say $n > 50$), i.e., when the acting load is many times larger than the critical load.

It was assumed in the previous discussion that the frame is absolutely rigid and that lateral displacements of the edges of the compressed plate are entirely suppressed. The problem can be discussed in an analogous manner for the case where the vertical bars of the frame, keeping the lateral edges of the plate straight, can move freely laterally when the plate is compressed. Taking

$$w = f \cos \frac{\pi x}{2b} \cos \frac{\pi y}{2a}$$

$$v = C_1 \sin \frac{\pi y}{a} \cos \frac{\pi x}{2b} - ey \qquad (m)$$

$$u = C_2 \sin \frac{\pi x}{b} \cos \frac{\pi y}{2a} + \alpha x$$

we determine α from the condition that the sum of normal stresses along the vertical edges of the plate is equal to zero. This gives

$$\alpha = -\frac{\pi^2 f^2}{16b^2} + \frac{2C_2}{b} + \nu e \qquad (n)$$

Substituting this in the third of expressions (m) and calculating the strain energy, we again use Eqs. (e) for determining f, C_1, and C_2. In the case of a square plate, with $\nu = 0.3$, we obtain from the first two of these equations

$$C_1 = 0.144 \frac{f^2}{a} \qquad C_2 = 0.1215 \frac{f^2}{a} \qquad (o)$$

and the third of these equations becomes

$$f(0.411h^2 - 0.455ea^2 + 0.320f^2) = 0$$

The deflection of the buckled plate is

$$f = \sqrt{\frac{0.455ea^2 - 0.411h^2}{0.320}} \qquad (p)$$

By putting this deflection equal to zero, we obtain for the critical value of the longitudinal compression of the plate

$$e_{cr} = \frac{0.411}{0.455} \frac{h^2}{a^2} = 0.904 \frac{h^2}{a^2}$$

which coincides exactly with the value from Eq. (9-6) for a longitudinally compressed square plate. Using again the notation $n = e/e_{cr}$, Eq. (p) gives[1]

$$f = 1.133h \sqrt{n - 1} \qquad (q)$$

Substituting this in Eqs. (o) and using Eqs. (m), we obtain expressions for the stresses σ_x and σ_y in the plate. For the boundary these stresses are

$$(\sigma_x)_{x=a} = 1.41(n - 1)E\frac{h^2}{a^2}\left[\left(1.234 \cos \frac{\pi y}{2a} - 0.382\right)\cos \frac{\pi y}{2a} - 0.374\right]$$

$$(\sigma_y)_{y=a} = E\frac{h^2}{a^2}\left[1.74(n - 1)\left(\cos \frac{\pi x}{2a} - 0.366\right)\cos \frac{\pi x}{2a} - 1.062n + 0.158\right] \qquad (r)$$

Distribution of these stresses for the particular case where $n = 59.7$ is shown in Fig. 9-52.

The total load taken by the plate at a unit compression ne_{cr} is

$$P = -h \int_{-a}^{a} \sigma_y \, dx = 2ah\left(0.661 + \frac{0.339}{n}\right)n\sigma_{cr}$$

---

[1] This formula is in satisfactory agreement with experiments; see W. L. Howland and P. E. Sandorff, *J. Aeronaut. Sci.*, vol. 8, p. 261, 1941.

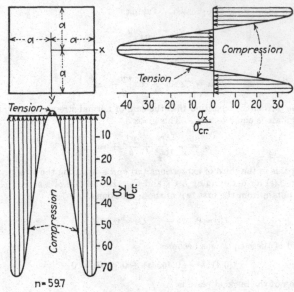

FIG. 9-52

and the effective width for a plate which is free to expand laterally is

$$c = 2a \left(0.661 + \frac{0.339}{n} \right) \tag{9-32}$$

Assuming that expressions (r) for stresses are accurate enough up to complete failure of the plate and that complete failure occurs when the maximum shearing stress reaches the value τ_{YP}, we can determine the ultimate load that the plate can carry from the condition

$$\tau_{\max} = \tfrac{1}{2}(\sigma_z - \sigma_y)_{\max} = \tfrac{1}{2}\sigma_{YP} \tag{s}$$

The maximum value of the shearing stress occurs at the points $x = \pm b$, $y = 0$ (Fig. 9-50) and is

$$\tau_{\max} = \frac{1}{2}\,(\sigma_z - \sigma_y)_{\max} = E\,\frac{h^2}{2a^2}\,(1.38n - 0.473)$$

With this value of $\tau_{\max}$ we obtain from Eq. (s)

$$n = \frac{0.725}{E}\,\frac{a^2}{h^2}\,\sigma_{YP} + 0.34$$

Substituting this value of n in the second of the expressions (r), we obtain for the ultimate load the value

$$\begin{aligned}
P_{\text{ult}} &= 2h \int_0^a \sigma_y\, dx = 0.867ah\sigma_{YP} + 1.02E\,\frac{h^3}{a} \\
&= 2ah\sigma_{YP}\left[0.434 + 2.04\,\frac{E}{\sigma_{YP}}\,\frac{h^2}{(2a)^2} \right] \\
&= 2ah\sigma_{YP}\left(0.434 + 0.566\,\frac{\sigma_{cr}}{\sigma_{YP}} \right)
\end{aligned} \tag{9-33}$$

The above calculated values of the reduced width are satisfactory[1] if n is not large, say $n < 5$. However, in the case of very thin plates, as in aircraft construction, experiments give for the reduced width smaller values than those obtained theoretically. The reason for this difference is contained in assumptions (a) and (b), which do not define accurately enough the deformation of the plate and give exaggerated values for its rigidity. It must be stated also that in experiments with thin plates various kinds of imperfections may reduce considerably the values of the reduced width and the ultimate strength.

Further investigations, based on the theory of large deflections of plates, showed[2] that the solution of the problem can be simplified if we assume that the shearing strain at the boundary of the plate (Fig. 9-50) vanishes. Then, again taking expression (a) for deflections w and substituting it into Eq. (8-38), the stress function and the stresses σ_x and σ_y, satisfying the boundary conditions, can be found. The corresponding values of reduced width found in this way are

$$c = \frac{9}{16} 2a \left(1 + \frac{8}{9} \frac{1}{n} \right)$$

for the immovable longitudinal edges of the plate and

$$c = \frac{1}{2} 2a \left(1 + \frac{1}{n} \right)$$

for the freely moving longitudinal edges of the plate. These values are somewhat smaller than those calculated above.

By using three trigonometric terms in the expression for w it was shown that the effective width of the plate can be represented (within the limits $\sigma_{cr} < \sigma < 60\sigma_{cr}$) by the following approximate formula:

$$c = 1.54h \sqrt{\frac{E}{\sigma}} + 0.19 (2a)$$

which seems to be in a satisfactory agreement with experimental results.[3]

The problem of the postbuckling behavior of thin rectangular plates under pure shear is also of great practical importance in aircraft design and was investigated by many engineers.[4] In investigations of this type an expression for w with several parameters is selected so as to represent the buckled shape of the plate with simply supported edges. When this expression is substituted into Eq. (8-38), the stress function can be found such that the edge conditions for the displacements u and v are satisfied approximately. With the expressions for u and v determined, the expression for strain energy of the plate is obtained. This expression will be in terms of the parameters used in the expression for w. These parameters are found from the conditions for minimum strain energy. This method of analysis requires a large amount of calculation for each particular case; hence, practical applications of the theory are

[1] See paper by R. Lahde and H. Wagner, *Luftfahrt-Forsch.*, vol. 13, p. 214, 1936. (For English translation, see *NACA Tech. Mem.* 814, 1936.)

[2] See K. Marguerre, *Luftfahrt-Forsch.*, vol. 14, p. 121, 1937. (For English translation, see *NACA Tech. Mem.* 833, 1937.)

[3] See Lahde and Wagner, *op. cit.*

[4] A. Kromm and K. Marguerre, *Luftfahrt-Forsch.*, vol. 14, p. 627, 1937; W. T. Koiter, *Natl. Luchtvaartlaboratorium, Amsterdam, Rept.* S. 295, 1944; S. Levy, K. L. Fienup, and R. M. Woolley, *NACA Tech. Note* 962, 1945, and S. Levy, R. M. Woolley, and J. N. Corrick, *ibid.*, no. 1009, 1946; S. G. A. Bergman, "Behavior of Buckled Rectangular Plates under the Action of Shearing Forces," Kungl. Tekniska Hogskola, Stockholm, 1948.

possible only if the results of calculations for a series of selected cases are represented in the form of tables or diagrams.[1] All these calculations assume perfect elasticity. For the calculation of the ultimate strength of buckled plates recourse should be made to empirical formulas which will be discussed later (see p. 425).

The problem of the postbuckling behavior of uniformly compressed circular plates was also studied.[2] Considering a symmetrical form of buckling, it was shown that with an increase of the ratio n of the compressive stress to its critical value the circumferential compressive stresses concentrate more and more in the outer ring of the plate and in the inner portion of the plate tensile stresses are produced.

9.14. Ultimate Strength of Buckled Plates.

It is general practice in the design of metallic structures to determine the proportions of thin plates in such a way as to eliminate all possibility of buckling under service conditions. However, as we have seen from the discussion of the previous article, a plate after buckling may, in some cases, carry without failure a load many times larger than the critical load at which buckling begins. Thus it is logical in cases where the question of economy in weight is of primary importance, as in aircraft construction, to consider not only the critical load but also the ultimate load that a plate may carry without complete failure.

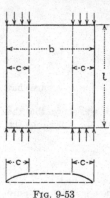

Fig. 9-53

In the case of compression of a rectangular plate with simply supported edges, it can be assumed for a rough calculation of the ultimate load that the load transmitted to the plate by a rigid block (Fig. 9-50) is carried finally by two strips of width c, one on each side of the sheet, and that the load distribution is uniform across these strips (Fig. 9-53). Then the middle portion of the sheet can be disregarded and the two strips can be handled as a long simply supported rectangular plate of width $2c$.[3] The critical stress for such a plate is (see p. 355)

$$\sigma_{cr} = \frac{4\pi^2 D}{h(2c)^2} = \frac{\pi^2 E h^2}{12(1 - \nu^2)c^2} \tag{a}$$

Assuming that the ultimate load is reached when σ_{cr} becomes equal to the yield-point stress σ_{YP} of the material, we find, from (a),

$$c = \frac{\pi h}{\sqrt{12(1 - \nu^2)}} \sqrt{\frac{E}{\sigma_{YP}}} \tag{b}$$

[1] A very extensive work of this kind, based on the theoretical work of Koiter, *op. cit.*, was made in the National Aeronautical Research Institute in Amsterdam; see W. K. G. Floor and T. J. Burgerhout, *Rept.* S. 370, Amsterdam, 1951; see also W. K. G. Floor, *Rept.* S. 427, Amsterdam, 1953.

[2] See K. O. Friedrichs and J. J. Stoker, *J. Appl. Mech.*, vol. 9, p. 7, 1942.

[3] Such an assumption was proposed by T. Von Kármán; see paper by Kármán, E. E. Sechler, and L. H. Donnell, *Trans. ASME*, vol. 54, p. 53, 1932.

and the ultimate load is

$$P_{ult} = 2ch\sigma_{YP} = \frac{\pi h^2}{\sqrt{3(1 - \nu^2)}} \sqrt{E\sigma_{YP}} \tag{9-34}$$

It is seen that the ultimate load is independent of the width b of the sheet and is proportional to the square of its thickness. This result is different from that obtained in the previous article by assuming that the edges of the plate are kept straight during buckling. If there is not such a restraint, experiments (see Art. 9.15) are in satisfactory agreement with formula (9-34). A better agreement can be obtained by using in this formula instead of a constant factor $\pi/\sqrt{3(1 - \nu^2)} = 1.90$ (for $\nu = 0.3$) a factor C varying with the proportions of the plate and given by the curve in Fig. 9-61, p. 425.

For comparison of formulas (9-33) and (9-34) with tests, we give in Table 9-24 the results obtained with three kinds of duralumin plates.[1] It is seen that the formula (9-34) gives results which are in satisfactory agreement with experiments. Formula (9-33) gives exaggerated values for P_{ult}, especially for small values of the ratio h/b.

TABLE 9-24. COMPARISON OF ULTIMATE LOADS FOR THREE COMPRESSED DURALUMIN PLATES

Thick-ness, h in.	Width, b in.	b/h	Length, a in.	E, psi	σ_{YP}, psi	Ultimate load, lb		
						Formula (9-33)	Formula (9-34)	Tests
0.0893	4.00	44.8	24	$10.6 \cdot 10^6$	41,000	10,200	9,990	7,300
0.0356	3.515	98.75	9	$10 \cdot 10^6$	45,000	2,700	1,620	1,175
0.0322	10.01	311	21	$10 \cdot 10^6$	45,000	6,375	1,310	1,270

It should be noted, however, that the latter formula has been developed on the assumption that the edges of the plate remain straight during buckling, while no such constraints existed in the tests. In actual structures where rigid ribs are placed along the edges of the plates, the conditions may be closer to those assumed in the derivations of the previous article and formula (9-33) may prove more satisfactory.

In the case of compression tests of thin tubes of square cross section, the edges of the flat sides of the tubes remain straight during buckling of the sides and actual conditions approach those assumed in the derivation of formula (9-33). In Table 9-25 is given a comparison of ultimate loads as calculated by using formulas (9-33) and (9-34) and as obtained by direct tests of brass tubes.[2] It is seen that in this case formula (9-33) is in better agreement with tests than in the case of compression of plates in V grooves. It is seen also that the agreement becomes less satisfactory as

[1] Data for the first plate are taken from the Bureau of Standards series of tests (see Art. 9.15) and data for the other two from the paper by E. E. Sechler, *Guggenheim Aeronaut. Lab.*, *Publ.* 27, California Institute of Technology, Pasadena, 1933.

[2] These experimental data have been communicated to the author by L. H. Donnell and were obtained at the Guggenheim Aeronautics Laboratory, California Institute of Technology, Pasadena, 1933.

TABLE 9-25. COMPARISON OF ULTIMATE LOADS OF COMPRESSED BRASS
SQUARE TUBES

Thickness, h in.	Width, b in.	Ratio, b/h	E, psi	σ_{YP}, psi	Total ultimate load of tube, lb		
					Formula (9-33)	Formula (9-34)	Tests
0.0065	1.0	154	$16.4 \cdot 10^6$	31,400	391	231	290
0.0056	1.0	178.5	$16.0 \cdot 10^6$	28,600	300	161	210
0.0065	2.0	308	$16.4 \cdot 10^6$	31,400	726	231	340

the ratio h/b decreases. For very small values of h/b formula (9-33) always gives exaggerated values for ultimate loads.

In discussing the ultimate strength of thin sheets submitted to the action of shearing stresses, let us consider a system consisting of three absolutely rigid bars with hinges at the joints (Fig. 9-54a).[1] The field $abcd$ between the bars consists of a very

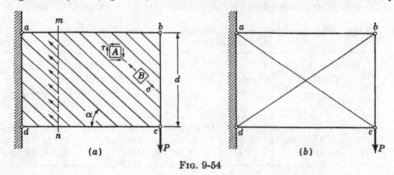

(a)   (b)

FIG. 9-54

thin plate which cannot resist bending or compression. Under the action of the load P the condition will be analogous to that of a system with flexible diagonals (Fig. 9-54b). Under the action of the load P the diagonal bd in compression buckles sideways and only the diagonal ac in tension is working. Since the thin plate cannot sustain shearing stress, it buckles sideways. This results in the formation of wrinkles, as shown in Fig. 9-55. Instead of pure shear τ as shown at point A, Fig. 9-54a, we will have simple tension in the wrinkles, with direction as shown at point B. Considering portion $mncb$ of the structure as a free body and denoting by σ the tensile stress in the wrinkles, we obtain the equation

$$\sigma hd \cos \alpha \sin \alpha = P$$

from which
$$\sigma = \frac{2P}{hd \sin 2\alpha} \qquad (c)$$

where d is the depth of the structure and h is the thickness of the plate.

Let us consider next a plate girder with a thin web (Fig. 9-56). The tensile forces

[1] This approximate theory is due to H. Wagner, Z. Flugtech. Motorluftsch., vol. 20, p. 200, 1929.

Fig 9-55. Diagonal-tension beam. NACA, Langley Field, Va. (*Taken from P. Kuhn, "Stresses in Aircraft and Shell Structures," Fig. 3.1, p. 48, McGraw-Hill Book Company, Inc., New York, 1956.*)

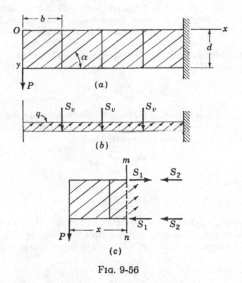

FIG. 9-56

transmitted from the wrinkled web to the lower flange are shown in Fig. 9-56b, and we see that the intensity of the distributed inclined forces is

$$q = \sigma h \sin \alpha \qquad (d)$$

Resolving this into horizontal and vertical components, we obtain

$$q_x = \sigma h \sin \alpha \cos \alpha = \frac{P}{d} \qquad q_v = \sigma h \sin^2 \alpha = \frac{P}{d} \tan \alpha \qquad (e)$$

The force q_x represents the addition per unit length to the axial compressive force in the flange, and q_v is the intensity of the transverse load acting on the flange. The compressive force S_v in the verticals, as seen from Fig. 9-56b, is

$$S_v = bq_v = P \frac{b}{d} \tan \alpha \qquad (f)$$

The forces F in the flanges (Fig. 9-56c) consist of two parts: (1) forces $S_1 = \pm M/d$ balancing the bending moment $M = Px$ which acts in the cross section mn of the girder and (2) the forces $S_2 = -\frac{1}{2}\sigma h d \cos^2 \alpha = -(P/2) \cot \alpha$, balancing the diagonal tension. Thus we obtain

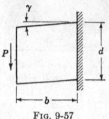

Fig. 9-57

$$F = \pm \frac{M}{d} - \frac{P}{2} \cot \alpha \qquad (g)$$

The angle α of the wrinkles (Fig. 9-56a) will be found from the condition of minimum strain energy. Assuming first that the flanges and the verticals are absolutely rigid, we have to consider only the strain energy of the web. Since the web is in the condition of simple diagonal tension [see Eq. (c)] we conclude that for minimum energy we have to take $\sin 2\alpha = 1$, i.e., $\alpha = 45°$. In actual cases the flanges and the verticals will deform also. Taking into account their strain energy, we obtain somewhat smaller values for α. For the proportions of girders used in aircraft construction, the angle α is usually not smaller than 38°.

Shear deformation of the girder (Fig. 9-56), in the case when the web does not buckle, can be found from the condition that the shearing strain is

$$\gamma = \frac{\tau}{G} = \frac{P}{dhG} \qquad (h)$$

In the case when the web is very thin and acts in simple tension only, the shear deflection γb (Fig. 9-57) of the portion of the girder between two stiffeners will be found from energy considerations, which gives the equation

$$\frac{P\gamma b}{2} = \frac{\sigma^2}{2E} bhd$$

From this we obtain

$$\gamma = \frac{4\tau}{E} = \frac{4P}{hdE} \qquad (i)$$

Equations (h) and (i) give the shear deformation of the girder in the two extreme cases when the web acts in pure shear only and in simple tension only. In actual cases the deformation of the flanges and the verticals must be taken into consideration, and we shall get

$$\gamma = \frac{\tau}{G_r}$$

where G_r depends on the proportions of the structure and lies between the extreme values G and $E/4$, as calculated above.

The above calculations for the diagonal tensile stress σ and for the forces in the verticals and flanges are based on the assumptions that the web is absolutely flexible and that the flanges and verticals are absolutely rigid. Experiments show that such conditions are approached only in the case of very thin webs and at loads near the ultimate.[1] To obtain a better approximation in the analysis, the deformation of the elements of the structure must be considered in more detail. Considering the action of the forces q_v on the flanges (Fig. 9-58a), we see that bending of the flanges as continuous beams will be produced. Because of this bending the diagonal tension in the web will not be uniformly distributed. The diagonal strips which are attached near the middle of the panel will be somewhat relieved of stress, while the diagonals attached near the verticals (Fig. 9-58b) have to carry higher stresses than given by Eq. (c). The force S_v also will be somewhat different from the value given by Eq. (f). Semiempirical formulas have been developed to take care of all these irregularities in stress distribution. The fact that, in the case of plates which are not very thin, the web will act partially in shear and partially in diagonal tension is considered in aircraft design, and on the basis of extensive experimental work empirical formulas have been established for the proper subdivision of the shearing force. The so-called "engineering stress theory of plane-

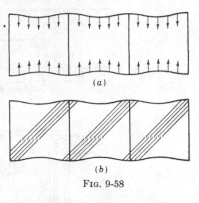

(a)

(b)

FIG. 9-58

web systems" is now well developed[2] and with its use the stresses in a plate girder, as well as its ultimate strength, can be predicted with sufficient accuracy. The stresses in the buckled plates and their deflections calculated in this way are in satisfactory agreement with the theory of large deflections of thin plates.[3]

9.15. Experiments on Buckling of Plates. *Compression of Rectangular Plates.* Experiments with plates having simply supported edges were the first to be made, since the theoretical solution for this case is a very simple one. Experimentally, however, this case is very complicated, since it is difficult in the experiment to realize the conditions of a simply supported edge. To permit, as nearly as possible, complete freedom of rotation of the edges, frames with V notches are used[4] (Fig. 9-59). However, an unrounded edge of the plate is not entirely free to rotate in such a notch and any rotation during buckling is accompanied by some displacement of the middle plane perpendicular to the surface of the plate

---

[1] See R. Lahde and H. Wagner, *Luftfahrt-Forsch.*, vol. 13, p. 262, 1936.

[2] See Paul Kuhn, "Stresses in Aircraft and Shell Structures," McGraw-Hill Book Company, Inc., New York, 1956. This book also gives a bibliography on the subject.

[3] See papers by S. Levy, et al., *loc. cit.*

[4] Figure 9-59 represents the testing apparatus used at the Bureau of Standards; see L. Schuman and G. Back, *NACA Tech. Rept.* 356, 1930.

(Fig. 9-60). The loaded edges of the plate are sometimes made semi-circular in order to realize the application of the load in the middle plane of the plate.[1]

Owing to initial curvature of plates and eccentricities in application of the load, lateral buckling usually begins at a very small load. To determine from the observed deflections the critical value of the load, it is advantageous to apply the method suggested by R. V. Southwell in testing columns (see p. 191). In discussing bending of compressed plates with an

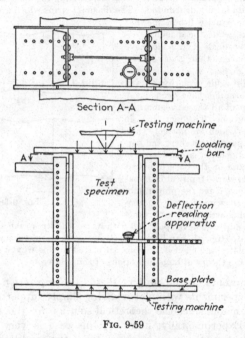

Fig. 9-59

initial curvature (see Art. 8.6) it was shown that, when the load P approaches its critical value, the corresponding term in the double trigonometric series representing the deflections becomes preponderant. This term increases with the load in such a way that a straight line is obtained when observed deflections δ are plotted against values of the ratio δ/P. The slope of this line gives the true value of the critical load. The values obtained in this manner are usually in very satisfactory agreement with the theoretical values. The wave formation obtained during buckling of compressed plates is also usually in good agreement with the theory.[2]

[1] See paper by Sechler, *loc. cit.*
[2] See Schuman and Back, *op. cit.*

Most experiments on buckling of plates have been made with thin plates such as those used in aircraft structures, the principal interest being to establish the ultimate load that a compressed plate can carry. These experiments show that a thin sheet of metal of considerable width can carry a much larger load than the critical load predicted theoretically. This ultimate value of the load can be calculated with sufficient accuracy from the formula [see formula (9-34)]

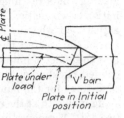

$$P_{\text{ult}} = Ch^2 \sqrt{E\sigma_{\text{YP}}} \qquad (9\text{-}35)$$

in which h is the thickness of the plate, E the modulus of elasticity of the material, σ_{YP} its yield-point stress, and C a factor depending on the properties of the material and on the proportions

FIG. 9-60

of the plate. Experimental values[1] for this factor are shown in Fig. 9-61 and there is also given a curve for C which can be used in practical calculations of ultimate strength of thin compressed sheets of metal.

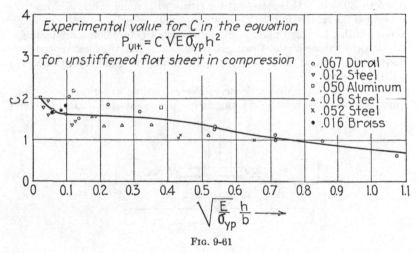

FIG. 9-61

Compression of Angles.[2] A compressed strut of angle section may fail in two entirely different ways: (1) as a column (see Art. 5.5) or (2) owing to buckling of the flanges. Tests made with struts of symmetrical angle sections show that, when the flange widths are small as compared with

[1] These values are taken from Sechler, *loc. cit.*

[2] See Bridget, Jerome, and Vosseller, *Trans. ASME*, Applied Mechanics Division, vol. 56, p. 569, 1934. Buckling of compressed angles beyond the yield point is described in the dissertation by C. F. Kollbrunner, Zürich, 1935.

the length, the struts buckle as columns, while for wide flanges plate buckling occurs first (see Fig. 9-7).   In order to obtain definite conditions for buckling of the strut and of its flanges, such end supports as shown in Fig. 9-62 were used.   A $\frac{5}{8}$-in. steel ball was inserted in each end block, making the strut hinged about the centers of the balls.   To eliminate initial eccentricities, a special arrangement was used providing for an adjustment whereby the ball could be moved along a line parallel to the axis of the greatest moment of inertia of the cross section of the strut, so that the effect of initial eccentricities could be removed and a very sharply defined buckling of the strut could be obtained.   To obtain definite edge conditions for the flanges, each side of the angle section was chamfered along its width at the ends to an angle of 60° and the end blocks each had two 120° V grooves in which the ends of the strut were supported, thus ensuring hinged edge conditions for the flanges.

The experimental results obtained thus with struts of 24 SRT aluminum alloy are shown in Fig. 9-63.   The specimens tested were 22 in. in length and of 0.025-in. thickness; the width of the sides varied from 0.405 to 2.025 in.   It is seen that for struts with small width of sides the experimental results follow Euler's curve very closely.   In the case of wider widths, in which plate buckling occurred before column buckling, the critical values of the load, given in the figure, were calculated by Southwell's method, and we have again a very satisfactory agreement with the theory.

Buckling of Plates under Shearing Forces.   Most experiments of this kind have been made with very long strips, the longitudinal sides of which

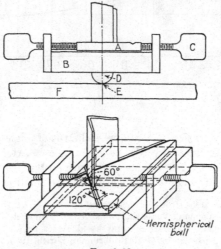

Fig. 9-62

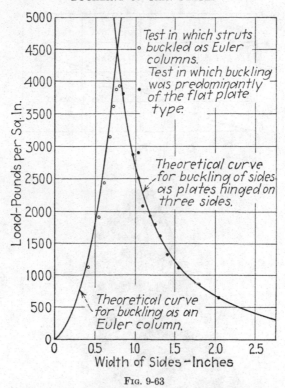

Test in which struts buckled as Euler columns.

Test in which buckling was predominantly of the flat plate type.

Theoretical curve for buckling of sides as plates hinged on three sides.

Theoretical curve for buckling as an Euler column.

Width of Sides –Inches

Load–Pounds per Sq. In.

FIG. 9-63

were clamped.[1] For this case we have an exact solution (see Art. 9.7), and experiments show that, regarding wave formation, there is satisfactory agreement with the theory.   The magnitude of the critical load is also obtained with sufficient accuracy if the Southwell method is applied.

In Fig. 9-64 the results of postbuckling deflection tests of a square duralumin plate (0.2 by 150 by 150 mm) tested[2] in pure shear are shown. The edges of the plate were free to rotate but could not move in the plane of the plate so that the stresses in this plane could be developed in full. Because of some initial curvature the deflection δ started at a load below the critical load, as shown in the figure.   For comparison the theoretical curve is also shown, calculated[3] for the deflection of a plate having an initial deflection $\delta_0 = 0.5h$.   This curve follows satisfactorily the experimental curve.

[1] See F. Bollenrath, *Luftfahrt-Forsch.*, vol. 6, p. 1, 1929. See also H. J. Gough and H. L. Cox, *Proc. Roy. Soc., London*, series A, vol. 137, p. 145, 1932.

[2] See E. Seydel, *Z. Flugtech. u. Motorluftsch.*, vol. 24, p. 78, 1933.

[3] This curve was calculated by S. G. A. Bergman, *loc. cit.*

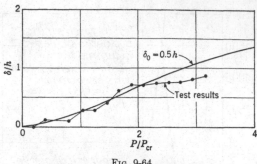

FIG. 9-64

FIG. 9-65

A very extensive series of tests of the postbuckling resistance of thin plates in shear was made in connection with the development of the engineering theory of plane-web systems mentioned in the preceding article.[1]

Some experiments have been made also with corrugated rectangular plates[2] subjected to the action of shear. The shearing forces were applied by the use of a hinged rectangular frame along one diagonal of which external tensile forces were acting (Fig. 9-65). The wave formation and

---

[1] A description of these tests and a bibliography are given in Kuhn, *loc. cit.*

[2] See "Deutsche Versuchsanstalt für Luftfahrt E. V.," 1931.

the magnitude of the critical load are in satisfactory agreement with the theory developed on the assumption that a corrugated plate can be considered as a plate of nonisotropic material.[1] It will be noted that in this case the nodal lines make only a small angle with the direction of the sides parallel to corrugations instead of approximately 45° as in the case of unstiffened plates.

9.16. Practical Applications of the Theory of Buckling. of Plates. *Applications in Design of Compression Members.* In discussing buckling of columns it was shown that for practical proportions of compression members failure occurs when, in the weakest cross section, the maximum combined direct and bending stress becomes equal to the yield point stress of the material. Only in the case of very slender struts can sufficient buckling occur within the elastic limit to be equivalent to complete failure. This fact must be kept in mind in designing compression members built up of comparatively thin sheets of metal. Take, for instance, the case of a compression member, the cross section of which is shown in Fig. 9-66. The two vertical web plates of width b and thickness h, with angles along the edges, are connected by diagonals and battens in horizontal planes as shown by the dotted lines. To be on the safe side, we neglect the resistance to twist of the angles and consider the web plates as uniformly compressed rectangular plates with simply supported edges. Since the length of these plates is large in comparison with their width, the critical stress within the elastic limit is determined, for $\nu = 0.3$, by the formula (see p. 355)

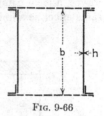

FIG. 9-66

$$\sigma_{cr} = \frac{\pi^2 E}{2.73} \frac{h^2}{b^2}$$

In the case of very slender struts, say $l/r > 150$, it is justifiable to take for the ratio b/h such a value that the critical stress for the plate is equal to the critical stress for the entire strut. Then, using Euler's formula for the strut, the equation for determining the required value of the ratio b/h is

$$\frac{\pi^2 E}{2.73} \frac{h^2}{b^2} = \frac{\pi^2 E r^2}{l^2} \tag{a}$$

in which l/r is the slenderness ratio for the strut. From this equation we obtain

$$\frac{b}{h} \approx 0.60 \frac{l}{r} \tag{b}$$

Thus, to have the factor of safety for buckling of the web equal to the

---

[1] See the paper by Bergmann and Reissner, *loc. cit.*

factor of safety for buckling of a slender strut, the proportions of the web must satisfy condition (b).

If a compression member has proportions usual in structural engineering, failure occurs as a result of local yielding of the material in the weakest cross section. Owing to various kinds of inaccuracies, this yielding may be produced while the average compressive stress in the strut is far below the yield-point stress. Under this condition it is evident that a built-up strut can be expected to behave as a solid one only if the proportions of the plates are such that they will not buckle at a stress below the yield point of the material. To eliminate the possibility of buckling at stresses below the yield point, long compressed plates of structural steel with simply supported edges must be of such proportions that $b/h \leq 36$ if the yield-point stress is 34,000 psi. Since long compressed plates buckle in many waves, a local compression along one half-wave will have practically the same effect as a uniform compression of the entire plate. Thus the same

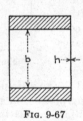

Fig. 9-67

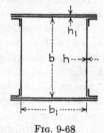

Fig. 9-68

ratio, $b/h \leq 36$, can be recommended also in the case of a compression member like the one shown in Fig. 9-66. It should be noted also that, owing to the local character of buckling of the web, the transverse diaphragms, which are usually placed at certain intervals along the length of compression members, will not increase the stability of the web.

If the vertical web plates of a compression member are rigidly connected with heavy horizontal plates (Fig. 9-67), it is justifiable to consider the vertical plates as built in at the longitudinal edges. In such case we can take $b/h = 48$ as the limiting ratio.

The conditions at the edges of the vertical webs of a compression member, the cross section of which is shown in Fig. 9-68, are intermediate between simply supported and clamped edge conditions. Assuming that the degree of restraint depends approximately on the magnitude of the ratio

$$\frac{h^3 b_1}{h_1^3 b}$$

we obtain for partial restraint, by interpolation between the ratios 36 and 48,

$$\frac{b}{h} = 48 \left(1 - \frac{h^3 b_1}{4 h_1{}^3 b} \right) \tag{c}$$

from which the required ratio b/h for any proportions of plates with $h_1{}^3 b > h^3 b_1$ can be calculated. When $h_1{}^3 b = h^3 b_1$ the critical compressive stress for horizontal plates is the same as for vertical webs; thus the latter are in the condition of plates with simply supported edges. Equation (c) in such a case gives the value 36.

In the cases shown in Fig. 9-69 each flange can be considered as a plate simply supported along one edge and entirely free along the other. In such cases, to eliminate the possibility of buckling at a stress below the yield point, we can take for structural steel ($\sigma_{YP} = 34{,}000$ psi) the limiting ratio $b/h = 12$. In the case shown in Fig. 9-70, assuming that the upper edges of the vertical web plates are built in, the limiting ratio is $b/h = 21$.

Applications in Design of Plate Girders.[1] The proportioning of plate girders is based at present to a great extent on empirical rules. Being the result of long experience, these rules usually give satisfactory proportions

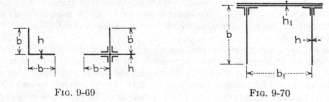

FIG. 9-69   FIG. 9-70

and at the same time are sufficiently flexible so that they leave considerable freedom for individual judgment. As a result of this we have a variety of dimensions of plate girders designed for the same span and the same load.

In designing girders of exceptionally large dimensions the proper selection of the web thickness and stiffener spacing becomes an important problem which can be solved satisfactorily only on the basis of a rational theory or by using model tests.

The problem of determining the proper web thickness and stiffening of the web is essentially a stability problem. It is well-known that a web, if not sufficiently thick or not satisfactorily stiffened, may buckle sideways and act as a tie while the stiffeners are working as struts. To take into account the possibility of such buckling, several engineers have recommended[2] that a narrow strip of web making an angle of 45° with the longitudinal axis of the girder be considered as a column carrying a compressive stress over its cross section equal to the shearing stress at the

[1] See Timoshenko, *Engineering*, vol. 138, p. 207, 1934.

[2] See, for instance, the paper by F. E. Turneaure, *J. Western Soc. Eng.*, vol. 12, p. 788, 1907.

neutral axis. The length of this column is taken as $h \sqrt{2}$, in which h is the unsupported depth of the web, the ends of the column being considered as built in. This assumption, as we shall see later, underestimates considerably the stability of the unstiffened webs.

To get a more reliable conclusion regarding the stability of the web, experiments on plate girders, together with the theory of buckling of a thin plate under the action of normal and shearing stresses in its plane, should be considered.

The first experiments with buckling of thin webs transmitting shearing and bending stresses were made by Fairbairn in connection with the construction of the famous Britannia and Conway tubular bridges.[1] Even up to the present time these classical experiments have held great interest for engineers working with thin-walled structures. The Britannia bridge is of the tubular type with a rectangular cross section. The larger tubes have a span of 450 ft and cross-sectional dimensions of 27 by 16 ft. As this was an unusually large structure for that time, it was decided to make experiments with models to determine the safe dimensions of the tube and the most favorable distribution of material. After a considerable amount of preliminary experimenting, it was decided to test large models, one-sixth the linear dimensions of the intended bridge. The sides of these model tubes consisted of sheets 3 ft 9 in. deep and only 0.1 in. thick. The first experiments showed that at a comparatively small load undulations in the sides appeared which formed angles of about 45° with the line of the bottom.

"It was evident, from these experiments, that the tension throughout the bottom and the compression throughout the top stood in the relation of action and reaction to each other, the diagonal strain in the sides being the medium of communication.

"A diagonal wave of puckering clearly exposed the line of severest strain. It was evident that the sides were exposed to unfair strain from the change of shape consequent on the tendency of the top and bottom to come together, the plates being strong enough, if they could but be kept in shape; and it was therefore determined, in this experiment, to modify the construction of the sides. This was done by the addition of pillars of angle-iron throughout, of the whole height of the sides, riveted to them, having the effect of stiffening them, and at the same time of keeping the top and bottom in place. They were prototypes of the T-iron pillars used in the large tubes."

Further experiments illustrated the importance of the pillars in the sides, for, with a small addition of metal to the weight of the tube, the top and bottom remained precisely the same as before while the ultimate strength was increased considerably. From these experiments it was learned that "as the depth of a web increased, the precautions requisite for maintaining the sides in shape become very formidable." The T irons, gussets, and stiffening plates for this purpose in one of the Britannia tubes weigh 215 tons, or upward of one-third of the whole weight of the sides.

The experimental tubes were submitted to a concentrated load at the middle, and the shearing force was constant along the length of the span. In the design of the actual bridge it was taken into account that the maximum shearing force diminishes toward the middle, and the web was taken $\frac{1}{2}$ in. thick in the middle portion and $\frac{5}{8}$ in. thick at the ends.[2]

[1] See William Fairbairn, "An Account of the Construction of the Britannia and Conway Tubular Bridges," London, 1849.

[2] For further description of these bridges, see Timoshenko, "History of Strength of Materials," McGraw-Hill Book Company, Inc., New York, 1953.

Some experimental work on plate girders was also done at that time.[1] The thickness of the web of the model girder was $\frac{1}{4}$ in. throughout; the over-all depth was 10 ft at the center, and 6 ft at the ends, and the distance between the bearings was 66 ft. The girder failed by buckling of the web. Later on the girder was repaired and the vertical web stiffened by the addition of angle-iron pillars at each joint in the vertical plates of the web. In this way the strength of the girder was considerably increased, and finally it failed at a larger load by a simultaneous collapse of the top and the bottom.

Further experiments with plate girders were made by the Belgian engineer Houbotte.[2] Two plate girders, 1.50 m in span length, 0.5 cm in thickness of the web, and 30 and 49 cm in depth, were tested. Loaded at the middle, both these girders failed by buckling of the web, which had no stiffeners. The girder of larger depth failed at smaller load although its section modulus was twice as great as that of the girder with smaller depth.

Additional work with plate girders was done by Lilly.[3] A plate girder of the following dimensions was constructed: depth $9\frac{1}{2}$ in.; length, 5 ft. 3 in. The flanges were made up of two plates 2 by $\frac{3}{8}$ in. and two angles $1\frac{1}{4}$ by $1\frac{1}{4}$ by $\frac{1}{4}$ in. The framework of the girder was made in separate halves and was bolted together to the web. This construction allowed different thicknesses of web to be used in the experiments. A large number of tests were then carried out with different thicknesses of the web and spacing of stiffeners. Applying the load at the middle the wave formation in the web was obtained.

"It was found that the wave-length of the wave formation is nearly independent of the thickness, if the stiffeners are of great strength compared with the web. The angle of inclination of the wave depends upon the distance apart of the stiffeners and the depth of the girder. The stiffeners prevent the formation of the waves, and severe local stresses are set up around the ends of the stiffeners, causing a crumpling up of this part of the web."

Experiments with a plate girder of larger size were made by Turneaure.[4] The principal conclusions of these experiments were: (1) The stresses in web plates with stiffeners, when stressed within the elastic limit, agree closely with the theoretical stresses, and as a necessary result the axial stresses in vertical stiffeners, not subjected to local loads, are practically zero. (2) The elastic limit strength of a web plate without stiffeners is about twice the ultimate strength given by Euler's column formula applied to a diagonal column element (see p. 431).

A series of tests on rolled I beams and built-up plate girders was made by Moore and Wilson.[5] One example of a tested plate girder with a buckled web is shown in Fig. 9-71. These experimenters came to the conclusion "that the ability to resist buckling of thin webs without intermediate stiffeners had been underestimated" and "that it may be safe to build girders without intermediate stiffeners if the ratio of the unsupported width to the thickness of the web does not exceed 60. However, it is necessary to decrease the working stress allowable in the web as this ratio becomes greater."

[1] See Edwin Clark, "The Britannia and Conway Tubular Bridges," 2 vols., London, 1850.

[2] See M. Houbotte, *Der Civilingenieur*, vol. 4, 1856.

[3] See W. E. Lilly, *Engineering*, vol. 83, p. 136, 1907; see also his book "The Design of Plate Girders and Columns," 1908.

[4] Turneaure, *loc. cit.*

[5] See H. F. Moore, *Univ. Illinois Bull.* 68, 1913, and H. F. Moore and W. M. Wilson, *Univ. Illinois Bull.* 86, 1916.

FIG. 9-71

An extensive series of tests of plate girders with thin webs (1,000 by 2,000 by 3.5 mm) was made at the Institute of Technology, Stockholm.[1] It was found that, owing to various imperfections, the web begins to buckle at a very small load, and in most cases no change in behavior of the web plates was noticed at the critical load. On the basis of these tests it was concluded "that the theoretical critical load of plane web plates bears no direct relation to the ultimate load, and that the ratio of the ultimate load to the theoretical critical load increases with the slenderness of the web." It was found also that "the load-bearing capacity of the web is not exhausted until yielding begins in a comparatively large portion of the web."

An experimental investigation of the necessary rigidity of stiffeners for plate girders was made by Massonnet at the University of Liége.[2]

From the experiments that have been made, it can be seen that a plate girder can transmit the shearing force to the bearings in two different ways: (1) if the load is not sufficient to produce wave formation, the web of the girder transmits the shearing force by working in shear, and (2) in the case of larger loads, which produce wave formation, one part of the shearing force is transmitted by shearing stresses in the web, as before, and the other part as in a truss, in which the web plate acts as a series of ties and the stiffeners as struts. The magnitude of the load at which wave formation begins depends on the thickness of the web and on the spacing and dimensions of stiffeners. In the case of a sufficient thickness of the web and satisfactory stiffening, a plate girder can carry the total load for which it is designed without any buckling of the web. We usually have such proportions in bridges. On the other hand there are constructions with very thin webs which buckle at the very beginning of loading and the total load is transmitted practically as in a truss. We have examples of such girders in aircraft construction.[3]

Although buckling of the web does not mean an immediate failure of the girder, the dimensions in the case of bridges are usually taken so as to eliminate buckling under service conditions. A common procedure is to

[1] See report by G. Wästlund and S. G. A. Bergman, "Buckling of Webs in Deep Steel I-Girders," Institution of Structural Engineering and Bridge Building, Stockholm, 1947. An extensive bibliography is given in this paper.

[2] See C. Massonnet, *Publ. Intern. Assoc. Bridge Structural Eng.*, vol. 14, p. 125, 1954.

[3] See H. Wagner, *Z. Flugtech. Motorluftsch.*, vol. 20, p. 200, 1919. See also *Eng. News*, vol. 40, p. 154, 1899.

adopt a certain value of the working stress in shear and on this basis to decide upon the web thickness.   Then the spacing of stiffeners is determined so as to enable the web to transmit shearing stresses without buckling.   Observing that in railway girders the total load varies approximately as the span and assuming the ratio of the depth to the span to be constant, it can be seen that the above procedure would result in nearly the same thickness for all spans.   Assuming that this thickness is satisfactory for small bridges, it certainly will be insufficient for larger spans, and some increase in the thickness for eliminating the possibility of buckling of the web becomes necessary.   This is prov'ded for in specifications. For example, the 1950 specifications of the Americ un Railway Engineering Association (AREA) and the 1949 specifications of the American Institute of Steel Construction (AISC) require that the thickness of the web shall be not less than $d/170$, where d represents the (clear) distance between the flanges in inches.

Another limitation for thickness is usually obtained from a consideration of corrosion and from the fact that very thin plates, if deep and long, are very awkward to handle in construction.   A thickness of $\frac{3}{8}$ in. is usually considered as the minimum thickness to protect against corrosion and ensure a satisfactory handling of material during construction and shipping.

To obtain a rational basis for design, it is necessary to make a study of the elastic stability of thin webs.   In discussing buckling of the web we must consider three cases: (1) near the supports the shearing force is the most important factor and the part of the web between two stiffeners can be considered as a rectangular plate subjected to the action of uniform shear (Fig. 9-72a); (2) at the middle of the span the shearing stresses can

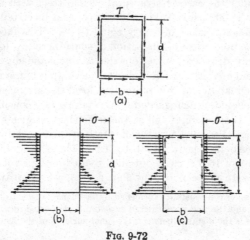

Fig. 9-72

be neglected in comparison with normal stresses and the part of the web between two stiffeners is in the condition of pure bending (Fig. 9-72b); (3) in the intermediate cross sections there is a combination of normal and shearing forces, as shown in Fig. 9-72c. These three cases have been discussed in Arts. 9.6 and 9.7.

In the application of the theoretical formulas obtained previously, we begin with the determination of the thickness of the web at the middle of the span, where only bending stresses need be taken into consideration. Since a long rectangular plate submitted to bending subdivides at buckling into comparatively short waves (see Table 9-6), the constraint due to the vertical stiffeners usually will be negligible and we can assume that

$$\sigma_{cr} = 23.9 \frac{\pi^2 E}{12(1 - \nu^2)} \frac{h^2}{d^2} \tag{d}$$

From this formula the required value of the ratio of the thickness of the web to its depth[1] can be calculated for any value of σ_{cr}. To eliminate any possibility of buckling of the web under service conditions, the value of σ_{cr} must be larger than the maximum bending stress in the web. In choosing the necessary factor of safety we proceed here differently than in the case of plates of compression members. In the case of columns any local buckling of the plates usually means a complete failure of the structure, and it was recommended in our previous discussion to take for plates such proportions that no buckling will occur below the yield-point stress. In the case of plate girders, however, buckling of the web does not represent any immediate danger to the structure but simply indicates that the web no longer takes its full share of compressive bending stresses. Under such circumstances it seems reasonable to use a lower factor of safety.

Let us consider the case of structural carbon steel with a yield-point stress equal to 34,000 psi. Taking 16,000 psi as a basic tensile stress in this case, we have a factor of safety equal to $2\frac{1}{8}$. This factor not only provides for the possibility of the action of some increased loads but also takes care of any stress concentrations due to rivets and sharp changes in cross section and any fatigue effect due to fluctuation of the stresses. In considering the stability of web plates, the local stress concentration and fatigue effects should be disregarded. Also taking into consideration that any constraint at the flanges and at the stiffeners is neglected in formula (d), it seems logical to take a factor of safety against buckling lower than $2\frac{1}{8}$. We suggest taking it equal to 1.5. Then, assuming that the maximum stress of 16,000 psi is obtained by deducting rivet holes, which may constitute 15 per cent of the flange area, we substitute for σ_{cr} in

---

[1] In this discussion the resistance furnished by the flange angles to buckling of the web is neglected and the web is considered as a simply supported rectangular plate of width d, where d is the distance between the inner surfaces of the flanges.

formula (d) the value of $16{,}000\ (0.85)(1.5) = 20{,}400$ psi. In this way we obtain, with $E = 30 \cdot 10^6$ psi and $\nu = 0.3$,

$$\frac{d}{h} = \sqrt{23.9\,\frac{\pi^2 E}{12(1 - \nu^2)16{,}000 \cdot 0.85 \cdot 1.5}} \approx 180 \qquad (e)$$

As mentioned previously, the AREA and AISC specifications limit this ratio to

$$\frac{d}{h} \le 170 \qquad (f)$$

Considering now the stiffening of the web by ribs, let us determine first the limiting value of the ratio d/h at which such stiffening will not be necessary except, of course, at the points of application of heavy concentrated loads.   In such case the portions of the web between these loads can be considered as rectangular plates which, near the supports, will be submitted principally to the action of shearing stresses.   Neglecting the constraint at the flanges and assuming that the plate is long, we take for the critical value of shearing stresses Eq. (9-9).   Then

$$\tau_{cr} = 5.35\,\frac{\pi^2 E}{12(1 - \nu^2)}\,\frac{h^2}{d^2} \qquad (g)$$

Substituting for τ_{cr} the yield-point stress,[1] we obtain the limiting value for the ratio d/h. Taking again structural steel with $\sigma_{YP} = 34{,}000$ psi and assuming that

$$\tau_{YP} = 0.58\sigma_{YP}$$

we obtain $\qquad \dfrac{d}{h} = \sqrt{\dfrac{5.35\pi^2 E}{12(1 - \nu^2)\tau_{YP}}} \approx 86 \qquad (h)$

For materials with higher values of the yield-point stress we will get from (h) smaller values for the ratio d/h. Taking, for instance, silicon steel with $\tau_{YP} = 26{,}000$ psi and nickel steel with $\tau_{YP} = 29{,}000$ psi, we obtain $d/h = 75$ and $d/h = 71$, respectively.

In calculating the required distances between the vertical stiffeners near the supports, we consider a portion of the web between two stiffeners as a simply supported plate of length equal to depth d and of width b equal to the distance between the axes of the stiffeners.[2]   For calculating critical values of the shearing stress we use the curve in Fig. 9-24.   Then

$$\tau_{cr} = \left(5.35 + 4\,\frac{b^2}{d^2}\right)\frac{\pi^2 E}{12(1 - \nu^2)}\,\frac{h^2}{b^2} \qquad (i)$$

[1] Since buckling of the web does not mean an immediate failure of the girder, we use Eq. (g) up to the yield-point stress, neglecting some permanent set of the material in the regions of transition between the proportional limit and the yield point.

[2] We assume that the distance between stiffeners is smaller than the depth d.

Taking again a factor of safety equal to 1.5, we obtain

$$\frac{b}{h} = \frac{1.16}{\sqrt{0.415(\tau/E) - (h^2/d^2)}} \tag{j}$$

where τ is the gross shearing stress in the panel at the support. Assuming that $E = 30 \times 10^6$ psi and $d/h = 170$, we obtain from formula (j)

$$\frac{b}{h} = \frac{9,850}{\sqrt{\tau - 2,500}} \tag{k}$$

The AREA specifications limit the clear distance between vertical stiffeners to 72 in. or the distance given by the formula

$$\frac{b}{h} = \frac{10,500}{\sqrt{\tau}} \tag{l}$$

The maximum allowable gross shearing stress is taken as 11,000 psi. The AISC specifications give the formula

$$\frac{b}{h} = \frac{11,000}{\sqrt{\tau}} \tag{m}$$

These specifications limit the distance b to 84 in. and give the allowable gross shearing stress as 13,000 psi. If the gross shearing stress is equal to 11,000 psi, we obtain from formulas (k), (l), and (m) the values $b/h = 107$, 100, and 105, respectively.

When the thickness of the web and the distance between the stiffeners at the supports are determined, some intermediate panels can be considered and the stability of the web can be checked by using the curves shown in Fig. 9-27.

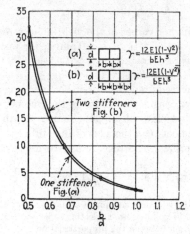

FIG. 9-73

For determining the required flexural rigidity of stiffeners, Tables 9-19 and 9-20 should be used. The values of γ are represented by curves[1] in Fig. 9-73. It is seen that, in the case of three panels, the required γ for the two intermediate stiffeners is larger than in the case of two panels, and γ should increase somewhat when the number of panels increases. Assuming that in all practical cases the required rigidity will not be larger than twice that given by Table 9-19, we arrive at the values given in Table 9-26

TABLE 9-26. REQUIRED MOMENT OF INERTIA (IN.[4]) FOR STIFFENERS
(b = 60 in.)

	$h = \frac{3}{8}$ in.	$h = \frac{7}{16}$ in.	$h = \frac{1}{2}$ in.	$h = \frac{9}{16}$ in.
d = 60 in. $\frac{b}{d} = 1$ $\gamma = 3.30$	0.96	1.52	2.27	3.23
d = 80 in. $\frac{b}{d} = \frac{3}{4}$ $\gamma = 11.6$	3.36	5.34	8.00	11.4
d = 96 in. $\frac{b}{d} = \frac{5}{8}$ $\gamma = 25.2$	7.30	11.6	17 3	24.7
d = 120 in. $\frac{b}{d} = \frac{1}{2}$ $\gamma = 60.0$	17.4	27.6	41.3	58.8

for the required moment of inertia of the cross section of the stiffeners for various depths and thicknesses of web and for a stiffener spacing of b = 60 in.[2]

[1] Note that in the figure the notation $\gamma = EI/Db$ is used. Hence values for γ given in Table 9-19 must be multiplied by 2 and the values in Table 9-20 by 3.

[2] For the application of stability theory to the design of the webs of plate girders, see F. Bleich, *Prelim. Publs. 1st Congr. Intern. Assoc. Bridge Structural Eng.*, Paris, 1932, and "Buckling Strength of Metal Structures," McGraw-Hill Book Company, Inc., New York, 1952; E. Chwalla, *Prelim. Rept. 2d Congr. Intern. Assoc. Bridge Structural Eng.*, Berlin, 1936; P. P. Bijlaard, *Publ. Intern. Assoc. Bridge Structural Eng.*, Zürich, vol. 8, 1947; and J. M. Young and R. E. Landau, *Proc. Inst. Civil Engrs.*, vol. 4, p. 299, 1955.

CHAPTER 10

BENDING OF THIN SHELLS

10.1. Deformation of an Element of a Shell. Let $ABCD$ (Fig. 10-1) represent an infinitely small element cut out from a shell by two pairs of adjacent planes normal to the middle surface of the shell and containing its principal curvatures. Taking the coordinate axes x and y tangent at O to the lines of principal curvatures and the axis z normal to the middle surface, as shown in the figure, we denote by r_x and r_y the radii of principal curvatures in the xz and yz planes, respectively. The thickness of the shell, which is assumed constant, we denote by h.

In considering bending of the shell, we assume that linear elements such as AD and BC, which are normal to the middle surface of the shell, remain straight and become normal to the deformed middle surface of the shell. Let us begin with a simple case in which, during bending, the lateral faces of the element $ABCD$ rotate only with respect to their lines of intersection with the middle surface. If r_x' and r_y' are the values of the radii of curvature after deformation, the unit elongations of a thin lamina at a distance z from the middle surface (see Fig. 10-1) are

$$\epsilon_x = -\frac{z}{1 - z/r_x}\left(\frac{1}{r_x'} - \frac{1}{r_x}\right) \qquad \epsilon_y = -\frac{z}{1 - z/r_y}\left(\frac{1}{r_y'} - \frac{1}{r_y}\right) \qquad (a)$$

If, in addition to rotation, the lateral sides of the element are displaced parallel to themselves, owing to stretching of the middle surface, and if the corresponding unit elongations of the middle surface in the x and y directions are denoted by ϵ_1 and ϵ_2, respectively, the elongation ϵ_x of the lamina considered above, as seen from Fig. 10-2, is

$$\epsilon_x = \frac{l_2 - l_1}{l_1}$$

Substituting

$$l_1 = ds\left(1 - \frac{z}{r_x}\right) \qquad l_2 = ds(1 + \epsilon_1)\left(1 - \frac{z}{r_x'}\right)$$

we obtain

$$\epsilon_x = \frac{\epsilon_1}{1 - z/r_x} - \frac{z}{1 - z/r_x}\left[\frac{1}{(1 - \epsilon_1)r_x'} - \frac{1}{r_x}\right] \qquad (b)$$

In the same manner, we find for the elongation in the y direction

$$\epsilon_y = \frac{\epsilon_2}{1 - z/r_y} - \frac{z}{1 - z/r_y} \left[\frac{1}{(1 - \epsilon_2)r_y'} - \frac{1}{r_y} \right] \qquad (c)$$

In our further discussion the thickness h of the shell always will be assumed small in comparison with the radii of curvature. In such a case the quantities z/r_x and z/r_y can be neglected in comparison with unity. We shall neglect also the effect of the elongations ϵ_1 and ϵ_2 on the curvature. Then, instead of expressions (b) and (c), we obtain

$$\epsilon_x = \epsilon_1 - z \left(\frac{1}{r_x'} - \frac{1}{r_x} \right) = \epsilon_1 - \chi_x z$$

$$\epsilon_y = \epsilon_2 - z \left(\frac{1}{r_y'} - \frac{1}{r_y} \right) = \epsilon_2 - \chi_y z \qquad (10\text{-}1)$$

where χ_x and χ_y are the changes of curvature. Using these expressions for the components of strain of a lamina and assuming that there are no

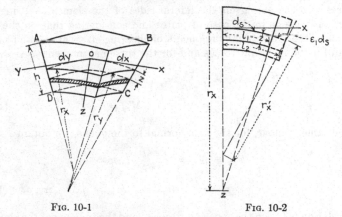

Fig. 10-1   Fig. 10-2

normal stresses between laminae ($\sigma_z = 0$), we obtain the following expressions for the components of stress:

$$\sigma_x = \frac{E}{1 - \nu^2} [\epsilon_1 + \nu\epsilon_2 - z(\chi_x + \nu\chi_y)]$$

$$\sigma_y = \frac{E}{1 - \nu^2} [\epsilon_2 + \nu\epsilon_1 - z(\chi_y + \nu\chi_x)] \qquad (10\text{-}2)$$

On each side of the element $ABCD$ the corresponding forces can be replaced by a normal force applied at the centroid of the side and by a bending moment. Since the thickness of the shell is very small, the lateral sides of the element can be considered as rectangles; hence the resultant forces will act in the middle surface of the shell. Using for these resultant forces and for the bending moments per unit length the same notations as in the case of plates (see Arts. 8.1 and 8.2), we obtain[1]

$$N_x = \int_{-h/2}^{+h/2} \sigma_x \, dz = \frac{Eh}{1 - \nu^2} (\epsilon_1 + \nu\epsilon_2)$$

$$N_y = \int_{-h/2}^{+h/2} \sigma_y \, dz = \frac{Eh}{1 - \nu^2} (\epsilon_2 + \nu\epsilon_1)$$

(10-3)

and also

$$M_x = \int_{-h/2}^{+h/2} z\sigma_x \, dz = -D(\chi_x - \nu\chi_y)$$

$$M_y = \int_{-h/2}^{+h/2} z\sigma_y \, dz = -D(\chi_y + \nu\chi_x)$$

(10-4)

where D has the same meaning as in the case of plates [see Eq. (8-3)] and denotes the *flexural rigidity* of the shell.

A more general case of deformation of the element in Fig. 10-1 is obtained if we assume that, in addition to normal stresses, shearing stresses also are acting on the lateral sides of the element. Using the same notations as in the case of plates and considering that on the side normal to the x axis the components of shearing stress are τ_{xy} and τ_{xz}, we obtain for the resultant forces and for the torsional moment

$$Q_x = \int_{-h/2}^{+h/2} \tau_{xz} \, dz \qquad N_{xy} = \int_{-h/2}^{+h/2} \tau_{xy} \, dz$$

$$M_{xy} = -\int_{-h/2}^{+h/2} z\tau_{xy} \, dz \quad (10\text{-}5)$$

In the same manner, for the side normal to the y axis, we obtain

$$Q_y = \int_{-h/2}^{+h/2} \tau_{yz} \, dz \qquad N_{yx} = N_{xy} = \int_{-h/2}^{+h/2} \tau_{zy} \, dz$$

$$M_{yx} = -M_{xy} = \int_{-h/2}^{+h/2} z\tau_{xy} \, dz \quad (10\text{-}6)$$

The relation between the shearing stress τ_{xy} and the twisting of the element $ABCD$ (Fig. 10-1) can be established exactly in the same manner as in the case of an element cut out from a plate (see p. 324); in this way we obtain

$$\tau_{xy} = -2Gz\chi_{xy} \qquad M_{xy} = D(1 - \nu)\chi_{xy} \qquad (10\text{-}7)$$

where χ_{xy} takes the place of $\partial^2 w/\partial x \, \partial y$ in the case of a plate and represents the twist of the element $ABCD$ during bending of the shell, so that $\chi_{xy} \, dx$ is the rotation[2] of the edge BC relative to Oz with respect to the x axis.

[1] The positive directions of moments and forces are the same as shown in Figs. 8-5 and 8-11 for bending of plates.

[2] Rotations with respect to the x, y, and z axes are taken positive in conformity with the right-hand screw rule.

If, in addition to twist, there is a shearing strain γ in the middle surface of the shell, we obtain

$$\tau_{xy} = (\gamma - 2z\chi_{xy})G \qquad N_{xy} = \int_{-h/2}^{+h/2} \tau_{xy}\, dz = \frac{\gamma hE}{2(1 + \nu)}$$

$$M_{xy} = - \int_{-h/2}^{+h/2} \tau_{xy}z\, dz = D(1 - \nu)\chi_{xy} \qquad (10\text{-}8)$$

Thus assuming that during bending of a shell the linear elements normal to the middle surface remain straight and become normal to the deformed middle surface, we can express the resultant forces N_x, N_y, and N_{xy} and the moments M_x, M_y, and M_{xy} in terms of six quantities: the three components of strain ϵ_1, ϵ_2, and γ of the middle surface of the shell and the three quantities χ_x, χ_y, and χ_{xy} representing the changes of curvature and the twist of the middle surface.

The strain energy of a deformed shell consists of two parts: (1) the strain energy due to bending and (2) the strain energy due to stretching of the middle surface. For the first part of this energy we can use Eq. (8-29). Substituting in it the changes of curvatures χ_x, χ_y, and χ_{xy}, instead of curvatures $\partial^2 w/\partial x^2$, $\partial^2 w/\partial y^2$, and $\partial^2 w/\partial x\, \partial y$, we obtain

$$U_1 = \tfrac{1}{2}D\iint[(\chi_x + \chi_y)^2 - 2(1 - \nu)(\chi_x\chi_y - \chi^2_{xy})]\, dA \qquad (10\text{-}9)$$

where the integration should be extended over the entire surface of the shell.

That part of the energy due to stretching of the middle surface is

$$U_2 = \iint\tfrac{1}{2}(N_x\epsilon_1 + N_y\epsilon_2 + N_{xy}\gamma)\, dA$$

or, by using Eqs. (10-3) and (10-8),

$$U_2 = \frac{Eh}{2(1 - \nu^2)} \iint\left[(\epsilon_1 + \epsilon_2)^2 - 2(1 - \nu)\left(\epsilon_1\epsilon_2 - \frac{1}{4}\gamma^2\right)\right]\, dA \qquad (10\text{-}10)$$

The total energy of deformation is obtained by adding together expressions (10-9) and (10-10). Applications of these expressions in discussing bending and buckling of shells will be shown later.

10.2. Symmetrical Deformation of a Circular Cylindrical Shell. There are many practical cases in which the load acting on a cylindrical shell is symmetrically distributed with respect to the axis of the cylinder. A tube submitted to the action of uniformly distributed internal pressure, a vertical cylindrical reservoir containing a liquid, or a rotating drum submitted to the action of centrifugal forces are examples of such symmetrical loading. Since in these cases all points of the middle surface of the shell, lying in the same cross section perpendicular to the axis of symmetry, have the same displacements, it is sufficient to consider one elemental strip mn

(Fig. 10-3), of unit width,[1] cut out from the shell by two axial[2] sections. An element dx of this strip (Fig. 10-3c) is submitted to the action of forces N_x and $N_y\, dx$ in the middle surface of the shell and of a force $q\, dx$ normal to the surface, where q is the intensity of the load acting on the shell. Also there will be bending moments acting on the sides of the element.

Let us assume that the forces N_x are constant, i.e., that the cylindrical shell is submitted to the action of a uniform axial tension or compression. The forces N_y will depend on the radial displacements of the points of the strip during deformation of the shell. Denoting these displacements in the z direction by w, we find that the strain of the middle surface of the

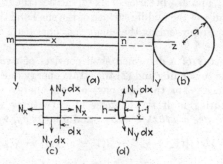

FIG. 10-3

shell in the circumferential direction is $-w/a$, where a is the radius of the middle surface of the shell. By using Eqs. (10-3) we obtain

$$N_x = \frac{Eh}{1 - \nu^2}\left(\epsilon_1 - \nu\,\frac{w}{a}\right) \qquad N_y = \frac{Eh}{1 - \nu^2}\left(-\frac{w}{a} + \nu\epsilon_1\right)$$

from which

$$N_y = \nu N_x - \frac{w}{a}\,Eh \tag{a}$$

Considering bending of the strip mn, the forces (a) give a component in the radial direction (Fig. 10-3d), the magnitude of which per unit length is

$$\frac{N_y}{a} = \frac{1}{a}\left(\nu N_x - \frac{w}{a}\,Eh\right)$$

Owing to curvature of the strip in the xz plane, the longitudinal forces N_x also give a radial component, the magnitude of which is

$$N_x\,\frac{d^2w}{dx^2}$$

---

[1] This width will be assumed as very small in comparison with the radius a, and the cross section of the strip will be considered rectangular.

[2] This term will be used hereafter to designate a section through the axis of the cylinder.

Summing up all transverse loads per unit length of the strip, we obtain

$$q + \frac{1}{a}\left(\nu N_x - \frac{w}{a} Eh\right) + N_z \frac{d^2w}{dx^2} \qquad (b)$$

and the differential equation for the bending of the strip is

$$D \frac{d^4w}{dx^4} = q + \frac{1}{a} \nu N_x - \frac{w}{a^2} Eh + N_z \frac{d^2w}{dx^2} \qquad (10\text{-}11)$$

The quantity D is taken for the flexural rigidity of the strip, since distortion of the cross section is prevented by the action of adjacent strips.

If the load q and the forces N_x are given, the deflection of the shell is found from Eq. (10-11). Application of this equation in studying buckling of shells will be shown later (see Art. 11.1).

10.3. Inextensional Deformation of a Circular Cylindrical Shell.[1] In the discussion of deformations of a circular ring (Art. 7.2), it was pointed out that a simplification in the analysis can be obtained if the extension of the center line of the ring may be neglected. The same kind of simplification can be obtained also in the case of *inextensional deformation* of a circular cylindrical shell. Let us consider the limitations that must be imposed on the displacements of the points of the middle surface of the shell in order to have inextensional deformation. Taking the origin of coordinates at any point in the middle surface of the shell, directing the axes as shown in Fig. 10-4, and denoting by u, v, and w the components of the dis-

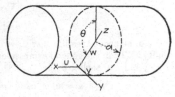

Fig. 10-4

placement of that point, we find that the condition that there is no stretching of the middle surface in the x direction is

$$\epsilon_1 = \frac{\partial u}{\partial x} = 0 \qquad (a)$$

The condition of inextensibility in the circumferential direction will be written as in the case of a ring [see Eq. (b), p. 282]

$$\epsilon_2 = \frac{\partial v}{a\,\partial\theta} - \frac{w}{a} = 0 \qquad (b)$$

The condition that there is no shearing strain in the middle surface is

$$\gamma = \frac{\partial u}{a\,\partial\theta} + \frac{\partial v}{\partial x} = 0 \qquad (c)$$

which is the same as in the case of small deflections of plates except that $a\,d\theta$ takes the place of dy. The three conditions (a), (b), and (c) will be satisfied by taking for the components of displacement the same expressions as in the case of a circular ring (see Art. 7.2). These components can be put in the following form:

[1] The theory of inextensional deformations of shells is due to Lord Rayleigh, *Proc. London Math. Soc.*, vol. 13, 1881, and *Proc. Roy. Soc., London*, vol. 45, 1889.

$$u_1 = 0$$

$$v_1 = \sum_{n=1}^{\infty} a(a_n \cos n\theta - a'_n \sin n\theta)$$

$$w_1 = - \sum_{n=1}^{\infty} na(a_n \sin n\theta + a'_n \cos n\theta)$$

$$(d)$$

where a is the radius of the middle surface of the shell, θ the central angle, and a_n and a'_n constants which must be calculated for each particular case of loading. The displacements (d) represent the case in which all cross sections of the shell deform identically. On these displacements we can superpose displacements which vary along the length of the cylinder and which are given by the series

$$u_2 = - \sum_{n=1}^{\infty} \frac{a}{n} (b_n \sin n\theta + b'_n \cos n\theta)$$

$$v_2 = x \sum_{n=1}^{\infty} (b_n \cos n\theta - b'_n \sin n\theta)$$

$$(e)$$

$$w_2 = -x \sum_{n=1}^{\infty} n(b_n \sin n\theta + b'_n \cos n\theta)$$

It can be proved readily that these expressions also satisfy the conditions of inextensibility. Then the general expressions for displacements in inextensional deformation of a cylindrical shell are

$$u = u_1 + u_2 \qquad v = v_1 + v_2 \qquad w = w_1 + w_2 \qquad (f)$$

In calculating inextensional deformations of a cylindrical shell under the action of a given system of forces, it is advantageous to use the expression for strain energy of bending [Eq. (10-9)]. The changes of curvature χ_x, χ_y, and χ_{xy}, which enter in this expression, can be calculated in the following way. The quantity χ_x, representing the change of curvature in the direction of the generator, is equal to zero, since the generators, as seen from expressions (d) and (e), remain straight. The quantity χ_y, representing the change of curvature of the circumference, can be determined as in the case of a ring (see Art. 7.1) and we obtain

$$\chi_y = \frac{1}{a^2} \left(w + \frac{\partial^2 w}{\partial \theta^2} \right)$$

or, by using condition (b),

$$\chi_y = \frac{1}{a^2} \left(\frac{\partial v}{\partial \theta} + \frac{\partial^2 w}{\partial \theta^2} \right)$$

$$(g)$$

In calculating twist we note that an element of a generator, during deformation, rotates[1] with respect to the y axis through an angle equal to $-\partial w/\partial x$ and with respect to z axis through an angle equal to $\partial v/\partial x$. Considering now a similar element of a generator at a circumferential distance $a\, d\theta$ from the first one, we see that its rotation about the y axis, corresponding to displacement w, is

$$- \frac{\partial w}{\partial x} - \frac{\partial^2 w}{\partial \theta\, \partial x}\, d\theta$$

$$(h)$$

[1] In determining the sign of rotation the right-hand-screw rule is used.

Rotation of the same element in the plane tangent to the shell is

$$\frac{\partial v}{\partial x} + \frac{\partial(\partial v/\partial x)}{\partial \theta}\, d\theta$$

Owing to the central angle $d\theta$ between the two elements, the latter rotation has a component with respect to the y axis equal to

$$-\frac{\partial v}{\partial x}\, d\theta \tag{i}$$

From the results (h) and (i) we conclude that the total angle of twist between the two elements under consideration is

$$-\left(\frac{\partial^2 w}{\partial \theta\, \partial x} + \frac{\partial v}{\partial x}\right) d\theta$$

Hence
$$\chi_{xy} = \frac{1}{a}\left(\frac{\partial^2 w}{\partial \theta\, \partial x} + \frac{\partial v}{\partial x}\right) \tag{j}$$

Substituting the calculated changes of curvatures in expression (10-9) for the strain energy of bending[1] and using for displacements expressions (f), we finally obtain for the total energy of deformation of a cylindrical shell of length $2l$ (Fig. 10-5) the following expression:

$$U = \pi Dl \sum_{n=2}^{\infty} \frac{(n^2 - 1)^2}{a^3} \left\{ n^2 \left[a^2(a_n{}^2 + a_n'^2) + \frac{1}{3} l^2(b_n{}^2 + b_n'^2) \right] \right.$$
$$\left. + 2(1 - \nu)a^2(b_n{}^2 + b_n'^2) \right\} \tag{10-12}$$

This expression does not contain a term with $n = 1$, since, as was pointed out in considering the deformation of a circular ring (Art. 7.2), the corresponding displacements are the same as for a rigid body and do not contribute to the strain energy.

Let us now apply the above expression for U in calculating the deformations produced in a cylindrical shell by two equal and opposite forces P acting along a diameter at a distance c from the middle (Fig. 10-5). These forces produce work only on radial displacements w of their points of application, and since the terms with coefficients a_n and b_n in the expressions for w_1

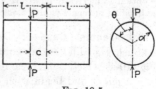

FIG. 10-5

and w_2 [see Eqs. (d) and (e)] vanish at these points, only terms with coefficients a_n' and b_n' will enter in the expressions for deformation. With the use of the principle of virtual displacements, the equations for calculating coefficients a_n' and b_n' become

$$\frac{\partial U}{\partial a_n'}\, \delta a_n' = -na\, \delta a_n'(1 + \cos n\pi)P$$

$$\frac{\partial U}{\partial b_n'}\, \delta b_n' = -nc\, \delta b_n'(1 + \cos n\pi)P$$

[1] The energy of deformation due to stretching of the middle surface of the shell is zero in this case, since the deformation is assumed inextensional.

Substituting expression (10-12) for U, we obtain, for the case where n is an even number,

$$a'_n = - \frac{a^2 P}{n(n^2 - 1)^2 \pi Dl}$$

$$b'_n = - \frac{ncPa^3}{(n^2 - 1)^2 \pi Dl \left[\dfrac{n^2 l^2}{3} + 2(1 - \nu)a^2 \right]}$$

If n is an odd number, we obtain

$$a'_n = b'_n = 0$$

Substituting the values of a'_n and b'_n in expressions (f) and also putting $a_n = b_n = 0$, we obtain rapidly convergent series for displacements u, v, and w. Although these expressions do not satisfy rigorously the conditions at the free edges of the cylindrical shell, the displacements calculated are in good agreement with experiments,[1] provided the thickness h of the shell is small in comparison with the radius a. The same method also can be used sometimes in calculating deformations of a portion of a cylindrical shell cut out from a complete cylinder of radius a by two axial sections making an angle α with each other (Fig. 10-6). Taking, for instance, for the displacements the series

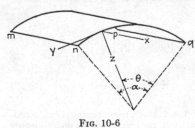

FIG. 10-6

$$u = - \sum \frac{\alpha a b_n}{\pi n} \sin \frac{n\pi\theta}{\alpha}$$

$$v = \sum a a_n \cos \frac{n\pi\theta}{\alpha} + x \sum b_n \cos \frac{n\pi\theta}{\alpha}$$

$$w = - \sum \frac{n\pi}{\alpha} a a_n \sin \frac{n\pi\theta}{\alpha}$$

$$\qquad - x \sum \frac{n\pi}{\alpha} b_n \sin \frac{n\pi\theta}{\alpha}$$

we obtain an inextensional deformation such that the displacements u and w and also the bending moments vanish along the edges mn and pq.

10.4. General Case of Deformation of a Cylindrical Shell.[2] To establish the differential equations for the displacements u, v, and w (Fig. 10-4) which define the deformation of a shell, we proceed as in the case of plates and begin with the equations of equilibrium of an element cut out from the cylindrical shell by two adjacent axial sections and by two adjacent sections perpendicular to the axis of the cylinder. The corresponding element of the middle surface of the shell, after deformation, is shown in Figs. 10-7a and 10-7b. In Fig. 10-7a the resultant forces, dis-

---

[1] Such experiments were made at the University of Michigan by I. A. Wojtaszak with brass tubes having a diameter of 6 in., a thickness of $\frac{1}{16}$ in., and lengths of 30 and 24 in. Deflection curves drawn for the generator for which $\theta = 90°$ (Fig. 10-5) deviated only slightly from straight lines, found by substituting a'_n and b'_n (found above) in the equation for the displacement w given by (f).

[2] A general theory of bending of thin shells has been developed by A. E. H. Love, "Mathematical Theory of Elasticity," 4th ed., chap. 24, p. 515, 1927.

cussed in Art. 10.1, are shown. Before deformation the axes x, y, and z at any point of the middle surface had the directions of the generator, the tangent to the circumference, and the normal to the middle surface of the shell, respectively. After deformation, which is assumed very small, these directions are slightly changed and we take the z axis normal to the deformed middle surface, the x axis in the direction of a tangent to the generator, which may become curved, and the y axis perpendicular to the xz plane. The directions of the resultant forces also will be changed slightly accordingly, and these changes must be considered in writing the equations of equilibrium of the element $OABC$.

Let us begin by establishing formulas for angular displacements of the sides BC and AB with respect to the sides OA and OC of the element, respectively. In these calculations we consider the displacements u, v, and w as very small; calculate the angular motions produced by each of

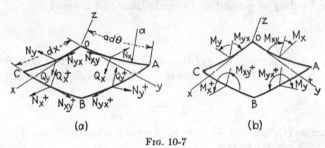

(a)   (b)

FIG. 10-7

these displacements; and obtain the resultant angular displacement by superposition. We begin with the rotation of the side BC with respect to the side OA. This rotation can be resolved into three component rotations with respect to the x, y, and z axes. Rotations of the sides OA and BC with respect to the x axis are due to the displacements v and w. Since the displacements v represent motion of the sides OA and BC in circumferential directions (see Fig. 10-4) and a is the radius of the middle surface of the cylinder, the corresponding rotation of the side OA, with respect to the x axis, is v/a and that of the side BC is

$$\frac{1}{a}\left(v + \frac{\partial v}{\partial x}\,dx\right)$$

Thus, owing to displacements v, the relative angular motion of BC with respect to AO about the x axis is

$$\frac{1}{a}\frac{\partial v}{\partial x}\,dx \tag{a}$$

Owing to displacements w, the side OA rotates with respect to the x axis by the angle $\partial w / a\, \partial\theta$ and the side BC by the angle

$$\frac{\partial w}{a\, \partial\theta} + \frac{\partial}{\partial x}\frac{\partial w}{a\, \partial\theta}\, dx$$

Thus, because of displacements w, the relative angular displacement is

$$\frac{\partial}{\partial x}\frac{\partial w}{a\, \partial\theta}\, dx \tag{b}$$

Summing up (a) and (b), the relative angular displacement about the x axis of the side BC with respect to the side OA is

$$\frac{1}{a}\left(\frac{\partial v}{\partial x} + \frac{\partial^2 w}{\partial x\, \partial\theta}\right) dx \tag{c}$$

The rotation about the y axis of the side BC with respect to the side OA is due to bending of generators in axial planes and is equal to[1]

$$-\frac{\partial^2 w}{\partial x^2}\, dx \tag{d}$$

The rotation about the z axis of the side BC with respect to the side OA is due to bending of the generators in tangent planes and is equal to

$$\frac{\partial^2 v}{\partial x^2}\, dx \tag{e}$$

The formulas (c), (d), and (e) give the three components of rotation of the side BC with respect to the side OA.

Let us establish now the corresponding formulas for the angular displacement of the side AB with respect to the side OC. Owing to curvature of the cylindrical shell, the initial angle between these lateral sides of the element $OABC$ is $d\theta$. However, because of displacements v and w, this angle will be changed and rotation of the lateral side OC with respect to the x axis becomes

$$\frac{v}{a} + \frac{\partial w}{a\, \partial\theta} \tag{f}$$

The corresponding rotation for the lateral side AB is

$$\frac{v}{a} + \frac{\partial w}{a\, \partial\theta} + \frac{\partial}{\partial\theta}\left(\frac{v}{a} + \frac{\partial w}{a\, \partial\theta}\right) d\theta$$

Hence, instead of the initial angle $d\theta$, we must take

$$d\theta + d\theta\left(\frac{\partial v}{a\, \partial\theta} + \frac{\partial^2 w}{a\, \partial\theta^2}\right) \tag{g}$$

[1] The signs of the angular displacements with respect to coordinate axes x, y, and z are taken in accordance with the right-hand screw rule.

In calculating the angle of rotation about the y axis of the side AB with respect to the side OC, we use for the twist expression (j) of the preceding article (see p. 447); then the required angular displacement is

$$-\left(\frac{\partial^2 w}{\partial\theta\,\partial x} + \frac{\partial v}{\partial x}\right) d\theta \qquad (h)$$

Rotation about the z axis of the side AB with respect to OC is due to displacements v and w. Owing to displacement v, the angle of rotation of the side OC is $\partial v/\partial x$ and that of the side AB is

$$\frac{\partial v}{\partial x} + \frac{\partial}{a\,\partial\theta}\frac{\partial v}{\partial x}\,a\,d\theta$$

so that the relative angular displacement is

$$\frac{\partial}{a\,\partial\theta}\frac{\partial v}{\partial x}\,a\,d\theta \qquad (i)$$

Because of displacement w, the side AB rotates in the axial plane by the angle $\partial w/\partial x$. The component of this rotation with respect to the z axis is

$$-\frac{\partial w}{\partial x}\,d\theta \qquad (j)$$

Summing up (i) and (j), the relative angular displacement about the z axis of the side AB with respect to the side OC is

$$\left(\frac{\partial^2 v}{\partial\theta\,\partial x} - \frac{\partial w}{\partial x}\right) d\theta \qquad (k)$$

Having the above formulas[1] for the angles, we may now obtain the three equations of equilibrium of the element $OABC$ (Fig. 10-7a) by projecting all forces on the x, y, and z axes. Beginning with those forces parallel to the resultant forces N_x and N_{yx}, and projecting them on the x axis, we obtain $(\partial N_x/\partial x)\,dx\,a\,d\theta$ and $(\partial N_{yx}/\partial\theta)\,d\theta\,dx$. Owing to the angle of rotation given by expression (k), the forces parallel to N_y give in the x direction a component

$$-N_y\left(\frac{\partial^2 v}{\partial\theta\,\partial x} - \frac{\partial w}{\partial x}\right) d\theta\,dx$$

Because of the rotation given by expression (e), the forces parallel to the resultant forces N_{xy} give in the x direction a component

$$-N_{xy}\frac{\partial^2 v}{\partial x^2}\,dx\,a\,d\theta$$

[1] These formulas can be obtained readily for a cylindrical shell from the general formulas given by Love, *ibid.*, p. 523.

Finally, owing to the angles given by expressions (d) and (h), the forces parallel to Q_x and Q_y give in the x direction the components

$$-Q_z \frac{\partial^2 w}{\partial x^2} dx\, a\, d\theta - Q_y \left(\frac{\partial^2 w}{\partial \theta\, \partial x} + \frac{\partial v}{\partial x} \right) d\theta\, dx$$

Regarding external forces acting on the element, we assume that there is only a normal pressure of intensity q, the projection of which on the x axis is zero.

Summing up all the projections calculated above, we obtain

$$\frac{\partial N_z}{\partial x} dx\, a\, d\theta + \frac{\partial N_{yz}}{\partial \theta} d\theta\, dx - N_y \left(\frac{\partial^2 v}{\partial \theta\, \partial x} - \frac{\partial w}{\partial x} \right) d\theta\, dx$$

$$-N_{xy} \frac{\partial^2 v}{\partial x^2} dx\, a\, d\theta - Q_z \frac{\partial^2 w}{\partial x^2} dx\, a\, d\theta - Q_y \left(\frac{\partial^2 w}{\partial \theta\, \partial x} + \frac{\partial v}{\partial x} \right) d\theta\, dx = 0$$

In the same manner, two other equations of equilibrium can be written. After simplification, all three equations can be put in the following form:

$$a \frac{\partial N_z}{\partial x} + \frac{\partial N_{yz}}{\partial \theta} - a Q_z \frac{\partial^2 w}{\partial x^2} - a N_{xy} \frac{\partial^2 v}{\partial x^2} - Q_y \left(\frac{\partial v}{\partial x} + \frac{\partial^2 w}{\partial x\, \partial \theta} \right)$$
$$- N_v \left(\frac{\partial^2 v}{\partial x\, \partial \theta} - \frac{\partial w}{\partial x} \right) = 0$$

$$\frac{\partial N_y}{\partial \theta} + a \frac{\partial N_{zy}}{\partial x} + a N_z \frac{\partial^2 v}{\partial x^2} - Q_z \left(\frac{\partial v}{\partial x} + \frac{\partial^2 w}{\partial x\, \partial \theta} \right)$$
$$+ N_{yz} \left(\frac{\partial^2 v}{\partial x\, \partial \theta} - \frac{\partial w}{\partial x} \right) - Q_y \left(1 + \frac{\partial v}{a\, \partial \theta} + \frac{\partial^2 w}{a\, \partial \theta^2} \right) = 0 \tag{10-13}$$

$$a \frac{\partial Q_z}{\partial x} + \frac{\partial Q_y}{\partial \theta} + N_{xy} \left(\frac{\partial v}{\partial x} + \frac{\partial^2 w}{\partial x\, \partial \theta} \right) + a N_z \frac{\partial^2 w}{\partial x^2}$$
$$+ N_y \left(1 + \frac{\partial v}{a\, \partial \theta} + \frac{\partial^2 w}{a\, \partial \theta^2} \right) + N_{yz} \left(\frac{\partial v}{\partial x} + \frac{\partial^2 w}{\partial x\, \partial \theta} \right) + qa = 0$$

In the derivation of these equations the change in size of the element due to stretching of the middle surface was not considered. In the solution of problems of stability a further refinement is sometimes introduced and the strains ϵ_1 and ϵ_2 of the middle surface are taken into account in writing the equations of equilibrium of the element. Since ϵ_1 and ϵ_2 are small quantities expressed by the derivatives of the displacements u, v, and w [see Eqs. (10-15)], they must be introduced only in those terms of Eqs. (10-13) which are not multiplied by the derivatives of the displacements. Considering, for instance, the case of buckling of a cylindrical shell under lateral pressure (Art. 11.5), we shall find that the stress resultant N_y is very large in comparison with the other stress resultants; thus $N_y(1 + \epsilon_1)$ should be introduced, instead of N_y, in the second and the third of Eqs. (10-13) and $q(1 + \epsilon_1)(1 + \epsilon_2)$ should be substituted for q in the third equation to take into account the stretching of the middle surface. In the case of buckling of a cylindrical shell under torsion (Art. 11.11), the stress resultants N_{xy} and N_{yx} become the most important; considering the effect of stretching of the middle surface, $N_{yx}(1 + \epsilon_1)$ and $N_{xy}(1 + \epsilon_2)$ instead of N_{yx} and N_{xy} should be substituted in the first and second of Eqs. (10-13). This question of taking account of stretching of the middle surface will be discussed later in considering particular problems.

Considering now the three equations of moments with respect to the x, y, and z axes (Fig. 10-7b) and again taking into consideration the small angular displacements of the sides BC and AB with respect to OA and OC, respectively, we obtain the following equations:

$$a \frac{\partial M_{xy}}{\partial x} - \frac{\partial M_y}{\partial \theta} - a M_x \frac{\partial^2 v}{\partial x^2} - M_{yx} \left(\frac{\partial^2 v}{\partial x \, \partial \theta} - \frac{\partial w}{\partial x} \right) + a Q_y = 0$$

$$\frac{\partial M_{yx}}{\partial \theta} + a \frac{\partial M_x}{\partial x} + a M_{xy} \frac{\partial^2 v}{\partial x^2} - M_y \left(\frac{\partial^2 v}{\partial x \, \partial \theta} - \frac{\partial w}{\partial x} \right) - a Q_x = 0$$

$$M_x \left(\frac{\partial v}{\partial x} + \frac{\partial^2 w}{\partial x \, \partial \theta} \right) + a M_{xy} \frac{\partial^2 w}{\partial x^2} + M_{yx} \left(1 + \frac{\partial v}{a \, \partial \theta} + \frac{\partial^2 w}{a \, \partial \theta^2} \right)$$
$$- M_y \left(\frac{\partial v}{\partial x} + \frac{\partial^2 w}{\partial x \, \partial \theta} \right) + a(N_{xy} - N_{yx}) = 0$$

(10-14)

By using the first two of these equations we can eliminate Q_x and Q_y from Eqs. (10-13) and obtain in this way three equations containing the resultant forces N_x, N_y, and N_{xy} and the moments M_x, M_y, and M_{xy}. By using the formulas of Art. 10.1, we can express all these quantities in terms of the three strain components ϵ_1, ϵ_2, and γ of the middle surface and the three curvature changes χ_x, χ_y, and χ_{xy}, all of which are represented in terms of the displacements u, v, and w as follows (see Art. 10.3):

$$\epsilon_1 = \frac{\partial u}{\partial x} \qquad \epsilon_2 = \frac{\partial v}{a \, \partial \theta} - \frac{w}{a} \qquad\qquad \gamma = \frac{\partial u}{a \, \partial \theta} + \frac{\partial v}{\partial x}$$
$$\chi_x = \frac{\partial^2 w}{\partial x^2} \qquad \chi_y = \frac{1}{a^2} \left(\frac{\partial v}{\partial \theta} + \frac{\partial^2 w}{\partial \theta^2} \right) \qquad \chi_{xy} = \frac{1}{a} \left(\frac{\partial v}{\partial x} + \frac{\partial^2 w}{\partial x \, \partial \theta} \right)$$

(10-15)

Thus we finally obtain the three differential equations for determining the displacements u, v, and w.

10.5. Symmetrical Deformation of a Spherical Shell. Let us assume that the vertical diameter is the axis of symmetry of the deformation of a spherical shell (Fig. 10-8) and let us consider an element $OABC$ cut out from the shell by two meridional sections an angle $d\psi$ apart and by two conical surfaces normal to the meridians and inclined to the axis of symmetry by the angles θ and $\theta + d\theta$. Taking the x and y axes as tangents at O to the meridian and to the parallel circle, respectively, and the z axis in the radial direction, as shown in the figure, we denote the corresponding components of the displacement at O by u, v, and w. In the case of a symmetrical deformation, $v = 0$ and u and w are functions of the angle θ only. Between the meridional faces OA and BC of the element there is a small angle $d\psi$ which can be obtained by rotating the meridional plane OA with respect to the x and z axes by the angles $d\psi \sin \theta$ and $d\psi \cos \theta$, respectively. The angle between the lateral faces OC and AB of the element is equal to $d\theta$, and the direction of the face AB is obtained by rotation

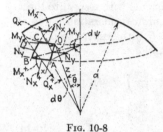

of the face OC with respect to the y axis by the angle[1] $-d\theta$. By using these initial values of the angles between the faces of the element $OABC$ and by denoting the resultant forces and the moments as shown in the figure, we can write readily the differential equations of equilibrium of the element.

Fig. 10-8

In the case of symmetrical deformation we have only three equations to consider: projections of forces on the x and z axes and moments of forces with respect to the y axis. Projecting all forces on the x axis and assuming that any external load is normal to the shell, we obtain

$$\left(N_x + \frac{\partial N_x}{\partial \theta} \, d\theta \right) a \sin \left(\theta + d\theta \right) d\psi - N_x a \sin \theta \, d\psi$$
$$- N_y a \, d\theta \cos \theta \, d\psi - Q_x a \sin \theta \, d\psi \, d\theta = 0$$

In the same manner the other two equations of equilibrium can be written. After simplification the three equations of equilibrium become

$$\frac{\partial N_x}{\partial \theta} + (N_x - N_y) \cot \theta - Q_x = 0$$

$$\frac{\partial Q_x}{\partial \theta} + Q_x \cot \theta + N_x + N_y + qa = 0 \qquad (10\text{-}16)$$

$$\frac{\partial M_x}{\partial \theta} + (M_x - M_y) \cot \theta - Q_x a = 0$$

where q is the intensity of the external load. These equations should be used in investigating deformation of a spherical shell under the action of a normal load symmetrically distributed with respect to a diametral axis.

In writing the equations of equilibrium for the case of the buckled surface of a shell, which is assumed symmetrical with respect to a diametral axis, we should take into account small changes of the angles between the faces of an element such as $OABC$, due to deformation. Considering the change of the angle between the faces OC and AB of the element, we con-

[1] The right-hand-screw rule is used in determining the sign of rotation.

clude from the assumed symmetry of deformation that there will be rotation only with respect to the y axis. This angle of rotation for the face OC is

$$\frac{u}{a} + \frac{dw}{a\,d\theta}$$

Thus the angle between the faces OC and AB after deformation becomes

$$d\theta + \frac{d}{d\theta}\left(\frac{u}{a} + \frac{dw}{a\,d\theta}\right)d\theta \qquad (a)$$

Considering now the change of the angle between the faces AO and BC, we observe that, owing to symmetry of deformation, these faces rotate only in their own planes by the angle

$$-\left(\frac{u}{a} + \frac{dw}{a\,d\theta}\right)$$

Such a rotation in the plane of the face BC has components with respect to the x and z axes equal to

$$\left(\frac{u}{a} + \frac{dw}{a\,d\theta}\right)\cos\theta\,d\psi \qquad \text{and} \qquad -\left(\frac{u}{a} + \frac{dw}{a\,d\theta}\right)\sin\theta\,d\psi$$

respectively. Thus, after deformation, the direction of the face BC with respect to the face AO can be obtained by the rotation of the face AO with respect to the x and z axes through the angles

$$\sin\theta\,d\psi + \left(\frac{u}{a} + \frac{dw}{a\,d\theta}\right)\cos\theta\,d\psi \qquad (b)$$

and

$$\cos\theta\,d\psi - \left(\frac{u}{a} + \frac{dw}{a\,d\theta}\right)\sin\theta\,d\psi \qquad (c)$$

respectively.

Using the angles given by expressions (a), (b), and (c), instead of the initial angles $d\theta$, $\sin\theta\,d\psi$, and $\cos\theta\,d\psi$, the equations of equilibrium of the element $OABC$ become

$$\frac{dN_x}{d\theta} + (N_x - N_y)\cot\theta - Q_x + N_y\left(\frac{u}{a} + \frac{dw}{a\,d\theta}\right)$$
$$- Q_x\left(\frac{d^2w}{a\,d\theta^2} + \frac{w}{a}\right) = 0$$

$$\frac{dQ_x}{d\theta} + Q_x\cot\theta + N_x + N_y + qa + N_x\left(\frac{d^2w}{a\,d\theta^2} + \frac{du}{a\,d\theta}\right) \qquad (10\text{-}17)$$
$$+ N_y\left(\frac{u}{a} + \frac{dw}{a\,d\theta}\right)\cot\theta = 0$$

$$\frac{dM_x}{d\theta} + (M_x - M_y)\cot\theta - Q_x a + M_y\left(\frac{u}{a} + \frac{dw}{a\,d\theta}\right) = 0$$

Eliminating Q_x from these equations, we obtain two equations containing N_x, N_y, M_x, M_y. All these quantities can be expressed in terms of the displacements u and w by using Eqs. (10-3) and (10-4) of Art. 10.1. The quantities ϵ_1 and ϵ_2, entering in Eqs. (10-3), can be readily determined from geometrical considerations. In this case they are

$$\epsilon_1 = \frac{du}{a\,d\theta} - \frac{w}{a} \qquad \epsilon_2 = \frac{u\cos\theta}{a\sin\theta} - \frac{w}{a} \qquad (d)$$

By using expressions (a) and (b), we obtain for the changes of curvature

$$\chi_x = \frac{d^2w}{a^2\,d\theta^2} + \frac{du}{a^2\,d\theta} \qquad \chi_y = \left(\frac{u}{a^2} + \frac{dw}{a^2\,d\theta}\right)\cot\theta \qquad (e)$$

With these values of ϵ_1, ϵ_2, χ_x, and χ_y we obtain finally, from Eqs. (10-17), two equations containing only u and w. The application of these equations in discussing stability of a compressed spherical shell will be shown in Art. 11.13.

The differential equations of equilibrium developed in the last two articles are based on Love's general theory of small deflections of thin shells[1] which neglects stresses normal to the middle surface of the shell and assumes that the normals to the undeformed middle surface remain normal to the deformed middle surface. In recent years a considerable literature has been developed dealing with various improvements of Love's theory.[2] Some refinements of the theory were also obtained by taking into account the effect of shearing forces on bending of shells.[3] In the solution of many practical problems, the tendency has been to further simplify Love's equations by omitting certain terms, so that the mathematical treatment becomes simpler. Such a simplification was applied first in the case of torsion of thin cylindrical shells[4] and will be discussed in the next chapter. A similar simplification was developed later for other cases of buckling of cylindrical shells.[5]

[1] *Ibid.*

[2] A bibliography of this subject is given in the paper by P. M. Naghdi, *Appl. Mech. Rev.*, vol. 9, p. 365, 1956.

[3] See E. Reissner, *J. Appl. Mech.*, vol. 12, p. A69, 1945, and *J. Math. Phys.*, vol. 31, p. 109, 1952; P. M. Naghdi, *Quart. Appl. Math.*, vols. 14 and 15, 1956 and 1957; and V. L. Salerno and M. A. Goldberg, *J. Appl. Mech.*, vol. 27, p. 54, 1960.

[4] L. H. Donnell, *NACA Tech. Rept.* 479, 1933. The accuracy of the simplified equations was discussed by N. J. Hoff, *J. Appl. Mech.*, vol. 22, p. 329, 1955.

[5] S. B. Batdorf, *NACA Tech. Note* 1341, 1947.

CHAPTER 11

BUCKLING OF SHELLS

11.1. Symmetrical Buckling of a Cylindrical Shell under the Action of Uniform Axial Compression.[1]

If a cylindrical shell is uniformly compressed in the axial direction, buckling symmetrical with respect to the axis of the cylinder (Fig. 11-1) may occur at a certain value of the compressive load.[2] The critical value of the compressive force N_{cr} per unit length of the edge of the shell can be obtained by using the energy method. As long as the shell remains cylindrical, the total strain energy is the energy of axial compression. However, when buckling begins, we must consider, in addition to axial compression, the strain of the middle surface in the circumferential direction and also bending of the shell. Thus the strain energy of the shell is increased; at the critical value of the load this increase in energy must be equal to the work done by the compressive load as the cylinder shortens owing to buckling.

We assume for radial displacements during buckling the expression

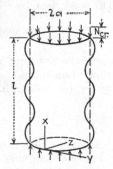

Fig. 11-1

$$w = -A \sin \frac{m\pi x}{l} \qquad (a)$$

where l is the length of the cylinder. The strains ϵ_1 and ϵ_2 in the axial and circumferential directions after buckling will be found from the condition that the axial compressive force during buckling remains constant. Using for the axial strain before buckling the notation

$$\epsilon_0 = -\frac{N_{cr}}{Eh} \qquad (b)$$

where h is the thickness of the shell, we obtain

$$\epsilon_1 + \nu\epsilon_2 = (1 - \nu^2)\epsilon_0$$

[1] For nonsymmetrical buckling, see Art. 11.3.

[2] See Timoshenko, Z. Math. Physik, vol. 58, p. 378, 1910. See also R. Lorenz, Z. Ver. deut. Ingr., vol. 52, p. 1766, 1908, and Physik. Z., vol. 13, p. 241, 1911.

Observing that

$$\epsilon_2 = -\nu\epsilon_0 - \frac{w}{a} = -\nu\epsilon_0 + \frac{A}{a}\sin\frac{m\pi x}{l} \qquad (c)$$

we find that

$$\epsilon_1 = \epsilon_0 - \nu\frac{A}{a}\sin\frac{m\pi x}{l} \qquad (d)$$

The change of curvature in the axial plane is

$$\chi_z = \frac{\partial^2 w}{\partial x^2} = A\frac{m^2\pi^2}{l^2}\sin\frac{m\pi x}{l} \qquad (e)$$

Substituting expressions (c), (d), and (e) in Eqs. (10-9) and (10-10) for strain energy and noting that, owing to symmetry of deformation,

$$\gamma = \chi_y = \chi_{zy} = 0$$

we find for the increase of strain energy during buckling the following expression:

$$\Delta U = -2\pi h E\nu\epsilon_0\int_0^l A\sin\frac{m\pi x}{l}\,dx + \frac{\pi A^2 Ehl}{2a} + A^2\frac{\pi^4 m^4}{2l^4}\pi alD \qquad (f)$$

The work done by compressive forces during buckling is

$$\Delta T = 2\pi N_{cr}\left(\nu\int_0^l A\sin\frac{m\pi x}{l}\,dx + \frac{a}{4}A^2\frac{m^2\pi^2}{l}\right) \qquad (g)$$

where the first term in the parentheses is due to the change $\epsilon_1 - \epsilon_0$ of the axial strain and the second term is due to bending of the generators given by Eq. (a). Equating expressions (f) and (g), we obtain

$$\sigma_{cr} = \frac{N_{cr}}{h} = D\left(\frac{m^2\pi^2}{hl^2} + \frac{E}{a^2 D}\frac{l^2}{m^2\pi^2}\right) \qquad (h)$$

Assuming that there are many waves formed along the length of the cylinder during buckling and considering σ_{cr} as a continuous function of $m\pi/l$, we find that the minimum value of expression (h) is

$$\sigma_{cr} = \frac{2}{ah}\sqrt{EDh} = \frac{Eh}{a\sqrt{3(1-\nu^2)}} \qquad (11\text{-}1)$$

and occurs at

$$\frac{m\pi}{l} = \sqrt[4]{\frac{Eh}{a^2 D}}$$

Thus the length of half-waves into which the shell buckles, for $\nu = 0.3$, is

$$\frac{l}{m} = \pi\sqrt[4]{\frac{a^2 D}{Eh}} = \pi\sqrt[4]{\frac{a^2 h^2}{12(1-\nu^2)}} \approx 1.72\sqrt{ah} \qquad (11\text{-}2)$$

It is seen that the results obtained for the symmetrical buckling of a cylindrical shell are similar to those obtained for buckling of a bar in an elastic medium (see Art. 2.10), and the discussion given there regarding the number of waves for the bar can be applied here also. It is seen also that a symmetrical buckling, which we are considering, may occur within the elastic limit only in the case of very thin shells. Taking, for instance, a steel shell with $E = 30 \cdot 10^6$ psi, $\sigma_{PL} = 60,000$ psi, and $\nu = 0.3$, we find from Eq. (11-1) that $a/h = 303$ and from Eq. (11-2) we conclude that the length of half-waves is less than one-tenth of the radius and that for cylinders whose length is not smaller than the diameter the number of half-waves is larger than 20. Our assumption that m is a large number is accurate enough in such cases.

Instead of the energy method the differential equation for symmetrical deflection of a cylindrical shell [Eq. (10-11)] can be used in calculating the critical load. In applying this equation we take $q = 0$ and measure the displacement w, not from the unstrained middle surface of the shell as was assumed in the derivation of the equation, but from the middle surface after uniform compression is applied. This requires that we replace w in Eq. (10-11) by $w + (\nu N_x a)/(Eh)$ and consider N_x positive when it is compression. Then the differential equation for symmetrical buckling of a cylindrical shell becomes

$$D \frac{d^4w}{dx^4} + N_x \frac{d^2w}{dx^2} + Eh \frac{w}{a^2} = 0 \qquad (11\text{-}3)$$

Substituting for w the previous expression (a) and equating to zero the coefficient of $\sin m\pi x/l$, we obtain from this equation the critical stress given by expression (h).

When the shell is not very thin and buckling occurs at a stress that is beyond the proportional limit, the critical load can be obtained again from Eq. (11-3), it being necessary only to introduce in the expression for the flexural rigidity D the tangent modulus E_t instead of E.[1] Then, proceeding as before, we obtain from Eq. (11-3)

$$\sigma_{cr} = \frac{h \sqrt{EE_t}}{a \sqrt{3(1 - \nu^2)}} \qquad (11\text{-}4)$$

By taking a series of values of σ_{cr} and determining the values of E_t from the compression-test diagram (see Art. 3.3), we can calculate the corresponding values of the ratio a/h from Eq. (11-4).

In the case of buckling beyond the proportional limit, the expression for the length of a half-wave becomes

$$\frac{l}{m} = \pi \sqrt[4]{\frac{a^2 h^2}{12(1 - \nu^2)}} \sqrt[4]{\frac{E_t}{E}} \approx 1.72 \sqrt{ah} \sqrt{\frac{E_t}{E}} \qquad (11\text{-}5)$$

---

[1] The value of E in the last term on the left-hand side of Eq. (11-3) is assumed unchanged.

Thus the length of waves becomes shorter for buckling beyond the proportional limit.

If we assume that the mechanical properties of the material beyond the proportional limit are the same in the axial and circumferential directions and if we introduce E_t instead of E in the first and the third terms of Eq. (11-3), we find that

$$\sigma_{cr} = \frac{E_t h}{a \sqrt{3(1 - \nu^2)}} \tag{11-6}$$

and that beyond the proportional limit the length of waves remains unchanged.

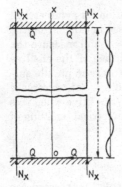

FIG. 11-2

In experimenting with cylindrical shells, the compression is usually applied by the rigid blocks of the testing machine and lateral expansion of the shell is prevented by friction. Then, instead of a stability problem, we have a problem involving the simultaneous action of compressive and bending forces[1] as shown in Fig. 11-2. Assuming that the compressive stress is smaller than the critical stress given by Eq. (11-3) and using the notation

$$\frac{N_x}{N_{cr}} = t$$

the general solution of Eq. (11-3) can be put in the following form:

$$w = C_1 e^{\alpha x} \sin (\beta x + \gamma_1) + C_2 e^{-\alpha x} \sin (\beta x + \gamma_2) \tag{i}$$

where $\alpha = \sqrt{1 - t} \sqrt[4]{\dfrac{Eh}{4a^2 D}} \qquad \beta = \sqrt{1 + t} \sqrt[4]{\dfrac{Eh}{4a^2 D}}$

and where C_1, C_2, γ_1, and γ_2 are four constants of integration which are to be determined in each particular case from the conditions at the edges. Assuming that the edges are simply supported, we have

$$\frac{\partial^2 w}{\partial x^2} = 0 \qquad \text{for } x = 0 \text{ and } x = l \tag{j}$$

$$w = \frac{\nu N_x a}{Eh} \qquad \text{for } x = 0 \text{ and } x = l \tag{k}$$

The second of these conditions states that the friction forces Q suppress entirely the lateral expansion of the shell at the edges. If the cylinder is not short and the load is not close to its critical value, we can put, in the general solution (i), $C_1 = 0$,[2] when considering the end $x = 0$ of the cylinder. Then we obtain a deflection in the form of waves which are rapidly damped out owing to the presence of the factor $e^{-\alpha x}$. The length of waves is

$$L = \frac{2\pi}{\beta} = \frac{2\pi}{\sqrt{1 + t}} \sqrt[4]{\frac{4a^2 D}{Eh}} \tag{l}$$

This length is somewhat larger than that obtained for the case of buckling [Eq. (11-2)] and approaches the latter when t approaches unity. When the load approaches its

[1] This problem of simultaneous compression and bending has been discussed by L. Föppl, *Sitzsber. math.-physik. Kl. bayer. Akad. Wiss., München,* 1926, p. 27, and by J. W. Geckeler, *Z. angew. Math. u. Mech.,* vol. 8, p. 341, 1928.

[2] See Timoshenko, "Strength of Materials," 3d ed., part 2, p. 2, D. Van Nostrand Company, Inc., Princeton, N.J., 1956.

critical value and t approaches unity, the factor $e^{-\alpha z}$ approaches unity also. The waves are no longer rapidly damped out; we cannot treat the edge conditions separately for each end and must consider all four constants in the expression (t). The deflection form is as shown in Fig. 11-2. The maximum deflection increases rapidly as the load approaches its critical value and failure occurs owing to yielding of the material at the crests of the waves nearest to the blocks of the testing machine. When the first half-wave, owing to plastic deformation, flattens out, the second half-wave begins to grow rapidly, and so on. We finally get the kind of deformation shown in the photograph[1] (Fig. 11-3). Such deformation is usually obtained with thicker tubes in which buckling occurs beyond the proportional limit. In the case of thin tubes, buckling which is nonsymmetrical with respect to the axis usually occurs (see Art. 11.4).

11.2. Inextensional Forms of Bending of Cylindrical Shells Due to Instability. If the edges of a uniformly axially compressed cylindrical shell are free to move

FIG. 11-3

laterally, an inextensional form of lateral bending may occur. Using expressions (e) of Art. 10.3, we take the following displacements[2] for buckling:

$$u = - \sum_{n=2}^{\infty} \frac{a}{n} b_n \sin (n\theta + \beta_n)$$

$$v = x \sum_{n=2}^{\infty} b_n \cos (n\theta + \beta_n) \tag{a}$$

$$w = -x \sum_{n=2}^{\infty} n b_n \sin (n\theta + \beta_n)$$

The critical value of compressive forces N_{cr} can be obtained now by using the energy method. During deformation as given by expressions (a) the strain energy of the shell increases by the amount [see Eq. (10-9)]

$$\Delta U = \pi Dl \sum_{n=2}^{\infty} b_n{}^2 \frac{(n^2 - 1)^2}{a^3} \left[\frac{n^2 l^2}{3} + 2(1 - \nu) a^2 \right] \tag{b}$$

---

[1] The photograph is taken from Geckeler, *op. cit.*

[2] The coefficients b_n in these expressions are equal to $\sqrt{b_n{}^2 + (b_n')^2}$ of Art. 10.3 and $\tan \beta_n = b_n'/b_n$

In calculating the work done by the compressive forces during bending, we note that, owing to displacements (a), the generators of the cylindrical shell become inclined to the x axis by the angle

$$\frac{1}{x}\sqrt{v^2 + w^2}$$

Then the work done is

$$\Delta T = N_{cr}\frac{l}{2}\int_0^{2\pi}\frac{v^2 + w^2}{x^2}\, a\, d\theta = \frac{\pi}{2}\, alN_{cr}\sum_{n=2}^{\infty} b_n^2(1 + n^2) \qquad (c)$$

Equating this work to the increase in strain energy (b), we obtain

$$N_{cr} = \frac{2D}{a^4}\frac{\displaystyle\sum_{n=2}^{\infty} b_n^2(n^2 - 1)^2\left[\frac{n^2l^2}{3} + 2(1 - \nu)a^2\right]}{\displaystyle\sum_{n=2}^{\infty} b_n^2(1 + n^2)}$$

The smallest value of N_{cr} is obtained by taking only one term, with $n = 2$, in the series, i.e., by assuming that the cross section of the shell flattens to an elliptical shape. Then

$$\sigma_{cr} = \frac{N_{cr}}{h} = \frac{2D}{a^4h}\frac{9}{5}\left[\frac{4}{3}l^2 + 2(1 - \nu)a^2\right] = \frac{3Eh^2}{10(1 - \nu^2)a^2}\left[\frac{4}{3}\frac{l^2}{a^2} + 2(1 - \nu)\right]$$

It can be seen that this critical stress is smaller than that obtained for a symmetrical form of buckling [Eq. (11-1)] provided that the ratio h/a is small and that the ratio l/a is not large.[1]

11.3. Buckling of a Cylindrical Shell under the Action of Uniform Axial Pressure. *General Case.* Although the case of symmetrical buckling of an axially compressed cylindrical shell has been discussed before (see Art. 11.1), we shall consider here a more general case by using Eqs. (10-13) and (10-14).[2] Assuming in this case that all resultant forces except N_z (Fig. 11-1) are very small and neglecting the products of these forces with the derivatives of the displacements u, v, and w, which are also small, we obtain from Eqs. (10-13)

$$a\frac{\partial N_z}{\partial x} + \frac{\partial N_{yz}}{\partial \theta} = 0$$

$$\frac{\partial N_y}{\partial \theta} + a\frac{\partial N_{xy}}{\partial x} + aN_z\frac{\partial^2 v}{\partial x^2} - Q_y = 0 \qquad (a)$$

$$a\frac{\partial Q_x}{\partial x} + \frac{\partial Q_y}{\partial \theta} + aN_z\frac{\partial^2 w}{\partial x^2} + N_y = 0$$

From Eqs. (10-14), neglecting the products of moments and derivatives

---

[1] See Timoshenko, *Z. Math. Physik*, vol. 58, p. 378, 1910.

[2] See R. Lorenz, *Physik. Z.*, vol. 13, p. 241, 1911; R. V. Southwell, *Phil. Trans. Roy. Soc., London,* series A, vol. 213, p. 187, 1914; and Timoshenko, *Bull. Electrotech. Inst., St. Petersburg*, vol. 11, 1914. The following discussion is taken from the latter paper.

of the displacements u, v, and w, we obtain

$$Q_x = \frac{\partial M_x}{\partial x} + \frac{\partial M_{yx}}{a \, \partial \theta}$$

$$Q_v = \frac{\partial M_y}{a \, \partial \theta} - \frac{\partial M_{xy}}{\partial x} \qquad (b)$$

Substituting these in Eqs. (a), the three equations of equilibrium for buckling of an axially compressed cylindrical shell become

$$a \frac{\partial N_x}{\partial x} + \frac{\partial N_{yx}}{\partial \theta} = 0$$

$$\frac{\partial N_y}{\partial \theta} + a \frac{\partial N_{xy}}{\partial x} + a N_x \frac{\partial^2 v}{\partial x^2} + \frac{\partial M_{xy}}{\partial x} - \frac{\partial M_y}{a \, \partial \theta} = 0 \qquad (c)$$

$$a N_x \frac{\partial^2 w}{\partial x^2} + N_v + a \frac{\partial^2 M_x}{\partial x^2} + \frac{\partial^2 M_{yx}}{\partial x \, \partial \theta} + \frac{\partial^2 M_y}{a \, \partial \theta^2} - \frac{\partial^2 M_{xy}}{\partial x \, \partial \theta} = 0$$

All resultant forces and moments entering in these equations can be expressed in terms of the displacements u, v, and w [see definitions of Art. 10.1 and Eqs. (10-15)], the positive directions of which are shown in Fig. 10-4. By taking compressive stress as positive and using the notations

$$\frac{h^2}{12a^2} = \alpha \qquad \frac{N_x(1 - \nu^2)}{Eh} = \phi \qquad (d)$$

we finally obtain the following equations:

$$\frac{\partial^2 u}{\partial x^2} + \frac{1 + \nu}{2a} \frac{\partial^2 v}{\partial x \, \partial \theta} - \frac{\nu}{a} \frac{\partial w}{\partial x} + \frac{1 - \nu}{2} \frac{\partial^2 u}{a^2 \, \partial \theta^2} = 0$$

$$\frac{1 + \nu}{2} \frac{\partial^2 u}{\partial x \, \partial \theta} + \frac{a(1 - \nu)}{2} \frac{\partial^2 v}{\partial x^2} + \frac{\partial^2 v}{a \, \partial \theta^2} - \frac{\partial w}{a \, \partial \theta}$$

$$+ \alpha \left[\frac{\partial^2 v}{a \, \partial \theta^2} + \frac{\partial^3 w}{a \, \partial \theta^3} + a \frac{\partial^3 w}{\partial x^2 \, \partial \theta} + a(1 - \nu) \frac{\partial^2 v}{\partial x^2} \right] - a\phi \frac{\partial^2 v}{\partial x^2} = 0 \qquad (11\text{-}7)$$

$$- a\phi \frac{\partial^2 w}{\partial x^2} + \nu \frac{\partial u}{\partial x} + \frac{\partial v}{a \, \partial \theta} - \frac{w}{a}$$

$$- \alpha \left[\frac{\partial^3 v}{a \, \partial \theta^3} + (2 - \nu)a \frac{\partial^3 v}{\partial x^2 \, \partial \theta} + a^3 \frac{\partial^4 w}{\partial x^4} + \frac{\partial^4 w}{a \, \partial \theta^4} + 2a \frac{\partial^4 w}{\partial x^2 \, \partial \theta^2} \right] = 0$$

These equations will be satisfied by assuming that

$$v = 0 \qquad \nu \frac{\partial u}{\partial x} = \frac{w}{a} = \text{const} \qquad (e)$$

This solution represents the cylindrical form of equilibrium in which the compressed shell expands uniformly in the lateral direction.

Another solution is obtained by assuming that $v = 0$ and that u and w are functions of x only. In this manner we obtain the case of buckling symmetrical with respect to the axis of the cylinder which was discussed in Art. 11.1.

Considering now the general solution of Eqs. (11-7), we assume that u, v, and w in these equations represent very small displacements from the cylindrical compressed form of equilibrium mentioned above [Eqs. (e)]. With the origin of coordinates at one end of the shell, and using, as before, the notations a and l for the radius and length of the shell, respectively, we take the solution of Eqs. (11-7) in the following form

$$u = A \sin n\theta \cos \frac{m\pi x}{l}$$

$$v = B \cos n\theta \sin \frac{m\pi x}{l} \qquad (f)$$

$$w = C \sin n\theta \sin \frac{m\pi x}{l}$$

which assumes that during buckling the generators of the shell subdivide into m half-waves and the circumference into $2n$ half-waves. At the ends we have

$$w = 0 \qquad \text{and} \qquad \frac{d^2w}{dx^2} = 0$$

which are the conditions of simply supported edges. The results obtained in this way can be used also for other edge conditions, since these conditions have only a small effect on the magnitude of the critical load if the length of the cylinder is not small (say $l > 2a$).[1]

Substituting expressions (f) in Eqs. (11-7) and using the notation

$$\frac{m\pi a}{l} = \lambda \qquad (g)$$

we obtain the following equations:

$$A\left(\lambda^2 + \frac{1 - \nu}{2}n^2\right) + B\frac{n(1 + \nu)\lambda}{2} + C\nu\lambda = 0$$

$$A\frac{n(1 + \nu)\lambda}{2} + B\left[\frac{(1 - \nu)\lambda^2}{2} + n^2 + \alpha(1 - \nu)\lambda^2 + \alpha n^2 - \lambda^2\phi\right] \qquad (h)$$
$$+ C[n + \alpha n(n^2 + \lambda^2)] = 0$$
$$A\nu\lambda + Bn\{1 + \alpha[n^2 + (2 - \nu)\lambda^2]\} + C[1 - \lambda^2\phi + \alpha(\lambda^2 + n^2)^2] = 0$$

Equating the determinant of these equations to zero and neglecting small quantities of higher order containing α^2 and ϕ^2 as factors, we obtain

$$\phi = \frac{N_x(1 - \nu^2)}{Eh} = \frac{R}{S} \qquad (i)$$

[1] Experiments show that the effect of edge conditions remains small also for shorter cylinders; see L. H. Donnell, *Trans. ASME*, vol. 56, p. 795, 1934.

where

$$R = (1 - \nu^2)\lambda^4 + \alpha[(n^2 + \lambda^2)^4 - (2 + \nu)(3 - \nu)\lambda^4 n^2 \\ + 2\lambda^4(1 - \nu^2) - \lambda^2 n^4(7 + \nu) + \lambda^2 n^2(3 + \nu) + n^4 - 2n^6]$$

and

$$S = \lambda^2 \left\{ (n^2 + \lambda^2)^2 + \frac{2}{1 - \nu}\left(\lambda^2 + \frac{1 - \nu}{2}n^2\right)[1 + \alpha(n^2 + \lambda^2)^2] \\ - \frac{2\nu^2\lambda^2}{1 - \nu} + \frac{2\alpha}{1 - \nu}\left(\lambda^2 + \frac{1 - \nu}{2}n^2\right)[n^2 + (1 - \nu)\lambda^2] \right\}$$

Experiments show (see Fig. 11-6) that thin cylindrical shells under compression usually buckle into short longitudinal waves so that λ^2 is a large number. Then, by keeping only the first term in the brackets of the numerator R and the first term in the denominator S, we can represent expression (i) in the following simplified form:

$$\phi = \frac{N_x(1 - \nu^2)}{Eh} = \alpha \frac{(n^2 + \lambda^2)^2}{\lambda^2} + \frac{(1 - \nu^2)\lambda^2}{(n^2 + \lambda^2)^2} \qquad (11\text{-}8)$$

For $n = 0$, Eq. (11-8) coincides with Eq. (h), Art. 11.1, which was obtained for the symmetrical type of buckling.

We obtain the smallest value of expression (11-8) when

$$\frac{(n^2 + \lambda^2)^2}{\lambda^2} = \sqrt{\frac{1 - \nu^2}{\alpha}} = \frac{2a}{h}\sqrt{3(1 - \nu^2)} \qquad (j)$$

in which case Eq. (11-8) gives

$$\phi = \frac{N_x(1 - \nu^2)}{Eh} = 2\sqrt{\alpha(1 - \nu^2)}$$

from which

$$\sigma_{cr} = \frac{(N_x)_{cr}}{h} = \frac{Eh}{a\sqrt{3(1 - \nu^2)}} \qquad (11\text{-}9)$$

This shows that the critical stress does not depend on the length l of the cylinder and is of the same magnitude as in the case of symmetrical buckling [see Eq. (11-1)]. The number of lobes into which the circumference subdivides during buckling remains indefinite as long as we consider λ^2 and n^2 as large numbers and determine the minimum value of expression (11-8) by treating it as a continuous function of $(n^2 + \lambda^2)^2/\lambda^2$. For any value of λ^2 smaller than $2a\sqrt{3(1 - \nu^2)}/h$ we find, from Eq. (j), the corresponding value of n^2.

In the case of shorter cylinders we cannot assume that λ is varying continuously; some additional discussion of expression (11-8) is necessary. If the cylindrical shell is so short that

$$\left(\frac{\pi a}{l}\right)^2 > \frac{2a}{h}\sqrt{3(1 - \nu^2)}$$

there will be formed during buckling only one half-wave in the axial direction and the smallest value of expression (11-8) is obtained by taking $n = 0$. That is, in such a case the form of buckling of the shell is symmetrical with respect to the axis of the cylinder. By taking the length of the cylinder shorter and shorter, we shall find that the second term in expression (11-8) becomes smaller and smaller in comparison with the first; by neglecting it we obtain, with $n = 0$,

$$\phi = \alpha\lambda^2 \qquad (k)$$

from which, by using notations (d), we obtain

$$\sigma_{cr} = \frac{\pi^2 E h^2}{12(1 - \nu^2)l^2} \qquad (l)$$

This is Euler's formula for an elemental strip.

When the length of the cylinder is such that $(\pi a/l)^2$ is somewhat smaller than $2a\sqrt{3(1 - \nu^2)}/h$, we shall continue to have one half-wave in the axial direction but n will no longer be zero and several lobes will appear along the circumference. The number of lobes, making expression (11-8) a minimum, will increase with the length of the cylinder up to the limit when two half-waves in the axial direction will be formed and the form of buckling again becomes symmetrical with respect to the axis. With further increase of the length, circumferential lobes again appear, and so on. For longer cylinders there will be only a small fluctuation in the critical value of compressive stress between two consecutive values of λ, and we can assume that this stress always remains equal to that found for a long cylinder with symmetrical buckling.

This discussion is of practical interest only in the case of very small values of the ratio h/a, in which case the critical stress (11-9) is below the proportional limit of the material. For thicker tubes failure will occur owing to yielding of the material rather than instability.

In the case of long cylindrical shells, we may expect the generators to buckle into long waves. In such case the quantity λ may become small. Neglecting in the numerator of expression (i) all terms containing products of α with powers of λ higher than the second and in the denominator all powers of λ higher than the second, we can represent this expression in the following form:

$$\phi = \frac{(1 - \nu^2)\lambda^4 + \alpha\{(n^4 - n^2)^2 + \lambda^2[4n^6 - (7 + \nu)n^4 + (3 + \nu)n^2]\}}{\lambda^2 n^2(n^2 + 1)} \qquad (m)$$

By taking $n = 1$, we obtain from this expression:

$$\phi = \frac{1 - \nu^2}{2}\lambda^2$$

or, by using notations (d),

$$\sigma_{cr} = \frac{\pi^2 E a^2}{2l^2}$$

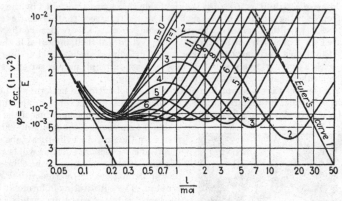

Fig. 11-4

This is Euler's formula for a strut, since $a^2/2$ is the square of the radius of gyration for the cross section of a thin tube. For $n = 1$ the cross sections remain circular (see p. 283) and the tube buckles as a strut.

For $n > 1$, neglecting additional small terms in expression (m), we obtain

$$\phi = \frac{(1 - \nu^2)\lambda^2}{n^2(n^2 + 1)} + \frac{\alpha n^2(n^2 - 1)^2}{\lambda^2(n^2 + 1)} \qquad (n)$$

The value of λ^2 which makes this expression a minimum is

$$\lambda^2 = \frac{\sqrt{\alpha}\, n^2(n^2 - 1)}{\sqrt{1 - \nu^2}} = \frac{hn^2(n^2 - 1)}{2a\sqrt{3(1 - \nu^2)}} \qquad (o)$$

and the corresponding value of the critical stress, from expression (m), is

$$\sigma_{cr} = \frac{Eh}{a\sqrt{3(1 - \nu^2)}} \frac{n^2 - 1}{n^2 + 1} \qquad (p)$$

This stress is smaller than that obtained for symmetrical buckling and its smallest value, for $n = 2$, is

$$\sigma_{cr} = \frac{3}{5} \frac{Eh}{a\sqrt{3(1 - \nu^2)}} \qquad (11\text{-}10)$$

As can be seen from expression (o), this buckling is characterized by comparatively long waves in the axial direction.[1]

If the ratio h/a is given, i.e., if α is known, we can, by using expression (i) and choosing a value for n representing the number of lobes of the buckled shell, obtain a curve representing the relation between λ and ϕ. Several curves[2] of this kind are shown in Fig. 11-4. The logarithmic values of $l/ma = \pi/\lambda$ are taken as abscissas, and the logarithmic values of $\phi = \sigma_{cr}(1 - \nu^2)/E$ are taken as ordinates. On the left side the curves approach, asymptotically, the inclined line representing the condition

---

[1] These waves were indicated by Southwell, loc. cit.

[2] These curves are taken from W. Flügge, "Statik und Dynamik der Schalen," Berlin, 1934. They are calculated for $\alpha = h^2/12a^2 = 10^{-5}$ and $\nu = \frac{1}{6}$ from an expression slightly different from expression (i).

of buckling of a strip [Eq. (l)].   On the right side the curves are limited by the curve $n = 1$ representing buckling of the shell as a strut.   It is seen that for shorter cylindrical shells the critical values of the compressive stress are always close to the value calculated for a symmetrical buckling of a long cylindrical shell, indicated in the figure by the dotted horizontal line.   For longer cylindrical shells, buckling into long waves with a comparatively small number of lobes [Eq. (p)] may occur at a stress which is smaller than that for a symmetrical buckling.

Noting the fact that a complete cylindrical shell submitted to uniform axial compression subdivides into a large number of small waves, we can conclude that in the case of a noncircular cylindrical shell buckling will start in those portions of the shell where the curvature is the smallest.   For instance, in the case of a cylindrical tube of an elliptical cross section, buckling will begin at the ends of the minor axis of the ellipse and an approximate value for the critical stress is obtained by substituting in formula (11-9) the larger radius of curvature instead of the radius a of a circular cylinder.

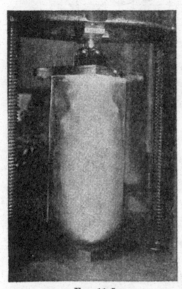

Fig. 11-5

11.4. Experiments with Cylindrical Shells in Axial Compression. From the discussion of the preceding article it is seen that only in the case of very thin shells will buckling occur within the elastic range in which the theoretical formulas can be applied.   Quite naturally, practically all the early experiments were made with comparatively thick tubes which fail, if longitudinally compressed, owing to yielding of the material and not to buckling.[1]   Later, in connection with the use of thin shells in aircraft structures, experiments were made with very thin cylindrical shells under axial pressure.[2]   Figure 11-5 shows a thin cylindrical shell in a compression-testing machine.   To realize a central application of the load, steel balls are used as shown.   The edges of the shell are welded to the end plates.   Owing to this additional constraint the edges are stiffened and buckling

---

[1] The first experiments with buckling of tubes were made by W. Fairbairn, "An Account of the Construction of the Britannia and Conway Tubular Bridges," London, 1849; see also E. Clark, "The Britannia and Conway Tubular Bridges," 2 vols., London, 1850.

[2] The first experiments of this kind and a comparison with theoretical formulas were made by Andrew Robertson, *Proc. Roy. Soc., London*, series A, vol. 121, p. 558, 1928.   Extensive experiments with thin cylindrical shells were made by E. E. Lundquist, *NACA Rept.* 473, 1933, and by L. H. Donnell, *Trans. ASME*, vol. 56, 1934. The results given in our discussion are taken from the latter paper.

usually occurs at some distance from the ends. Several examples of buckling of thin cylindrical shells of steel and brass are shown in Fig. 11-6. As should be expected from the theory, shells of such proportion as shown in the figure buckle in comparatively small waves. Usually the length of these waves in the axial and circumferential directions is about the same.

The results obtained in these experiments are shown in Fig. 11-7. The ratios of the radius to the wall thickness a/h are taken as abscissas, and

Fig. 11-6

the ratios of the ultimate compressive stress to the stress $Eh/[a\sqrt{3(1-\nu^2)}]$ calculated for short waves are taken as ordinates. The results obtained on duralumin shells[1] are denoted with crosses and those on steel and brass shells[2] with circles. It is seen that in all cases failure has occurred at a stress much lower than the theory predicts. In not one case was the ultimate stress more than 60 per cent of the theoretical. It is seen clearly that the ratio of the ultimate stress to the theoretical stress decreases as

[1] Tests by Lundquist, *op. cit.*
[2] Tests by Donnell, *op. cit.*

the ratio a/h increases; i.e., the discrepancy between experiment and theory is larger for thinner shells. To explain this discrepancy, a theory was advanced[1] which takes into account the initial deviations from the ideal cylindrical surface and considers bending of the shell due to this initial imperfection, assuming that deflections are not small. It was assumed also that the shells collapse when yielding of the material begins. Taking initial imperfections in the form of waves of equal length in the axial and circumferential directions in combination with waves of buckling symmetrical with respect to the axis,[2] it was found that the ultimate load, for a given value of the ratio $E/[\sigma_{YP} \sqrt{12(1 - \nu^2)}]$, can be represented

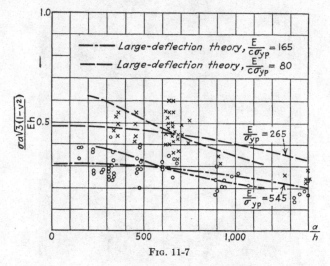

FIG. 11-7

as a function of the radius-thickness ratio a/h. The corresponding curves calculated for duralumin shells with $E/[\sigma_{YP} \sqrt{12(1 - \nu^2)}] = 165$ and for steel and brass shells with $E/[\sigma_{YP} \sqrt{12(1 - \nu^2)}] = 80$ are shown in Fig. 11-7 by dotted lines.[3] It can be seen that there is a satisfactory agreement between these curves and the test results.

On the basis of the existing experimental data, an empirical formula for calculating the ultimate strength of cylindrical shells under axial compression was developed. This formula takes into consideration the ratios a/h and E/σ_{YP} and gives

[1] See *ibid.*

[2] Donnell has chosen such a combination from a consideration of the energy of deformation. It should be noted that waves symmetrical with respect to the axis always are present owing to constraint at the ends preventing lateral expansion of the shell during compression (see Fig. 11-2).

[3] Letter c is used for $\sqrt{12(1 - \nu^2)}$ in the figure. The two steeper curves of the figure are here considered.

$$\sigma_{\text{ult}} = E \frac{0.6 \dfrac{h}{a} - 10^{-7} \dfrac{a}{h}}{1 + 0.004 \dfrac{E}{\sigma_{\text{YP}}}} \qquad (11\text{-}11)$$

For the shells tested, with $E/\sigma_{\text{YP}} = 545$ for steel and brass and

$$\frac{E}{\sigma_{\text{YP}}} = 265$$

for duralumin, the curves representing this formula are also shown in Fig. 11-7.

A further investigation of the reasons for the discrepancy between theory and experiment was made[1] by studying the postbuckling behavior of ideal compressed cylindrical shells. Proceeding as in the case of post-buckling behavior of plates (see Art. 9.13), the equation

$$\frac{\partial^4 F}{\partial x^4} + 2 \frac{\partial^4 F}{\partial x^2 \, \partial y^2} + \frac{\partial^4 F}{\partial y^4} = E \left[\left(\frac{\partial^2 w}{\partial x \, \partial y} \right)^2 - \frac{\partial^2 w}{\partial x^2} \frac{\partial^2 w}{\partial y^2} - \frac{1}{a} \frac{\partial^2 w}{\partial x^2} \right] \qquad (a)$$

was used, which differs from Eq. (8-38) for plates only by the last term on the right-hand side. Assuming buckling and taking an expression for w with several parameters and substituting it in Eq. (a), we can find the stress function F and calculate the stresses and strains of the middle surface (see Art. 8.7). Substituting them into Eqs. (10-15) we find the changes in curvatures, and using Eqs. (10-9) and (10-10) we find the total strain energy of the buckled shell. The energy of the external compressive force also can be written readily, since the axial strain of the shell is found already. Having the total energy U of the system, we now select the parameters, entering into the assumed expression for w, in such a way as to make this energy a minimum. In actual calculations[2] it was assumed that the buckled shell was subdivided into a number of circumferential and longitudinal waves and one of these waves was considered. On the basis of test results the shape of one wave was taken in the following form, the origin of coordinates being in the wave center:

$$\frac{w}{a} = \left(f_0 + \frac{1}{4} f_1 \right) + \frac{1}{2} f_1 \left(\cos \frac{mx}{a} \cos n\theta + \frac{1}{4} \cos \frac{2mx}{a} + \frac{1}{4} \cos 2n\theta \right)$$
$$+ \frac{1}{4} f_2 \left(\cos \frac{2mx}{a} + \cos 2n\theta \right) \qquad (b)$$

This expression contains five parameters, namely, the three quantities f_0, f_1, f_2 defining the radial displacements and the quantities m and n

[1] See T. V. Kármán and H. S. Tsien, *J. Aeronaut. Sci.*, vol. 8, p. 303, 1941. A discussion of the same problem by using the Ritz method of integration was made by P. Cicala, *Quart. Appl. Math.*, vol. 9, p. 273, 1951. See also P. M. Finkelstein, *Bull. Acad. Sci. Div. Tech. Sci.*, no. 7, p. 37, 1956 (Russian).

[2] Kármán and Tsien, *op. cit.*

defining the wave dimensions, $2\pi a/m$ in the longitudinal direction and $2\pi a/n$ in the circumferential direction. Equating to zero the derivative of the energy U with respect to f_0, we find that f_0 is such that the average of the circumferential stress σ_θ is equal to zero. To simplify the further calculations, the values of m/n, the *aspect ratio* of the waves, and the value of $\eta = n^2 h/a$, defining the number of waves in the circumferential direction, were assumed; then from the equations

$$\frac{\partial U}{\partial f_1} = 0 \qquad \frac{\partial U}{\partial f_2} = 0 \qquad (c)$$

the values of the parameters f_1 and f_2, as functions of the compressive force, can be calculated. In Fig. 11-8 the results of such calculations for

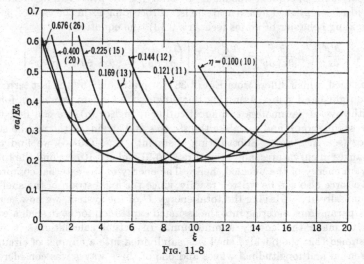

FIG. 11-8

the particular case $m = n$ and for several values of η are shown. The quantity ξ, equal to the ratio δ/h of the deflections to the shell thickness, is taken as abscissa, and the quantity $\sigma a/Eh$, proportional to the average axial compressive stress σ, is the ordinate. The numbers in parentheses give, for various values of η, the number of circumferential waves calculated for the case $a/h = 1,000$. When ξ approaches zero, Eqs. (c) give $f_2 = -\frac{1}{2}f_1$ and expression (b) gives the same wave pattern[1] as was assumed in Art. 11.3. We find also that for this case

$$\left(\frac{\sigma a}{Eh}\right)_{\min} = \frac{1}{\sqrt{3(1 - \nu^2)}} \qquad (d)$$

which again coincides with the result of the preceding article.

[1] Note that the origin of coordinates is different from that in Art. 11.3.

We see that to originate buckling in an ideal case we must have a compressive stress given by Eq. (d), but with an increase of the deflection the load required to keep the cylindrical shell in a buckled condition rapidly diminishes and approaches about one-third of the theoretically required buckling load. This explains why experimental results in the case of axial compression of cylindrical shells do not agree with the theory. In the case of thin experimental shells there always exist some initial deflections and other imperfections. For this reason bending starts when the load is small and the deflections reach the values at which further continuation of buckling requires a much smaller load than the theoretical buckling load for the ideal case.

In a further study[1] of the influence of imperfections on the process of buckling it was assumed that the cylindrical shell initially has waves of the form

$$w_0 = a_0 h \left(\cos \frac{mnx}{a} \cos n\theta + b \cos \frac{2mnx}{a} + c \cos 2n\theta + d \right) \qquad (e)$$

and the unit axial compression ϵ of the cylinder as a function of the average axial compressive stress σ was calculated. In these computations the constants b, c, d, m, n were selected so as to minimize the energy. The constant a_0 defining the magnitude of the initial imperfection was assumed to have the form

$$a_0 = \lambda_1 \frac{a^2}{h^2 m^{1.5} n^2}$$

where λ_1 is a small numerical factor. The results of the calculations for several values of λ_1 are given by curves in Fig. 11-9. In this figure σ_{cr} is the critical stress given by Eq. (11-9) of the preceding article and ϵ_{cr} is the corresponding simple unit compression. It is seen that the smallest imperfection reduces considerably the maximum load which the compressed cylindrical shell can carry. If, on the basis of actual measurements, the value of λ_1 is established, the corresponding curve in Fig. 11-9 can be used for determining the allowable compressive load.

All these investigations showed the great influence of various imperfections on the value of ultimate load which a compressed cylinder can carry. They showed also that theoretical investigations developed for an ideal cylindrical shell can be used only as a guide in developing empirical formulas based on numerous experiments.[2]

---

[1] See L. H. Donnell and C. C. Wan, *J. Appl. Mech.*, vol. 17, p. 73, 1950.

[2] A review of the corresponding publications can be found in Handbook of Structural Stability, *NACA Tech. Notes* 3781–3786, 1957–1958. See also Y. C. Fung and E. E. Sechler, Instability of Thin Elastic Shells, *Proc. First Symposium on Naval Structural Mechanics*, pp. 115–168, edited by J. N. Goodier and N. J. Hoff, Pergamon Press, New York, 1960.

11.5. Buckling of a Cylindrical Shell under the Action of Uniform External Lateral Pressure. In discussing buckling of a uniformly compressed circular ring (Art. 7.4), it was pointed out that the formula for the critical load obtained for a ring can be applied also in the case of cylindrical shells with free edges submitted to a uniform lateral pressure. The same formula can be applied also in the case of a shell with some constraint at the edges if the length of the shell is so large that the stiffening effect of any constraint at the edges can be neglected. If the length of a cylinder

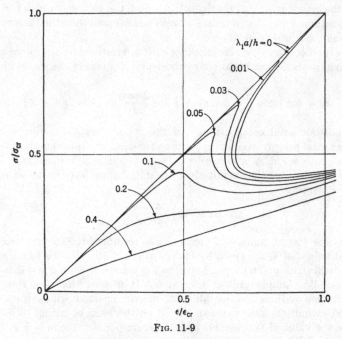

Fig. 11-9

is not very large in comparison with its diameter, we can no longer disregard the end conditions, and in calculating the intensity of lateral pressure at which buckling occurs[1] we must have recourse to the general equations of deformation of a cylindrical shell (Art. 10.4). Considering Eqs. (10-13), we assume that in the case of uniform lateral pressure all resultant forces, except N_y, are small and we neglect the terms containing the products of these resultants with the derivatives of the displacements

[1] The first investigations of this kind were made by R. Lorenz, *loc. cit.*, and R. V. Southwell, *Phil. Mag.*, vol. 25, p. 687, 1913, and *Phil. Trans. Roy. Soc., London*, series A, vol. 213, p. 187, 1914. A more accurate formula for the critical load was developed by R. von Mises, *Z. Ver. deut. Ingr.*, vol. 58, p. 750, 1914.

u, v, and w. In this manner we obtain

$$a \frac{\partial N_x}{\partial x} + \frac{\partial N_{yx}}{\partial \theta} - N_y \left(\frac{\partial^2 v}{\partial x \, \partial \theta} - \frac{\partial w}{\partial x} \right) = 0$$

$$\frac{\partial N_y}{\partial \theta} + a \frac{\partial N_{xy}}{\partial x} - Q_y = 0 \qquad (a)$$

$$a \frac{\partial Q_x}{\partial x} + \frac{\partial Q_y}{\partial \theta} + N_y \left(1 + \frac{\partial v}{a \, \partial \theta} + \frac{\partial^2 w}{a \, \partial \theta^2} \right) + qa = 0$$

In Eqs. (10-14) we assume that the bending and twisting moments are small and we neglect the products of these moments with the derivatives of the displacements u, v, and w. Then the first two of these equations give

$$Q_x = \frac{\partial M_{yx}}{a \, \partial \theta} + \frac{\partial M_x}{\partial x}$$

$$Q_y = \frac{\partial M_y}{a \, \partial \theta} - \frac{\partial M_{xy}}{\partial x} \qquad (b)$$

Substituting in Eqs. (a), we obtain

$$a \frac{\partial N_x}{\partial x} + \frac{\partial N_{yx}}{\partial \theta} - N_y \left(\frac{\partial^2 v}{\partial x \, \partial \theta} - \frac{\partial w}{\partial x} \right) = 0$$

$$\frac{\partial N_y}{\partial \theta} + a \frac{\partial N_{xy}}{\partial x} - \frac{\partial M_y}{a \, \partial \theta} + \frac{\partial M_{xy}}{\partial x} = 0$$

$$\frac{\partial^2 M_{yx}}{\partial x \, \partial \theta} + \frac{a \, \partial^2 M_x}{\partial x^2} + \frac{\partial^2 M_y}{a \, \partial \theta^2} - \frac{\partial^2 M_{xy}}{\partial x \, \partial \theta} \qquad (c)$$

$$+ N_y \left(1 + \frac{\partial v}{a \, \partial \theta} + \frac{\partial^2 w}{a \, \partial \theta^2} \right) + qa = 0$$

A particular solution of Eqs. (c) is obtained by assuming that under the action of uniform external pressure the circular cylindrical shell remains circular and undergoes only a uniform compression in the circumferential direction so that

$$v = 0 \qquad w = \frac{a^2 q}{E'h}$$

$$N_x = 0 \qquad N_y = -qa \qquad M_x = M_y = M_{xy} = 0$$

In discussing buckling of the shell we shall consider only small deflections from this uniformly compressed form of equilibrium, so that N_y in Eqs. (c) will differ but little from the value $-qa$, and we can put

$$N_y = -qa + N_y'$$

where N_y' is a small change in the resultant force $-qa$ corresponding to small displacements u, v, and w from the uniformly compressed cylindrical form of the shell.

We shall take into account also stretching of the middle surface of the shell during buckling and substitute $N_y(1 + \epsilon_1)$ and $q(1 + \epsilon_1)(1 + \epsilon_2)$ for N_y and q, respectively, in the second and third of Eqs. (c). Observing then that

$$\epsilon_1 = \frac{\partial u}{\partial x} \qquad \epsilon_2 = \frac{\partial v}{a\, \partial \theta} - \frac{w}{a}$$

we represent Eqs. (c) in the following form

$$a\,\frac{\partial N_x}{\partial x} + \frac{\partial N_{yx}}{\partial \theta} + qa\left(\frac{\partial^2 v}{\partial x\, \partial \theta} - \frac{\partial w}{\partial x}\right) = 0$$

$$\frac{\partial N_y'}{\partial \theta} + a\,\frac{\partial N_{xy}}{\partial x} - \frac{\partial M_y}{a\, \partial \theta} + \frac{\partial M_{xy}}{\partial x} = 0 \qquad (d)$$

$$\frac{\partial^2 M_{yx}}{\partial x\, \partial \theta} + a\,\frac{\partial^2 M_x}{\partial x^2} + \frac{\partial^2 M_y}{a\, \partial \theta^2} - \frac{\partial^2 M_{xy}}{\partial x\, \partial \theta} + N_y' - q\left(w + \frac{\partial^2 w}{\partial \theta^2}\right) = 0$$

By using the formulas of Art. 10.1 and Eqs. (10-15) we express now all resultant forces and moments in terms of the displacements u, v, and w. Substituting such expressions in Eqs. (d) and using the notations

$$\frac{qa(1 - \nu^2)}{Eh} = \phi \qquad \text{and} \qquad \frac{h^2}{12a^2} = \alpha \qquad (e)$$

we obtain

$$a^2\,\frac{\partial^2 u}{\partial x^2} + \frac{1 + \nu}{2}\,\frac{a\, \partial^2 v}{\partial x\, \partial \theta} - \nu a\,\frac{\partial w}{\partial x} + a\phi\left(\frac{\partial^2 v}{\partial x\, \partial \theta} - \frac{\partial w}{\partial x}\right)$$
$$+ \frac{1 - \nu}{2}\,\frac{\partial^2 u}{\partial \theta^2} = 0$$

$$\frac{1 + \nu}{2}\,\frac{a\, \partial^2 u}{\partial x\, \partial \theta} + \frac{1 - \nu}{2}\,a^2\,\frac{\partial^2 v}{\partial x^2} + \frac{\partial^2 v}{\partial \theta^2} - \frac{\partial w}{\partial \theta}$$
$$+ \alpha\left[\frac{\partial^2 v}{\partial \theta^2} + \frac{\partial^3 w}{\partial \theta^3} + a^2\,\frac{\partial^3 w}{\partial x^2\, \partial \theta} + a^2(1 - \nu)\,\frac{\partial^2 v}{\partial x^2}\right] = 0 \qquad (f)$$

$$a\nu\,\frac{\partial u}{\partial x} + \frac{\partial v}{\partial \theta} - w - \alpha\left[\frac{\partial^3 v}{\partial \theta^3} + (2 - \nu)a^2\,\frac{\partial^3 v}{\partial x^2\, \partial \theta} + a^4\,\frac{\partial^4 w}{\partial x^4}\right.$$
$$\left. + \frac{\partial^4 w}{\partial \theta^4} + 2a^2\,\frac{\partial^4 w}{\partial x^2\, \partial \theta^2}\right] = \phi\left(w + \frac{\partial^2 w}{\partial \theta^2}\right)$$

Thus the problem of determining the critical value of the lateral pressure is reduced to solving the above three differential equations and satisfying the boundary conditions. If the edges of the shell are simply supported, the boundary conditions require that w and $\partial^2 w/\partial x^2$ become zero at the ends. Assuming that the length of the cylinder is l and that x is measured from the middle cross section of the shell, we obtain a solution of Eqs. (f) satisfying the boundary conditions by taking for the displacements during buckling the following expressions:

$$u = A \sin n\theta \sin \frac{\pi x}{l}$$

$$v = B \cos n\theta \cos \frac{\pi x}{l} \qquad (g)$$

$$w = C \sin n\theta \cos \frac{\pi x}{l}$$

which show that during buckling the generators of the shell deflect to one half-wave of a sine curve while the circumference is subdivided into $2n$ half-waves. At the ends the displacements w and the derivative $\partial^2 w/\partial x^2$ both are zero, which represents the conditions of simply supported edges.

In actual cases the edges of the shell will usually be fastened to the supports before the uniform load q is applied; thus, in addition to uniform compression of the shell, assumed in our discussion, local bending at the edges will be produced. Assuming that the supports are absolutely rigid, the local bending will be of the type shown in Fig. 11-10. From the discussion of Art. 11.1 we conclude that the deflection curve of a generator is

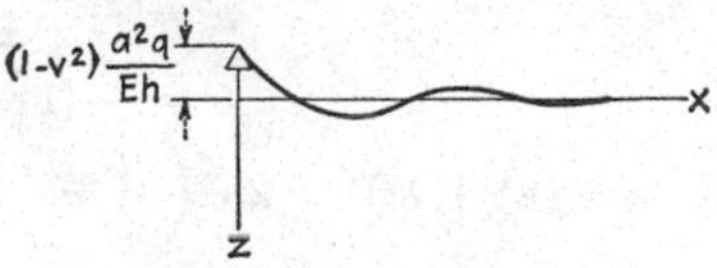

$$w = \frac{a^2 q}{Eh}(1 - \nu^2)e^{-\beta x} \cos \beta x$$

where $\beta = \sqrt[4]{\dfrac{3(1 - \nu^2)}{a^2 h^2}}$

Fig. 11-10

It has a wavy form in which the waves are rapidly damped out. The length of waves is

$$L = \frac{2\pi}{\beta} = 4.90 \sqrt{ah}$$

In the case of thin shells, h is small in comparison with a and the wavelength is usually much smaller than the radius, so that for a cylinder, the length of which is several times its radius, bending at the edges can be considered as a local factor that has no serious effect on the magnitude of the critical load. In the same manner, in the case of built-in edges, bending produced in the shell by bending moments M_x can be discussed and it can be shown that for longer cylinders there will not be much difference in the critical loads for simply supported or for built-in edges.

Substituting expressions (g) for the displacements in Eqs. (f) and using the notation $\lambda = \pi a/l$, we obtain the following equations:

$$A\left(-\lambda^2 - \frac{1 - \nu}{2}n^2\right) + B\left(\frac{1 + \nu}{2}n\lambda + n\lambda\phi\right) + C(\nu + \phi)\lambda = 0$$

$$A\left(\frac{1 + \nu}{2}n\lambda\right) - B\left[\frac{1 - \nu}{2}\lambda^2 + n^2 + n^2\alpha + \alpha(1 - \nu)\lambda^2\right]$$

$$- C(n + \alpha n^3 + \alpha n\lambda^2) = 0 \qquad (h)$$

$$A(\nu\lambda) - B[n + \alpha n^3 + (2 - \nu)\alpha n\lambda^2] - C[1 + \alpha\lambda^4 + \alpha n^4$$

$$+ 2\alpha n^2\lambda^2 + \phi(1 - n^2)] = 0$$

These equations can be satisfied by putting A, B, and C equal to zero, which corresponds to a uniformly compressed circular form of equilibrium of the shell. A buckled form of equilibrium becomes possible only if Eqs. (h) yield for A, B, and C solutions different from zero; this requires that the determinant of these equations becomes zero. In this manner the equation for determining the critical load is obtained. This equation has the form

$$\phi(D + E\alpha + F\phi) = G + H\alpha + K\alpha^2 \tag{i}$$

in which $D, E, \ldots , K$ have the following meanings:

$$D = (1 - n^2)(n^2 + \lambda^2)^2 - \nu\lambda^4$$

$$E = (1 - n^2)\left(n^2 + \frac{2\lambda^2}{1 - \nu}\right)[n^2 + (1 - \nu)\lambda^2] + \frac{1 + 3\nu}{1 - \nu}\, n^4\lambda^2$$

$$+ \frac{2 + 3\nu - \nu^2}{1 - \nu}\, n^2\lambda^4 - \frac{2\nu n^2\lambda^2}{1 - \nu} - 2\nu\lambda^4 - \frac{1 + \nu}{1 - \nu}\, n^2\lambda^2\,(\lambda^2 + n^2)^2$$

$$F = -\frac{1 + \nu}{1 - \nu}\,(1 - n^2)n^2\lambda^2$$

$$G = -(1 - \nu^2)\lambda^4$$

$$H = -(n^2 + \lambda^2)^4 + 2n^2\left(n^2 + \frac{3 - \nu}{2}\,\lambda^2\right)[n^2 + (2 + \nu)\lambda^2]$$

$$- [n^2 + (1 - \nu)\lambda^2][n^2 + 2(1 + \nu)\lambda^2]$$

$$K = -\lambda^4(n^2 + \lambda^2)[n^2(1 - \nu) + 2\lambda^2]$$

After omitting the small terms, which have very little effect on the magnitude of the critical pressure, and substituting for α, ϕ, and λ their expressions, we can put Eq. (i) in the following form:[1]

$$\frac{(1 - \nu^2)q_{cr}a}{Eh} = \frac{1 - \nu^2}{(n^2 - 1)(1 + n^2l^2/\pi^2a^2)}$$

$$+ \frac{h^2}{12a^2}\left(n^2 - 1 + \frac{2n^2 - 1 - \nu}{1 + n^2l^2/\pi^2a^2}\right) \tag{11-12}$$

When the shell is very long, l/a is a large number; neglecting, in Eq. (11-12), the terms containing the square of this ratio in the denominator, we obtain

$$q_{cr} = \frac{Eh^3(n^2 - 1)}{12a^3(1 - \nu^2)}$$

which, with $n = 2$, coincides with our previous result [see Eq. (7-15)]. In calculating the critical load for shorter cylinders, using our previous notations (e), Eq. (11-12) represents a linear relation between the quantities α and ϕ. For a given value of l/a, a chosen value of the number n, and with α and ϕ as coordinates, we obtain a straight line. By taking $n = 2, 3, 4, \ldots$, a system of lines is obtained which, for each value of

[1] This is the equation obtained by von Mises, *op. cit.*

a/l, forms a broken line. Several such lines are shown in Fig. 11-11. Having these lines, the magnitude of the critical load can be readily determined in each particular case. It is seen that the number of waves n, into which the shell buckles, increases as the length and the thickness of the shell decrease.

The results of calculations, by using Eq. (11-12), can be represented in another way by taking as abscissas the values of the ratio l/a and as ordinates the quantities $(1 - \nu^2)a(q_{cr}/Eh)$. Then, for each value of

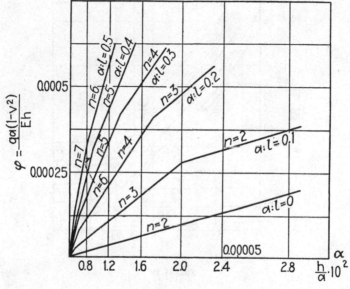

FIG. 11-11

$h^2/12a^2$, a line is obtained which is formed by portions of curves constructed for various values of n. Several lines of this kind are shown in Fig. 11-12.[1] This manner of representation of the results is similar to that used before in discussing buckling of plates (see Art. 9.2). It is seen that for shorter tubes the critical load increases rapidly as the ratio l/a decreases. For long tubes, say for $l/a > 50$, the critical load does not depend on the length and is equal to the value given by Eq. (7-15) derived for the case of an infinitely long tube.

[1] These curves are taken from Flügge, *loc. cit.* They are calculated by using an equation which is somewhat different from Eq. (11-12) and by taking $\nu = \frac{1}{6}$. Calculations show that the difference between the results obtained from the two equations mentioned is small and can be disregarded in practical applications.

The results obtained from Eq. (11-12) represent the correct values of the critical load only as long as the calculated compressive stresses are within the elastic region. Beyond the proportional limit Eq. (11-12) gives exaggerated values for q_{cr}; to obtain satisfactory results the tangent modulus E_t, instead of E, should be used. When the compressive stress approaches the yield-point stress, the value of E_t decreases rapidly so that the load producing a yield-point stress must be considered as the ultimate load for materials with a pronounced yield point.

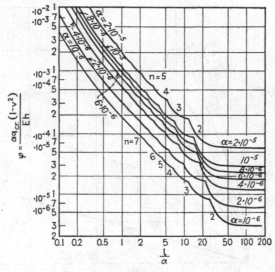

Fig. 11-12

In the case of steel cylindrical shells, the curves given in Fig. 11-13 can be used[1] for determining the critical value of the load or for determining the necessary thickness of the wall, if the ratio of length to diameter $l/2a$ is given and if a suitable factor of safety is chosen. The right-hand portion of these curves is calculated by using Eq. (11-12) and by taking $E = 30 \cdot 10^6$ psi and $\nu = 0.3$. On the left-hand side we have a system of parallel lines corresponding to compressive stresses equal to the yield-point stress, which in this case was taken very low (26,000 psi). Analogous curves can be constructed for any other material.

These curves can be used, not only in the case of shells with simply supported edges, but also in the case of built-in edges, since the manner of edge constraint is not of much influence on q_{cr}, and again in the case of a long shell stiffened by rings, provided the flexural rigidity of these rings is such that they can carry the lateral load alone without buckling. In such case l is the distance between rings.

The case in which a cylindrical shell with supported edges has some initial curva-

[1] These curves have been prepared by the ASME Research Committee on the Strength of Vessels under External Pressure, December, 1933.

ture is also of practical importance. One particular case of this kind has been discussed by Westergaard.[1]

A considerable number of experiments have been made with buckling of cylindrical shells under uniform lateral pressure.[2] One experimental series of this kind, together with a comparison of the experimental results with the theory developed by R. V. Southwell,[3] is given in the paper by G. Cook.[4] These experiments show that Southwell's theory, considering buckling of shorter tubes into a number of lobes, is in better

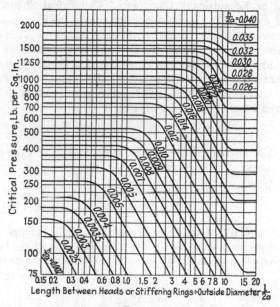

FIG. 11-13

agreement with experiments than the older formula developed by Carman,[5] which states that the critical pressure for shorter tubes is

$$q_{cr} = \frac{L}{l} (q_{cr})_0$$

In this formula l is the length of the tube, L is the so-called critical length, which by Carman is assumed equal to six diameters of the tube, and $(q_{cr})_0$ is the critical pressure for an infinitely long tube.

[1] H. M. Westergaard, *Reports 4th Intern. Congr. Appl. Mech.*, 1934, p. 274. See also the paper by R. D. Johnson presented before the Engineering Institute of Canada, Feb. 9, 1935.

[2] A bibliography on this subject was published by Gilbert Cook, *Brit. Assoc. Advancement Sci., Repts.*, Birmingham, 1913.

[3] Southwell, *loc. cit.*

[4] See G. Cook, *Phil. Mag.*, 6th series, vol. 28, p. 51, 1914.

[5] A. P. Carman, *Phys. Rev.*, vol. 21, p. 381, 1905.

A study of the effect of imperfections on the magnitude of the critical pressure was made by Donnell.[1] Using the same method as in the case of axially compressed cylindrical shells it was shown that imperfections do not produce so great a reduction in critical pressure as was found for the case of axial compression. This explains why a better agreement between theory and experiments is obtained for the case of lateral pressure.

11.6. Bent or Eccentrically Compressed Cylindrical Shells. In the discussion of Art. 11.3 it was assumed always that the force compressing the shell is centrally applied and that the resultant force N_x is constant. If the compressive force is applied with some eccentricity, we obtain a combined compression and bending of a cylindrical shell. Denoting by θ the angle that an axial plane makes with the plane in which bending occurs, the resultant forces N_x in the axial direction are no longer constant and can be represented by the formula[2]

$$N_x = -(N_0 + N_1 \cos \theta) \tag{a}$$

in which N_0/h is the uniform compressive stress and N_1/h is the maximum compressive stress due to bending. Then the maximum compressive stress is

$$\sigma = \frac{1}{h}(N_0 + N_1) \tag{b}$$

Considering Eqs. (11-7) of Art. 11.3 and noting that N_x is no longer constant, it is seen that we cannot use, as a solution of these equations, the expressions (f) of Art. 11.3 but must have recourse to the more general expressions

$$u = \cos \frac{m\pi x}{l} \sum_0^\infty A_n \cos n\theta$$

$$v = \sin \frac{m\pi x}{l} \sum_0^\infty B_n \sin n\theta \tag{c}$$

$$w = \sin \frac{m\pi x}{l} \sum_0^\infty C_n \cos n\theta$$

which imply that in a circumferential direction we have no longer simple sinusoidal waves, and that a more complicated form of buckling occurs. Because of the presence of the infinite series, the calculation of the critical

---

[1] L. H. Donnell, *J. Appl. Mech.*, vol. 23, p. 569, 1956. See also G. D. Galletly and R. Bart, *J. Appl. Mech.*, vol. 23, p. 351, 1956. References to new experiments with ring-stiffened cylindrical shells are given in the latter paper.

[2] N_x is taken positive in tension.

value of the maximum compressive stress (b) becomes much more complicated than before. This calculation[1] shows that we can get a satisfactory
approximation and are on the safe side by assuming, for any ratio N_0/N_1,
that buckling occurs when the maximum compressive stress becomes equal
to the critical stress calculated for symmetrical buckling. Thus

$$\frac{1}{h}\,(N_0 + N_1)_{cr} = \frac{Eh}{a\,\sqrt{3(1 - \nu^2)}} \tag{d}$$

In Fig. 11-14 a comparison of exact and approximate values of the critical
stresses is shown for the particular case where

$$\alpha = \frac{h^2}{12a^2} = 10^{-6} \qquad \nu = \frac{1}{6}$$

The quantities $N_0(1 - \nu^2)/Eh$ are plotted as abscissas and $N_1(1 - \nu^2)/Eh$
as ordinates. The results of exact calculations are given by the full line,
and the results obtained by using Eq. (d) are shown by the dotted line.

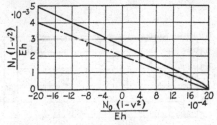

FIG. 11-14

It is seen that for pure bending ($N_0 = 0$) the exact solution gives for the
critical compressive stress a value which is about 30 per cent higher than
that obtained from Eq. (d).

An interesting case of instability of a thin cylindrical shell in bending
has been discussed by Brazier.[2] It is known that in the bending of thin
curved tubes a flattening of the cross sections takes place and such tubes
are more flexible than would be expected from the usual theory of curved
bars.[3] This phenomenon of flattening occurs also in the case of initially
straight cylinders, where it is due to curvature produced by bending.
During bending, the cross section becomes more and more oval until a

[1] Such calculations were made by W. Flügge, Ingr.-Arch., vol. 3, p. 463, 1932.

[2] See L. G. Brazier, Proc. Roy. Soc., London, series A, vol. 116, p. 104, 1927. The
same question has been discussed more rigorously by E. Chwalla, Z. angew. Math. u.
Mech., vol. 13, p. 48, 1933. See also E. Reissner, J. Appl. Mech., vol. 26, pp. 386–392,
1959.

[3] T. V. Kármán, Z. Ver. deut. Ingr., vol. 55, 1911, and Timoshenko, Trans. ASME,
vol. 45, p. 135, 1923.

point is reached at which the resistance to bending starts to decrease, after which, of course, complete collapse takes place. This type of failure occurs in comparatively thick tubes made of a material with a low modulus of elasticity, such as rubber tubes, and in thick metal tubes

FIG. 11-15

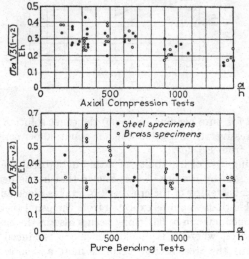

FIG. 11-16

stressed above the yield point. For thin metal tubes with built-in edges, such as used, for instance, in previously described experiments, failure always occurred as a result of buckling in small waves of the same kind as in the case of uniformly compressed cylindrical shells. Several examples of failure under bending are shown in Fig. 11-15. In Fig. 11-16 the results of Donnell's axial-compression and bending tests[1] on thin tubes of steel and brass are given. The ratios a/h are taken as abscissas, and the ratios of the maximum compressive stress (b) to the stress $Eh/a \sqrt{3(1 - \nu^2)}$ for a symmetrical buckling are taken as ordinates. The results show exactly the same decrease of σ_{cr} with the increase of a/h as are shown by the axially loaded specimens. The values found for σ_{cr} are about 1.4 times those found in axial-compression tests for all values of a/h.

11.7. Axial Compression of Curved Sheet Panels. In the case of a cylindrical shell supported along two generators and along two circular edges normal to the axis of the cylinder and uniformly compressed in the direction of the generators (Fig. 11-17), the same method as in the case of a circular cylindrical tube axially compressed (Art. 11.3) can be used[2] in calculating critical stresses. If β is the central angle of the shell, a the radius, and l the length in the direction of generators, then, with coordinate axes directed as shown in the figure, we can satisfy the boundary conditions by taking for the displacements during buckling of the shell the following expressions:

$$u = A \sin \frac{n\pi\theta}{\beta} \cos \frac{m\pi x}{l}$$

$$v = B \cos \frac{n\pi\theta}{\beta} \sin \frac{m\pi x}{l} \qquad (a)$$

$$w = C \sin \frac{n\pi\theta}{\beta} \sin \frac{m\pi x}{l}$$

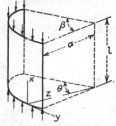

It can be seen that the radial displacements w and the bending moments become zero along the edges of the shell, i.e., for $x = 0$ and $x = l$ and for $\theta = 0$

Fig. 11-17

and $\theta = \beta$, as is required for simply supported edges.[3] Substituting expressions (a) in Eqs. (11-7), we obtain three homogeneous linear equations, and the equation for determining the critical value of the compressive stress is obtained, as before, by equating to zero the determinant of these equations. This equation is similar to Eq. (i) (see p. 464) obtained for cylindrical tubes, the only change being that the quantity $n\pi/\beta$ takes the place of the quantity n. If the central angle β is not small and the length l is of the same order of magnitude as βa, we can expect the

[1] L. H. Donnell, *Trans. ASME*, vol. 56, 1934. A similar series of tests with thin cylindrical shells of duralumin were made by E. E. Lundquist, *NACA Tech. Note* 479, 1933.

[2] This problem was discussed by Timoshenko, "Theory of Elasticity," vol. 2, p. 395, St. Petersburg, 1916.

[3] Note that v is not zero along the generators $\theta = 0$ and $\theta = \beta$ and that there are some circumferential displacements in the planes tangent to the middle surface of the shell. This corresponds to the case when the edges of the shell are supported in V grooves with some clearance.

shell to buckle into a large number of circumferential and longitudinal waves. Then by using notations (g) of Art. 11.3 and Eq. (11-8), we obtain for calculating critical stresses the following equation:

$$\phi = \alpha \frac{\left(\dfrac{n^2\pi^2}{\beta^2} + \lambda^2\right)^2}{\lambda^2} + \frac{(1 - \nu^2)\lambda^2}{\left(\dfrac{n^2\pi^2}{\beta^2} + \lambda^2\right)^2} \qquad (11\text{-}13)$$

from which we can conclude, as before in the case of thin cylindrical tubes, that the critical stress is the same as for the case of buckling of a tube into a form symmetrical with respect to the axis of the cylinder, and we can take

$$\sigma_{cr} = \frac{Eh}{a\sqrt{3(1 - \nu^2)}} \qquad (b)$$

If the angle β is very small, the conditions of buckling of the shell will approach those of a longitudinally compressed rectangular plate; the critical value of the compressive stress is obtained by taking $n = 1$ in Eq. (11-13). Then

$$\phi = \alpha \frac{\left(\dfrac{\pi^2}{\beta^2} + \lambda^2\right)^2}{\lambda^2} + \frac{(1 - \nu^2)\lambda^2}{\left(\dfrac{\pi^2}{\beta^2} + \lambda^2\right)^2} \qquad (c)$$

As the radius a of the shell becomes larger and larger, λ increases also and for a very large value of a we can omit the second term in expression (c). Then we obtain

$$\phi = \alpha \frac{\left(\dfrac{\pi^2}{\beta^2} + \lambda^2\right)^2}{\lambda^2} = \alpha\left(\frac{\pi^2}{\beta^2\lambda} + \lambda\right)^2 \qquad (d)$$

The smallest value of this expression occurs when

$$\lambda = \frac{\pi}{\beta} \qquad \text{or} \qquad \frac{l}{m} = \beta a$$

i.e., when the length of longitudinal half-waves is equal to the width of the curved panel. Then

$$\phi = 4\,\frac{\alpha\pi^2}{\beta^2}$$

Noting that

$$\phi = \frac{(1 - \nu^2)\sigma_{cr}}{E} \qquad \text{and} \qquad \alpha = \frac{h^2}{12a^2}$$

we obtain

$$\sigma_{cr} = \frac{\pi^2 E h^2}{3(1 - \nu^2)(\beta a)^2} \qquad (e)$$

This is the same value of the critical stress that we obtained for long rectangular plates.

Keeping both terms in expression (c), we find that it becomes a minimum for

$$\frac{\left(\frac{\pi^2}{\beta^2} + \lambda^2\right)^2}{\lambda^2} = \sqrt{\frac{1 - \nu^2}{\alpha}} \tag{f}$$

if
$$\beta a \geq 2\pi \sqrt[4]{\frac{a^2 h^2}{12(1 - \nu^2)}} \tag{g}$$

i.e., if the circumferential dimension of the panel is at least equal to twice the half-wave length for symmetrical buckling of the shell. Then

$$\phi = 2\sqrt{\alpha(1 - \nu^2)}$$

By substituting for ϕ its value $(1 - \nu^2)\sigma_{cr}/E$, we again obtain for the critical stress the value given by expression (b). It indicates that, for any value of the width βa of the curved panel satisfying condition (g), we can find from (f) such a length l/m of longitudinal half-waves that the critical stress becomes equal to that found before for symmetrical buckling of a thin tube. This value should be used in design of curved panels uniformly compressed along the generators.

If the circumferential dimension of the shell is smaller than that required by (g), expression (c) becomes a minimum if

$$\lambda^2 = \frac{\pi^2}{\beta^2}$$

This means that for a considerable length l such a narrow shell, like a long narrow compressed plate, will subdivide during buckling into squares. From Eq. (c) the magnitude of the critical stress in such a case is

$$\sigma_{cr} = \frac{\pi^2 E h^2}{3(1 - \nu^2)(a\beta)^2} + \frac{E\beta^2}{4\pi^2} \tag{11-14}$$

The first term on the right-hand side gives the stress calculated as for a plate and the second term gives the increase of the critical stress due to curvature of the shell.[1]

Experiments with axial compression of cylindrical panels show[2] that

---

[1] The effect of a central stiffener in the circumferential direction was investigated by S. B. Batdorf and M. Schildcrout, *NACA Tech. Note* 1661, 1948.

[2] A large number of such experiments with simply supported duralumin curved panels were made by J. S. Newell and W. H. Gale; see "A Report on Aircraft Materials Research at the Massachusetts Institute of Technology," 1931–1932. Some experimental data are given in the paper by S. C. Redshaw, *Aeronaut. Research Comm. Rept. and Mem.* 1565, 1933, and H. L. Cox and W. J. Clenshaw, *ibid.*, no. 1894, 1941.

formula (*b*), which can be put in the form

$$\sigma_{cr} = 0.6E \frac{h}{a} \qquad (11\text{-}15)$$

is in satisfactory agreement with test results providing the axial and circumferential dimensions of the panel are about equal and the central angle β is small, say $\beta < \frac{1}{2}$. With an increase of this angle, σ_{cr} diminishes; for $\beta > 2$ it is only about half of what we obtain from formula (11-15) and approaches the value given by experiments on thin cylindrical tubes (Art. 11.4).

If β is very small the conditions for buckling of a cylindrical shell approach those of buckling of plane panels, and an increase in the ultimate

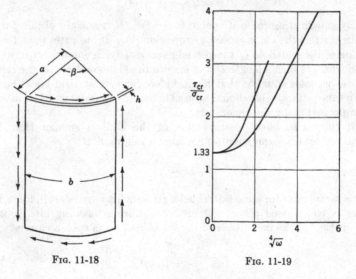

FIG. 11-18   FIG. 11-19

load, similar to that discussed in the case of buckling of plates (Art. 9.14), should be considered.

11.8. Curved Sheet Panels under Shear or Combined Shear and Axial Stress. The problem of buckling of curved sheet panels under pure shear (Fig. 11-18) is of practical interest in aircraft structures. Using the same general method as in the preceding articles, the critical values of shearing stress τ at which buckling begins were calculated[1] for long curved panels. These values are presented in Fig. 11-19, in which abscissas are the values of

[1] See D. M. A. Leggett, *Proc. Roy. Soc.*, series A, vol. 162, p. 62, 1937.

$$\sqrt[4]{\omega} = \frac{b}{\pi a} \sqrt{\frac{a}{h}} \sqrt[4]{12(1 - \nu^2)} \qquad (a)$$

where a is the radius of the cylindrical shell, h is its thickness, and $b = \beta a$ is the width of the panel. The ordinates are the ratios of τ_{cr} to σ_{cr_s} the latter value being calculated for a long compressed rectangular plate of width b and thickness h, as follows [see Eq. (9-7)]:

$$\sigma_{cr} = \frac{\pi^2}{3} \frac{E}{1 - \nu^2} \frac{h^2}{b^2} \qquad (b)$$

There are two curves in the figure. The upper one is calculated on the assumption that the displacements v in the circumferential direction vanish at the longitudinal edges of the panel. The lower one is calculated on the assumption that the normal stress σ_θ in the circumferential direction vanishes at the longitudinal edges of the panel. In both cases it is assumed that the radial displacements w and the bending moments at the longitudinal edges of the panel vanish. At $\omega = 0$ or $a \to \infty$, we have the case of a flat panel. Both curves have the same horizontal tangent with the ordinate 1.33 corresponding to the ratio τ_{cr}/σ_{cr} for long rectangular strips with simply supported edges [see Eqs. (9-7) and (9-8)]. At larger values of ω the lower curve approaches asymptotically the straight line for which $\tau_{cr}/\sigma_{cr} = 0.80 \sqrt[4]{\omega}$ and the value of τ_{cr}, for larger values of ω and for $\nu = 0.3$, can be calculated from the approximate formula[1]

$$\tau_{cr} = 4.82E \frac{h^2}{b^2} \sqrt[4]{1 + 0.0145 \frac{b^4}{a^2 h^2}}$$

$$= 1.67E \frac{h}{b} \sqrt{\frac{h}{a}} \sqrt{1 + 68.7 \frac{a^2 h^2}{b^4}} \qquad (11\text{-}16)$$

The problem of combined shear and axial compression was also investigated.[2] The results obtained are given by curves in Fig. 11-20. Each curve corresponds to a definite value of the ratio of the compressive stress σ_x to σ_{cr}, calculated from Eq. (b). Negative values of the ratio, indicated in the figure, correspond to axial compression, and positive values correspond to axial tension. The curve $\sigma_x/\sigma_{cr} = 0$ corresponds to the case of pure shear, discussed above. It was assumed in this calculation that the longitudinal edges of the panel are simply supported and the circumferential normal stress σ_θ vanishes along these edges. The points of intersection of the curves with the horizontal axis $\sqrt[4]{\omega}$ [see Eq. (a)] give the values of ω for which buckling is produced by compressive stresses

[1] For the case of a panel under shear with a central stiffener in either the circumferential or axial direction, see M. Stein and D. J. Yaeger, NACA Tech. Note 1972, 1949.

[2] See A. Kromm, Jahrb. deut. Luftfahrt-Forsch., p. 1832, 1940.

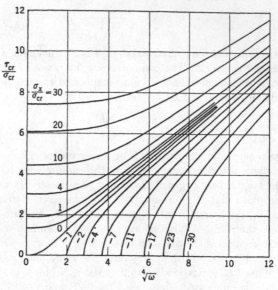

FIG. 11-20

σ_z acting alone ($\tau = 0$). The points of intersection of the curves with the vertical axis give τ_{cr} for a long flat strip for various amounts of the longitudinal tensile stress σ_z.

11.9. Buckling of a Stiffened Cylindrical Shell under Axial Compression. In discussing buckling of a compressed panel (Art. 11.7), it was assumed that the stiffeners and frames were absolutely rigid, so that the results obtained [Eq. (11-14)] can be used for the investigation of local stability of a stiffened compressed cylinder. In the case of large cylinders with comparatively flexible stringers and frames there is the possibility of buckling in which not only the skin but also the stringers and frames buckle and participate in wave formation. Such types of problems become of practical importance in the structural design of large aircraft.

Let us begin the discussion of the problem with the simplest case of buckling symmetrical with respect to the axis of the cylinder (Fig. 11-21). In such a case we have to consider buckling of only one stringer in the axial plane. The frames remain circular during buckling and act only in tension or compression[1] during buckling of the stringers. Denoting by w the deflection of the stringer at a frame, we find the unit compression of the frame to be w/a_θ, where a_θ is the radius of the center line of the frame. The compressive force S in the frame will be equal to $A_\theta E w/a_\theta$, where A_θ is the cross-sectional area of the frame. Then denoting by b the distance between the stringers, we find the angle between the forces S (Fig. 11-21b) to be b/a, and these forces will give a resultant $A_\theta E b w/a a_\theta$ opposing the deflection of the stringer. If the frames are close together, we can replace this resultant by a reaction, continuously

---

[1] It is assumed that the distance b between the stringers is small in comparison with the radius a of the cylinder.

distributed along the stringer, of intensity

$$q = \frac{A_\theta Ebw}{aa_\theta c}$$

where c denotes the distance between the frames. We see that the buckled stringer is in the same condition as a bar on an elastic foundation (Art. 2.10) the modulus of which is

$$\beta = \frac{A_\theta Eb}{aa_\theta c}$$

If such a bar has a large length, it subdivides during buckling into many half-waves, the length of which is [see Eq. (2-40)]

$$l_x = \sqrt[4]{\frac{\pi^4 EI}{\beta}}$$

and the critical load is [see Eq. (2-41)]

$$P_{cr} = \frac{2\pi^2 EI}{l_x^2}$$

Applying these equations in the case of the buckled stringer and denoting by I_s the centroidal moment of inertia of the stringer together with the active portion of the skin, we find the half-wave length of the buckled stringer:

$$l_x = \pi \sqrt[4]{\frac{I_s aa_\theta c}{A_\theta b}} \tag{11-17}$$

and the value of the critical load of the stringer is

$$P_{cr} = 2E \sqrt{\frac{I_s A_\theta b}{aa_\theta c}} \tag{11-18}$$

In our further discussion it is advantageous to replace the stiffened cylindrical shell by an equivalent unstiffened one, having different flexural and extensional rigidities in the longitudinal and circumferential directions. In discussing the properties of such

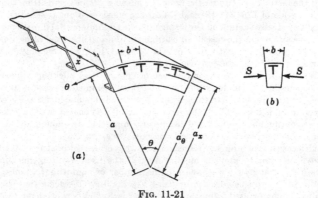

FIG. 11-21

an anisotropic shell in the axial direction, we introduce for the thickness and the moment of inertia per unit length of the circle of radius a_x, where a_x is the distance of the axis of the stringer from the axis of the cylinder (Fig. 11-21), the following notation:

$$h_x = \frac{A_s a}{b a_x} \tag{a}$$

$$i_x = \frac{I_s a}{b a_x} \tag{b}$$

where A_s is the cross-sectional area of the stringer together with the active portion of the skin. Similarly, for the circumferential direction we use the notation

$$h_\theta = \frac{A_\theta}{c} \tag{c}$$

$$i_\theta = \frac{I_\theta}{c} \tag{d}$$

Using this notation we represent Eqs. (11-17) and (11-18) in the following form:

$$l_x = \pi \sqrt[4]{\frac{i_x a_x a_\theta}{h_\theta}} \tag{11-19}$$

$$p_{cr} = \frac{P_{cr}}{b} \frac{a}{a_x} = 2E \sqrt{\frac{i_x h_\theta}{a_x a_\theta}} \tag{11-20}$$

where p_{cr} denotes the critical load per unit length of the circle of radius a_x.

To apply these equations in the case of a thin isotropic cylindrical shell of thickness h and radius a we have to put

$$h_\theta = h_x = h \qquad a_x = a_\theta = a \qquad i_x = \frac{h^3}{12}$$

Then we obtain

$$l_x = \pi \sqrt[4]{\frac{a^2 h^2}{12}} \qquad \sigma_{cr} = \frac{p_{cr}}{h} = \frac{Eh}{a \sqrt{3}}$$

Comparing these results with the previously obtained formulas (11-2) and (11-1), we find that the factor $1 - \nu^2$ under the radical is missing. This results from the fact that in the derivation of Eqs. (11-19) and (11-20) we considered one isolated stringer and neglected any stresses acting on the stringer in the circumferential direction. To take care of these stresses in the case of an isotropic cylindrical shell and bring Eqs. (11-19) and (11-20) into agreement with Eqs. (11-2) and (11-1), we have only to use for i_x the expression $h^3/[12(1 - \nu^2)]$ instead of $h^3/12$.

The results obtained above for the case of symmetrical buckling are satisfactory only if the length l_x calculated from Eq. (11-19) is large in comparison with the frame distance c, so that not less than three frames are included in one half-wave of a stringer. If this condition is not fulfilled, the replacement of the concentrated reactions of the frames by continuously distributed reactions is not accurate enough and we have to consider the stringers as bars on elastic supports (see Art. 2.6).

In such a case it is of interest to consider first what cross-sectional area A_θ the frames must have in order to prevent any bending of the frames during buckling of the stringers. Assuming that we have a large number of equidistant frames, we shall obtain the answer to this question by using Eq. (2-30). Observing that the reaction of a uniformly compressed frame on one stringer is $A_\theta E b w / a a_\theta$, we have to substitute in that equation

$$\alpha = \frac{A_\theta E b}{a a_\theta}$$

It is also necessary to substitute the frame distance c for l/m, $\pi^2 EI/c^2$ for P, and $\beta = \frac{1}{4}$. In this way we obtain the required cross-sectional area of the frame:

$$A_\theta = \frac{4\pi^2 I_s a a_\theta}{bc^3} = \frac{4\pi^2 i_x a_x a_\theta}{c^3} \qquad (11\text{-}21)$$

If the cross-sectional area of the frames is smaller than this value, we have to consider the stringer as a bar on elastic supports. Assuming that the buckled stringer has an inflection point at each second frame, we shall get the conditions represented in Fig. 2-26, and the value of P_{cr} is obtained from the curve in that figure. If we have an inflection point at every third frame, the conditions will be such as shown in Fig. 2-27, and for every value of α we can find P_{cr} from the curves AB and BC of that figure.

It should be noted also that in the above method of analysis the influence on the critical load of the skin of the cylinder is taken care of by including the active portion of the skin, i.e., the effective width of the skin, in the cross-sectional areas A_s and A_θ. This procedure is satisfactory in the case of symmetrical buckling, but a more elaborate investigation is required in the general case of buckling.

To find the critical compressive force in the general case of buckling, we again use the notion of anisotropic cylindrical shells. Proceeding as in the case of an isotropic shell (Art. 11.3) we shall arrive at three differential equations of equilibrium similar to Eqs. (11-7). In solving these equations we assume that the length of the cylinder is large in comparison with its diameter and that the edges at the ends are simply supported. These conditions are satisfied by taking expressions for the displacements similar to Eqs. (f), p. 464. Substituting them into the equations of equilibrium we obtain three equations for determining the constants defining the shape of the buckled cylinder. Proceeding as before and equating to zero the determinant of these equations, we obtain the equation for calculating the critical load.[1] This equation is a very complicated one, but it can be simplified for various particular cases of buckling. In the case of symmetrical buckling it gives the equation for p_{cr} derived above.

In the case of buckling in which the half-wave length l_x in the axial direction is of the same order as the half-wave length l_θ in the circumferential direction, this equation gives the critical load per unit length of the circle of radius a_x in the following form:

$$p_{cr} = \frac{(p_{cr})_{sym}}{Y + A + 1/Y} \left[\sqrt{(Y + A + 1/Y)(Y + B + C/Y) + F^2} - F \right] \qquad (11\text{-}22)$$

in which A, B, C, F are constants depending on the dimensions of the structure and its elastic properties (see below) and Y is proportional to $(l_\theta/l_x)^2$. In each particular case we have to select for Y such a value as to make the right-hand side of Eq. (11-22) a minimum. This minimum is the required value of p_{cr}. It is seen that this value is proportional to $(p_{cr})_{sym}$ given by Eq. (11-20). The constant A in Eq. (11-22) is defined by the following equation:

$$A = \frac{E}{G_r} \sqrt{\frac{h_\theta h_x a_x}{h^2 a_\theta}} \qquad (e)$$

[1] Such an investigation was made by A. Van der Neut, see *Natl. Luchtvaartlaboratorium Rept.* S. 314, Amsterdam, 1947. The results given below are taken from this paper.

where h is the thickness of the skin and h_x and h_θ are defined by Eqs. (a) and (c). The expression for G_r is given on p. 422. The constant B is given by the expression

$$B = \frac{G}{E} \frac{i_{x\theta}a_x{}^2 + i_{\theta x}aa_x}{a_\theta{}^2 i_x \sqrt{h_\theta a_x{}^3/h_x a_\theta a^2}} \tag{f}$$

where $Gi_{x\theta}$ and $Gi_{\theta x}$ are the torsional rigidities of the stringers and frame, together with the effective portions of the skin, per unit length of the circle of radius a_x and per unit length along the distance c between the frames, respectively. The constant C is

$$C = \frac{i_\theta a^3 h_x}{i_x a_\theta{}^3 h_\theta} \tag{g}$$

where i_θ is the moment of inertia of the cross section of a frame, together with the effective portion of the skin, per unit length of the distance c between frames. Finally, the constant F and the quantity Y in Eq. (11-22) are obtained from the formulas

$$F = \frac{(a - a_x)a_\theta + (a - a_\theta)a}{\sqrt{i_x a_\theta{}^2/h_x}} \tag{h}$$

$$Y = \left(\frac{l_\theta}{l_x}\right)^2 \sqrt{\frac{h_x a_\theta a^2}{h_\theta a_x{}^3}} \tag{i}$$

In each particular case we begin by calculating the constants $A, \ldots, F$. Substituting them into Eq. (11-22), taking several values for Y, and calculating the corresponding values of p, we can construct a curve representing p as a function of Y. The minimum ordinate of this curve gives the value for p_{cr}. As a guide in selecting Y we observe that for $F > 0$, as is usually the case in aircraft structures, the true value of Y lies between the value $\sqrt{C}$ and that value of Y which makes the right-hand side of Eq. (11-22) a minimum for the case $F = 0$. The latter value can be readily calculated.

In applying Eq. (11-22) to the case of a thin isotropic shell we have

$$a_x = a_\theta = a \qquad h_x = h_\theta = h \qquad \frac{E}{G_r} = \frac{E}{G} = 2(1 + \nu)$$

$$i_{x\theta} = i_{\theta x} = \frac{h^3}{3} \qquad i_x = \frac{h^3}{12}$$

Then, from Eqs. (e) to (i), we obtain

$$A = 2(1 + \nu) \qquad B = \frac{4}{1 + \nu} \qquad C = 1$$

$$F = 0 \qquad Y = \left(\frac{l_\theta}{l_x}\right)^2$$

and Eq. (11-22) becomes

$$p_{cr} = (p_{cr})_{sym} \left[\frac{Y + 4/(1 + \nu) + 1/Y}{Y + 2(1 + \nu) + 1/Y} \right]^{\frac{1}{2}}$$

The right-hand side of this equation becomes a minimum for $Y = 1$, and we obtain

$$p_{cr} = (p_{cr})_{sym} \left[\frac{1 + 2/(1 + \nu)}{2 + \nu} \right]^{\frac{1}{2}}$$

For $\nu = 0.3$ this value is about 2 per cent larger than the value obtained before (see Eq. 11-9). This discrepancy is understandable if we observe that in both cases the original equation was simplified by omitting terms which were considered as nonessential.

Let us consider now the case in which the length l_θ is large in comparison with l_x. In such a case the basic equation for calculating p_{cr} can be represented in the following form:

$$p_{cr} = (p_{cr})_{sym}\left[1 - \frac{n^2}{2}\sqrt{\frac{i_x a_\theta^2}{a^4 h_x}}(A - B + 2F)\right] \tag{11-23}$$

where A, B, F have the same meaning as before and n denotes the number of waves in the circumferential direction and cannot be large, since l_θ is assumed large. The factor

$$\sqrt{\frac{i_x a_\theta^2}{a^4 h_x}}$$

is very small and of the order h/a. Thus the second term in the brackets is usually a small positive number, and subtracting it from unity we obtain for p_{cr} a value only slightly smaller than $(p_{cr})_{sym}$.

If the longitudinal waves are many times longer than the circumferential waves, the quantity Y becomes small and the basic equation for calculating p_{cr} becomes

$$p_{cr} = \frac{n^2 - 1}{n^2 + 1}(p_{cr})_{sym}\sqrt{C} \tag{11-24}$$

To obtain short circumferential waves the number n must be large, and if C approaches unity, the value of p_{cr} approaches the value $(p_{cr})_{sym}$.

All these results assume that the edges of the cylindrical shell are simply supported and that the radial displacements w vanish at the ends. In the case of free edges the value of p_{cr} can be very different.

Since the above-calculated critical values of p correspond usually to short circumferential waves of buckling, these values can be used with sufficient accuracy also in the case of bending of circular cylindrical shells (see Art. 11.6).

11.10. Buckling of a Cylindrical Shell under Combined Axial and Uniform Lateral Pressure. There are cases in machine design and shipbuilding in which thin cylindrical shells are submitted to the simultaneous action of axial compression and uniform lateral pressure. Under the action of such forces the shell may retain its cylindrical form, but at certain critical values of the pressures this form of equilibrium may become unstable and the shell may buckle.[1] If u, v, and w denote, as before, small displacements from the compressed cylindrical form during buckling of the shell, the three differential equations of equilibrium for determining these displacements can be written by using Eqs. (f), Art. 11.5, for the case of lateral pressure and Eqs. (11-7) for the case of axial compression of the shell. Using the notation

$$\frac{qa(1 - \nu^2)}{Eh} = \phi_1 \quad \text{and} \quad \frac{N_x(1 - \nu^2)}{Eh} = -\phi_2 \tag{a}$$

[1] This problem has been solved by R. von Mises, "Stodola-Festschrift," p. 418, Zürich, 1929. It was also discussed by K. von Sanden and F. Tölke, *Ingr.-Arch.*, vol. 3, p. 24, 1932, and by W. Flügge, *ibid.*, vol. 3, p. 463, 1932.

these equations become

$$a^2 \frac{\partial^2 u}{\partial x^2} + \frac{1+\nu}{2} \frac{a \, \partial^2 v}{\partial x \, \partial \theta} - \nu a \frac{\partial w}{\partial x} + a\phi_1 \left(\frac{\partial^2 v}{\partial x \, \partial \theta} - \frac{\partial w}{\partial x} \right)$$
$$+ \frac{1-\nu}{2} \frac{\partial^2 u}{\partial \theta^2} = 0$$

$$\frac{1+\nu}{2} \frac{a \, \partial^2 u}{\partial x \, \partial \theta} + \frac{1-\nu}{2} a^2 \frac{\partial^2 v}{\partial x^2} + \frac{\partial^2 v}{\partial \theta^2} - \frac{\partial w}{\partial \theta} + \alpha \left[\frac{\partial^2 v}{\partial \theta^2} + \frac{\partial^3 w}{\partial \theta^3} \right.$$
$$\left. + a^2 \frac{\partial^3 w}{\partial x^2 \, \partial \theta} + a^2 (1-\nu) \frac{\partial^2 v}{\partial x^2} \right] - a^2 \phi_2 \frac{\partial^2 v}{\partial x^2} = 0 \qquad (b)$$

$$\nu a \frac{\partial u}{\partial x} + \frac{\partial v}{\partial \theta} - w - \alpha \left[\frac{\partial^3 v}{\partial \theta^3} + (2-\nu) a^2 \frac{\partial^3 v}{\partial x^2 \, \partial \theta} + a^4 \frac{\partial^4 w}{\partial x^4} \right.$$
$$\left. + \frac{\partial^4 w}{\partial \theta^4} + 2a^2 \frac{\partial^4 w}{\partial x^2 \, \partial \theta^2} \right] = \phi_1 \left(w + \frac{\partial^2 w}{\partial \theta^2} \right) + \phi_2 a^2 \frac{\partial^2 w}{\partial x^2}$$

Assuming that the edges of the shell are simply supported, we use for displacements the same expressions as in the case of an axially compressed shell [see Eqs. (f), p. 464] and take

$$u = A \sin n\theta \cos \frac{m\pi x}{l}$$
$$v = B \cos n\theta \sin \frac{m\pi x}{l} \qquad (c)$$
$$w = C \sin n\theta \sin \frac{m\pi x}{l}$$

Substituting these expressions in Eqs. (b), we obtain for A, B, and C three homogeneous linear equations. The equation for calculating the critical value of pressure is obtained by equating to zero the determinant of these equations. After simplifications this equation can be put in the following form:[1]

$$C_1 + C_2\alpha = C_3\phi_1 + C_4\phi_2 \qquad (d)$$

in which

$$C_1 = (1-\nu^2)\lambda^4$$
$$C_2 = (\lambda^2 + n^2)^4 - 2[\nu\lambda^6 + 3\lambda^4 n^2 + (4-\nu)\lambda^2 n^4 + n^6]$$
$$\quad + 2(2-\nu)\lambda^2 n^2 + n^4$$
$$C_3 = n^2(\lambda^2 + n^2)^2 - (3\lambda^2 n^2 + n^4)$$
$$C_4 = \lambda^2(\lambda^2 + n^2)^2 + \lambda^2 n^2$$
$$\alpha = \frac{h^2}{12a^2} \qquad \lambda = \frac{m\pi a}{l} \qquad (e)$$

---

[1] The equation is given in this form in the paper by Flügge, *op. cit.* Flügge used a system of equations slightly different from Eqs. (b), but this difference affects only terms of minor importance in the final equation.

If the dimensions of the shell are given and if certain values are assumed for the numbers m and $2n$, which indicate, respectively, the number of half-waves in the axial direction and in the circumferential direction, Eq. (d) represents a certain linear relation between the quantities ϕ_1 and ϕ_2, determining external pressures. Taking ϕ_1 and ϕ_2 as rectangular coordinates, a straight line is defined by Eq. (d). If we keep m constant and

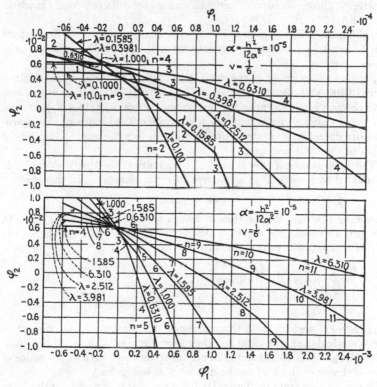

Fig. 11-22

give to n the values 2, 3, 4, . . . , a system of such straight lines is obtained. The portions of these lines which for a given abscissa have the smallest ordinates form a broken line which can be used for determining the critical values of pressures. In Fig. 11-22 such lines are constructed[1] for the case where $\alpha = 10^{-5}$, $\nu = \frac{1}{6}$ and for various values of λ.

Taking the points of intersection of these lines with the horizontal axis ($\phi_2 = 0$), we obtain the critical values of ϕ_1 when lateral pressures alone

[1] This figure is taken from *ibid.*

are acting. It is seen that ϕ_1 and the critical pressure increase as λ increases. This indicates that in the case of lateral pressure acting alone $m = 1$; i.e., the buckled shell has one half-wave in the axial direction, and the critical pressure increases as the length of the cylinder decreases. The number of waves in the circumferential direction, indicated on the sides of the polygons, also increases as the length of the cylinder becomes shorter. These conclusions coincide with statements previously made in Art. 11.5.

Taking the points of intersection of the same polygons with the vertical axis ($\phi_1 = 0$), we obtain the critical values of ϕ_2 providing axial pressure alone is acting.

For any given value of the ratio ϕ_1/ϕ_2 we draw through the origin a straight line with the slope ϕ_1/ϕ_2. The points of intersection of this line with the polygons determine the corresponding critical values of ϕ_1 and ϕ_2. It is seen that any axial pressure makes the critical value of the lateral pressure decrease; any lateral pressure produces a decrease in the critical value of the axial pressure.

The most common case is that of a cylindrical shell with closed ends submitted to the action of a uniform external pressure. In this case

$$\phi_2 = \tfrac{1}{2}\phi_1$$

Assuming that the shell is thin and keeping only the principal terms in Eq. (d), we obtain for this case the following simplified formula for the critical value of the lateral pressure:[1]

$$q_{cr} = \frac{Eh}{a}\,\frac{1}{n^2 + \tfrac{1}{2}(\pi a/l)^2}\left\{\frac{1}{[n^2(l/\pi a)^2 + 1]^2} + \frac{h^2}{12a^2(1 - \nu^2)}\left[n^2 + \left(\frac{\pi a}{l}\right)^2\right]^2\right\} \quad (11\text{-}25)$$

The values of n which make this expression a minimum and which must be used in calculating q_{cr} should be taken from Fig. 11-23, in which the ratios of the length l to the diameter $2a$ are taken as abscissas and the ratios of the thickness of the wall to the diameter as ordinates.[2]

It can be seen from the figure that the number of circumferential lobes into which the shell buckles increases as the length and thickness of the cylinder decrease. Experiments made with thin shells, which buckle within the elastic region of the material, are in satisfactory agreement with the theory.[3]

---

[1] This simplified formula was given by von Mises, loc. cit.

[2] This figure is taken from D. F. Windenburg and C. Trilling, Trans. ASME, vol. 56, p. 819, 1934. Note that numbers on the vertical axis are multiplied by 10^{-3}.

[3] Some experimental results are given in the paper by H. E. Saunders and D. F. Windenburg, Trans. ASME, vol. 54, p. 263, 1932, and in the paper by Windenburg and Trilling, loc. cit. See also the paper by T. Tokugawa, Proc. World Eng. Congr., Tokyo, vol. 29, p. 249, 1929.

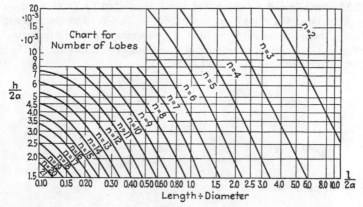

FIG. 11-23

The case of buckling of a cylindrical shell stiffened by equidistant circumferential rings also has been investigated.[1] Replacing the rings by equivalent increases in the circumferential direction of the flexural rigidity of the shell and of the thickness of the wall, and using the equation for non-isotropic shells, we obtain, for calculating critical pressures, an equation similar to Eq. (d). If A_v is the cross-sectional area of the ring, EI_v is the flexural rigidity of one ring, together with the corresponding portion of the wall of the cylinder,[2] and b is the distance between the rings, we use EI_v/b instead of $Eh^3/12(1 - \nu^2)$ in considering bending in the circumferential direction; we also use the equivalent thickness $h_y = h + (A_y/b)$ instead of h in considering the circumferential compression. Using the notations

$$\frac{I_y(1 - \nu^2)}{bha^2} = \alpha_1 \qquad \text{and} \qquad \frac{h_y(1 - \nu^2)}{h} = s \qquad (f)$$

the equation for determining the critical values of pressures becomes

$$C_1 + C_2\alpha + C_3\alpha_1 = C_4\phi_1 + C_5\phi_2 \qquad (g)$$

in which α, ϕ_1, and ϕ_2 have the same meaning as before (see notations a and e) and the coefficients C_1, C_2, . . . are as follows:

$$C_1 = s\lambda^4$$
$$C_2 = \lambda^6(\lambda^2 + 2n^2) + s\lambda^2n^2[2(\lambda^2 - 1)^2 + 2(n^2 - 1)^2 + 5\lambda^2n^2 - 2]$$
$$C_3 = (n^2 - 1)^2[\lambda^4 + s(2\lambda^2 + n^2)n^2]$$
$$C_4 = \lambda^4n^2 + s(2\lambda^2 + n^2)n^4 - s(3\lambda^2 + n^2)n^2$$
$$C_5 = \lambda^6 + s\lambda^2n^2(2\lambda^2 + n^2 + 1)$$

Again, for given dimensions of the shell and ribs and for assumed values for m and n we obtain a straight line giving the relation between ϕ_1 and ϕ_2. Keeping m constant and taking $n = 2, 3, . . .$, polygons similar to those shown in Fig. 11-22 can be obtained and the critical pressure can be calculated.

[1] Flügge, loc. cit.; Sanden and Tölke, loc. cit.; and P. P. Bijlaard, J. Aeronaut. Sci., vol. 24, p. 437, 1957.

[2] It is assumed here that the cross section of the ring is symmetrical with respect to the middle surface of the shell; i.e., half of the ring is projecting inside the shell.

11.11. Buckling of a Cylindrical Shell Subjected to Torsion. In the case of buckling of a cylindrical shell under the action of torque applied at the ends, the resultant shearing force N_{xy} becomes of primary importance. Taking this into account, the general equations of equilibrium (10-13) can be put in the following simplified form:[1]

$$a\,\frac{\partial N_x}{\partial x} + \frac{\partial N_{yx}}{\partial \theta} - aN_{xy}\,\frac{\partial^2 v}{\partial x^2} = 0$$

$$\frac{\partial N_y}{\partial \theta} + a\,\frac{\partial N_{xy}}{\partial x} + N_{yx}\left(\frac{\partial^2 v}{\partial x\,\partial\theta} - \frac{\partial w}{\partial x}\right) - Q_y = 0 \qquad (a)$$

$$a\,\frac{\partial Q_x}{\partial x} + \frac{\partial Q_y}{\partial \theta} + (N_{xy} + N_{yx})\left(\frac{\partial v}{\partial x} + \frac{\partial^2 w}{\partial x\,\partial\theta}\right) + N_y = 0$$

We substitute in these equations

$$N_{xy} = N_{yx} = \frac{M}{2\pi a^2} + N'_{xy}$$

where M is the applied torque, $M/2\pi a^2$ is the resultant shearing force due to this torque, and N'_{xy} is the small change in this force due to buckling. We take into consideration also the stretching of the middle surface during buckling and substitute $N_{yx}(1 + \epsilon_1)$ and $N_{xy}(1 + \epsilon_2)$ instead of N_{yx} and N_{xy} in the second terms of the first and the second equations of the system (a). Then, neglecting the products of the derivatives of N'_{xy} with the displacements u, v, and w, we obtain

$$a\,\frac{\partial N_x}{\partial x} + \frac{\partial N'_{yx}}{\partial \theta} + \frac{M}{2\pi a^2}\left(\frac{\partial^2 u}{\partial x\,\partial\theta} - a\,\frac{\partial^2 v}{\partial x^2}\right) = 0$$

$$\frac{\partial N_y}{\partial \theta} + a\,\frac{\partial N'_{xy}}{\partial x} + 2\,\frac{M}{2\pi a^2}\left(\frac{\partial^2 v}{\partial x\,\partial\theta} - \frac{\partial w}{\partial x}\right) - Q_y = 0$$

$$a\,\frac{\partial Q_x}{\partial x} + \frac{\partial Q_y}{\partial \theta} + N_y + 2\,\frac{M}{2\pi a^2}\left(\frac{\partial v}{\partial x} + \frac{\partial^2 w}{\partial x\,\partial\theta}\right) = 0$$

Substituting in these equations [see Art. 10.1 and Eqs. (10-15)],

$$N_x = \frac{Eh}{1 - \nu^2}\left[\frac{\partial u}{\partial x} + \frac{\nu}{a}\left(\frac{\partial v}{\partial\theta} - w\right)\right]$$

$$N_y = \frac{Eh}{1 - \nu^2}\left[\nu\,\frac{\partial u}{\partial x} + \frac{1}{a}\left(\frac{\partial v}{\partial\theta} - w\right)\right]$$

$$N'_{xy} = \frac{Eh}{2(1 + \nu)}\left(\frac{\partial v}{\partial x} + \frac{1}{a}\,\frac{\partial u}{\partial\theta}\right)$$

[1] Here, as before, u, v, and w are small displacements from the cylindrical form of equilibrium of the twisted shell. The products of all moments and all forces, except N_{xy}, with the derivatives of u, v, and w are neglected.

$$M_x = -\frac{Eh^3}{12(1-\nu^2)}\left[\frac{\partial^2 w}{\partial x^2} + \frac{\nu}{a^2}\left(\frac{\partial^2 w}{\partial \theta^2} + \frac{\partial v}{\partial \theta}\right)\right]$$

$$M_y = -\frac{Eh^3}{12(1-\nu^2)}\left[\nu\frac{\partial^2 w}{\partial x^2} + \frac{1}{a^2}\left(\frac{\partial^2 w}{\partial \theta^2} + \frac{\partial v}{\partial \theta}\right)\right]$$

$$M_{xy} = \frac{Eh^3}{12(1+\nu)a}\left(\frac{\partial v}{\partial x} + \frac{\partial^2 w}{\partial x\,\partial\theta}\right)$$

and using the notations

$$\alpha = \frac{h^2}{12a^2} \qquad \phi = \frac{M(1-\nu^2)}{2\pi a^2 Eh} = \frac{\tau(1-\nu^2)}{E} \tag{b}$$

we obtain

$$a^2\frac{\partial^2 u}{\partial x^2} + \frac{1-\nu}{2}\frac{\partial^2 u}{\partial \theta^2} + \frac{a(1+\nu)}{2}\frac{\partial^2 v}{\partial x\,\partial\theta} - \nu a\frac{\partial w}{\partial x}$$

$$+ \phi a\left(\frac{\partial^2 u}{\partial x\,\partial\theta} - a\frac{\partial^2 v}{\partial x^2}\right) = 0$$

$$\frac{\partial^2 v}{\partial \theta^2} + \frac{a^2(1-\nu)}{2}\frac{\partial^2 v}{\partial x^2} + \frac{a(1+\nu)}{2}\frac{\partial^2 u}{\partial x\,\partial\theta} - \frac{\partial w}{\partial \theta} + \alpha\left[\frac{\partial^2 v}{\partial \theta^2}\right.$$

$$\left. + a^2(1-\nu)\frac{\partial^2 v}{\partial x^2} + a^2\frac{\partial^3 w}{\partial x^2\,\partial\theta} + \frac{\partial^4 w}{\partial \theta^3}\right] + \phi a\left(\frac{\partial^2 v}{\partial x\,\partial\theta} - \frac{\partial w}{\partial x}\right) = 0 \tag{c}$$

$$\frac{\partial v}{\partial \theta} + a\nu\frac{\partial u}{\partial x} - w - \alpha\left[a^4\frac{\partial^4 w}{\partial x^4} + 2a^2\frac{\partial^4 w}{\partial x^2\,\partial\theta^2} + \frac{\partial^4 w}{\partial \theta^4}\right.$$

$$\left. + (2-\nu)a^2\frac{\partial^3 v}{\partial x^2\,\partial\theta} + \frac{\partial^3 v}{\partial \theta^3}\right] + 2\phi a\left(\frac{\partial v}{\partial x} + \frac{\partial^2 w}{\partial x\,\partial\theta}\right) = 0$$

Thus the problem of buckling of a cylindrical shell under the action of torque is reduced to the integration of these three equations.[1]   There is a substantial difference between these equations and those obtained previously in discussing lateral or axial compression of a cylinder.   The difference lies in the fact that in the same equation we encounter both odd and even orders of derivatives of a displacement with respect to the same independent variable.   This indicates that we can no longer satisfy the equations by using solutions in the form of products of sines or cosines, which means physically that there are no generators which remain straight during buckling and which form a system of straight nodal lines for a buckled surface.   In the case of torsion we can expect nodal lines of helical form, and this condition will be fulfilled if we take for the displacements the following expressions:

[1] The first investigation of buckling of cylindrical shells under torsion is due to E. Schwerin; see *Repts. Intern. Congr. Appl. Mech.*, Delft, 1924, and also *Z. angew. Math. u. Mech.*, vol. 5, p. 235, 1925.

$$u = A \cos \left(\frac{\lambda x}{a} - n\theta \right)$$

$$v = B \cos \left(\frac{\lambda x}{a} - n\theta \right) \qquad (d)$$

$$w = C \sin \left(\frac{\lambda x}{a} - n\theta \right)$$

where, as before, a is the radius of the cylinder, l its length, n the number of waves in the circumferential direction, and

$$\lambda = \frac{m\pi a}{l}$$

The corresponding form of buckling has n circumferential waves which spiral along the cylinder; the pitch L of the corresponding helical lines is found from the condition

$$\frac{\lambda L}{a} = 2\pi n$$

from which

$$L = \frac{2a\pi n}{\lambda}$$

Regarding conditions at the ends, we assume first that the cylinder is so long that constraints at the edges do not affect greatly the magnitude of the critical stress and can therefore be disregarded. Substituting expressions (d) in Eqs. (c), we obtain

$$-A \left(\lambda^2 + \frac{1-\nu}{2} n^2 - \lambda n\phi \right) + B \left(\frac{1+\nu}{2} \lambda n + \lambda^2 \phi \right) - C\nu\lambda = 0$$

$$A \frac{1+\nu}{2} \lambda n - B \left[n^2(1+\alpha) + \frac{1-\nu}{2} \lambda^2(1+2\alpha) - 2\phi\lambda n \right]$$
$$+ Cn \left(1 + \alpha n^2 + \alpha\lambda^2 - 2\phi \frac{\lambda}{n} \right) = 0 \qquad (e)$$

$$A\lambda n - Bn \left[1 + \alpha n^2 + (2-\nu)\lambda^2\alpha - 2\phi \frac{\lambda}{n} \right]$$
$$+ C[1 + \alpha(\lambda^2 + n^2)^2 - 2\phi\lambda n] = 0$$

These three homogeneous linear equations can yield for A, B, and C solutions different from zero only if their determinant is zero. Equating the determinant to zero and neglecting[1] terms containing α^2, $\alpha\phi$, α^3, ϕ^2, and ϕ^3, we find that

$$\phi = \frac{R}{S} \qquad (f)$$

---

[1] From notations (b) it can be seen that α and ϕ are small quantities, since buckling within the elastic limit may occur only in the case of very thin shells so that h^2/a^2 is very small. The quantity τ/E is also very small.

where

$$R = \lambda^4(1 - \nu^2) + \alpha[2\lambda^4(1 - \nu^2) + (\lambda^2 + n^2)^4 + (3 + \nu)\lambda^2 n^2$$
$$- (2 + \nu)(3 - \nu)\lambda^4 n^2 - (7 + \nu)\lambda^2 n^4 - 2n^6 + n^4]$$

and $$S = 2\lambda n^5 - 2\lambda n^3 + 4\lambda^3 n^3 - 2\lambda^3 n + 2\lambda^5 n$$

Taking for m and n certain integers, we can calculate the corresponding value of ϕ from this expression.

We begin with $n = 1$. In this case, as can be seen from expressions (d), a cross section of the tube remains circular and moves, during buckling, only in its plane (see p. 283). For $n = 1$, expression (f) becomes

$$\phi = \frac{\lambda^4(1 - \nu^2) + \alpha\lambda^4[\lambda^4 + 4\lambda^2 + (2 + \nu)(1 - \nu)]}{2\lambda^3(\lambda^2 + 1)} \quad (g)$$

Neglecting, in the numerator, the terms with the factor α as very small, we obtain

$$\phi = \frac{\lambda(1 - \nu^2)}{2(\lambda^2 + 1)} \quad (h)$$

To obtain such buckling within the elastic limit, ϕ must be very small [see notation (b)]; hence λ must also be small[1] and we neglect λ^2 in comparison with unity. Then

$$\phi = \tfrac{1}{2}\lambda(1 - \nu^2) \quad (i)$$

We obtain the smallest value for λ when, during buckling, the cylinder forms only one complete wave in the axial direction, i.e., from expressions (d), when

$$\frac{\lambda l}{a} = 2\pi \quad \text{and} \quad \lambda = \frac{2\pi a}{l} \quad (j)$$

Substituting in (i), we obtain

$$\phi = \frac{\pi a(1 - \nu^2)}{l}$$

and, by using notation (b),

$$M_{cr} = \frac{2\pi^2 a^3 E h}{l} \quad (11\text{-}26)$$

Noting that $2\pi a^3 h$ is the polar moment of inertia of the cross section of the tube, we find that the result obtained is in complete agreement with Greenhill's solution obtained for sideways buckling of a long thin rod under torsion [see Eq. (2-74)].

[1] The case of buckling which we are considering requires large values for l/a; thus large values of λ are excluded from this discussion.

If we take $n = 2$ in expression (f), we obtain

$$\phi = \frac{\lambda^4(1 - \nu^2) + \alpha[\lambda^8 + 16\lambda^6 + \lambda^4(74 - 4\nu + 2\nu^2) + \lambda^2(156 - 12\nu) + 144]}{4(\lambda^5 + 7\lambda^3 + 12\lambda)}$$

$$(k)$$

By taking $\lambda = 1$ in this expression, we obtain a very high value for ϕ. Much smaller values can be obtained by assuming λ very small or very large. Calculations show that the first of these two assumptions gives smaller values for ϕ. Adopting this assumption and retaining only the important terms in expression (k), we obtain

$$\phi = \frac{\lambda^4(1 - \nu^2) + 144\alpha}{48\lambda} \qquad (l)$$

We now determine λ from the condition that ϕ becomes a minimum. Then

$$\frac{\partial \phi}{\partial \lambda} = \frac{3\lambda^2(1 - \nu^2)}{48} - \frac{3\alpha}{\lambda^2} = 0$$

from which

$$\lambda = \sqrt[4]{\frac{48\alpha}{1 - \nu^2}} = \sqrt{\frac{2h}{a\sqrt{1 - \nu^2}}}$$

Substituting in Eq. (l), we obtain

$$\phi = 2\sqrt[4]{\frac{\alpha^3}{3}}(1 - \nu^2)$$

and, by using notations (b),

$$M_{cr} = \frac{\pi\sqrt{2}\, E}{3(1 - \nu^2)^{\frac{3}{4}}}\sqrt{ah^5}$$

and

$$\tau_{cr} = \frac{M_{cr}}{2\pi a^2 h} = \frac{E}{3\sqrt{2}\,(1 - \nu^2)^{\frac{3}{4}}}\left(\frac{h}{a}\right)^{\frac{3}{2}} \qquad (11\text{-}27)$$

Calculations with $n > 2$ always give[1] for τ_{cr} values larger than those we have just obtained with $n = 2$. Thus formula (11-27) must always be used in calculating critical stresses for a long cylindrical shell in twist.

In the case of shorter cylinders the conditions at the ends can no longer be disregarded and the problem of determining critical stress becomes more complicated. The general procedure will be as follows: We consider Eq. (f) as an equation of eighth degree in λ. Assuming certain values for ϕ, α, and n, we find eight roots $\lambda_1, \lambda_2, \ldots, \lambda_8$ of this equation, and substituting them in expressions (d), we obtain eight corresponding elementary solutions. By superposing these solutions, the expressions for the displacements become

[1] Such calculations have been made by Schwerin, loc. cit., and by Flügge, "Statik und Dynamik der Schalen," Berlin, 1934.

$$u = \sum_{i=1}^{8} A_i \cos\left(\frac{\lambda_i x}{a} - n\theta\right)$$

$$v = \sum_{i=1}^{8} B_i \cos\left(\frac{\lambda_i x}{a} - n\theta\right) \qquad (m)$$

$$w = \sum_{i=1}^{8} C_i \sin\left(\frac{\lambda_i x}{a} - n\theta\right)$$

Regarding the constants A_i, B_i, C_i it should be noted that for any root λ_i of Eq. (f) we obtain definite ratios for A_i/C_i and B_i/C_i from Eqs. (e). Thus there are only eight independent constants in expressions (m). For determining these constants we have eight boundary conditions, four at each end of the cylinder. Assuming, for instance. that the ends are simply supported, we have for each end

$$u = v = w = \frac{\partial^2 w}{\partial x^2} + \nu \frac{\partial^2 w}{a^2 \, \partial\theta^2} = 0 \qquad (n)$$

In the case of clamped edges, we have at each end

$$u = v = w = \frac{\partial w}{\partial x} = 0 \qquad (o)$$

Substituting expressions (m) in the boundary conditions, we obtain eight linear homogeneous equations. Equating to zero the determinant of these equations, we finally obtain an equation from which can be calculated the length l of the cylinder, to which the assumed values of ϕ, α, and n correspond. Repeating such calculations for a considerable number of assumed values of ϕ, α, and n, we can construct curves representing critical stresses as functions of the geometrical dimensions of the shell.

Such calculations, however, would require an immense amount of work; to make the problem less intricate, several simplifications have been proposed by Donnell.[1] He showed that in the first two equations of the system (c) all terms containing α or ϕ as a factor can be omitted. In the third equation of the same system only such terms with factors α or ϕ which contain the derivatives of w will be retained. In this manner Eq. (f) is simplified considerably also. Further simplification is introduced in the boundary conditions by omitting the requirement that $u = 0$ at the ends of the shell. Then the rest of the end conditions can be

[1] L. H. Donnell, *NACA Rept.* 479, 1933. For discussion of the accuracy of the simplified equation, see J. Kempner, *J. Appl. Mech.*, vol. 22, p. 117, 1955, and N. J. Hoff, *ibid.*, vol. 22, p. 329, 1955. For use of the simplified equation in other cases of loading see the paper by S. B. Batdorf, *NACA Tech. Notes* 1341 and 1342.

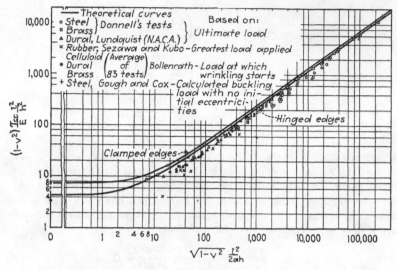

F$_{IG}$. 11-24

satisfied by keeping only four terms in the summations in expressions (m) and by replacing the simplified Eq. (f) of eighth degree in λ with an approximate equation of the fourth degree. In such a manner the boundary conditions and the determinant for calculating the critical stress are brought to the same form as in the problem of buckling of an infinitely long strip under shear, as discussed by Skan and Southwell.[1] As a result of all these simplifications, Donnell obtained the following formulas for short and moderately long shells:

$$(1 - \nu^2)\frac{\tau_{\mathrm{cr}}}{E}\frac{l^2}{h^2} = 4.6 + \sqrt{7.8 + 1.67\left(\sqrt{1 - \nu^2}\,\frac{l^2}{2ha}\right)^{\frac{3}{2}}} \quad (11\text{-}28)$$

for clamped edges, and

$$(1 - \nu^2)\frac{\tau_{\mathrm{cr}}}{E}\frac{l^2}{h^2} = 2.8 + \sqrt{2.6 + 1.40\left(\sqrt{1 - \nu^2}\,\frac{l^2}{2ah}\right)^{\frac{3}{2}}} \quad (11\text{-}29)$$

for simply supported edges.

To check these formulas some 50 tests were made, the results of which, together with those obtained by other experimenters,[2] are plotted in Fig. 11-24. It can be seen that all tests give values for the failure stress some-

[1] S. W. Skan and R. V. Southwell, *Proc. Roy. Soc., London*, series A, vol. 105, p. 582, 1924.

[2] See E. E. Lundquist, *NACA Tech. Note* 427, 1932; Katsutada Sezawa and Kei Kubo, *Tokyo Imp. Univ. Rept.* 76, vol. 6, 1931. The experiments by F. Bollenrath,

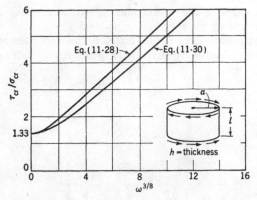

FIG. 11-25

what lower than the values of critical stress calculated from theoretical formulas.

A more detailed investigation[1] in the case of simply supported edges gives, instead of formula (11-28), the following equation:

$$\tau_{cr} = 4.39 \frac{E}{1 - \nu^2} \frac{h^2}{l^2} \sqrt{1 + 0.0257(1 - \nu^2)^{\frac{1}{2}}\left(\frac{l}{\sqrt{ah}}\right)^3} \quad (11\text{-}30)$$

A comparison of this equation with Eq. (11-28) is shown in Fig. 11-25. In the figure, the quantities σ_{cr} and ω are defined as follows:

$$\sigma_{cr} = \frac{\pi^2}{3} \frac{E}{1 - \nu^2} \frac{h^2}{l^2} \quad (p)$$

$$\omega = \frac{12(1 - \nu^2)}{\pi^4} \frac{l^4}{a^2 h^2} \quad (q)$$

It is seen that Eq. (11-30) gives somewhat smaller values for τ_{cr} than Eq. (11-28).

For longer cylinders, namely, where

$$\frac{1}{\sqrt{1 - \nu^2}} \frac{l^2 h}{(2a)^3} > 7.8 \quad \text{for clamped edges}$$

and

$$\frac{1}{\sqrt{1 - \nu^2}} \frac{l^2 h}{(2a)^3} > 5.5 \quad \text{for simply supported edges}$$

Luftfahrt-Forsch., vol. 6, p. 1, 1929, and by Gough and Cox, *Proc. Royal Soc., London*, series A, vol. 137, p. 145, 1932, refer to buckling of flat strips. See also W. A. Nash, *Soc. Exptl. Stress Anal.*, vol. 16, p. 55, 1959.

[1] See Kromm, *loc. cit.*

it is recommended to use, in calculating critical stresses, the formulas developed previously for very long cylinders.

Regarding the number of waves formed in the circumferential direction during buckling of tubes, the experiments are in very satisfactory agreement with the simplified theory.

FIG. 11-26

FIG. 11-27

Figure 11-26 represents the machine used by Donnell in his torsion and also combined torsion and bending and torsion and compression tests. The three types of load are applied by three conveniently located cranks. The amount of the load is read directly on the dial gauges. Figure 11-27 shows a cylindrical shell buckled under torsion.

The problem of combined torsion and longitudinal tension or compression was also investigated theoretically.[1] The results obtained are shown in Fig. 11-28. The curves in this figure are calculated for various values of $\sqrt[4]{\omega}$, where ω is defined by expression (q). The case $\omega = 0$ corresponds to a long rectangular strip subjected to shear and uniform lateral compression or tension.[2] The abscissas are the ratios σ_x/σ_0 where [see Eq. (11-1)]

$$\sigma_0 = \frac{E}{\sqrt{3(1 - \nu^2)}} \frac{h}{a}$$

The ordinates are the ratios τ/τ_0, where τ_0 is the critical stress for pure shear as calculated from Eq. (11-30).

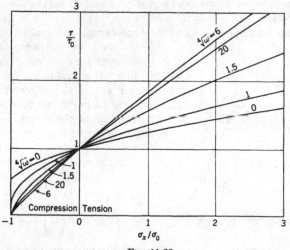

FIG. 11-28

Experiments with torsion of cylindrical shells do not show sudden buckling or great sensitivity to various initial imperfections as in the case of axial compression. Theoretical investigations similar to those made for axial compression[3] do not show a rapid diminishing of torque during postbuckling deformation.

11.12. Buckling of Conical Shells. The general case of buckling of compressed conical shells represents a complicated mathematical problem.[4] A solution has been obtained for the particular case in which the

---

[1] See Kromm, *loc. cit.* See also the paper by S. B. Batdorf, M. Stein, and M. Schildcrout, *NACA Tech. Note* 1345, 1947.

[2] This case was discussed by C. Schmieden, *Z. angew. Math. u. Mech.*, vol. 15, p. 278, 1935.

[3] Tsu-Tau Loo, *Proc. 2d U.S. Natl. Congr. Appl. Mech.*, Ann Arbor, p. 345, 1954.

[4] See A. Pflüger, *Ingr.-Arch.*, vol. 8, p. 151, 1937.

thickness h of the shell is proportional to the radius of curvature r (Fig. 11-29) so that

$$h = h_1 \frac{r}{r_1} \qquad (a)$$

where h_1 is the thickness of the shell at the bottom of the cone. Assuming that the cone is subjected to the action of uniformly distributed vertical load q, we shall have for membrane forces the values

$$N_x = -\frac{qr}{2 \cos \alpha} \qquad N_\theta = -qr \cos \alpha \qquad (b)$$

where N_x acts in the direction of a generator and N_θ acts in the circumferential direction. Denoting by u and v the displacements in those directions during buckling and by w the displacements normal to the surface

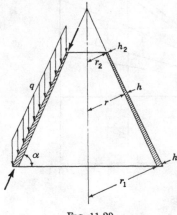

FIG. 11-29

of the shell, we can write, as before, three equations of equilibrium containing three unknown functions u, v, and w. If the edges of the shell are free to move, we take a particular solution of those equations in the following form:

$$u = A \left(\frac{r}{r_1}\right)^m \cos n\theta$$

$$v = B \left(\frac{r}{r_1}\right)^m \sin n\theta \qquad (c)$$

$$w = C \left(\frac{r}{r_1}\right)^m \cos n\theta$$

in which m and n are parameters which have to be selected so as to make the buckling load a minimum.

Substituting solutions (c) into the equations of equilibrium we get three homogeneous linear equations for calculating the constants A, B, and C. These equations can yield solutions different from zero only if their determinant is zero. This gives us the equation for calculating the buckling load. This equation can be put in the following simplified form:

$$aq_1 = b + ck_1 \qquad (d)$$

in which

$$q_1 = q \frac{1 - \nu^2}{E} \frac{r_1}{h_1} \qquad k_1 = \frac{h_1^2}{12 r_1^2} \qquad (e)$$

and the quantities a, b, c denote expressions which can be calculated if the geometrical dimensions and elastic constants of the shell are known and if values of the parameters m and n are assumed. The preliminary considerations show that, to make the buckling load a minimum, we have to take $m = 2$. Taking then various values of n and making calculations for various proportions of shells, we obtain, from Eq. (d), the results represented by curves[1] in Fig. 11-30. Using these curves we get, for

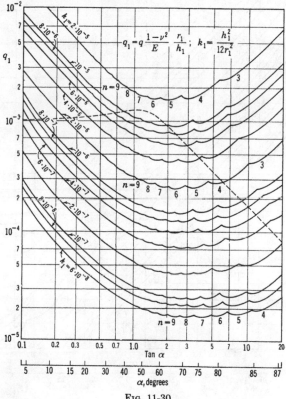

$$q_1 = q \frac{1-\nu^2}{E} \frac{r_1}{h_1} ; \quad k_1 = \frac{h_1^2}{12 r_1^2}$$

Fig. 11-30

given values of the angle α and the quantity k_1, the corresponding buckling values of q_1 and q. The region of elastic buckling is indicated by the dotted line for the case of structural steel with $\sigma_{YP} \approx 28,000$ psi.

In a similar way the case of uniformly distributed pressure normal to the conical surface can be treated. The corresponding values of the critical pressure can be obtained from the curves in Fig. 11-31. For $\alpha = 90°$ these curves give $n = 2$, and for q_{cr} the same magnitude is obtained as for long cylindrical shells.

[1] These curves are taken from Pflüger, *loc. cit.*

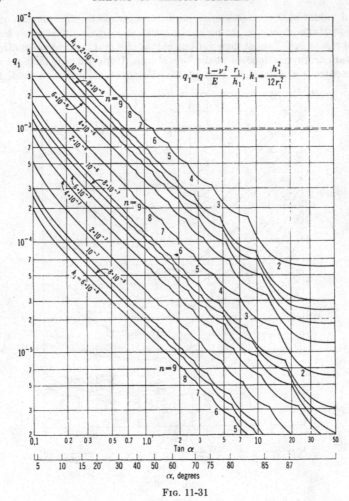

$$q_1 = q\,\frac{1-\nu^2}{E}\,\frac{r_1}{h_1};\ k_1 = \frac{h_1^2}{12 r_1^2}$$

Fig. 11-31

11.13. Buckling of Uniformly Compressed Spherical Shells. If a spherical shell is submitted to uniform external pressure, it may retain its spherical form and undergo only a uniform compression. The magnitude of the uniform compressive stress in this case is

$$\sigma = \frac{qa}{2h} \qquad (a)$$

where q is the pressure per unit area of the middle surface, a is the radius of the sphere, and h is the thickness of the shell. If the pressure increases beyond a certain limit, the spherical form of equilibrium of the compressed shell may become unstable and buckling then occurs.[1] In calculating the critical value of pressure at which buckling occurs, we assume that the buckled surface is symmetrical with respect to a diameter of the sphere and we use Eqs. (10-17) derived for the case of symmetrical deformation of a symmetrically loaded shell. In the derivation of equations for the buckled surface of the sphere, we proceed as before and assume that the quantities u, v, and w (Fig. 10-8) represent components of small displacements during buckling from the compressed spherical form. Then N_x and N_y in Eqs. (10-17) differ but little from the uniform compressive forces $qa/2$, and we can put

$$N_x = -\frac{qa}{2} + N_x' \quad \text{and} \quad N_y = -\frac{qa}{2} + N_y' \tag{b}$$

where N_x' and N_y' are the resultant forces due to small displacements u, v, and w. Substituting these expressions in Eqs. (10-17), using[2] $q(1 + \epsilon_1 + \epsilon_2)$ instead of q, and neglecting small terms such as products of N_x', N_y', and Q_x with the derivatives of u, v, and w, we obtain

$$\frac{dN_x'}{d\theta} + (N_x' - N_y') \cot\theta - Q_x - \frac{qa}{2}\left(\frac{u}{a} + \frac{dw}{a\,d\theta}\right) = 0$$

$$\frac{dQ_x}{d\theta} + Q_x \cot\theta + N_x' + N_y' + qa\left(\frac{du}{a\,d\theta} + \frac{u}{a}\cot\theta - \frac{2w}{a}\right)$$

$$- \frac{qa}{2}\left(\frac{du}{a\,d\theta} + \frac{d^2w}{a\,d\theta^2}\right) - \frac{qa}{2}\cot\theta\left(\frac{u}{a} + \frac{dw}{a\,d\theta}\right) = 0 \tag{c}$$

$$\frac{dM_x}{d\theta} + (M_x - M_y)\cot\theta - Q_x a = 0$$

Eliminating Q_x from the first two equations by the use of the third one and substituting in these equations (see p. 456)

$$N_x' = \frac{Eh}{1 - \nu^2}\left[\frac{du}{a\,d\theta} - \frac{w}{a} + \nu\left(\frac{u\cot\theta}{a} - \frac{w}{a}\right)\right]$$

$$N_y' = \frac{Eh}{1 - \nu^2}\left[\frac{u\cot\theta}{a} - \frac{w}{a} + \nu\left(\frac{du}{a\,d\theta} - \frac{w}{a}\right)\right]$$

---

[1] The buckling of spherical shells was discussed by R. Zoelly, Dissertation, Zürich, 1915, and E. Schwerin, *Z. angew. Math. u. Mech.*, vol. 2, p. 81, 1922. A general solution of the problem, considering also unsymmetrical buckling, has been given by A. Van der Neut, Dissertation, Delft, 1932. See also K. Marguerre, *Proc. 5th Intern. Congr. Appl. Mech.*, 1938, pp. 93–101.

[2] This accounts for a small change of pressure on an element of the surface due to stretching of the surface. The corresponding terms in the equations are very small and could be neglected. We keep them only to bring the calculations into agreement with existing derivations in which the above-mentioned stretching of the surface has been considered.

$$M_x = -\frac{D}{a^2}\left[\frac{du}{d\theta} + \frac{d^2w}{d\theta^2} + \nu\left(u + \frac{dw}{d\theta}\right)\cot\theta\right]$$

$$M_y = -\frac{D}{a^2}\left[\left(u + \frac{dw}{d\theta}\right)\cot\theta + \nu\left(\frac{du}{d\theta} + \frac{d^2w}{d\theta^2}\right)\right]$$

we obtain for determining u and w two equations which, with the notations

$$\alpha = \frac{D(1-\nu^2)}{a^2Eh} = \frac{h^2}{12a^2} \qquad \phi = \frac{qa(1-\nu^2)}{2Eh} \qquad (d)$$

can be represented in the following form:

$$(1+\alpha)\left[\frac{d^2u}{d\theta^2} + \cot\theta\,\frac{du}{d\theta} - (\nu + \cot^2\theta)u\right] - (1+\nu)\frac{dw}{d\theta}$$
$$+ \alpha\left[\frac{d^3w}{d\theta^3} + \cot\theta\,\frac{d^2w}{d\theta^2} - (\nu + \cot^2\theta)\frac{dw}{d\theta}\right] - \phi\left(u + \frac{dw}{d\theta}\right) = 0 \quad (e)$$

$$(1+\nu)\left(\frac{du}{d\theta} + u\cot\theta - 2w\right) + \alpha\left[-\frac{d^3u}{d\theta^3} - 2\cot\theta\,\frac{d^2u}{d\theta^2}\right.$$
$$\left. + (1+\nu+\cot^2\theta)\frac{du}{d\theta} - \cot\theta(2-\nu+\cot^2\theta)u - \frac{d^4w}{d\theta^4}\right.$$
$$\left. - 2\cot\theta\,\frac{d^3w}{d\theta^3} + (1+\nu+\cot^2\theta)\frac{d^2w}{d\theta^2} - \cot\theta(2-\nu+\cot^2\theta)\frac{dw}{d\theta}\right]$$
$$- \phi\left(-u\cot\theta - \frac{du}{d\theta} + 4w + \cot\theta\,\frac{dw}{d\theta} + \frac{d^2w}{d\theta^2}\right) = 0 \quad (f)$$

We begin with Eq. (e), which can be simplified by omitting α in comparison with unity in the first factor.[1] We note also that the expressions in the brackets become identical if we replace u with $dw/d\theta$. Thus a simplification can be obtained if we introduce a new variable ψ by putting

$$u = \frac{d\psi}{d\theta} \qquad (g)$$

Equation (e) then becomes

$$\frac{d^3\psi}{d\theta^3} + \cot\theta\,\frac{d^2\psi}{d\theta^2} - (\nu + \cot^2\theta)\frac{d\psi}{d\theta} - (1+\nu)\frac{dw}{d\theta}$$
$$+ \alpha\left[\frac{d^3w}{d\theta^3} + \cot\theta\,\frac{d^2w}{d\theta^2} - (\nu + \cot^2\theta)\frac{dw}{d\theta}\right] - \phi\left(\frac{d\psi}{d\theta} + \frac{dw}{d\theta}\right) = 0$$

Using the symbol H for the operation

$$\frac{d^2(\,\cdots\,)}{d\theta^2} + \cot\theta\,\frac{d(\,\cdots\,)}{d\theta} + 2(\,\cdots\,) \qquad (h)$$

---

[1] From notation (d) it is seen that α is a very small quantity, since buckling within the elastic limit may occur only in the case of very thin shells where h/a is small.

we can write this equation as follows:

$$\frac{d}{d\theta}[H(\psi) + \alpha H(w) - (1 + \nu)(\psi + w) - \alpha(1 + \nu)w - \phi(\psi + w)] = 0$$

In this equation the fourth term, containing the factor α, can be neglected in comparison with the third. Integrating the equation after this simplification and assuming the constant of integration equal to zero,[1] we obtain

$$H(\psi) + \alpha H(w) - (1 + \nu)(\psi + w) - \phi(\psi + w) = 0 \qquad (i)$$

Proceeding in the same manner with Eq. (f), we obtain

$$\alpha HH(\psi + w) - (1 + \nu)H(\psi) - (3 + \nu)\alpha H(w) + 2(1 + \nu)(\psi + w)$$
$$+ \phi[-H(\psi) + H(w) + 2(\psi + w)] = 0 \qquad (j)$$

Thus the investigation of buckling of a spherical shell reduces to the integration of the two Eqs. (i) and (j).

We can solve these equations by using Legendre functions of the orders $0, 1, 2, \ldots$, which are[2]

$$P_0(\cos \theta) = 1 \qquad P_1(\cos \theta) = \cos \theta \qquad P_2(\cos \theta) = \tfrac{1}{4}(3 \cos 2\theta + 1)$$
$$P_3(\cos \theta) = \tfrac{1}{8}(5 \cos 3\theta + 3 \cos \theta) \qquad P_4(\cos \theta) = \tfrac{1}{64}(35 \cos 4\theta$$
$$+ 20 \cos 2\theta + 9)$$

$\cdots \cdots \cdots \cdots \cdots \cdots \cdots \cdots \cdots \cdots \cdots \cdots \cdots$

$$P_n(\cos \theta) = 2\frac{1 \cdot 3 \cdot 5 \cdots (2n - 1)}{2^n \cdot n!}\left[\cos n\theta\right.$$
$$+ \frac{1}{1}\frac{n}{2n - 1}\cos (n - 2)\theta + \frac{1 \cdot 3}{1 \cdot 2}\frac{n(n - 1)}{(2n - 1)(2n - 3)}\cos (n - 4)\theta$$
$$\left. + \frac{1 \cdot 3 \cdot 5}{1 \cdot 2 \cdot 3}\frac{n(n - 1)(n - 2)}{(2n - 1)(2n - 3)(2n - 5)}\cos (n - 6)\theta + \cdots\right] \qquad (k)$$

All these functions satisfy the equation

$$\frac{d^2P_n}{d\theta^2} + \cot \theta \frac{dP_n}{d\theta} + n(n + 1)P_n = 0 \qquad (l)$$

Hence, performing the operation indicated in (h), we obtain

$$H(P_n) = -\lambda_n P_n \qquad (m)$$
where
$$\lambda_n = n(n + 1) - 2 \qquad (n)$$
and
$$HH(P_n) = \lambda_n^2 P_n \qquad (o)$$

---

[1] From Eq. (g) it is seen that an addition of a constant to ψ does not affect the value of u.

[2] Numerical tables and graphs of these functions can be found in Jahnke and Emde, "Tables of Functions," 4th ed., Dover Publications, New York, 1945.

Proceeding with Legendre functions as we did before with trigonometric functions and using series, we obtain general expressions for any symmetrical buckling of a spherical shell by assuming that

$$\psi = \sum_{n=0}^{\infty} A_n P_n$$
$$w = \sum_{n=0}^{\infty} B_n P_n$$

(p)

Substituting these expressions in Eqs. (i) and (j) and using Eqs. (m) and (o), we obtain

$$\sum_{n=0}^{\infty} \{A_n[\lambda_n + (1 + \nu) + \phi] + B_n[\alpha\lambda_n + (1 + \nu) + \phi]\}P_n = 0$$

$$\sum_{n=0}^{\infty} \{A_n[\alpha\lambda_n{}^2 + (1 + \nu)(\lambda_n + 2) + \phi(\lambda_n + 2)]$$
$$+ B_n[\alpha\lambda_n{}^2 + (3 + \nu)\alpha\lambda_n + 2(1 + \nu) - \phi(\lambda_n - 2)]\}P_n = 0$$

The series on the left-hand sides of these equations can vanish only if each term by itself vanishes. Thus we obtain for each value of n the following two homogeneous equations:

$$A_n[\lambda_n + (1 + \nu) + \phi] + B_n[\alpha\lambda_n + (1 + \nu) + \phi] = 0$$
$$A_n[\alpha\lambda_n{}^2 + (1 + \nu)(\lambda_n + 2) + \phi(\lambda_n + 2)] + B_n[\alpha\lambda_n{}^2 + (3 + \nu)\alpha\lambda_n$$
$$+ 2(1 + \nu) - \phi(\lambda_n - 2)] = 0 \quad (q)$$

Buckling of the shell becomes possible if these equations, for some value of n, yield for A_n and B_n a solution different from zero, which requires that the determinant of these equations becomes equal to zero. In this manner we obtain the following equation for calculating the critical value of the external pressure:[1]

$$(1 - \nu^2)\lambda_n + \alpha\lambda_n[\lambda_n{}^2 + 2\lambda_n + (1 + \nu)^2] - \phi\lambda_n[\lambda_n + (1 + 3\nu)] = 0 \quad (r)$$

One solution of this equation is

$$\lambda_n = 0$$

The corresponding value of n, from Eq. (n), is

$$n = 1$$

and from the first of Eqs. (q) we conclude that

$$A_1 = -B_1$$

---

[1] The quantities α and ϕ are very small and are neglected in comparison with unity in the derivation of this equation.

The corresponding displacements, from Eqs. (p), are

$$u = \frac{d\psi}{d\theta} = -A_1 \sin \theta$$

$$w = -A_1 \cos \theta$$

They represent a displacement of the sphere as a rigid body by the amount A_1 along the axis of symmetry.

To get displacements corresponding to buckling of the shell, we must assume in Eq. (r) that λ_n is different from zero. Then

$$\phi = \frac{(1 - \nu^2) + \alpha[\lambda_n{}^2 + 2\lambda_n + (1 + \nu)^2]}{\lambda_n + (1 + 3\nu)} \tag{s}$$

For any value of n we can calculate ϕ, and by using notation (d) we obtain the corresponding value of the external pressure. To find the smallest value of ϕ and q at which buckling may occur, we consider expression (s) as a continuous function of λ_n and determine its minimum from the condition that

$$\frac{d\phi}{d\lambda_n} = 0$$

This gives, after neglecting small terms,

$$\lambda_n{}^2 + 2(1 + 3\nu)\lambda_n - \frac{1 - \nu^2}{\alpha} = 0$$

and, approximately,

$$\lambda_n = -(1 + 3\nu) + \sqrt{\frac{1 - \nu^2}{\alpha}} \tag{t}$$

Substituting in (s), we obtain

$$\phi_{\min} = 2\sqrt{(1 - \nu^2)\alpha} - 6\nu\alpha$$

By using notations (d),

$$q_{cr} = \frac{\phi_{\min} 2Eh}{a(1 - \nu^2)} = \frac{2Eh}{a(1 - \nu^2)}\left(\sqrt{\frac{1 - \nu^2}{3}}\frac{h}{a} - \frac{\nu h^2}{2a^2}\right)$$

or, by neglecting the second term in the parenthesis,

$$q_{cr} = \frac{2Eh^2}{a^2 \sqrt{3(1 - \nu^2)}} \tag{11-31}$$

and from Eq. (a)

$$\sigma_{cr} = \frac{Eh}{a \sqrt{3(1 - \nu^2)}} \tag{11-32}$$

This stress has the same magnitude as the critical stress for an axially compressed cylindrical shell of radius a and of thickness h [see Eq. (11-9)].

In the above derivation a continuous variation of λ_n has been assumed. But λ_n is defined from Eq. (n) in which n is an integer. Hence, to get a more accurate value for σ_{cr}, two adjacent integers, as obtained from Eq. (n), instead of the value (t), should be substituted in Eq. (s) and the value of λ_n which gives the smaller value for ϕ_n should be used in calculating critical stresses. From Eq. (t) it may be seen that λ_n is a large number; hence the numbers n are also large and the result of this more accurate calculation of σ_{cr} will differ but little from that given by formula (11-32).

Until now we have considered only symmetrical buckling of the shell, but a more general investigation shows[1] that, owing to symmetry of the uniformly compressed spherical shell with respect to any diameter, Eq. (s), derived on the assumption of symmetry, gives all possible values of ϕ_{cr} and that formula (11-32) always can be used for calculating the critical stress.

Experiments with thin spherical shells subjected to uniform external pressure[2] show that buckling occurs at pressures much smaller than that given by Eq. (11-31) and that the collapse of the buckled shell occurs suddenly as in the previously discussed case of axial compression of cylindrical shells. Approximate calculations[3] show that in the vicinity of the spherical shape of equilibrium of a compressed shell there exist forms of equilibrium slightly deviated from the spherical shape which require pressures much smaller than that given by Eq. (11-31). This indicates, as discussed in the case of cylindrical shells (see p. 471), that a very small disturbance during loading may produce buckling at pressures much smaller than required by the classical theory. This also explains the suddenness of the buckling phenomenon of compressed spherical shells and the wide scatter of experimental values of critical pressures.

For practical applications we can use the following empirical formula[4] for calculating q_{cr}:

$$q_{cr} = \left(1 - 0.175 \frac{\theta° - 20°}{20°}\right)\left(1 - \frac{0.07a/h}{400}\right)(0.3E)\left(\frac{h}{a}\right)^2$$

[1] See paper by Van der Neut, *loc. cit.*; also Flügge, *loc. cit.*

[2] See K. Klöppel and O. Jungbluth, *Der Stahlbau*, vol. 22, p. 121, 1953.

[3] See T. V. Kármán and Hsue-Shen Tsien, *J. Aeronaut. Sci.*, vol. 7, pp. 43 and 276, 1940. See also K. O. Friedrichs, "T. V. Kármán Anniversary Volume," p. 258, 1941; V. I. Feodosiev, *Appl. Math. Mech.*, vol. 18, p. 35, 1954 (Russian); and E. L. Reiss, H. J. Greenberg, and H. B. Keller, *J. Aeronaut. Sci.*, vol. 24, p. 533, 1957.

[4] See Klöppel and Jungbluth, *op. cit.* There is given also in that paper some information about stiffened spherical shells.

which gives satisfactory results for (see Fig. 11-32) $400 \leq a/h \leq 2,000$ and $20° \leq \theta \leq 60°$.

In the case of very small values of the angle θ (Fig. 11-32) we shall have a slightly bent circular plate which under external pressure buckles in a direction opposite to the initial curvature, similar to buckling of a slightly bent bar (see Art. 7.8).

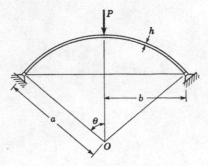

FIG. 11-32

An approximate solution[1] for the case of a load P acting along the symmetry axis (Fig. 11-32) gives the following equations for calculating the critical load:

$$\mu = \sqrt{0.152(\lambda + 74.9)} - 2.88 \qquad \text{for } 20 < \lambda < 100$$
$$\mu = \sqrt{0.093(\lambda + 11.5)} - 0.94 \qquad \text{for } 100 < \lambda < 500 \qquad (11\text{-}33)$$

where
$$\lambda = \frac{b^4}{a^2 h^2} \qquad \mu = \frac{Pa}{Eh^3}$$

and in which b is the radius of the circular plate, a is the radius of the sphere, and h is the thickness of the plate.

[1] This phenomenon was discussed by C. B. Biezeno, *Z. angew. Math. u. Mech.*, vol. 15, p. 10, 1935. For a further discussion of similar problems see H. Nylander, *Osterr. Ingr. Arch.*, vol. 9, p. 181, 1955.

APPENDIX

TABLE A-1. TABLE OF THE FUNCTIONS $\phi(u)$, $\psi(u)$, $\chi(u)$

$$\phi(u) = \frac{3}{u}\left(\frac{1}{\sin 2u} - \frac{1}{2u}\right) \qquad \psi(u) = \frac{3}{2u}\left(\frac{1}{2u} - \frac{1}{\tan 2u}\right) \qquad \chi(u) = \frac{3(\tan u - u)}{u^3}$$

$2u = kl$	$\phi(u)$	$\Delta\phi$	$\psi(u)$	$\Delta\psi$	$\chi(u)$	$\Delta\chi$	$2u = kl$
0	1.0000		1.0000		1.0000		0
		0.0008		0.0007		0.0010	
0.10	1.0008		1.0007		1.0010		0.10
		0.0039		0.0020		0.0030	
0.20	1.0047		1.0027		1.0040		0.20
		0.0059		0.0034		0.0051	
0.30	1.0106		1.0061		1.0091		0.30
		0.0083		0.0047		0.0072	
0.40	1.0189		1.0108		1.0163		0.40
		0.0110		0.0063		0.0094	
0.50	1.0299		1.0171		1.0257		0.50
		0.0138		0.0078		0.0117	
0.60	1.0437		1.0249		1.0374		0.60
		0.0166		0.0094		0.0142	
0.70	1.0603		1.0343		1.0516		0.70
		0.0198		0.0111		0.0168	
0.80	1.0801		1.0454		1.0684		0.80
		0.0232		0.0131		0.0198	
0.90	1.1033		1.0585		1.0882		0.90
		0.0271		0.0152		0.0231	
1.00	1.1304		1.0737		1.1113		1.00
		0.0151		0.0085		0.0128	
1.05	1.1455		1.0822		1.1241		1.05
		0.0162		0.0090		0.0138	
1.10	1.1617		1.0912		1.1379		1.10
		0.0175		0.0097		0.0148	
1.15	1.1792		1.1009		1.1527		1.15
		0.0187		0.0105		0.0159	
1.20	1.1979		1.1114		1.1686		1.20
		0.0201		0.0111		0.0170	
1.25	1.2180		1.1225		1.1856		1.25
		0.0216		0.0120		0.0183	
1.30	1.2396		1.1345		1.2039		1.30
		0.0232		0.0128		0.0196	
1.35	1.2628		1.1473		1.2235		1.35
		0.0250		0.0137		0.0210	
1.40	1.2878		1.1610		1.2445		1.40
		0.0268		0.0147		0.0226	
1.45	1.3146		1.1757		1.2671		1.45
		0.0288		0.0158		0.0243	
1.50	1.3434		1.1915		1.2914		1.50

TABLE A-1. TABLE OF THE FUNCTIONS $\phi(u)$, $\psi(u)$, $\chi(u)$ (Continued)

$2u = kl$	$\phi(u)$	$\Delta\phi$	$\psi(u)$	$\Delta\psi$	$\chi(u)$	$\Delta\chi$	$2u = kl$
1.50	1.3434		1.1915		1.2914		1.50
		0.0310		0.0169		0.0260	
1 55	1.3744		1.2084		1 3174		1.55
		0 0334		0.0182		0.0281	
1.60	1.4078		1.2266		1.3455		1.60
		0.0361		0 0196		0.0303	
1.65	1.4439		1 2462		1.3758		1.65
		0.0391		0 0211		0 0327	
1.70	1.4830		1 2673		1.4085		1.70
		0 0422		0 0228		0.0353	
1.75	1.5252		1 2901		1 4438		1.75
		0.0458		0 0246		0.0383	
1.80	1.5710		1.3147		1 4821		1.80
		0 0498		0 0267		0.0416	
1.85	1.6208		1.3414		1 5237		1.85
		0.0542		0 0290		0 0452	
1.90	1.6750		1.3704		1 5689		1 90
		0.0593		0.0316		0.0493	
1 95	1.7343		1.4020		1.6182		1 95
		0.0650		0 0345		0.0540	
2 00	1.7993		1.4365		1.6722		2.00
		0.0137		0.0073		0 0114	
2.01	1 8130		1.4438		1.6836		2.01
		0.0140		0.0074		0 0117	
2.02	1.8270		1.4512		1.6953		2 02
		0.0143		0 0075		0.0118	
2.03	1 8413		1.4587		1.7071		2 03
		0.0145		0.0077		0.0121	
2 04	1.8558		1.4664		1.7192		2 04
		0.0148		0.0078		0.0122	
2.05	1.8706		1.4742		1 7314		2 05
		0.0152		0 0080		0.0126	
2.06	1.8858		1.4822		1 7440		2 06
		0.0154		0.0082		0 0128	
2.07	1.9012		1.4904		1.7568		2 07
		0.0156		0.0083		0.0130	
2.08	1.9168		1.4987		1.7698		2 08
		0.0161		0.0084		0.0134	
2.09	1 9329		1.5071		1.7832		2 09
		0.0164		0.0087		0.0135	
2.10	1.9493		1.5158		1 7967		2.10
		0 0168		0.0088		0 0139	
2.11	1.9661		1.5246		1 8106		2.11
		0 0170		0.0090		0.0141	
2.12	1.9831		1 5336		1 8247		2.12
		0 0174		0 0091		0.0145	
2.13	2.0005		1 5427		1 8392		2 13
		0 0179		0.0094		0 0147	
2.14	2.0184		1.5521		1 8539		2.14
		0.0182		0.0095		0 0150	
2.15	2.0366		1.5616		1 8689		2.15

TABLE A-1. TABLE OF THE FUNCTIONS $\phi(u)$, $\psi(u)$, $\chi(u)$ (Continued)

$2u = kl$	$\phi(u)$	$\Delta\phi$	$\psi(u)$	$\Delta\psi$	$\chi(u)$	$\Delta\chi$	$2u = kl$
2.15	2.0366		1.5616		1.8689		2.15
		0.0186		0.0097		0.0154	
2.16	2.0552		1.5713		1.8843		2.16
		0.0189		0.0100		0.0157	
2.17	2.0741		1.5813		1.9000		2.17
		0.0194		0.0101		0.0160	
2.18	2.0935		1.5914		1.9160		2.18
		0.0198		0.0104		0.0163	
2.19	2.1133		1.6018		1.9323		2.19
		0.0203		0.0106		0.0168	
2.20	2.1336		1.6124		1.9491		2.20
		0.0207		0.0109		0.0172	
2.21	2.1543		1.6233		1.9663		2.21
		0.0211		0.0110		0.0174	
2.22	2.1754		1.6343		1.9837		2.22
		0.0218		0.0114		0.0179	
2.23	2.1972		1.6457		2.0016		2.23
		0.0222		0.0115		0.0183	
2.24	2.2194		1.6572		2.0199		2.24
		0.0228		0.0118		0.0187	
2.25	2.2422		1.6690		2.0386		2.25
		0.0200		0.0122		0.0192	
2.26	2.2654		1.6812		2.0578		2.26
		0.0237		0.0124		0.0197	
2.27	2.2891		1.6936		2.0775		2.27
		0.0244		0.0126		0.0201	
2.28	2.3135		1.7062		2.0976		2.28
		0.0249		0.0130		0.0205	
2.29	2.3384		1.7192		2.1181		2.29
		0.0256		0.0133		0.0211	
2.30	2.3640		1.7325		2.1392		2.30
		0.0262		0.0136		0.0216	
2.31	2.3902		1.7461		2.1608		2.31
		0.0269		0.0140		0.0222	
2.32	2.4171		1.7601		2.1830		2.32
		0.0277		0.0143		0.0227	
2.33	2.4448		1.7744		2.2057		2.33
		0.0283		0.0147		0.0233	
2.34	2.4731		1.7891		2.2290		2.34
		0.0291		0.0150		0.0239	
2.35	2.5022		1.8041		2.2529		2.35
		0.0298		0.0154		0.0245	
2.36	2.5320		1.8195		2.2774		2.36
		0.0305		0.0159		0 0251	
2.37	2.5625		1.8354		2.3025		2.37
		0.0314		0.0162		0.0259	
2 38	2.5939		1.8516		2.3284		2.38
		0.0323		0.0167		0.0266	
2.39	2.6262		1.8683		2.3550		2.39
		0.0334		0.0171		0.0272	
2.40	2.6596		1.8854		2.3822		2.40

TABLE A-1. TABLE OF THE FUNCTIONS $\phi(u)$, $\psi(u)$, $\chi(u)$ (Continued)

$2u = kl$	$\phi(u)$	$\Delta\phi$	$\psi(u)$	$\Delta\psi$	$\chi(u)$	$\Delta\chi$	$2u = kl$
2.40	2.6596		1.8854		2.3822		2.40
		0.0339		0.0177		0.0281	
2.41	2.6935		1.9031		2.4103		2.41
		0.0352		0.0181		0.0288	
2.42	2.7287		1.9212		2.4391		2.42
		0.0362		0.0186		0.0296	
2.43	2.7649		1.9398		2.4687		2.43
		0.0372		0.0191		0.0306	
2.44	2.8021		1.9589		2.4993		2.44
		0.0382		0.0197		0.0313	
2.45	2.8403		1.9786		2.5306		2 45
		0 0395		0 0203		0.0324	
2.46	2.8798		1.9989		2 5630		2.46
		0.0406		0 0209		0.0334	
2.47	2.9204		2.0198		2 5964		2.47
		0.0420		0.0215		0.0343	
2.48	2.9624		2.0413		2 6307		2.48
		0.0432		0.0222		0.0355	
2.49	3.0056		2.0635		2 6662		2.49
		0.0446		0.0229		0.0365	
2.50	3.0502		2.0864		2.7027		2.50
		0.0461		0.0236		0.0378	
2.51	3.0963		2 1100		2.7405		2.51
		0.0475		0 0243		0.0389	
2.52	3.1438		2.1343		2.7794		2.52
		0.0493		0.0252		0.0403	
2.53	3.1931		2.1595		2 8197		2.53
		0.0506		0.0260		0.0415	
2.54	3.2437		2.1855		2.8612		2.54
		0.0526		0.0269		0.0431	
2.55	3.2963		2.2124		2.9043		2.55
		0.0545		0.0278		0.0445	
2.56	3.3508		2.2402		2.9488		2.56
		0.0564		0 0288		0.0461	
2.57	3.4072		2.2690		2 9949		2.57
		0.0585		0 0298		0.0478	
2.58	3.4657		2.2988		3.0427		2.58
		0.0605		0.0309		0.0495	
2.59	3.5262		2.3297		3.0922		2.59
		0.0628		0.0321		0.0513	
2.60	3.5890		2.3618		3.1435		2.60
		0.0652		0.0332		0.0533	
2.61	3.6542		2.3950		3.1968		2.61
		0.0678		0.0345		0.0554	
2.62	3.7220		2.4295		3.2522		2.62
		0.0705		0.0359		0.0575	
2.63	3.7925		2.4654		3.3097		2.63
		0.0734		0.0373		0.0599	
2.64	3.8659		2.5027		3.3696		2.64
		0.0762		0.0388		0.0623	
2.65	3.9421		2.5415		3.4319		2.65

TABLE A-1. TABLE OF THE FUNCTIONS $\phi(u)$, $\psi(u)$, $\chi(u)$ (*Continued*)

$2u = kl$	$\phi(u)$	$\Delta\phi$	$\psi(u)$	$\Delta\psi$	$\chi(u)$	$\Delta\chi$	$2u = kl$
2.65	3.9421		2.5415		3.4319		2.65
		0.0797		0.0404		0.0650	
2.66	4.0218		2.5819		3.4969		2.66
		0 0829		0.0422		0.0677	
2.67	4.1047		2.6241		3.5646		2.67
		0.0867		0·0439		0.0707	
2.68	4.1914		2.6680		3.6353		2.68
		0.0906		0.0460		0.0739	
2.69	4.2820		2.7140		3.7092		2.69
		0.0946		0.0479		0.0771	
2 70	4.3766		2.7619		3.7863		2.70
		0.0991		0.0502		0 0808	
2.71	4.4757		2.8121		3.8671		2.71
		0.1038		0.0527		0 0846	
2.72	4.5795		2.8648		3.9517		2.72
		0.1090		0.0551		0.0888	
2.73	4.6885		2.9199		4.0405		2.73
		0.1144		0.0579		0.0932	
2.74	4.8029		2.9778		4.1337		2.74
		0.1204		0.0608		0.0980	
2.75	4.9233		3.0386		4.2317		2.75
		0.1266		0 0641		0.1032	
2.76	5.0499		3.1027		4.3349		2.76
		0.1336		0 0675		0.1087	
2.77	5.1835		3 1702		4.4436		2.77
		0.1410		0.0712		0.1148	
2.78	5 3245		3.2414		4.5584		2.78
		0.1491		0.0752		0.1213	
2.79	5.4736		3 3166		4.6797		2.79
		0.1579		0.0797		0.1285	
2.80	5.6315		3.3963		4.8082		2.80
		0.1675		0.0844		0.1362	
2.81	5.7990		3 4807		4.9444		2.81
		0.1780		0.0897		0.1448	
2.82	5.9770		3 5704		5.0892		2.82
		0.1894		0.0955		0.1540	
2.83	6.1664		3 6659		5.2432		2.83
		0.2021		0.1017		0.1643	
2.84	6.3685		3.7676		5.4075		2.84
		0.2160		0.1088		0.1757	
2.85	6 5845		3 8764		5.5832		2.85
		0.2315		0.1164		0.1881	
2.86	6.8160		3.9928		5.7713		2.86
		0.2486		0.1251		0.2020	
2.87	7.0646		4 1179		5.9733		2.87
		0.2676		0.1346		0 2174	
2.88	7.3322		4.2525		6.1907		2.88
		0.2890		0.1452		0.2348	
2.89	7.6212		4.3977		6.4255		2.89
		0.3131		0.1573		0.2543	
2.90	7.9343		4.5550		6.6798		2.90

TABLE A-1. TABLE OF THE FUNCTIONS $\phi(u)$, $\psi(u)$, $\chi(u)$ (*Continued*)

$2u = kl$	$\phi(u)$	$\Delta\phi$	$\psi(u)$	$\Delta\psi$	$\chi(u)$	$\Delta\chi$	$2u = kl$
2.90	7.9343		4.5550		6.6798		2.90
		0.3402		0.1709		0.2763	
2.91	8.2745		4.7259		6.9561		2.91
		0.3710		0.1862		0.3012	
2.92	8.6455		4.9121		7.2573		2.92
		0.4061		0.2039		0.3298	
2.93	9.0516		5.1160		7.5871		2.93
		0.4466		0.2241		0.3625	
2.94	9.4982		5.3401		7.9496		2.94
		0.4933		0.2474		0.4004	
2.95	9.9915		5 5875		8.3500		2.95
		0.5478		0 2747		0.4446	
2.96	10.5393		5 8622		8.7946		2.96
		0.6117		0 3066		0.4964	
2.97	11.1510		6 1688		9 2910		2.97
		0.6876		0 3446		0.5579	
2.98	11.8386		6.5134		9.8489		2.98
		0.7785		0.3901		0 6315	
2.99	12.6171		6.9035		10.4804		2.99
		0.8886		0.4451		0.7209	
3.00	13.5057		7.3486		11.2013		3.00
		1.0238		0 5127		0.8304	
3.01	14.5295		7.8613		12.0317		3.01
		1.1924		0.5970		0.9671	
3.02	15.7219		8.4583		12 9988		3.02
		1.4063		0 7040		1.1405	
3.03	17.1282		9.1623		14.1393		3.03
		1.6834		0.8426		1.3651	
3.04	18.8116		10 0049		15 5044		3.04
		2.0513		1 0265		1.6633	
3.05	20.8629		11.0314		17.1677		3.05
		2 5547		1.2782		2.0711	
3.06	23.4176		12.3096		19.2388		3.06
		3.2684		1 6350		2.6498	
3.07	26.6860		13.9446		21 8886		3.07
		4.3300		2.1659		3.5103	
3.08	31.0160		16.1105		25.3989		3.08
		6.0084		3.0051		4.8712	
3.09	37.0244		19.1156		30.2701		3.09
		8.8990		4.4503		7.2138	
3.10	45.9234		23.5659		37.4839		3.10
		14.5332		7.2675		11.7808	
3.11	60.4566		30.8334		49.2647		3.11
		27.9956		13.9987		22.6930	
3.12	88 4522		44.8321		71.9577		3.12
		76.2965		38.1491		61.8440	
3.13	164 7487		82.9812		133.8017		3.13
		1034 4142		517.2088		838.4545	
3.14	1199 1629		600.1900		972.2562		3.14
		∞		∞		∞	
π	$\pm\infty$		$\pm\infty$		$\pm\infty$		π

TABLE A-1. TABLE OF THE FUNCTIONS $\phi(u)$, $\psi(u)$, $\chi(u)$ (Continued)

$2u = kl$	$\phi(u)$	$\Delta\phi$	$\psi(u)$	$\Delta\psi$	$\chi(u)$	$\Delta\chi$	$2u = kl$
π	$\pm\infty$		$\pm\infty$		$\pm\infty$		π
		∞		∞		∞	
3.15	-227.1668		-112.9747		-183.8716		3.15
		123.4092		61.7055		100.0325	
3.16	-103.7576		-51.2692		-83.8391		3.16
		36.5228		18.2624		29.6049	
3.17	-67.2348		-33.0068		-54.2342		3.17
		17.5035		8.7527		14.1884	
3.18	-49.7313		-24.2541		-40.0458		3.18
		10.2713		5.1365		8.3263	
3.19	-39.4600		-19.1176		-31.7195		3.19
		6.7537		3.3778		5.4750	
3.20	-32.7063		-15.7398		-26.2445		3.20
		4.7787		2.3903		3.8742	
3.21	-27.9276		-13.3495		-22.3703		3.21
		3.5593		1.7807		2.8858	
3.22	-24.3683		-11.5688		-19.4845		3.22
		2.7541		1.3779		2.2330	
3.23	-21.6142		-10.1909		-17.2515		3.23
		2 1940		1.0980		1.7790	
3.24	-19.4202		-9.0929		-15.4725		3.24
		1 7890		0.8954		1 4507	
3.25	-17.6312		-8.1975		-14.0218		3.25
		1.4865		0.7443		1.2057	
3.26	-26.1447		-7.4532		-12.8161		3.26
		1.2548		0.6284		1.0178	
3.27	-14.8899		-6.8248		$-11 7983$		3.27
		1.0733		0.5376		0.8707	
3.28	-13.8166		-6.2872		$-10 9276$		3.28
		0 9285		0.4652		0.7533	
3.29	-12.8881		$-5 8220$		-10.1743		3.29
		0.8111		0.4066		0.6581	
3.30	-12.0770		-5.4154		$-9 5162$		3.30
		4.6522		2.3367		3.7784	
3.40	-7.4248		-3.0787		-5.7378		3.40
		2.0479		1.0354		1.6681	
3.50	-5.3769		-2.0433		-4.0697		3.50
		1.1477		0.5861		0.9389	
3.60	-4.2292		-1.4572		-3.1308		3.60
		0.7302		0.3785		0.6016	
3.70	-3.4990		-1.0787		-2.5292		3.70
		0.5029		0.2659		0.4170	
3.80	-2.9961		-0.8128		-2.1113		3.80
		0.3647		0.1981		0.3070	
3.90	-2.6314		-0.6147		-1.8043		3.90
		0.2744		0.1544		0.2349	
4.00	-2.3570		-0.4603		-1.5694		4.00

Table A-1. Table of the Functions $\phi(u)$, $\psi(u)$, $\chi(u)$ (Continued)

$2u = kl$	$\phi(u)$	$\Delta\phi$	$\psi(u)$	$\Delta\psi$	$\chi(u)$	$\Delta\chi$	$2u = kl$
4.00	−2.3570		−0.4603		−1.5694		4.00
		0.2116		0.1248		0.1854	
4.10	−2 1454		−0.3355		−1.3840		4.10
		0.1662		0.1038		0.1498	
4.20	−1.9792		−0.2317		−1.2342		4.20
		0.1317		0.0887		0.1237	
4.30	−1.8475		−0.1430		−1.1105		4.30
		0.1046		0.0778		0.1036	
4.40	−1.7429		−0.0652		−1.0069		4.40
		0.0826		0.0696		0.0881	
4.50	−1.6603		0.0044		−0.9188		4.50
		0.0641		0 0638		0.0757	
4.60	−1 5962		0.0682		−0.8431		4.60
		0.0810		0.1169		0.1235	
4.80	−1.5152		0.1851		−0.7196		4.80
		0.0238		0.1124		0.0962	
5.00	−1.4914		0.2975		−0 6234		5.00
		0.0568		0.1520		0 0938	
5.25	−1.5482		0.4495		−0 5296		5.25
		0.1964		0.1975		0.0733	
5.5	−1 7446		0.6470		−0 4563		5.5
		0.4898		0.3277		0.0589	
5.75	−2.2344		0.9747		−0 3974		5.75
		1.5111		0.8268		0.0482	
6 0	−3 7455		1.8015		−0.3492		6.0
		25.3412		12.7331		0.0404	
6 25	−29.0867		14.5346		−0.3088		6.25
		∞		∞		0.0048	
2π	± ∞		± ∞		−0.3040		2π
		∞		∞		0.0295	
6.5	4.1490		−2.0242		−0.2745		6.5

Note: The functions $\phi(u)$, $\psi(u)$, and $\chi(u)$ are known as Berry functions. Table A-1 was compiled by Niles and Newell, "Airplane Structures," 3d ed., vol. II, pp. 72–78, John Wiley & Sons, Inc., New York, 1943. The first table of this type was published by A. P. Van der Fleet, *Bull. Polytechnic Inst.*, *St. Petersburg*, 1904.

TABLE A-2. TABLE OF THE FUNCTIONS $\eta(u)$ AND $\lambda(u)$

$$\eta(u) = \frac{12(2 \sec u - 2 - u^2)}{5u^4} \qquad \lambda(u) = \frac{2(1 - \cos u)}{u^2 \cos u}$$

$2u = kl$	$\eta(u)$	$\lambda(u)$
0	1.000	1.000
0.20	1.004	1.004
0.40	1.016	1.016
0.60	1.037	1.038
0.80	1.070	1.073
1.00	1.114	1.117
1.20	1.173	1.176
1.40	1.250	1.255
1.60	1.354	1.361
1.80	1.494	1.504
2.00	1.690	1 704
2.20	1.962	1 989
2.40	2.400	2.441
2.60	3.181	3.240
2.80	4.822	4 938
2.90	6.790	6.940
3.00	11.490	11.670
π	∞	∞

TABLE A-3. PROPERTIES OF SECTIONS
O = shear center J = torsion constant C_w = warping constant

	$J = \dfrac{2bt_f{}^3 + ht_w{}^3}{3}$ $C_w = \dfrac{t_f h^2 b^3}{24}$	If $t_f = t_w = t$: $J = \dfrac{t^3}{3}(2b + h)$
	$e = h\dfrac{b_1{}^3}{b_1{}^3 + b_2{}^3}$ $J = \dfrac{(b_1 + b_2)t_f{}^3 + ht_w{}^3}{3}$ $C_w = \dfrac{t_f h^2}{12}\dfrac{b_1{}^3 b_2{}^3}{b_1{}^3 + b_2{}^3}$	If $t_f = t_w = t$: $J = \dfrac{t^3}{3}(b_1 + b_2 + h)$
	$e = \dfrac{3b^2 t_f}{6bt_f + ht_w}$ $J = \dfrac{2bt_f{}^3 + ht_w{}^3}{3}$ $C_w = \dfrac{t_f b^3 h^2}{12}\dfrac{3bt_f + 2ht_w}{6bt_f + ht_w}$	If $t_f = t_w = t$: $e = \dfrac{3b^2}{6b + h}$ $J = \dfrac{t^3}{3}(2b + h)$ $C_w = \dfrac{tb^3 h^2}{12}\dfrac{3b + 2h}{6b + h}$
	$J = \dfrac{2bt_f{}^3 + ht_w{}^3}{3}$ $C_w = \dfrac{b^3 h^2}{12(2b + h)^2}$ $\times [2t_f(b^2 + bh + h^2) + 3t_w bh]$	If $t_f = t_w = t$: $J = \dfrac{t^3}{3}(2b + h)$ $C_w = \dfrac{tb^3 h^2}{12}\dfrac{b + 2h}{2b + h}$
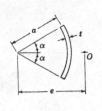	$e = 2a\dfrac{\sin\alpha - \alpha\cos\alpha}{\alpha - \sin\alpha\cos\cdot\alpha}$ $J = \dfrac{2a\alpha t^3}{3}$ $C_w = \dfrac{2ta^5}{3}$ $\times\left[\alpha^3 - \dfrac{6(\sin\alpha - \alpha\cos\alpha)^2}{\alpha - \sin\alpha\cos\alpha}\right]$	If $2\alpha = \pi$: $e = \dfrac{4a}{\pi} \quad J = \dfrac{\pi at^3}{3}$ $C_w = \dfrac{2ta^5}{3}\left(\dfrac{\pi^3}{8} - \dfrac{12}{\pi}\right)$ $= 0.0374ta^5$

NAME INDEX

531

SUBJECT INDEX

Allowable stresses, for beam-columns, 37–45
  for built-up columns, 206
  for columns, 192–211
  for lateral buckling, 277
  (*See also* Design)
Amplification factor for beam-columns, 15, 29
Angle section, column, 363, 425
  shear center, 217
  torsion constant, 212, 213
  warping, 217
Anisotropic plates, 403, 406
Anisotropic shells (*see* Shells)
Anticlastic bending, 321
Approximate calculation of critical loads, by energy method, 82–94
  by Ritz method, 92–94
  by successive approximations, 116–125, 267*n*.
    graphical, 123
    numerical, 120
Arches, bending, 278–289
  buckling, 297–310
    lateral, 313–318
      bridge, 317*n*.
  catenary, 304
  circular, 297
  experiments with, 298*n*., 303, 310
  flat, 305–310
  parabolic (*see* Parabolic arch)
  three-hinged, 300, 303
  with varying cross section, 302, 303

Bars, bimetallic, 310–313
  curved, bending, 278–289, 313
    buckling, 297–310
      lateral, 313–318
    inextensional deformation, 282
  dynamic buckling, 158
  elastic buckling, 46–162
    by approximate methods (*see* Approximate calculation)
    by beam-column theory, 59
    with built-in ends, 49, 54
    with distributed loads, 100, 130
    due to own weight, 101
    due to torque, 156
    dynamic methods for, 152
    on elastic foundation, 94, 107

Bars, elastic buckling, with elastic restraints, 60
    experiments on, 185–211
    fundamental case, 49, 52
      by energy method, 87, 91
      by successive approximations, 117
    with intermediate forces, 98
    with load through fixed point, 55
    nonprismatic, 98, 113, 120, 123, 125, 130
    with one end fixed, and other end free, 47, 52, 132, 153
      by energy method, 87, 88
      and other end guided, 49
      and other end pinned, 53, 59
    in pure torsion, 156
    with rounded ends, 57
    shear effect on, 132
    static methods for, 152
    with tangential load, 153
  energy method for (*see* Energy method)
  equilibrium method for, 83–86, 152
  inelastic buckling, 163–184
    with distributed loads, 183
    on elastic foundation, 183
    with elastic restraints, 182, 183
    Engesser-Kármán theory, 176
    experiments on, 179, 186
    fundamental case, 175
    lateral, 272–277
    nonprismatic, 184
    reduced modulus theory, 177
    tangent modulus theory, 178
    with various end conditions, 182
  large deflections, 76
  lateral buckling, 251–277
    (*See also* Beams, lateral buckling)
  of minimum weight, 131
  with nonconservative forces, 152
  nonprismatic (*see* Nonprismatic bars)
  rigid, with elastic supports, 83, 84
  tapered, 125, 130, 184
    buckling, inelastic, 184
    lateral, 262
  thin-walled (*see* Thin-walled bars)
  torsion, 212–224
    nonuniform, 218–224
    pure, 212–218
  torsion properties, table, 530
  torsional buckling, 225–250
    with bending, 229–236